The Science of Human Intelligence

In this revised and updated edition of Earl Hunt's classic textbook, *Human Intelligence*, two research experts explain how key scientific studies have revealed exciting information about what intelligence is, where it comes from, why there are individual differences, and what the prospects are for enhancing it. The topics are chosen based on the weight of evidence, so readers can evaluate what ideas and theories the data support. These topics include IQ testing, mental processes, brain imaging, genetics, population differences, sex, aging, and likely prospects for enhancing intelligence based on current scientific evidence. Readers will confront ethical issues raised by research data and learn how scientists pursue answers to basic and socially relevant questions about why intelligence is important in everyday life. Many of the answers will be surprising and stimulate readers to think constructively about their own views.

Richard J. Haier is Professor Emeritus at the University of California, Irvine, USA. He pioneered neuroimaging studies of intelligence, created *The Intelligent Brain* (The Great Courses), served as president of the International Society for Intelligence Research, and is editor in chief of *Intelligence*. He received the Lifetime Achievement Award from the International Society for Intelligence Research (2020), coedited *The Cambridge Handbook of Intelligence and Cognitive Neuroscience* (2021), and authored *The Neuroscience of Intelligence* (2017). Dr. Haier has done podcasts with Jordan Peterson, Scott Barry Kaufman, and Lex Fridman. His personal website is www.richardhaier.com.

Roberto Colom is Professor of Differential Psychology at Universidad Autónoma de Madrid, Spain. He has authored or edited twenty books and published 174 peer-reviewed articles. He has a wide network of scientific collaborations around the globe, and he is a member of the International Society for Intelligence Research. He was among the top five researchers in the latest bibliometric analysis of articles published in the journal *Intelligence*. For further information, visit https://sites.google.com/site/colomresearch/Home.

Earl Hunt was Professor Emeritus at the University of Washington, USA, where he was a faculty member since 1966. He also taught at Yale University; the University of California, Los Angeles; and the University of Sydney, Australia. His other books include *Concept Learning* (1962), *Experiments in Induction* (1966), *Artificial Intelligence* (1975), *Will We Be Smart Enough?* (1995), *Thoughts on Thought* (2002), *The Mathematics of Behavior* (2007), and *Human Intelligence* (2011). He received the International Society for Intelligence Research's Lifetime Achievement Award for his contributions to the study of intelligence.

The Science of Human Intelligence

Second Edition

RICHARD J. HAIER

University of California, Irvine

ROBERTO COLOM

Universidad Autónoma de Madrid

EARL HUNT

University of Washington

CAMBRIDGE
UNIVERSITY PRESS

CAMBRIDGE
UNIVERSITY PRESS

Shaftesbury Road, Cambridge CB2 8EA, United Kingdom

One Liberty Plaza, 20th Floor, New York, NY 10006, USA

477 Williamstown Road, Port Melbourne, VIC 3207, Australia

314–321, 3rd Floor, Plot 3, Splendor Forum, Jasola District Centre, New Delhi – 110025, India

103 Penang Road, #05-06/07, Visioncrest Commercial, Singapore 238467

Cambridge University Press is part of Cambridge University Press & Assessment, a department of the University of Cambridge.

We share the University's mission to contribute to society through the pursuit of education, learning and research at the highest international levels of excellence.

www.cambridge.org
Information on this title: www.cambridge.org/highereducation/isbn/9781108477154

DOI: 10.1017/9781108569576

First published 2011
Fourth printing 2019
Second edition 2024

A catalogue record for this publication is available from the British Library.

Library of Congress Cataloging-in-Publication Data
Names: Haier, Richard J., author. | Colom, Roberto, 1964- author. | Hunt, Earl B., author.
Title: The science of human intelligence / Richard J. Haier, University of California, Irvine, Roberto Colom, Universidad Autónoma de Madrid, Earl Hunt, University of Washington.
Description: Second edition. | Cambridge, United Kingdom ; New York, NY : Cambridge University Press, [2024] | Earlier edition authored by the late Earl B. Hunt as: Human intelligence, published in 2011. | Includes bibliographical references and index.
Identifiers: LCCN 2023006713 (print) | LCCN 2023006714 (ebook) | ISBN 9781108477154 (hardback) | ISBN 9781108701969 (paperback) | ISBN 9781108569576 (epub)
Subjects: LCSH: Intellect.
Classification: LCC BF431 .H257 2011 (print) | LCC BF431 (ebook) | DDC 153.9–dc23/eng/20230415
LC record available at https://lccn.loc.gov/2023006713
LC ebook record available at https://lccn.loc.gov/2023006714

ISBN 978-1-108-47715-4 Hardback
ISBN 978-1-108-70196-9 Paperback

Additional resources for this publication at www.cambridge.org/haier

Dedication

To the memory of our friend Earl (Buz) Hunt, who championed clear thinking and incisive but constructive skepticism about intelligence research

Contents

Preface

This book is the second edition to the classic textbook *Human Intelligence* (2011), written by our friend Earl (Buz) Hunt (1933–2016). Hunt's book provided a balanced and comprehensive presentation of the scientific evidence about human intelligence. We wrote this book to do the same with updated research data, but with this warning: we have our own points of view elaborated after seventy-five years of combined research experience between us, so this book is not blandly neutral on every issue. Not surprisingly, we do not agree about every point or the emphasis it should receive. Our disagreements mirror the field. The astute reader may even detect a bit of tension in passages that are written as compromises of our different views. We often channeled Buz as the tiebreaker.

Intelligence is a controversial subject for many reasons. This book presents empirical data that address complex ideas and the issues surrounding them. Many findings might be surprising and contrary to what you may have heard or what you believe. Misunderstandings and mistaken ideas abound, especially in the popular media,

about what intelligence is, where it comes from, its importance in everyday life, and the meaning of average differences among populations. This book is intended to inoculate you against erroneous information and to enable you to think and converse about complex topics with facts.

The weight of evidence is a key concept at the core of this book. Questions about intelligence do not have simple answers. No one study is definitive, and it takes many years to sort out inconsistent research results to establish a weight of evidence. Intelligence research has many mysteries, twists, and turns, and, most importantly, know it or not, it is highly relevant to your life.

Our job is to inform you with clear explanations of the best evidence. New data are coming rapidly, especially from neuroscience and genetic studies. Each chapter intends to promote critical thinking about what science already knows and what remains to be learned. There are hundreds of recent good studies we could have chosen to include. We want you to see the forest through the trees, the so-called big picture, so we have focused on some of the best, informative, and interesting research. We often will quote researchers directly when they interpret their findings to give you a feel for how research is communicated by the people who do it. For the same reason, we often reproduce the exact figures and tables of results from the original sources.

With this in mind, Chapter 1, "A Brief Voyage to the Past," presents historic milestones relevant for understanding the present state of research and preparing for future findings. Common misunderstandings about intelligence are debunked, including opinions that intelligence cannot be defined or measured for scientific research, that intelligence tests do not measure anything important for everyday life, or that intelligence is based exclusively on early environmental and social experiences. These views are not supported by compelling evidence, as we will see.

Chapter 2, "Basic Concepts," discusses the role of models and hypothesis testing and sets out major questions about intelligence. The need for quantitative assessments is underscored for answering questions at distinguishable but interrelated levels of analysis: psychological traits, cognitive processes, and biology (e.g., the brain, genomes, influences of experiences).

Chapter 3, "Psychometric Models of Intelligence," covers testing intelligence at the behavioral level, what test scores mean, and what they do not mean. The available evidence supports the conclusions that (1) standardized intelligence tests are among the most reliable and valid psychological assessment tools, (2) people can be reliably ordered on these test scores, and (3) the analysis of these scores supports the view that human cognitive abilities are organized in a pyramid-like structure with a prominent general factor of intelligence (g) common to all tests of mental abilities and that some mental abilities require more g than others.

Chapter 4, "Cognitive Models of Intelligence and Information Processing," focuses on the dynamics of intelligence within the brain. Perception, attention, learning, and memory have been studied as elements of intelligence. Can intelligence be explained by individual differences in attention and memory ability? Or is intelligence the factor that integrates all these elements into something psychologically meaningful that is more than the sum of its parts? We present evidence that favors explanations somewhere between psychometric and biology-based models by showing how individuals apply their information processing abilities.

Chapter 5, "Intelligence and the Brain," focuses on a neuroscience approach. Neuroimaging technologies have taken intelligence research deep into the brain. Structural and functional features of the cortex, specific neural circuits throughout the brain, and characteristics of neurons are related to intelligence differences among individuals as assessed by standardized tests. Using patterns of connectivity among brain areas (the connectome), we can now identify brain "fingerprints" that predict intelligence test scores. This is one of the most exciting areas of progress, but further advances are coming quickly, and they are tasty food for hungry minds.

Chapter 6, "The Genetic Basis of Intelligence," presents overwhelming evidence that genetic variation is related to intelligence differences. Classic behavioral genetics research (based on twin, adoption, and family studies) is now taken to another level with DNA assessment. Polygenic scores can predict intelligence differences among individuals to some degree. This opens a new and exciting field that seeks to understand the molecular biology of specific brain systems related to intelligence. These findings have the science fiction–like potential to evolve into learning how to change brain systems to enhance intelligence. Chapter 13 is fully devoted to the enhancement of intelligence.

Chapter 7, "Experience and Intelligence," examines the impact of nongenetic environmental factors on intelligence. We know genes have substantial influence, but we also know that they are clearly insufficient for explaining all manifestations of intelligence. Furthermore, the same environmental feature may have a differential impact on each person's unique genome and connectome.

Understanding how complex interactions among genetic and environmental factors contribute to intelligence differences is one of the most challenging topics in all science.

Chapter 8, "Intelligence and Everyday Life," explains why intelligence matters to you. Intelligence is critical for understanding numerous consequential social outcomes, such as academic achievement, occupational status and success, physical and mental health, longevity, creativity, eminence, income, socioeconomic status, and accident-proneness, just to name some of them. We challenge you to find any other psychological trait showing such a large set of social corre-lates. We explain here how the integrative nature of intelligence might help account for this wide network of findings.

Chapter 9, "Introduction to the Scientific Study of Population Differences," is intended to provide a context for some of the most complex findings in intelligence research that indicate average test score dif-ferences among some populations. We detail these findings in Chapter 10 (sex differences), Chapter 11 (age differences), and Chapter 12 (intelligence in the world). Here we present key points to keep in mind when considering population data. As a pre-view, here are three of them: (1) making sweeping assertions about population differ-ences must be regarded with skepticism; (2) it is critical to distinguish scientific findings and political recommendations; and (3) when population differences are observed, it may be unclear if they can be attributed to intelligence or to other personal and social/cultural variables.

Chapter 10, "Sex Differences and Intelligence," deals with our current under-standing of average cognitive ability differ-ences between men and women. There is a large research literature and compelling evi-dence that men and women do not differ on general intelligence (g), but there are aver-age differences in some specific mental abil-ities. We address structural and functional brain differences and whether these differ-ences could account for men/women dispar-ities across vocations and professions. An important issue is to what, if any, extent these findings might inform social/educa-tional policies.

Chapter 11, "Intelligence and Aging," addresses the stability of intelligence as peo-ple age. There are longitudinal studies show-ing that intelligence is the most stable psychological trait, but this does not mean there are no changes across the life span. Different aspects of intelligence age differ-ently, and the effects of age are variable among individuals. Research is trying to dis-cover how genes and environmental factors influence aging effects on intelligence. Since we all age, this research has especially per-sonal meaning.

Chapter 12, "Intelligence in the World," discusses average intelligence similarities and differences among countries, conti-nents, and ancestry. Devoting a chapter to this complex and delicate subject is neces-sary because there is much popular discus-sion and argument about it, especially on social media. Intelligence research is also relevant to what economists call human capital. The role of intelligence for eco-nomic development and well-being is an emerging area of interest. Therefore, we dis-cuss competing interpretations of the avail-able data. We acknowledge that we do not yet have a solid weight of evidence about why differences exist, but the issues are important, and we think skipping them would be irresponsible.

Chapter 13, "Enhancing Intelligence," addresses the most far-reaching and ulti-mate topic in all of intelligence research: how can we improve general cognitive abil-ity? There is evidence of a generational increase in average IQ scores (Flynn effect) and of a small average increase related to education. However, there have been many claims of large increases for individuals resulting from early childhood education, computer gaming, memory training, stimu-lation of the brain, and a host of others. So far, none of them produce replicated or last-ing effects. We think advances in neuroscience and genetic research may pro-vide a rational basis for enhancing intelli-gence through neurobiology, brain development, and perhaps gene expression.

We see a glimmer of what might be possible and the thought-provoking questions that will arise.

The epilogue, "A Final Word," summarizes ten key findings about intelligence that are strongly supported by currently available research evidence. There also is a challenge to think about what the future holds for understanding intelligence beyond test scores.

Let us begin this book with the words of Douglas K. Detterman, a longtime intelligence researcher:

> Intelligence is the most important thing of all to understand, more important than the origin of the universe, more important than climate change, more

important than curing cancer, more important than anything else. That is because human intelligence is our major adaptive function and only by optimizing it will we be able to save ourselves and other living things from ultimate destruction.

> (Detterman, 2016, p. v)

We strongly agree, and, perhaps by the time you finish reading, you will agree too.

Reference

Detterman, D. K. 2016. Was intelligence necessary? *Intelligence*, 55, v–viii.

Acknowledgments

This project has a long history, but it has come to fruition with the help of David Repetto and Rowan Groat at Cambridge University Press. We thank all the reviewers who offered a range of comments and suggestions. Many of our colleagues around the world read all or some of the chapters of this book. We thank James Lee, Natalia Goriounova, Anna-Lena Schubert, Kirsten Hilger, Matt Euler, and Sophie von Stumm for their feedback. We also thank all the people who contributed figures (listed in the main text). Of course, even with their input, any mistakes belong exclusively to us. Finally, we thank the Hunt family – they have never wavered in their support of this project.

CHAPTER 1

A Brief Voyage to the Past

1.1 Introduction: We Are Not All Equally Intelligent

This book is about the nature of intelligence, its causes and uses, and why it differs among people. Scientific psychology has much to say about intelligence, but unfortunately, much that has been said is misunderstood.

There is a reason for this. The findings derived from empirical studies of intelligence have important and sometimes uncomfortable social consequences. For example, school systems might use intelligence and cognitive ability tests to stream students into specialized programs. Colleges have used cognitive tests to screen applicants for admission to higher education opportunities, at least until recently. Scholastic tests are not called intelligence tests, but they typically show substantial correlations with them (Frey and Detterman, 2004; Kaufman et al., 2012; Pokropek et al., 2021). This is hard to accept when students are disappointed or adversely affected by test results. It is even more difficult to acknowledge average score differences among some populations, such as students from different countries completing the widely used standardized Program for International Student Assessment (PISA) tests of mathematical ability (Rindermann, 2018). Blame the test or misrepresent the concept of intelligence are common exhortations, which have had recent impact as many colleges and universities have abandoned the use of standardized tests as part of the admissions process.

Testing is not confined to the educational system. Volunteers for military services must obtain passing scores on a test of general mental competence. Job applicants often are tested for cognitive abilities. There are a variety of special assistance programs for people who do not have the cognitive competence to cope with the complexities of the modern world. Low intelligence test scores can be offered as evidence of diminished mental capacity during criminal trials (Oleson, 2016).

There is agreement that some people are smarter than others, but things become complex when we try to be precise about what this means. Every knowledgeable person, for instance, would agree that Pasteur and Michelangelo were both highly intelligent, but was either more intelligent than the other? How did their intelligence differences come about, and why did they differ so much from most people? Such questions began to be formalized back in sixteenth-century Europe.

1.1.1 *Spain, Sixteenth Century*

The Spaniard Huarte de San Juan (1529–1588) was a physician who is now recognized as the father of differential psychology. His book *The Examination of Wits* (1575) connected psychology and biology, discussing differences among people. The book was a huge bestseller at the time, and it was translated into the main European languages (English, French, German, and Italian). Huarte's ideas influenced authors such as Francis Bacon, David Hume, Immanuel Kant, Jean-Jacques Rousseau, Francis Galton, and Noam Chomsky, to name a few.

Huarte emerged in the middle of a rich intellectual environment. As described by Robert Goodwin, arts and sciences flourished in the so-called Spanish Golden Age (Goodwin, 2015). This period began in 1492 – when the Spaniards connected the Old with the New World – and lasted until approximately 1659. During the kingship of Philip II (1556–1598), Spain achieved its greatest international influence and power (it was known as the empire on which the sun never set). The broad social context had strong positive impact on the intellectual milieu within this first global empire connecting Europe, America, and Asia. Thus, for instance, the main residence of Philip II (El Escorial, Madrid) was the epicenter of a variety of scientific and cultural developments: (1) it included the greatest private library in the world at the time (14,000 volumes), (2) scientists from different countries were invited to discuss a variety of topics, (3) the academy of mathematics was funded in 1582, (4) intellectuals were protected by the monarch, and (5) the judicial system was greatly improved, influencing the rest of the world.

Organizing such a huge global empire required efficient managing of the available human capital across the multiple connected world regions, and Huarte applied his theoretical framework for matching humans with the explosion of new occupations. Huarte noted that different occupations require distinguishable mental abilities, and he thought it would be possible to analyze these requirements to achieve the goal of matching them to people's mental abilities. He argued that both individuals and society would benefit from this systematic approach.

Huarte described three mental faculties/abilities: (1) understanding, (2) memory, and (3) imagination. He argued that these faculties are present in unique combinations within individual brains and could be characterized by a number of physical features.

He thought that when people attack problems, some will use their imagination to envisage how a solution might work out, while others will rely on their memories of solutions that have worked in the past. Huarte also defined understanding (*entendimiento*) as a distinguishable ability. Huarte's distinction between problem solving by imagination or by memory is mirrored in contemporary models of intelligence that distinguish between abstract reasoning ability and the ability to apply previously learned solution methods, as we will see later in this book.

He also anticipated another contemporary idea: the need to have a biological explanation for intelligence. Huarte offered a theory based on the sixteenth-century notion that the body is governed by four humors – blood, bile, black bile, and phlegm. Although that old formulation is no longer viable, one of the most active areas of current intelligence research deals with the relation between intelligence and the brain (Haier, 2017), as we will detail in Chapter 5.

As noted, Huarte analyzed how to match people and occupations for the benefit of both individuals and the society in which they live. These attempts were the origins of modern occupational counseling and vocational guidance. Occupations can be characterized by distinguishable cognitive requirements, and people do have different cognitive ability profiles. You can use education for teaching people to do many things, but we can also discover the cognitive strengths (and weaknesses) of any individual and find which occupations might be most suitable for them.

Huarte asked questions such as: Why are children of the same parents so different? What are the key psychological features of the different professions/occupations? How can we match the wide variety of individuals with the psychological requirements of the professions? How can society promote the development of wits? He considered how different cognitive abilities relate to one another and also how intelligence interacts with treatment (e.g., programs to aid learning complex vs. simple information). Acknowledging this pioneering approach, Douglas Detterman wrote, "Huarte was not only the first to suggest a multifactorial model of intelligence, but also the first to describe aptitude by treatment interactions. ... Much of what Huarte says sounds strikingly modern. Indeed, his theory contains many of the aspects of current models of intelligence" (Detterman, 1982, pp. 100–101).

1.1.2 England and France, Nineteenth and Twentieth Centuries

In the nineteenth century, the Briton Sir Francis Galton (1822–1911) explored Africa, made major contributions to the development of statistics ("whenever you can, count" became his favorite dictum), and conducted research in psychology. He endorsed the theory of evolution proposed by his half cousin Charles Darwin, and he thought that human intelligence was inherited in the same sense as the rest of human traits. He considered that a person's intelligence could be assessed by examining brain size or measuring the efficiency of the nervous system by recording the speed of reaction to elementary signals. He also understood the value of studying twins.

Galton made contributions to several disciplines, not just psychology. Because of his explorations of remote regions in Africa (paying the costs of the expeditions himself), he was elected into the Royal Geographical Society at age thirty-four. He also discovered anticyclones, invented tools for registering weather data, and contributed to the establishment of the Meteorological Office in Britain. In 1890, he persuaded Scotland Yard to use fingerprints for identification purposes. He designed and built the Anthropometric Laboratory for the 1884 London International Exhibition, where he measured thousands of people, getting their heights, weights, physical abilities, and reaction times.

Perhaps his major contribution to psychology was the generalization of Darwin's framework for the scientific study of human mental traits. His cousin had demonstrated that there are widespread individual differences in the physical realm due to evolution processes, so Galton posited there must be individual differences at the mental or psychological level also due to evolution. He adopted Aldolphe Quetelet's bell-shaped curve for quantifying these differences, invented the correlation coefficient to quantify the strength of relationship between two variables, and promoted the use of percentiles for ranking people. In one of his most famous books, *Hereditary Genius* (Galton,

1869), Galton emphasized that eminence ran in families because it was due partly to genetic influences. He coined the term *eugenics* to describe a progressive movement that encouraged more procreation among the upper classes, a view sometimes called positive eugenics. This idea was taken to horrific opposite extremes by the Nazis, who distorted the original concept beyond recognition to include killing or sterilizing people they thought were not worthy of procreation. Ray Fuller's biography of Galton cited this passage about eminence from *Hereditary Genius*: "There is no escape from the conclusion that nature prevails enormously over nurture *when the differences of nurture do not exceed what is commonly found among persons of the same rank of society and in the same country*" (Fuller, 1995, p. 404, emphasis added). In other words, genetic influences (nature) were strongest when environments (nurture) were similar. Galton is often today criticized for his views on eugenics, especially when no positive/ negative distinction is made, but Fuller highlighted a more balanced view.

From a different but related perspective, a French experimental psychologist, Alfred Binet (1857–1911), developed the first intelligence tests to be used in schools. In 1903, he published a book, *The Experimental Study of Intelligence*, summarizing his research on intelligence, mainly based on the systematic observation of his two daughters (Binet, 1903). He was aware of Galton's conceptualization of intelligence, but he had another view. Binet's view regarding the measurement of human intelligence has dominated the assessment of intelligence, as we will see.

We will also discuss the contributions of another Briton, Charles Spearman. According to Ian Deary, Huarte's work, "if bound together with (Charles) Spearman's (1927) *The Abilities of Man*, the resulting volume would be a comprehensive review covering the period from antiquity to the establishment of scientific psychology, and would pose almost all the important questions currently being addressed in mental ability differences" (Deary, 2000, pp. 47–48).

1.2 Testing for Intelligence

Society requires methods for selecting candidates either into employment directly or into educational systems that serve as channels to future employment. Not everyone can do whatever they want independent of required ability. Students have to be selected, jobs have to be filled, and when behavioral problems arise, mental competence must be assessed. We often rely on formal testing to accomplish such selections objectively as much as possible to avoid unjustified biases. If you do not like testing, what is your alternative that can do better?

1.2.1 Testing before Psychological Science

Modern psychologists did not invent testing. In the early days of the Chinese empire, an elaborate series of local, regional, and nationwide tests were used to select officers for the imperial bureaucracy. Candidates had to write traditional poetry, explain the importance of fearing the will of heaven, and know the words of the sages.

What the Chinese tested, and what we today attempt to evaluate, is a collection of mental traits that we call intelligence or general cognitive ability. These traits define individual differences in abilities and skills with broad application in many settings. Some of the most important aspects of intelligence are the abilities to reason, plan, solve problems, and learn. You demonstrate this by showing that, after exposure to knowledge, you have learned something useful. For example, the skills needed to do well on a college entrance test are not exactly all the skills you need to acquire a bachelor's degree, but there is some overlap of a general factor. That is why both the classic Chinese and modern testing work. It is also why they work imperfectly.

1.2.2 Alfred Binet Invents Modern Intelligence Testing

Modern schooling is an unusual form of education. Before 1800, most humans were educated on the job – observing and then helping adults and serving as apprentices. Universal education, the requirement that every child learn by practicing seemingly esoteric exercises in a setting divorced from everyday life, is a late nineteenth-/early twentieth-century idea.

At the start of the twentieth century, France was committed to providing public education for all its citizens. However, the French Ministry of Education had a problem. The idea of universal public education had been adopted so that all children would have an opportunity to compete for desirable positions in society. Given different backgrounds and abilities, this goal was not easy to achieve.

By 1900, it was apparent to educators that some children had a great deal of trouble learning in this manner. The French educational administration needed a way of identifying such children so that they could either be dropped from the system or channeled into an educational program more suited to their abilities. It was also important to prevent children with behavioral problems instead of academic weakness to be sent to remedial programs by teachers eager to have discipline problems removed from their classrooms.

French educators needed an objective method to evaluate students' potential to learn so the goal of universal education could be achieved. So, from the very first, testing was intended to benefit society as well as individuals. To meet this challenge, the Education Ministry hired Alfred Binet, who began his task by making two assumptions:

1. Mental competence increases over the childhood years. The typical six-year-old can solve problems a four-year-old cannot; a four-year-old can solve problems a two-year-old cannot – and so on, at least from birth to the late teenage years. Therefore, it makes sense to talk about mental age – the level of mental competence at which a child is operating.

    Binet took a pragmatic approach to the measurement of mental age. He asked experienced teachers what sorts of problems children could solve at different ages. Once he had a set of problems typical of what most children could solve at age six, seven, eight, and so on, he could assess a person's mental age by finding the most difficult problems that a child could solve. Mental age could then be compared to chronological age, to determine whether a child has been performing below, at, or above the average cognitive level that would be expected based on chronological age.

2. Binet then made his second assumption: a child's relative standing in mental development, compared to their age group, will remain fairly constant as the child ages. If Claude and Pierre are both six years old, but Claude has a mental age of eight and Pierre has one of five, Binet assumed that four years later, when they were both ten, Claude would have a mental age higher than ten and Pierre a mental age lower than ten.

    Therefore, it follows that if you test children on entrance to school (age six), and you find that some are markedly behind (have mental ages in the three to four range), those children are likely to be behind their classmates at all ages and, therefore, are candidates for removal from the standard school program and entrance to remedial education.

That is just what the French education system wanted to know. The Education Ministry accepted Binet's approach, and the modern era of intelligence testing had begun.

1.2.3 *The Intelligence Quotient (IQ)*

Mental age was a meaningful concept for children since cognitive abilities increase as the brain, and the body itself, is developing. But as mental testing expanded to the evaluation of adolescents and adults with mature brain development, there was a need for a different measure of intelligence. The intelligence quotient (IQ) was born. Originally, IQ referred to mental age divided by chronological age, but it now refers to any

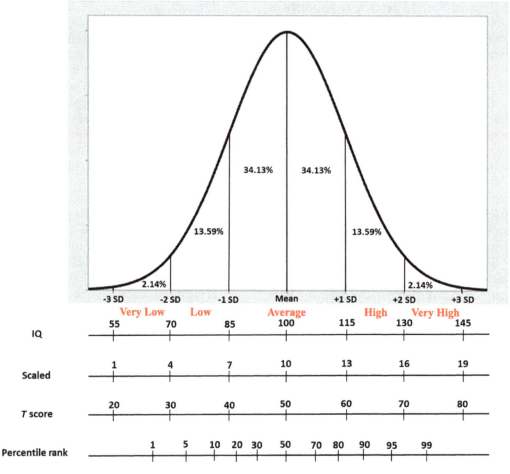

Figure 1.1 The bell curve for IQ. The area under this curve represents 100 percent of the population. The area under the curve and to the left of a given IQ value represents the proportion of people in a population who have IQs lower than the given IQ value. This can be seen in the percentile rank line. Conversely, the area to the right indicates the fraction of people who have this IQ or a higher one. For example, 50 percent of the area under the curve lies to the left of IQ = 100, indicating that half the population has an IQ of less than 100. Just over 68 percent have IQs between 85 and 115. To the right of an IQ of 130 (2 standard deviations above the mean of 100), only 2.14 percent is under the curve, indicating that only 2.14 percent of all people have IQs of 130 or higher (see Table 1.1 for more examples). The bell curve for IQ scores is a special example of the normal, or Gaussian, distribution. Some intelligence tests use scaled scores or a *T* score; both are equivalent to IQ and percentile rank, as shown.

standardized score on an intelligence test. Modern tests have been developed according to a scoring protocol where average intelligence receives a score of 100 and other scores are assigned so that the scores are distributed normally around 100, with a standard deviation of 15 (standard deviation is a measure of the scatter of scores around a mean).

As illustrated in Figure 1.1, in a normal distribution (also called a bell curve, because of its shape), approximately two-thirds of all scores lie between 85 and 115. Five percent of all scores are above 125, and 1 percent are above 135. Similarly, 5 percent are below 75, and 1 percent are below 65. Therefore, IQ, in the narrow sense, is a score indicating a person's relative performance on an intelligence test, compared to the performance of other people in an appropriately chosen comparison group.

Table 1.1 The distributions of standard (z) scores (based on standard deviation units; negative values are below the mean) and IQ scores in terms of the percentage of people above or below selected scores

Standard score (z)	IQ score	% Below	% Above
−2.33	65	0.982	99.018
−2.00	70	2.275	97.725
−1.00	85	15.866	84.134
−0.67	90	25.249	74.751
0.00	100	50.000	50.000
0.67	110	74.751	25.249
1.00	115	84.134	15.866
2.00	130	97.725	2.275
2.33	135	99.018	0.982
3.00	145	99.865	0.135
4.00	160	99.997	0.003

In the broader sense, the term *IQ* often is used as a synonym for intelligence, that is, as a shorthand term for individual differences in cognition. A person who has high intelligence will probably have a high IQ score, but the distinction between the two is important. The use of IQ as synonymous with intelligence causes much confusion and deserves careful consideration.

In interpreting IQ scores, it is often useful to think of percentiles, which indicate the percentage of people in the reference group whose scores are below or above a certain level. What that level is is shown by the IQ score and by the properties of the bell curve itself. Table 1.1 shows examples of reference scores. The properties of these scores follow from the assumption that IQ scores will fit the normal distribution illustrated in Figure 1.1. An IQ of 65 would, if accompanied by other indications of mental competence, be cause for considering a person mentally disabled. If IQ is distributed normally, about 1 percent of all people have IQ scores this low. Average IQ is, by definition, 100. Approximately half of

all scores lie between 90 and 110. About 16 percent of the scores lie above 115, 2 percent lie above 130, and 1 percent lie above 135.

MENSA, an international organization whose members have high IQ scores (ninety-eighth percentile plus 2 standard deviations, or IQs greater than 130, is the minimum for membership), defines the even higher 4-sigma group as people with IQs over 160. (*Sigma* is a term frequently used to refer to the standard deviation.) This latter level of score would be expected three times in every 100,000 observations.

If someone says that their child has an IQ of, say, 110, this does not mean that the child's mental age is 10 percent higher than their chronological age or that the child is 10 percent smarter than a child with an IQ of 100. It means that the child has a test score in the top 25 percent of test scores at the child's age.

Why are IQ scores distributed normally? IQ tests are constructed by choosing appropriate numbers of easy, intermediate, and hard cognitive problems or exercises (items). The total scores will be normally distributed in the population for which the test was intended.

There is no interpretation of IQ independent of the tests themselves. IQ scores are used to describe people *relative to each other*. This contrasts with a variable like height of an individual, which is defined independently of the height of other people. Height happens to be distributed approximately normally, within the populations of adult men and women.

The distribution of height is a fact of nature. The fact that IQ test scores are normally distributed is an outcome of the test construction procedure. Nevertheless, despite limitations, test scores are a reasonable and useful way to quantify differences among people. Importantly, these cognitive differences exist regardless of their formal measurement by IQ tests, which provide an excellent quantification of these cognitive differences; the former does not create the latter.

IQ scores are also used to make predictions and to indicate associations, as in predicting a student's likely academic progress or investigating the association between intelligence and income differences. There are technical reasons for wanting to deal with normally distributed scores when we apply the statistical methods used for making predictions and analyzing associations, as discussed at length in later chapters.

There is another, less technical reason for requiring that IQ scores be normally distributed. Many other human factors that can be measured on scales with physical interpretations, like height and weight, are distributed normally. If we could measure intelligence in some physical manner, such as measuring the efficiency of the nervous system, the number of neurons, or the integrity of white matter connections in the brain, these measures would likely turn out to be normally distributed. Therefore, it seems appropriate to require that IQ scores be normally distributed.

In the late nineteenth and early twentieth centuries, this reasoning seemed compelling, because the normal distribution itself was regarded as a law of nature. Today, there is still a good argument for assuming a normal distribution. If a person's intelligence is due to a large number of causes, each of which has a small effect, intelligence would be distributed normally across the population. In fact, as explained in Chapter 6, a large number of genes, each with small effects, apparently contribute to intelligence differences.

Binet's assumption that mental competence increases as children grow older was correct. He was also correct that there are marked individual differences in the rate at which mental competence increases across age. His second assumption was that relative standings remain constant as children age. This is true on the whole, although there are some exceptions. The smartest kid in grade school does not always become a Phi Beta Kappa in college or a wealthy CEO. However, after the age of about ten, indicators of relative cognitive competence are fairly stable. Stable refers to a person's rank in a group ranked by IQ scores. Even though scores change over time, the rank does not change much. The smartest eleven-year-olds will tend to be the smartest eighteen-year-olds. Variance (individual differences in cognitive ability) and mean levels (average cognitive ability) tell different stories, and it is extremely important to keep this in mind. The same happens with height: on average, eighteen-year-olds are taller than eleven-year-olds, but the tallest eleven-year-old will tend to be the tallest eighteen-year-old. The rank ordering over time allows estimating stability values in a given physical or psychological trait.

Evidence for the stability of intelligence is demonstrated in a countrywide study of Scottish schoolchildren, coordinated by Ian Deary (more details of this classic study are in Section 1.3.1). There were substantial correlations between intelligence test scores taken at age eleven and subsequent measures taken when the examinees were in their sixties and seventies, even though the average scores were higher in the older adults (Deary, 2014). Based on this and similar studies, most researchers regard intelligence as a *trait*, a characteristic of the individual that is stable over time and that is revealed in many situations.

1.2.4 The Stanford–Binet and Wechsler Tests

Lewis Terman, from Stanford University, translated and modified Binet's tests for use in the United States of America. The resulting test, the Stanford–Binet Intelligence Test, is still used today in updated form (SB-5). The Binet and Stanford–Binet tests were intended for use with schoolchildren. In the late 1930s, David Wechsler, a clinical psychologist working at New York City's Bellevue Hospital, created a similar test for adults, the Wechsler–Bellevue test. It has subsequently been modified into the Wechsler Adult Intelligence Scale (WAIS). It and a companion test for children, the Wechsler Intelligence Scale for Children (WISC), are the most widely individually administered intelligence tests today.

Both the Wechsler and the Stanford–Binet tests are individually administered. The examinee sits down with a trained examiner and attempts to solve a series of problems, divided pragmatically into problems that vary in the demands that they place on language and memory. Wechsler has described this as an opportunity for the examinees to display their cognitive abilities and skills during a standardized interview with an experienced observer.

The resulting IQ scores have proven to be highly useful in many domains (see Chapter 8 for details). For instance, the WAIS is widely used to evaluate a person who, for whatever reason, is of suspected mental disability. Examples of such use are the adjudication of legal competence and the analysis of status following brain injury.

Other applications of these tests for individuals are extensions of these ideas. The cost of testing is evaluated relative to the potential benefits of the results in making judgments about an individual case. Are the decisions made about this person improved by knowing test scores, and if they are, is the value of a typical decision enough to justify the costs of the test (see Box 1.1 and Chapter 8)? The Wechsler and

Stanford–Binet tests are not the only individually administered intelligence tests, but they have played an important role in the development of testing for intelligence.

1.2.5 Group Testing

The next major step in intelligence testing was a spin-off from a critical military need. When the United States entered World War I, the army had to make rapid evaluations of mental abilities of large numbers of incoming soldiers to help direct them into appropriate specialties, as was Huarte's goal noted earlier. The War Department sponsored development of a test that could be administered to large groups of recruits. Psychologists responded by developing the Army Alpha Test, a written test suitable for group administration, and the Beta test, a version which could be given to recruits who could not read.

The military tests are examples of successful personnel classification tests. Today,

Box 1.1 Is There Any Value in Knowing Your IQ?

Arthur Jensen, known for his pioneering studies of intelligence, discussed IQ testing on the *Phil Donahue* television talk show more than forty years ago. He received hundreds of letters with questions from viewers. In his book *Straight Talk about Mental Tests* (Jensen, 1981), chapter 7 was devoted to addressing some of these questions. Regarding whether it is important to know your own IQ, Jensen answered in part, "I can do what I try to do, with some effort, and I don't believe that knowing my IQ would ever have been of any use to me in the process of trying to achieve any of my goals. ... The best way to find out if you can achieve something is to try to achieve it. No person should approach a challenge as a statistic to be predicted by a test score in a regression equation." In

other words, it is not generally important to know your own IQ.

Here are some other interesting questions from the 1980s and Jensen's answers from his chapter 7 (they presage chapters later in this book):

Question: If group IQ tests are abandoned by the public school, what would take their place [note that group IQ testing used to be common in public schools, although this is no longer true]?

Answer (in part): Schools should focus on achievement instead of on the measurement of cognitive abilities. The exception is when students show unusual learning problems that require attention by qualified psychologists. ... The aim of IQ tests is not measuring specific skills and knowledge, but the general cognitive ability underlying observable performance. [See more about this general ability in our Chapter 3.]

Box 1.1 *(continued)*

Question: What can I do to raise my child's IQ [intelligence]?

Answer (in part): I don't know of any psychological prescription that will lead to the fulfillment of this parental wish. No such formula has been discovered. [This is still the case, as explained in our Chapter 13.]

Question: Isn't there a danger that scientific knowledge about such subjects as genetics, intelligence, and race might be misused by racists?

Answer: The place to stop the misuse of knowledge is not at the point of inquiry, but at the point of misuse. To avoid pursuing scientific inquiry for fear that racists will misuse it is to grant them the power of censorship of research. ... Already well-established findings in genetics [see our Chapter 6] and differential psychology clearly contradict the essential tenets of racism. ... The sound use and interpretation of mental testing can help reinforce the democratic ideal of treating every person according to his or her individual characteristics, rather than according to race, sex, social class, religion, or national origin. [See more in our Chapters 9 and 12.]

Jensen's views are widely accepted among today's intelligence researchers. More of Jensen's work will be discussed in Chapters 3, 7, and 13.

---

cognitive tests for personnel classification are widely used in the civilian sector as well as in the military (we present supporting data in Chapter 8). The costs and benefits of testing within a personnel classification system are not the same as the costs and benefits of testing intended for individual counseling and/or placement.

In a personnel classification system, correct classifications have a value and incorrect classifications have a cost, as seen from the perspective of the institution setting the test, rather than as seen from the perspective of the examinee. A classification test is economical if, on average, the cost of administering the test is less than the value of improved decision-making. This view shifts the focus from decisions about an individual to the average value of a decision, calculated over the population. The shift greatly affects the economics of testing. A cheap test, which makes only a moderate improvement in the accuracy of the selection decisions, administered to thousands or even millions of people, can be a valuable classification instrument.

Well over a hundred group-administered classification tests have been developed.

They include the Scholastic Ability Test (SAT) used in the college admissions process in the United States and the General Aptitude Test Battery (GATB), which for years was used by the US Department of Labor to provide a test score to guide in industrial hiring.

Describing all the tests in use today literally takes a volume, and the volume has to be updated annually. We discuss more details of using tests for selection in Chapter 8. If you are interested in traveling more through this rich forest, you can visit the highly recommended web page of the International Test Commission (www.intestcom.org).

The important point is that since the 1930s, intelligence testing has been widely used to make important decisions about academic and vocational careers (Detterman, 2014). Testing is also used as a guide in medical rehabilitation, such as evaluating the course of treatment following insults to the brain. The tests are also widely used in research on the description, causes, and consequences of being more or less intelligent, as we shall see throughout the rest of this book.

1.3 Do the Tests Work?

There has been debate about how well intelligence tests work. How accurate does a test have to be before we say that it works? How well do we understand why the tests work?

People might be uncomfortable using indicators that they do not understand. To the scientist interested in intelligence, though, this question ought to be a challenge. What are the consequences of basing important decisions about education, employment, and personal planning on test scores? Let's do three things: (1) explain the criterion by which tests are judged as being more or less accurate, (2) provide a few statistics, and (3) describe some key controversies.

1.3.1 *Tests' Accuracy*

Tests would be perfectly accurate if, whenever a test score was used as a predictor, there was some critical score such that everyone who had a score equal to or higher than the critical score succeeded, and everyone with a lower score failed.

No such intelligence test exists, and none ever will. Test scores are not perfect indicators of a person's cognitive power, because other things besides intellectual competence contribute to success (or failure). And of course, success (or failure) itself is not an either/or thing.

The question is not whether test scores can be used to make perfect predictions of success or failure; the question is whether using test scores reliably improves our ability to predict who will succeed or fail. Tests that do this are said to be valid, and validity is a matter of degree. The more using a test improves prediction, the higher the test's validity is.

Here is an example that illustrates the issue. In the 1930s, the Scottish psychologist Godfrey Thomson conceived a plan to test the entire nation of Scotland. With the support of the government, he managed to do it. In 1932, Thomson's Moray House intelligence test was given to all eleven-year-olds in Scotland (more than 80,000; see study details in Chapter 11 and more findings in Chapter 8, based on an impressive seventy-year follow-up of surviving participants [Whalley and Deary, 2001]). Figure 1.2 shows the fraction of the original respondents to the 1932 test who were still alive at different times from the 1930s until the start of the twenty-first century. The data, based on death records, are shown separately for women and men from the upper quartile (top 25 percent) and lower quartile (bottom 25 percent) of the distribution of intelligence test scores.

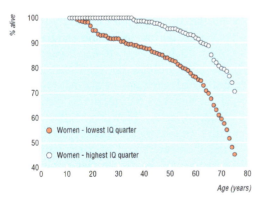
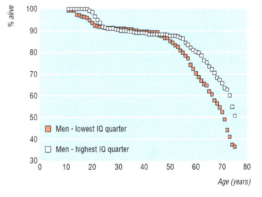

Figure 1.2 IQ and mortality. The percentage (*y*-axis) of individuals surviving to different ages (*x*-axis) for a cohort of approximately 2,200 Scottish children tested at age eleven, in 1932. The left panel shows high-IQ women (white circles) live longer than low-IQ women (red circles). The same finding is weaker for men (right panel).
From Whalley and Deary (2001), with permission.

The data show that 70 percent of the women with higher intelligence scores in childhood were still alive at age seventy-five, whereas 45 percent of the women with lower intelligence scores in childhood were still alive at age seventy-five. This is quite a substantial difference. Men showed a similar, albeit weaker pattern (and there was a notable minimized IQ influence during the years of World War II). This is an impressive and unexpected finding indicating that intelligence test scores obtained in childhood are a valid statistical predictor of length of life. Try to find another psychological factor for predicting longevity, and see what happens (Deary et al., 2010).

Now let's use this finding to illustrate some of the complications dealing with test scores as predictors and, by implication, intelligence itself.

1.3.2 A Few Statistics

We first must find out whether we are dealing with an artifact due to some relationship that is forced between test scores and the outcome of interest. For instance, in the United States, SAT scores of people who enter college are higher than test scores of people who do not enter college. That fact alone is uninteresting, because test scores are only one of the indicators used to decide who gets to go to college.

However, this sort of problem did not arise in the Scottish longevity findings, for it is unlikely that a score on a test taken at age eleven has any direct influence on survival at age seventy. The statistic indicates some meaningful relationship between mortality and whatever the test measures. But what? We can imagine three different reasons why a cognitive ability test score might predict longevity.

1. There could be a direct relationship. We know, today, that intelligence test scores can be partly attributed to the state of the brain. It could be that the test scores revealed (imperfectly) the state of the brain at age eleven, and this state carried forward over the years, producing an association between test scores and later mortality. This explanation would have appealed to both Huarte de San Juan and Galton.

2. There could be an indirect relationship. Perhaps intelligence, as revealed by the tests, makes people less likely to do unhealthful things, like drinking alcohol to excess, driving a motor vehicle after drinking to excess, declining to take vaccinations against influenza, and so on. This explanation – that intelligence is related to finding smart solutions as you navigate your way through life – would have appealed to Binet.

3. Finally, it could be that the relationship is not due to intelligence at all. It could be due to a third variable that influences test scores and also influences mortality. Parental socioeconomic status (SES) is a term used to refer to the general place in society occupied by a child's family. It includes such things as financial status, education, and generally beneficial lifestyle practices, such as having children inoculated against common diseases. It is also an indicator of the extent to which a child is likely to have family connections to help establish their own lifestyle. There is a positive statistical relationship between children's test scores and parental SES. It might be that childhood test scores predict longer life because parental SES assists in the development of a lifestyle that promotes longevity (or the opposite, for low test scores and parental SES). Richard Nisbett has argued that the relationships between test scores and social outcomes are actually relationships between SES and the outcomes (Nisbett, 2009). However, SES is influenced in part by intelligence, so there is an inherit confound of these variables, which we discuss in later chapters because of their relevance.

Which of these explanations do you think best accounts for the Scottish data?

There are probabilistic relationships between test scores and a variety of life outcomes in life. In general, the higher a person's test score, the more likely that person is to have good things happen. These include educational achievement, on-the-job performance, income, marital status, (physical and mental) health, and longevity. Most of the statistical relationships are high enough to be economically valuable evidence to guide decision-making, both in personal life planning and in personnel classification. But because the relationships are probabilistic, they have to be understood in statistical terms. This demands some knowledge of probability on the part of the person who tries to interpret these relationships, although, generally, we humans are not so good at using probabilities for decision-making (Gigerenzer, 2015).

Test scores are not only correlated with life outcomes; they also are correlated with a variety of other measures that are correlated with the outcomes. The parental SES example just given is illustrative. It is often difficult to know why there is a relationship between intelligence test scores and outcomes such as educational achievement, job performance, and health.

The fact that there is a correlation between an IQ score and variable X, where X is almost any life outcome, does not prove itself that intelligence causes X. This caveat is important. There might be several plausible interpretations for the relationship, and they might not be mutually exclusive. Keeping an open mind is mandatory for science and cannot hurt for society at large (Sagan, 1995).

In the intelligence–life expectancy example, three alternative explanations were offered, and none of them excluded the others. The questions "What causes intelligence?" and "What does intelligence cause?" are not easily answered. Do not expect simple answers about intelligence in this book. If you encounter simple answers somewhere else, please be suspicious.

1.3.3 *Objections to Testing for Intelligence*

Because test scores can be used to make major social decisions, it is hardly surprising that they have been controversial. Let's check five basic objections to the tests and

the responses that science can provide based on the best available evidence. (More details about the evidence for all five objections is found in Chapter 8.)

> *Objection 1. The tests cannot possibly be meaningful. It is unreasonable to believe that performance on an out-of-context snapshot test, made up by people who do not know the examinee, could possibly be an important measure of mental abilities.*

This is a common objection, but it is not supported by empirical evidence that test scores predict a variety of meaningful social outcomes. This is called *predictive validity*. Predictions from intelligence test scores are not perfect, but, in many circumstances, they provide cost-effective, reliable information not available from other sources. Few tests, if any, have stronger and more widespread predictive validity than do intelligence tests.

> *Objection 2. The tests don't work. There are people who have only modest test scores and do well and people who have high test scores and are not doing well.*

Predictions made based on test scores are never 100 percent accurate. Perfect prediction is an impossible goal, as noted, because intelligence is not the only factor that contributes to success or failure across the numerous aspects of life.

When people have to deal with probabilistic relationships, they try to get an intuitive feeling for the relationship, rather than analyzing numbers. This leads to an overinterpretation of exceptional cases. Because the tests work well on average, as overwhelming evidence indicates, failures stand out (dog bites human is not news; human bites dog is news). People remember the case of the person with low test scores who did brilliantly or the one with high test scores who did miserably. The fact that many people perform about as well as their test scores predict, with a small amount of variation, just does not have the same impact as the outliers.

On the other hand, many of us live in a cognitively stratified society. Children of parents with high, middle, or low SES go to schools with children mostly of the same background as their own. Academic testing mechanisms stratify college students on the basis of cognitive test scores. In the workplace, our coworkers generally tend to be about as intelligent as we are. When we deal with people of greatly varied intelligence, we tend to do so in a stereotyped way. A sales clerk, for instance, encounters customers of highly varied intelligence but does not interact with them in a way that would reveal their intelligence. Stratification also occurs in neighborhoods, because there is a great deal of residential segregation by SES. Neighborhood life is stratified by intelligence to a large degree (Murray, 2012).

Within a person's local society, there will not be a great deal of variation in intelligence, so other factors, such as variations in personality traits, will play a more apparent role in social success. Therefore, when people try to evaluate intelligence by referring to their personal experiences, they are likely to undervalue the role of intelligence in the general society. Nonetheless, scientists have established an impressive body of research that shows intelligence is an important part of everyday life.

Objection 3. The tests work only in academic settings.

Leaving aside a few spectacular but rare cases, such as billionaires heading companies that pioneer new technology, it is often difficult to establish criteria for who is succeeding in the workplace or nonacademic setting on an individual basis. This makes the statistical analyses more complex than they are in the case of analyzing academic success where the criteria are fairly straightforward (e.g., grades), and hence harder to explain to the nonstatistician.

Objection 4. The tests work only for certain demographic groups. The tests do not predict well for other groups.

This kind of bias has been studied extensively for decades (Jensen, 1980), and there are other technical kinds of test bias (Wicherts, 2016). This objection speaks to simple bias for or against groups where a test is biased in favor of a group if people with high test scores actually do not perform well on the predicted variable. Or, a test is biased against a group if people with low test scores actually perform well on the predicted variable. The evidence supports the conclusion that test scores have about the same power to predict achievement within all demographic groups. This is the case even if average scores differ between groups (Warne, 2020). There is more about test bias in Boxes 9.3 and 12.1.

Objection 5. The tests should not be used because they are prejudiced against some groups who tend to get low scores.

This objection raises one of the most troublesome topics in psychology: the possibility that there are intelligence differences among populations, especially groups of distinguishable ancestry. The fact is that there are differences in the average intelligence test scores of different groups, as acknowledged by (for instance) the American Psychological Association (APA; Neisser et al., 1996). The debate is not about the existence of these differences but about their meaning and their implications for social action. These are two separate issues that frequently are mixed in uninformative ways for reasons that have nothing to do with scientific research.

If the tests predict performance only within a group, discrepancies in test scores across groups could simply be disregarded. However, that is not the case. Test scores predict performance across different populations. Although many opinions have been expressed, the fact is that there is no consensus about the causes of group differences in IQ scores (see Chapters 9 and 12).

Should test scores be used in personnel screening if doing so would reduce the proportion of applicants from any group who are accepted for employment or education?

There is no justification for using a test that would have an adverse impact on one group or another if that test is not a valid predictor of performance (Sackett et al., 2021). If the test is a valid indicator – as intelligence tests usually are – then the policy maker is faced with a trade-off. Should the best people be selected, regardless of group membership, or should some effort be made to balance rates of acceptance across groups? This is a policy issue, not a scientific one. Scientific research can provide information about the costs and benefits of a policy, but the decision is up to the policy makers and to the society they represent.

Please keep this distinction between science and policy in mind at all times.

1.4 Debunking Myths before Moving Forward

Russell T. Warne and colleagues published a comprehensive analysis of how introductory psychology textbooks inform students about intelligence research (Warne et al., 2018). We will use their analysis here for highlighting some of the key issues regarding the scientific intelligence construct to help readers consider the rest of this book with an open mind.

The first issue noted by Warne and colleagues is that psychology courses invest less time to teach intelligence than the demonstrated importance of the construct deserves for a comprehensive understanding of human behavior. Upper-level intelligence courses are rare, and therefore, what undergraduates learn about intelligence is mostly taught in introductory psychology courses. The two key research questions addressed by the analysis of Warne and colleagues are (1) What are the most frequently discussed topics related to intelligence in introductory psychology textbooks? and (2) How accurate are introductory psychology textbooks in their discussion of intelligence?

They considered twenty-nine textbooks published between 2011 and 2017. For testing the accuracy levels of the information included in the textbooks, two reference documents were used: (1) the mainstream statement on intelligence coordinated by Linda Gottfredson (Gottfredson, 1997) and (2) the APA report on intelligence coordinated by Ulric Neisser (Neisser et al., 1996). These documents were chosen because they summarize widely replicated findings in intelligence research and are old enough to be known by the textbooks' authors. The documents agree on major points.

These were key findings derived from Warne et al.'s analyses:

1. The average textbook devotes 3.3 percent of the text to the intelligence concept.
2. Thirty-eight percent of the books include a complete chapter for intelligence.
3. These are the top ten topics (based on devoted number of pages): IQ, Charles Spearman and the general factor of intelligence (g), multiple intelligences, Sternberg's theory of intelligence, measurement of intelligence, reliability and validity, Alfred Binet, Stanford–Binet test, environment and intelligence, and intellectual disability.
4. On a scale ranging from 0 to 4, the average number of logical fallacies used to dismiss intelligence research was 1.8 (SD = 1.2). The fallacies considered are listed in Box 1.2.
5. Seventy-nine percent of the books included fallacies or inaccuracies about intelligence research.
6. The most common inaccurate statement relates to test bias as a key limitation of intelligence testing, albeit extensive research has shown that the issue of test bias is not supported (since scores in different populations predict outcomes equally well).
7. Multiple intelligences and Sternberg's theory of intelligence received much more attention than deserved according to mainstream research (Box 1.3). These nonmainstream models are covered more than the g-factor model, which has the most research support by far.
8. Politically correct statements and models are covered more than controversial ones.

Box 1.2 Gottfredson's Thirteen Logical Fallacies Used to Dismiss Intelligence Research (Gottfredson, 2009)

Fallacies are mistaken beliefs based on unsound and faulty argumentation and logic. They can be intentional (for cheating the audience) or not (due to a lack of the proper evidence). Here we briefly present the thirteen fallacies identified by Linda Gottfredson and used by some critics of the science of human intelligence.

1. *Yardstick mirrors construct.* This fallacy involves portraying the superficial appearance of an intelligence test as if these visible features mimic the essence of the phenomenon it measures. However, this is not the case. A thermometer's appearance does not provide any clue about the nature of heat. Everybody understands the second fact, but some people reject, with vehemence, the fact that exactly the same applies to the tests designed for measuring the construct of intelligence.

2. *Intelligence is a marble collection.* This fallacy argues that general intelligence (*g*) is just an aggregation of separate specific abilities, not a singular phenomenon in itself. Like marbles in a bag, intelligence is thought to be an aggregate of many separate abilities psychometricians choose to add. This mistake is because IQ scores typically are calculated by computing a person's score based on the various subtests included in a standardized battery. This fallacy takes for granted that the way scores are computed mirrors how general intelligence is constituted. However, intelligence is not the sum of several independent skills but its common core, the psychological factor that binds and integrates the remaining human cognitive abilities.

3. *Nonfixedness proves malleability.* People grow and learn. No doubt about that, although some people learn more stuff and faster than others. Developmental change within individuals, however, tells a different story to the fact that IQ levels change hard. IQ scores compare individuals within the same reference group, and the ordering of individuals across the life span is quite stable, especially after the childhood period.

4. *Improvability proves equalizability.* This fallacy states that because social interventions can raise mean cognitive levels, individual differences in cognitive ability can be eradicated. However, mean levels and variability point to independent facts, as we will highlight in this book.

5. *Gene–environment interaction nullifies heritability.* Genes and environment can work together to produce phenotypes (true). Therefore, it is argued, calling to this fallacy, we cannot separate the contribution of either one to individual differences in intelligence (false): "this is analogous to saying that it would be impossible to estimate whether differences in quality of Tango performances among Chinese couples is owing more to skill variation among the male partners than to skill variation among the female partners (genetic versus non-genetic variation) or to what extent differences among couples in their quality of performance depend on the chemistry between two partners (gene–environment interaction)" (Gottfredson, 2009, p. 38). It is crucial to understand that the typical course of human development and variations in development tell different stories. As underscored by Belsky and colleagues, "variation is the norm

Box 1.2 *(continued)*

instead of the exception" (Belsky et al., 2020, p. 3).

6. *Genetic similarity of 99.9 percent among humans negates differences.* Humans are more than 99 percent alike genetically. Therefore, that remaining less than 1 percent must be trivial. However, differences in 3 million base pairs (contained in the less than 1 percent figure) is hardly trivial, as demonstrated by scientists working in large-scale research projects, such as ENIGMA (Thompson et al., 2020) or the One Thousand Genomes Project.

7. *Contending definitions of intelligence negate evidence.* Because there are disparate definitions of intelligence held by experts in the field, no one really knows what (if anything) IQ tests really measure. However, using this argument, one can say the same of gravity or health. Furthermore, "competing verbal definitions do not negate either the existence of a suspected phenomenon or the possibility of measuring it" (Gottfredson, 2009, p. 45).

8. *Phenotype equals genotype.* It is argued that differences among humans, including their intelligence, are innate, genetically determined. This fallacy portrays phenotypic differences in intelligence as if they were exclusively genotypic, which is false by any means. IQ standardized tests measure phenotypes, and only genetically informative designs can help to separate the contribution of genetic and nongenetic factors to the measured differences.

9. *Biological equals genetic.* This fallacy assumes that a biological difference (in, say, cortical thickness or brain nerve conduction velocity) must be genetically caused. However, this is openly false. Genes contribute to our biology, but variables such as nutrition and disease also make substantial contributions.

10. *Environmental equals nongenetic.* The fallacy is based on the presumption that environmental influences on development are unaffected by individuals' genes. However, although environments are physically external to individuals, they are not independent of genes: "individuals select, create and reshape their personal environments according to their interests and abilities. ... Differences in personal circumstances are somewhat genetically shaped" (Gottfredson, 2009, pp. 50, 53–54).

11. *The imperfect measurement pretext.* This fallacy claims that IQ tests must perfectly measure intelligence and/or make predictions perfectly before they can be used or trusted. However, "testing is hardly the only useful source of information about students and employees, but few are as reliable, construct valid, and predictive in education and employment settings as IQ tests" (Gottfredson, 2009, p. 53–54).

12. *The dangerous thought trigger.* Socially acceptable ideas (whatever this means in different cultural settings) are the default belief or should be given less scrutiny: "the implicit premise seems to be that unsettling truths do no good and comforting lies no harm" (Gottfredson, 2009, p. 54).

13. *Happy thoughts leniency.* Gottfredson writes, "Mere theoretical possibility elevates the scientific credibility of a politically popular idea above that of an empirically plausible but unpopular conclusion" (p. 56). There are false assumptions that are almost never questioned: genetic, but not environmental, influences limit human freedom and equality. However, the historical record shows how environmental engineering can easily eliminate freedom in the blink of an eye. Janet R. Richards offered a perfect

Box 1.2 *(continued)*

example: "there is no reason at all to think, in general, that differences between people that result from differences of environment are easier to change than differences resulting from genes. . . . Nobody can unbake a baked potato" (Richards, 2000, p. 121).

Finding reliable knowledge about intelligence requires careful thinking and skepticism because, as underscored by Gottfredson, "the 13 fallacies seem to hold special power in the public media, academic journals, college textbooks, and the professions. . . . Fallacies are tricks of illogic to protect the false from refutation. . . . Sophistry is best dealt with by recognizing it for what it is: arguments whose power to persuade resides in their logical flaws" (p. 58).

Box 1.3 Robert Sternberg and Howard Gardner Argue for a Different Emphasis

Defining intelligence as a general ability (g) that applies to many diverse situations has its critics. Their views provide a constructive skepticism that helps refine our understanding of intelligence. The main criticism is that intelligence has numerous components and standardized tests capture only one slice of the cake representing human intelligence. It is true that the evidence shows other factors are relevant to some extent, but overly negative skepticism about the central importance of the general ability factor is unwarranted. What sort of evidence would make us want to either replace or expand the model of general intelligence adopted by most researchers?

Robert Sternberg distinguished three classes of intelligence (analytic, creative, and practical) that he thought were mostly independent. However, Nathan Brody (among others) demonstrated that this is not the case (Brody, 2003a, 2003b). Indeed, the predictive validity of these three types of intelligence are due to a common factor that is nearly indistinguishable from general intelligence (g). Nonetheless, the augmentation of conventional tests would be good for the field. His work on creative intelligence may represent a pragmatic advance, and the notion of practical intelligence is close to the classic concept of crystallized intelligence.

Howard Gardner starts with two premises: (1) there is no general trait of overall mental competence and (2) there are a variety of different and unrelated intelligences. The popularity of his framework of multiple intelligences, especially among educators, is based on circumstantial evidence and anecdotes rather than on the systematic data collection and analysis that is the mark of science. Despite the popularity of Gardner's views, there is virtually no objective supporting evidence (Ferrero et al., 2021). The negative attitude that Gardner expresses toward measurement does not lead to constructive science. Gardner's theory is not a map for scientific progress because science proceeds when it defines testable hypotheses, and so far, there are no measures of multiple intelligences that show independent factors consistent with his ideas. There is nothing wrong with advocating the expansion of intelligence concepts, but this must be data driven. Even if there are multiple "intelligences," this does not negate the utility of common IQ tests. If you want to test "kinesthetic intelligence," go ahead and do it, but don't confuse it with tests of

Box 1.3 *(continued)*

cognitive ability. If Gardner had substituted "talents" for "intelligences," the model might be less controversial within the scientific community.

Dissatisfaction with conventional testing is unlikely to go away completely, despite the compelling evidence in its favor after a century of research. Sternberg and Gardner made some reasonable criticisms, but their alternatives have not yet found compelling empirical support. Any claim that intelligence, as currently assessed, is not important simply defies the facts (Chapter 8).

Cognitive abilities, personality traits, and interests interact to enhance or depress influences of intelligence. A constructive approach depends on developing research that will elucidate these interactions and lead to novel insights and formal models. There are serious attempts to do so, but so far, their impact has been limited (Schmidt, 2014; Ackerman, 1996).

Warne and colleagues did a good job in their article debunking some widespread myths about intelligence, and Warne subsequently published a comprehensive book detailing thirty-five myths and the research that debunks them (Warne, 2020). Nevertheless, we think the best way for teaching intelligence is to show what the best available evidence supports. Clues for teaching intelligence properly can be found in several articles published in a special issue of the journal *Intelligence* in 2014 (Detterman, 2014; Brody, 2014; Haier, 2014; Mackintosh, 2014; Plucker and Esping, 2014; Sternberg, 2014; Hunt, 2014). The titles of some of these articles are interesting:

- "You Should Be Teaching Intelligence!" (Detterman, 2014)
- "The Universe, Dark Matter, and Streaming Intelligence" (Haier, 2014)
- "Teaching Intelligence: Why, Why It Is Hard and Perhaps How to Do It" (Hunt, 2014)

As we discuss in the next chapter, science can provide answers now, but future research efforts might show that what we consider reliable and valid knowledge today may turn out to be arguable or perhaps even incorrect tomorrow. Michael Crichton provided a lesson that we scientists must keep in mind, in the introduction section of his novel *Timeline* (Crichton, 1999, pp. ix–x):

> If you were to say to a physicist in 1899 that in 1999 moving images would be transmitted into homes all over the world from satellites in the sky; that bombs of unimaginable power would threaten the species; that antibiotics would abolish infectious disease but that disease would fight back; that women would have the vote, and pills to control reproduction; that millions of people would take to the air every hour in aircraft capable of taking off and landing without human touch; that you could cross the Atlantic at two thousand miles an hour; that humankind would travel to the moon, and then lose interest; that microscopes would be able to see individual atoms; that people would carry telephones weighing a few ounces, and speak anywhere in the world without wires; and that most of these miracles depended on devices the size of a postage stamp, which utilized a new theory called quantum mechanics – if you said all this, the physicist would almost certainly pronounce you mad. . . . The most informed scientist, standing on the threshold of the 20th Century, had no idea what was to come.

This is our perspective for the book you have in your hands. Keep your mind open.

We will explain what we know and how we achieved this knowledge. We also acknowledge that new research and insights will always challenge the status quo.

Before leaving this section on myths about intelligence, there are two more noteworthy examples of myth busting. The first is an old, discredited, but still widely read book published nearly forty years ago. In 1981, the Harvard paleontologist Stephen J. Gould wrote a scathing analysis of virtually all aspects of modern research on intelligence. He asserted as false the fundamental idea that intelligence was a meaningful term or could be quantified (Gould, 1981). He stated, incorrectly, that g was merely a statistical artifact of factor analysis. He concluded that there was no reliable evidence relating brain size to intelligence. He also claimed that attempts to show group differences in brain size were motivated by racial prejudice. Gould had previously achieved considerable public credibility as a commentator on science, so his views were widely accepted despite compelling negative reviews of his work in the technical literature (Jensen, 1982; Davis, 1983; Carroll, 1995), and these negative reviews are sustained and amplified by more recent critiques that include newer research (Warne et al., 2019; Lewis et al., 2011).

Gould had many facts just plain wrong and worse, and when confronted with detailed technical refutations of his key points and even with new MRI-based data on brain size and intelligence (Willerman et al., 1991), Gould declined to correct his mistakes or modify his opinions in the second edition of his book (Gould, 1996). Today, this book is still widely used in psychology courses to "debunk" intelligence research in general. The "Gould effect" has been coined to describe the deliberate practice of creating false controversy about a scientific question (Woodley et al., 2018). If Gould were alive today, it is hard to imagine how he might credibly refute brain/intelligence relationships based on many replicated studies using accurate neuroimaging methods and tests of intelligence, as detailed in Chapter 5.

The second critical myth is that genetics has no influence on intelligence differences among individuals because all differences are due to social/cultural influences. The compelling evidence against this view is detailed in Chapter 6, but here we note a seminal moment in the history of this belief. In 2017, a DNA study with a large sample showed correlations with educational attainment, which is highly correlated with general intelligence scores (Sniekers et al., 2017). This study, detailed in Chapter 6, was considered a breakthrough in the field, as underscored by the editorial that accompanied its publication in the preeminent scientific journal *Nature* (Editorial, 2017, p. 386): "It's the latest in a series of studies to probe the details of how genetics influences cognitive ability. Note the word 'how.' For, despite claims to the contrary – some well-meaning and some merely ignorant – it's well established and uncontroversial among geneticists that together, differences in genetics underwrite significant variation in intelligence between people. It's just that those differences seem to be many and of little consequence by themselves. As such, intelligence is a classic polygenic human trait – just like many other cognitive and physical features, from mental disorders to height. ... The more that researchers probe traits such as intelligence, and show how there is no genetic basis for discrimination, the more they distance themselves from the mistakes of the past. What most people know about intelligence must be updated." In our view, this acknowledgment was a historic moment for the science of human intelligence.

1.5 Summary

We are not all equally intelligent, and this fact is unrelated to the formal measurement of this psychological factor. Intelligence can be considered an integrative general cognitive ability (g). We all use language and numbers. We all can perceive, attend, memorize, plan, solve problems, and reason. However, all these cognitive activities must be integrated in some way before acting or before showing overt behaviors. Some people have better integration than others. Why this is so is a fundamental question for the science of human intelligence and for all cognitive science.

Standardized intelligence tests are designed to obtain estimates of this integrative general ability. Empirical evidence supports the conclusion that they do a good job, but there are inherent limitations to measuring intelligence by tests, and there is room for improvement. Science depends on measurement, so understanding key aspects of the history of intelligence testing in this chapter is key for understanding the research data presented in the subsequent chapters.

1.6 Questions for Discussion

1.1 Do you think Binet's motivation for developing a test of intelligence makes sense today?

1.2 Scores on intelligence tests are designed to be normally distributed. What would happen if this were not the case?

1.3 Would you want to know your IQ?

1.4 Which objections to intelligence testing do you find most convincing?

1.5 Can you spot any logical fallacies in your own views about intelligence?

References

Ackerman, P. L. 1996. A theory of adult intellectual development: Process, personality, interests, and knowledge. *Intelligence*, 22, 227–257.

Belsky, J., Caspi, A., Moffitt, T. E., & Poulton, R. 2020. *The origins of you: How childhood shapes later life*. Cambridge, MA: Harvard University Press.

Binet, A. 1903. *L'étude expérimentale de l'intelligence*. Paris: Schleicher.

Brody, N. 2003a. Construct validation of the Sternberg Triarchic Abilities Test: Comment and reanalysis. *Intelligence*, 31, 319–329.

Brody, N. 2003b. What Sternberg should have concluded. *Intelligence*, 31, 339–342.

Brody, N. 2014. A plea for the teaching of intelligence: Personal reflections. *Intelligence*, 42, 136–141.

Carroll, J. B. 1995. Reflections on Stephen Jay Gould's The Mismeasure of Man (1981): A retrospective review. *Intelligence*, 21, 121–134.

Crichton, M. 1999. *Timeline*. New York: Alfred A. Knopf.

Davis, B. D. 1983. Neo-Lysenkoism, IQ, and the press. *Public Interest*, 73, 41–59.

Deary, I. J. 2000. *Looking down on human intelligence: From psychometrics to the brain*. Oxford: Oxford University Press.

Deary, I. J. 2014. The stability of intelligence from childhood to old age. *Current Directions in Psychological Science*, 23, 239–245.

Deary, I. J., Weiss, A., & Batty, G. D. 2010. Intelligence and personality as predictors of illness and death: How researchers in differential psychology and chronic disease epidemiology are collaborating to understand and address health inequalities. *Psychological Science in the Public Interest*, 11, 53–79.

Detterman, D. K. 1982. Does G exist? *Intelligence*, 6, 99–108.

Detterman, D. K. 2014. You should be teaching intelligence! *Intelligence*, 42, 148–151.

Editorial. 2017. Intelligence research should not be held back by its past. *Nature*, 545, 385–386.

Ferrero, M., Vadillo, M. A., & León, S. P. 2021. A valid evaluation of the theory of multiple intelligences is not yet possible: Problems of methodological quality for intervention studies. *Intelligence*, 88, 101566.

Frey, M. C., & Detterman, D. K. 2004. Scholastic assessment or *g*? The relationship between the scholastic assessment test and general cognitive ability. *Psychological Science*, 15, 641–641.

Fuller, R. 1995. *Seven pioneers of psychology: Behaviour and mind*. London: Routledge.

Galton, F. 1869. *Hereditary genius: An inquiry into its laws and consequences*, London: Macmillan.

Gigerenzer, G. 2015. *Simply rational: Decision making in the real world*. Oxford: Oxford University Press.

Goodwin, R. 2015. *Spain: The centre of the world, 1519–1682*. New York: Bloomsbury Press.

Gottfredson, L. S. 1997. Mainstream science on intelligence: An editorial with 52 signatories, history, and bibliography. *Intelligence*, 24, 13–23.

Gottfredson, L. S. 2009. Logical fallacies used to dismiss the evidence on intelligence testing. In Phelps, R. P. (ed.), *Correcting fallacies about educational and psychological testing*. Washington, DC: American Psychological Association.

Gould, S. J. 1981. *The mismeasure of man*. New York: Norton.

Gould, S. J. 1996. *The mismeasure of man*. 2nd ed. New York: Norton.

Haier, R. J. 2014. The universe, dark matter, and streaming intelligence. *Intelligence*, 42, 152–155.

Haier, R. J. 2017. *The neuroscience of intelligence*. Cambridge: Cambridge University Press.

Hunt, E. 2014. Teaching intelligence: Why, why it is hard and perhaps how to do it. *Intelligence*, 42, 156–165.

Jensen, A. 1980. *Bias in mental testing*. New York: Free Press.

Jensen, A. R. 1981. *Straight talk about mental tests*. New York: Free Press.

Jensen, A. R. 1982. The debunking of scientific fossils and straw persons. *Contemporary Education Review*, 1, 121–135.

Kaufman, S. B., Reynolds, M. R., Liu, X., Kaufman, A. S., & McGrew, K. S. 2012. Are cognitive g and academic achievement *g* one and the same *g*? An exploration on the Woodcock–Johnson and Kaufman tests. *Intelligence*, 40, 123–138.

Lewis, J. E., Degusta, D., Meyer, M. R., et al. 2011. The mismeasure of science: Stephen Jay Gould versus Samuel George Morton on skulls and bias. *PLoS Biology*, 9, 1.

Mackintosh, N. J. 2014. Why teach intelligence? *Intelligence*, 42, 166–170.

Murray, C. 2012. Coming apart: The state of white America, 1960–2010. *New York Review of Books*, 59, 21–23.

Neisser, U., Boodoo, G., Bouchard, T. J., Jr. et al. 1996. Intelligence: Knowns and unknowns. *American Psychologist*, 51, 77–101.

Nisbett, R. E. 2009. *Intelligence and how to get it: Why schools and cultures count*. New York: Norton.

Oleson, J. C. 2016. *Criminal genius: A portrait of high-IQ offenders*. Berkeley: University of California Press.

Plucker, J. A., & Esping, A. 2014. Developing and maintaining a website for teaching and learning about intelligence. *Intelligence*, 42, 171–175.

Pokropek, A., Marks, G. N., & Borgonovi, F. 2021. How much do students' scores in PISA reflect general intelligence and how much do they reflect specific abilities? *Journal of Educational Psychology*, 114, 1121–1135.

Richards, J. R. 2000. *Human nature after Darwin: A philosophical introduction*. London: Routledge.

Rindermann, H. 2018. *Cognitive capitalism: Human capital and the wellbeing of nations*. Cambridge: Cambridge University Press.

Sackett, P. R., Zhang, C., Berry, C. M., & Lievens, F. 2021. Revisiting meta-analytic estimates of validity in personnel selection: Addressing systematic over-correction for restriction of range. *Journal of Applied Psychology*, 107, 2040–2068.

Sagan, C. 1995. *The demon-haunted world: Science as a candle in the dark*. New York: Random House.

Schmidt, F. L. 2014. A general theoretical integrative model of individual differences in interests, abilities, personality traits, and academic and occupational achievement: A commentary on four recent articles. *Perspectives on Psychological Science*, 9, 211–218.

Sniekers, S., Stringer, S., Watanabe, K., et al. 2017. Genome-wide association meta-analysis of 78,308 individuals identifies new loci and genes influencing human intelligence. *Nature Genetics*, 49, 1107–1112.

Sternberg, R. J. 2014. Teaching about the nature of intelligence. *Intelligence*, 42, 176–179.

Thompson, P. M., Jahanshad, N., Ching, C. R. K., et al. 2020. ENIGMA and global neuroscience: A decade of large-scale studies of the brain in health and disease across more than 40 countries. *Translational Psychiatry*, 10, 100.

Warne, R. T. 2020. *In the know: Debunking 35 myths about human intelligence*. Cambridge: Cambridge University Press.

Warne, R., Astle, M., & Hill, J. 2018. What do undergraduates learn about human intelligence? An analysis of introductory psychology textbooks. *Archives of Scientific Psychology*, 6, 32–50.

Warne, R. T., Burton, J. Z., Gibbons, A., & Melendez, D. A. 2019. Stephen Jay Gould's analysis of the army beta test in *The Mismeasure of Man*: Distortions and misconceptions regarding a pioneering mental test. *Journal of Intelligence*, 7, 1–22.

Whalley, L. J., & Deary, I. J. 2001. Longitudinal cohort study of childhood IQ and survival up to age 76. *BMJ*, 322, 819.

Wicherts, J. M. 2016. The importance of measurement invariance in neurocognitive ability testing. *Clinical Neuropsychologist*, 30, 1006–1016.

Willerman, L., Schultz, R., Rutledge, J. N., & Bigler, E. D. 1991. Invivo brain size and intelligence. *Intelligence*, 15, 223–228.

Woodley, M. A., Dutton, E., Figueredo, A. J., et al. 2018. Communicating intelligence research: Media misrepresentation, the Gould effect, and unexpected forces. *Intelligence*, 70, 84–87.

CHAPTER 2

Basic Concepts

2.1 Introduction: A Framework for Understanding Intelligence

In this chapter, we address several fundamental issues that are important for learning about the scientific concept of intelligence. Many of the arguments about intelligence are less informative than they might be. Often, disagreements are misdirected because concepts of intelligence differ, and there is confusion about the relation between test scores and intelligence. Here we provide a framework for constructively discussing theories, models, and facts about intelligence.

2.1.1 *Manifest and Latent Variables*

Figure 2.1 presents a simple model organizing relevant variables. Rectangles represent manifest variables, things that might be observed

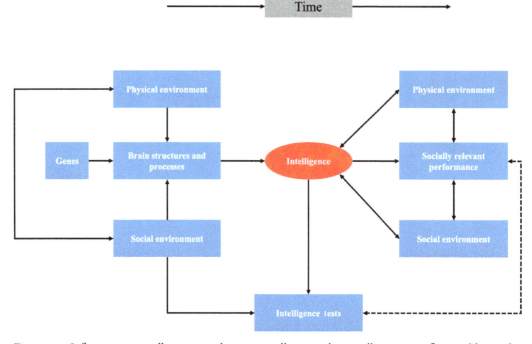

Figure 2.1 Influences on intelligence: a schematic to illustrate that intelligence is influenced by, and influences, a number of biological and environmental variables. These relationships change over time (arrows on top) such that some variables that influenced intelligence early in development (left side) subsequently are themselves influenced by intelligence (right side). Examples of the former are brain structures and processes and the physical and the social environments. Examples of the latter are performance on intelligence tests, socially relevant performance, and the physical and social environments.

and measured. Test scores, grades, and salaries are examples of manifest variables. Ellipses represent latent variables, concepts used in theories. Intelligence, culture, and socioeconomic status (SES) are examples of latent variables. They can be defined and assessed indirectly by their manifest indicators, but they are not measured directly. Their values are inferred from assessments of the appropriate manifest variables. For example, SES is inferred from education, wealth, and residence. Intelligence is inferred from test scores and other indices of cognitive performance.

Figure 2.1 also shows three kinds of relationships among variables: (1) causation, (2) reciprocal causation, and (3) correlation. Causation is indicated by a single-headed arrow. For instance, there is a single-headed arrow between genes and brain structure, because genetic makeup influences brain structure. Reciprocal causation, in which one variable causes another and then the

second feeds back into the first, is indicated by a double-headed arrow. The dashed double-headed arrow between intelligence and the social environment indicates that a person's intelligence may indirectly contribute to their social environment (not directly causal), which in turn influences the further development of their intelligence. Correlation, in which two variables tend to occur together, without any implication of causation, is indicated by a double-headed arrow with a dashed line between the arrowheads. An example is the correlation between physical and social environment.

2.1.2 *Causes of Intelligence*

Our understanding of intelligence starts with genomes. Although we talk about the human genome, in fact, each person has a unique genome (Ridley, 2003; Collins, 2010; Plomin, 2018).

A person's genetic makeup (genotype) underlies their potential for development of the brain structures and functions that support cognition to some degree. We are not clones. There are substantial individual differences in genetic makeup that have implications for the development of intelligence (detailed in Chapter 6). Although the genotype is established at conception, parts of the genotype may not be expressed until certain ages. For example, there are a number of medical conditions, such as Alzheimer's disease, that have some genetic basis but are not displayed until later in adult life. The same might happen for intelligence.

The extent to which genetic potential is realized depends in part on environmental factors that encourage or inhibit the development of intelligence. In developed societies, about 50 percent of the variance in IQ scores among people can be statistically associated with genetic variation. Importantly, this value is just an average that applies to a population, not to any individual. The contribution of genetic variability to individual differences in intelligence changes across the life span and may change in different societies or in different environmental settings. Although 50 percent is a substantial value, it is important to underscore that no one ever inherited a test score in the same sense that they inherited eye color. People differ in the extent to which they have inherited features (such as brain properties) that allow them to produce the mental abilities required to solve problems proposed on cognitive tests and to interact with their physical and social environments. As noted, genetic contribution unfolds throughout life. For example, the genetic variance associated with intelligence increases with age, a finding known as the Wilson effect (Bouchard, 2013). (More about this is detailed in Chapter 6.)

How does genetic potential interact with the physical and social environments? Here is a straightforward example. If a pregnant woman abuses alcohol, her child may be born with fetal alcohol syndrome, a serious form of mental retardation. This physical condition occurred, at least in part, because of the woman's social environment. Fetal alcoholism does not occur in societies that enforce abstinence. This condition can be a substantial problem in societies where alcohol is freely available as a recreational drug, especially if social stressors that lead to alcohol abuse are present and the individual shows some degree of psychological vulnerability (which itself might be influenced by genetics).

Social influences also act on the developing mind. At this point, we have to shift from the concept of brain to the concept of mind. It is useful to distinguish between mental abilities that are related to a person's capacity to process information in the abstract and abilities that are related to the possession of information, either about the world or about how to solve problems. Information and problem solving are associated with the society within which persons live.

Perhaps the most striking example is literacy. The ability to read appears to foster a willingness to evaluate abstract arguments and to take multiple perspectives, both behaviors evaluated by intelligence tests. These can be considered direct influences of literacy on intelligence. Literacy also has an indirect influence, for it opens the door to formal education, and education by definition is a major avenue for passing on culturally acquired knowledge. Literacy facilitates the transmission of cultural knowledge, and cultural knowledge allows us to behave more intelligently (with no guarantees of the outcome).

Medicine offers another example. Health workers wash their hands when they move from patient to patient. Prior to the mid-nineteenth century, this was not seen as necessary. Does this make today's physicians more intelligent than the physicians of Pasteur's day? In one sense, they are. Knowledge is cognitive power.

There is a two-way interaction between brain and mind. One of the major limitations on knowledge acquisition is the ability to concentrate mental effort on a topic, especially in the face of distractions. The ability to concentrate depends on how well certain brain structures work. It also depends on how efficiently information about the external world is coded inside the nervous system. The coding systems people use depend very much on their experience with the topic at hand.

For these reasons, researchers see intelligence as a complex trait produced by interactions between genetic influences and environmental experiences (see more details in Chapter 7). Where do IQ scores fit into this picture?

2.1.3 *Measurement of Intelligence*

The center portion of Figure 2.1 makes a point that should be obvious but is often forgotten. A person's intelligence test score is produced by two things: the decision to construct an intelligence test in a particular way based on the social environment and the examinee's intelligence ability to perform the test once it has been constructed.

Different societies might construct different tests, depending on the mental abilities that each society values. This does not mean that tests are narrowly culture-bound, since many mental abilities are seen as vital in all societies. For instance, all societies demand that their members learn their native language. On the other hand, societies may differ in the emphasis that they place on other aspects of cognition.

These points were illustrated by the research of Manuel de Juan-Espinosa, a psychologist from Spain who has studied conceptions of intelligence held by the Fang, a society of mixed agriculturalists and hunters in the African country of Equatorial Guinea (Juan-Espinosa and Palacios, 1996).

When you ask people in Western society to list the attributes of an intelligent person, you generally get statements about the ability to solve abstract problems and the ability to comprehend and use language. Spatial orientation, the ability to locate oneself in the physical environment, either is not mentioned or mentioned far down the list of other abilities. When the Fang list the qualities of an intelligent person, they say intelligent people do not get lost in the forest. This does not mean that the Fang devalue the sorts of verbal skills that Westerners

mention. In fact, they stress verbal skills in much the same way as Westerners. If the Fang were to construct intelligence tests, they would include tests of verbal skills, but the Fang might emphasize spatial orientation in more detail than do Westerners.

IQ and related standardized tests, such as PISA surveys (Program for International Student Assessment; intended primarily for fifteen-year-old high school students for worldwide country comparisons) are products of their cultures. They test some aspects of intelligence but not others. However, the tests are not arbitrary. Today, tests are constructed and evaluated with sophisticated statistical methods collectively called *psychometrics*. We discuss these methods in Chapter 3 because they apply to most research issues discussed in this book.

IQ tests would not have survived unless scores predicted socially important behaviors, such as academic and vocational achievement. Because the test scores do meet numerous diverse predictive validity criteria (detailed in Chapter 8), the tests must either evaluate mental skills that are important for the society or evaluate mental skills that are not important in themselves but whose possession is correlated with the possession of skills that are important.

Intelligence tests might be like an army physical examination, where the candidate is required to do push-ups. Soldiers are not going to do push-ups in combat, but the ability to do push-ups is correlated with the ability to move heavy objects (carrying artillery shells to a gun position), which soldiers may have to do. The same argument applies to the mental gymnastics required to perform well on intelligence tests. Note that the content of the exercises or problems included on intelligence tests is irrelevant. It is the abilities and skills required to solve these problems that are important. Recall the example of the thermometer described in the first fallacy in Box 1.2. Liquors are another example: their impact on our bodies, including our brains and our manifest behaviors, is unrelated with their color, scent, or flavor. Their effects on us depend on factors we cannot see directly, namely, their concentration of alcohol and our own metabolic features.

While all human societies are not identical, there is a core set of cognitive skills that all societies rely on. People in all cultures learn to speak the native language, control their attention, and remember events. For this reason, an intelligence test that is valid in one culture is unlikely to be invalid in another.

In fact, a primary reason intelligence tests are predictive of important real-world outcomes is that they are good estimates of a general factor of intelligence (g). The g-factor is found in diverse cultures throughout the world (Warne and Burningham, 2019). It likely reflects general information-processing capacity in the brain. One new development in the assessment of this general ability is the potential for using video game performance to derive a general cognitive score for individuals (Quiroga et al., 2015; Quiroga et al., 2019). The g-factor is a key concept in intelligence research and will be discussed in detail in Chapter 3.

2.1.4 *Uses of Intelligence*

We now come to Figure 2.1 (right), dealing with results of intelligence. People use their cognitive abilities to some extent to define their environments. For example, brighter children may seek out more opportunities to read, and more reading in turn may influence better reasoning. Brighter adults may seek companions who are also bright and thereby create a more stimulating environment, which in turn might influence greater knowledge.

Cognitive abilities are not the only abilities we have. Differences in behavior are also produced by individual differences in a variety of temperamental and motivational traits. These are often referred to as personality variables. It is not always clear where to draw the line between intelligence and personality. One monograph addresses the interplay between the concepts (Colom et al., 2019). For instance, conscientiousness, the tendency to fulfill obligations to others (including one's employers), is usually considered a personality trait. However, it is possible to see conscientiousness as an aspect of intelligence. It makes sense to be conscientious in fulfilling obligations to people who control resources that you want.

We can make a distinction between cognition and personality by asking if we are talking about whether a person *can do* or *will do* a certain behavior. To the extent that the answer is can do, the mental acts controlling the behavior are part of cognition, and hence individual differences in them are aspects of intelligence. To the extent that the answer is will do, the mental control is part of motivation, and individual differences are part of personality. Any particular action has both can-do and will-do considerations. It is surprising how often explanations of behavior focus on personality to the exclusion of intelligence, and vice versa. This is usually not helpful.

2.1.5 *Products of Intelligence*

Intelligence manifests in two general ways: test scores and socially relevant behaviors. Test scores are relatively easy to analyze; socially relevant behavior is harder. Nevertheless, socially relevant behaviors are important to consider.

There are statistical associations (correlations) between intelligence test scores and measures of socially relevant behaviors, including academic achievement, income, health, and prestige/reputation (see Chapter 8). Understanding why they occur is an important goal of research. The fact that intelligence test scores and measures of socially relevant behaviors are correlated suggests a number of possible causes, all of which are worthy of investigation. Note that the computation of correlation coefficients (designated by r) between two variables ranges from -1 (a perfect relationship where a high score on one variable goes with a low score on the other variable) to $+1$ (a perfect relationship where a high score on one variable goes with a high score on the other variable).

Performance on an intelligence test and performance in socially relevant situations might both depend on the same cognitive processes. Correlation coefficients between

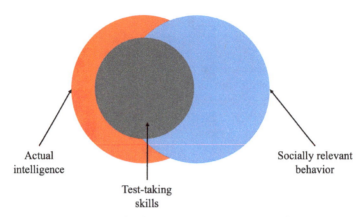

Figure 2.2 Conceptual relationships among actual intelligence, test-taking skills, and socially relevant behavior. People possess cognitive abilities and skills to varying degrees. Some of these abilities and skills are evaluated by the tests, and some of those are specific to the test situation (test-taking skills), while others are relevant to social behavior in general.

them generally are not large, although this does not mean they are meaningless (Funder and Ozer, 2019). Gilles Gignac and Eva Szodorai considered 708 correlations derived from eighty-seven meta-analyses (a statistical technique for combining results from different studies) (Gignac and Szodorai, 2016). They found that correlation coefficients in individual differences research can be considered low, medium, or high at these values: 0.10 (25th percentile), 0.20 (50th percentile), and 0.30 (75th percentile). Correlation values of 0.50 or higher were observed in only 3 percent of the published scientific studies.

Overall, the correlations observed in intelligence research are high enough to indicate that test scores and behaviors tap common traits but low enough to indicate that the traits that affect the tests and the behaviors are not the same. Figure 2.2 shows a conceptual example.

Socially relevant behaviors are partly influenced by cognitive skills. Some of the abilities and skills that define actual intelligence are reflected in test scores. Other socially important cognitive skills that are part of intelligence are not reflected in test scores, and test scores also reflect test-specific skills that are not related to socially relevant behavior. However, some important

cognitive abilities required for socially relevant behavior are not tapped by the tests. What might these be?

Most conventional intelligence testing includes a limited number of assessments that provide a snapshot of a person at the time of testing. This approach might not evaluate abilities that reveal themselves over a relatively long period of time, such as the ability to plan, to allocate time for extended courses of action, or to integrate information from multiple sources. It is difficult to tap these skills in a time-limited test where individual items take only a few minutes to solve. New approaches to intelligence testing that address such issues have not yet been developed, but this is an area of research ripe for the twenty-first century. Robert Mislevy, for example, refers to the usual snapshot approach as "Drop from the Sky" because it provides no context for snapshot results. (Mislevy, 2018).

2.1.6 Cause and Effect in Intelligence Research

The bidirectional arrows in Figure 2.1 highlight how hard it is to identify cause and effect when studying intelligence. If everything were simple, we would place causal variables on the left, intelligence in the center, and the effects of having intelligence on the right. To some extent, this can be done (Rohrer, 2018). Genetics and a child's physical and social environment influence intelligence, and a person's intelligence influences socially relevant behaviors and thus can alter the person's environment. Intelligence influences a person's environment, and feedback from the environment can alter intelligence. The main problem is understanding how the feedback interactions actually work in any detail.

Take the case of aging. As we grow older, two different processes influence intelligence. On one hand, there is a decline in brain function. But on the other hand, at the same time, people acquire better knowledge about the demands of their culture. There are large individual differences in both of these processes. Some people remain cognitively fit until old age; others noticeably lose key cognitive abilities during middle age. Some people infer relevant cultural knowledge as they pass through the world; others only have unrelated experiences that offer little cultural insight. In both these situations, intelligence appears to act as both cause and effect. Generally, more intelligent adults are more open to engaging with and extracting knowledge from the world about them. We discuss aging issues further in Chapter 11.

2.1.7 Reaction Ranges and the Challenge Hypothesis

Imagine Carmen at age fifty. We want to estimate Carmen's intelligence without actually measuring it. For achieving this goal, we can (1) obtain an initial estimate based on the average intelligence of persons at age fifty; (2) collect facts about Carmen's physiological status, including genetic background, medical history, nutrition habits, and so on, and use these to compute a biological correction factor to the initial estimate from (1); and (3) collect facts about Carmen's social environment, including education, marital status, hobbies, profession, and so forth, compute a social correction factor, and add it in to the initial estimate.

The twice-adjusted figure is the final estimate of Carmen's current intelligence. However, we would not know how that intelligence had been acquired, because estimation does not explain the dynamics of intelligence. A person's genetic makeup does not provide that person with a certain number of intelligence units, any more than it provides a person with a certain number of points on an IQ test. Genetic status provides a "reaction range," a range of probable levels of intelligence. Environmental factors define where the person operates within that range.

Figure 2.3 (left) depicts hypothetical reaction ranges. Cognitive functioning is shown on the ordinate (y-axis) and the level of favorableness of the environment on the abscissa (x-axis). Where a person actually functions is identified by the point at which a vertical line drawn from the environment's rating crosses the reaction range line. Observable cognitive performance is influenced by the combination of genetic reaction range and environmental quality.

In this model, the level of genetic inheritance cannot be inferred from IQ scores unless environmental quality is known, nor can environmental quality be inferred from IQ scores unless genetic inheritance is known. Most differences in the cognitive performance of identical (monozygotic or MZ) twins can be attributed to the environment because MZ twins, having identical genotypes, will have nearly identical genetically determined reaction ranges. Note that random development factors and interactions also can account for any differences and for variation not directly attributable to genes or environment.

Figure 2.3 shows hypothetical genetic reaction ranges as negatively accelerated curves that rise steeply at first, and then flatten out as they approach an asymptotic value (no further changes). This was an arbitrary choice, as both the cognitive performance and environmental quality axes are shown on arbitrary scales. Perhaps attempts to improve cognition will result in negatively accelerated curves. Why might this be so?

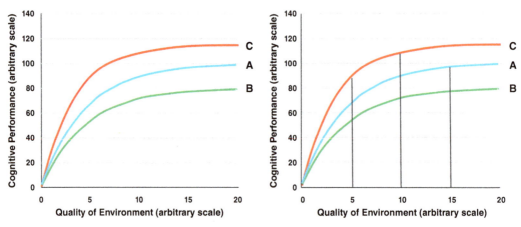

Figure 2.3 The concept of reaction range. (left) The y-axis represents the quality of cognitive performance, and the x-axis represents the extent to which the environment supports development of cognitive skills. A person's level of cognitive performance is influenced by the combination of reaction range and quality of the environment. Persons A, B, and C each have a unique genetic potential for cognitive performance, indicated by the three curved lines. Their actual performance will be defined by the quality of the environment. If the environment goes from very poor (at the far left of the figure) to moderate (toward the middle), there will be considerable improvement of performance because the curve starts out steep. Going from moderate to good environments (toward the right) does not result in much improvement because the curve flattens. (right) Suppose the environmental quality is fixed (arbitrarily) at a quality of 10 for all individuals. Differences in cognitive performance will occur due to individual differences in genetic potential. However, if individual A is placed in an environment of quality 15 and C in an environment of quality 5, A will outperform C, even though C has greater genetic potential.

There are many ways to restrict intelligence, but we know of relatively few ways to enhance it. This implies negatively accelerated growth curves. For example, there are a number of ways of producing physical environments that can constrain the development of intelligence. Prolonged famine is one; infection of the brain is another. Once these extreme states are avoided, the incremental effects of improving the physical environment are probably rather small (see Chapters 7 and 13). The same thing is true for establishing genetic potential. There are many catastrophic genetic conditions where the presence of a single anomaly greatly restricts intellectual development, but if there is no anomaly, genetic potential seems to depend on the combined effects of a large number of genes, no one of which contributes a great deal (see Chapter 6).

The reaction range does not mean it influences a particular individual. Cognitive skills are acquired by investing time and energy to develop cognitive abilities. Willingness to invest often depends on one's perception of the likely outcome of the investment. Here is one way to think about this.

THE CHALLENGE HYPOTHESIS

Given genetic potential, intelligence is developed by engaging in cognitively challenging activities. Environments vary in the extent to which they support such challenges, and individuals vary in the extent to which they seek them out. The development of intelligence depends on the extent to which the individual wishes, is allowed to, or is required to meet environmental challenges.

Consider language and literacy. Every human being is required to learn a spoken first language. Only in the last 200 years have societies begun to demand literacy, which is a much harder skill to acquire. It is worth the effort. A literate person has acquired a cognitive skill that influences intelligence more than for an illiterate person with an identical, but unrealized, biological potential for intelligence.

The example expands. Literate societies dominate the planet. They have developed formal education systems that stimulate further intellectual development. Mechanisms of education such as the school, the newspaper, and the internet are different from, and far more challenging than, the educational systems of the nineteenth century. As a result, today's children generally have better cognitive skills than children in the past. This may be one reason that IQ scores rose from generation to generation throughout the twentieth century, a phenomenon known as the Flynn effect (Flynn, 2018; Pietschnig and Voracek, 2015), although some evidence finds the increase already is apparent in infants, suggesting improved nutrition and health care also might be relevant (Lynn, 2009) (see Chapter 13).

Literacy is an example of a compulsory challenge; individuals have to meet it or suffer consequences. In other cases, there may be options. Learning about statistics and probability provides a good example. On a population basis, relatively few people study statistics. Students who complete elementary statistics courses can solve problems that Pascal and Gauss could not solve. The typical student in a modern statistics class does not have the biological potential for the intelligence that Pascal and Gauss had, even if they can use modern methods to solve problems.

Robert Sternberg has identified three ways a person has of responding to cognitive challenges (Sternberg, 2000):

1. *Adapting* – changing your own cognitive behaviors to meet the challenge. Studying is an example of adapting.
2. *Selecting* – finding a new environment that does not present the challenge you want to avoid. If you are a college student majoring in engineering, but math classes are difficult for you, consider switching majors.
3. *Shaping* – changing the environment to adapt it to your current abilities. If you have difficulty doing arithmetic, use the calculator on your smartphone.

Adapting might be the most important indicator of intelligence. Both adapting and shaping require a willingness to engage with the environment. This is in itself a reliable individual trait. Selecting can be rational in some situations, but it has the danger of becoming a way to avoid intellectual challenge and, hence, to restrict the development of one's biological potential for intelligence. Philip Ackerman has conducted research showing that willingness to engage in intellectual challenges is characteristic of the people who hold the occupations and vocations that we think of as requiring intelligence (Ackerman, 1996; Ackerman et al., 2005).

2.1.8 *Intelligence Is Part of a System*

Defining intelligence solely in terms of test performance is not ideal. It focuses our attention on explaining variations in test scores, which is important in some situations, but it is also important to study individual variations in socially relevant behavior.

In practice, there are two problems. First, variations in socially relevant performance are influenced by noncognitive as well as cognitive factors. As we noted, success and failure are based on both can do and will do. Second, the display of socially relevant behaviors depends on the opportunity to display them, which may be quite beyond a person's control, no matter what their personal characteristics are. Here are two examples.

In the early twentieth century, Joseph P. Kennedy, an American financier, amassed a considerable fortune. Subsequently, three of his sons – John, Robert, and Edward – became US senators. In 1960, John became president of the United States.

One can argue that Joseph Kennedy and his wife provided their sons the genetic capability to become intelligent. But his fortune certainly helped their political careers.

Now an example of restriction. The Declaration of Independence of the United States, written in 1776, contains the statement "All men are created equal." At that time, women were disenfranchised, and the

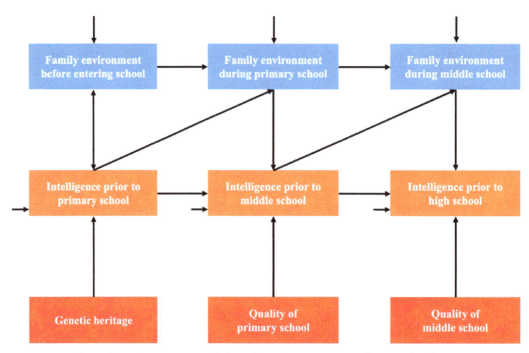

Figure 2.4 A hypothetical system view of the relationships among intelligence, family environment, and the quality of schooling. Intelligence and family environment are system variables. Genetic heritage and quality of school are external variables. Unknown sources that influence the system variables are indicated by a short arrow.

slavery of Africans was condoned by many. More than 200 years later, Condoleezza Rice, an African American woman, who had served in many high-level positions, including as Secretary of State, observed that when the Declaration was written, "They weren't talking about me." Dr. Rice's impressive achievements, which surely are attributed to a large extent to her intelligence, would have been impossible in 1776.

The point of these examples is that both the causes and effects of intelligence are embedded in a matrix of other variables. This presents challenges to research on intelligence. The task is difficult, but it is not impossible. While an ideal study may be practically impossible to implement, a great deal can be learned from the accumulation of findings from less-than-ideal studies. This is called the *weight of evidence*, and it is key to the presentations in this book. Progress can be made by investigating the issues that can be studied and by restricting excessive interpretations and conclusions where we do not have the best evidence.

As we have seen, intelligence is one of the multiple variables in the system defined by human society. A system is a set of interdependent variables, in which each variable is related to the others. In closed systems, the interdependence is complete; the value of each variable is completely determined by the other variables in the system. In open systems, some variables are influenced by conditions outside the system.

The real-world systems we study are always open. Therefore, it is important to distinguish between system variables, which exert measured reciprocal influences on each other, and external variables, which influence the system variables but are not influenced by the system variables. Consider a hypothetical study of the roles of genetics, family influence, and intelligence during primary school, middle school, and high school. Figure 2.4 diagrams the hypothetical system.

Intelligence and family environment are system variables, because they influence each other. Genetic potential and quality of schooling are external variables, because they exert influences on other measures but are not influenced by them. If we can obtain measurements of all these variables, we can use statistical methods to evaluate the relative influences of each variable on the others. Then we can address the external, unmeasured variables.

Figure 2.4 makes no provision for extrafamilial influences on the family environment, such as financial emergencies. Nor is there any provision for extrafamilial and extraeducational influences on intelligence, such as physical injury. Therefore, we have to allow for the influence of unmeasured (and often unknown) external variables. These are indicated by the short arrows. While we cannot identify these variables, statistical techniques allow us to estimate the size of their influence compared to the influence of the measured system variables on each other. Hopefully, the influence of the unknowns will be small. If it is large, analyses within the system will account for only a small part of what we need to know.

We can learn a great deal by comparing systems models to each other. The intelligence-education system displayed in Figure 2.4 treats genetic influences as a one-shot effect – genetics influences intelligence prior to entering school but has no direct influence subsequently. In fact, some genetic effects unfold over time as noted previously. For example, individual differences in the rate at which connections are developed in the brain during adolescence may result in differences in the ability to control impulsive behavior, which may influence how much a student learns in middle school. These influences could be modeled by extending arrows from the genetic inheritance box to the boxes representing measurements of intelligence at each time period, not just at the point of school entrance. The extension is needed only if the added arrows produce a system that explains more of the variation in intelligence than did the simpler system. System analyses of this sort allow researchers to move beyond disagreements over the meaning of a correlation between IQ scores and just one other variable, in isolation from the system in which both occur.

Still, problems remain. Systems incorporating human intelligence are so complicated that no one study can include adequate measures of all relevant variables. The size of the influence of unknown variables can be evaluated, but what those variables are and how they exert their influence will remain unknown. Therefore, we cannot explain all the causes and ramifications of intelligence at once. We can identify and analyze reasonably closed subsystems, dealing with a particular aspect of intelligence. No one such study will tell us all about intelligence, but taken together to form a weight of evidence, they can reveal many things, as we detail in subsequent chapters.

2.2 Scientific Theories

Theories and the testable hypotheses they generate are the glue that hold scientific observations together. They summarize the chains of cause and effect that scientists use to understand how the world works. The theory of evolution, for instance, is an explanation of why life on earth is so diverse. Theories of physical phenomena can be used to design aircraft.

Theory lets us look back to explain why something happened the way it did and to look forward to predict what will happen in the future. Theories are not unique to science. Many statements in philosophy, popular discourse, and religion are theoretical statements. Nevertheless, scientific theories differ from nonscientific theories (usually called opinions) in three ways. The first is that they must refer to variables that are measurable in principle. The variables in a theory do not need to be measurable at the time the theory is presented. The history of modern physics contains several instances in which theoreticians postulated the existence of subatomic particles before the technology to measure them had been developed. However, no variable can be admitted to a scientific theory if, in principle, that variable can never be measured.

The second distinction between scientific theories and opinions is that scientific theories must be testable and interpretable by objective means. Gurus are not permitted, and personal faith is never to be confused with evidence. This does not mean that anyone without training should be able to understand a scientific theory. The necessary technical material may be hard to master, but the rules of mastery have to be open and nonmystical. Consider the following two cases. Albert Einstein's theory of relativity was arguably the most important development in scientific theory in the twentieth century. When Einstein presented the theory of relativity, he had to present a chain of assumptions and deductions that other appropriately trained scientists could follow. The theory was accepted, or not, by physicists on the basis of their analysis of Einstein's ideas, not because they attributed any mystic properties to Einstein the person. Contrast this to the decision of Saul of Tarsus – to accept Jesus as the Messiah and begin a new life as St. Paul, the Apostle to the Gentiles. St. Paul never claimed to have analyzed Jesus's philosophy; he claimed to have received a revelation from God.

The third distinction between scientific theories and opinions is that scientific theories must account for empirical data. A scientific theory is vacuous unless it implies that certain observable events will happen or that certain patterns will occur in data. There is no requirement that the observations be easy to make, or even that they be possible at the time that the theory is presented. Several of the predictions of Einstein's theory of relativity have only recently been evaluated, using technologies that were not available in his lifetime. The requirement for prediction means that the theory must be stated with sufficient precision that, given the opportunity to observe, we can tell whether the prediction was confirmed or denied. Newton's theory of motion predicted that large and small objects will fall at the same speed in a vacuum, regardless of their mass – and we can observe that they do. By contrast, the prophecies of the Delphic Oracle,

Nostradamus, and other seers are notorious for their ambiguity.

2.2.1 *Choosing between Competing Theories*

When a theory correctly predicts an observation, as in a hypothesis, this is evidence in support of the theory, but evidence can never prove that a theory or a hypothesis is correct. If an observation is made that contradicts a prediction, then the theory or hypothesis may be disconfirmed, assuming the observation was based on sound research, showed unambiguous results, and addressed a precise prediction.

Precision in the physical sciences is more attainable than it is in the social sciences. In psychology and other social sciences, we depend more on the weight of evidence from multiple studies; a single study rarely can disconfirm a theory. This is why statistical analysis is key to intelligence research. If theories always made exact predictions, and if we could be absolutely certain that our observations were accurate, then science would proceed in the way just described. In practice, that is not what happens. Predictions are often made in general terms, and measurements are never exact. Therefore, empirical results seldom completely rule out a hypothesis or a theory. Instead, what we usually learn is that findings are more compatible with one hypothesis or theory than with another. This should modify our thinking about the winning and losing theories, but not necessarily cause us to accept the winner and reject the loser. Indeed, it often happens that as more evidence accumulates, our confidence in all theories under consideration falls, and we must develop new theories to accommodate new information.

One of the most debated topics in intelligence research, the nature–nurture issue, provides a good example. In its pure form, the nature side of this debate asserts that intelligence is mainly inherited through genes. The nurture side states that intelligence is obtained mainly through experience, with the corollary that early childhood experiences are particularly important.

Let us look at the logic of the argument. Francis Galton observed that the eminent people of his own generation were highly likely to have had eminent parents or grand-parents (Galton, 1869). He identified families who produced people of eminence generation after generation, concluding that intelligence, as other human traits, is partly inherited.

The hypothesis that intelligence is inherited to some extent implies that there will be a statistical association between the eminence of families from one generation to the next. The null hypothesis is that familial eminence varies randomly from one generation to the next; in other words, the null hypothesis is that there will be no difference or no association. Galton found that the generation-to-generation associations of eminence were stronger than would be expected by random variation. Accordingly, he accepted the hypothesis that eminence (and therefore intelligence) is partly inherited. Galton was correct. The data supported the nature hypothesis relative to the null hypothesis of no familial association. The position relative to the nurture (environmental) hypothesis is different.

People inherit both their genetic makeup and their social environment from their parents. Therefore, both the nature and nurture hypotheses predict associations of familial eminence from generation to generation. The association that Galton found actually was evidence for either the nature or the nurture hypothesis against the null hypothesis that families don't matter in achieving eminence, but it was not valid evidence to decide against either nature or nurture.

This example shows that, in general, simply testing a hypothesis against the null hypothesis is not always informative. Proponents of the nurture hypothesis can point to studies that show that children raised in unfavorable environments perform poorly in school and on intelligence tests, at least during the early school years. On its face, this evidence seems to support the importance of the environment. However, proponents of the nature hypothesis can point out, correctly, that such children are not a random sample of children in their generation. Indeed, there is reason to believe that many of them are the children of parents who are not highly intelligent to begin with. Thus, these data do not discriminate between the hypotheses.

By the middle of the twentieth century, the debate became more refined because the studies became more complex. Scientists learned that both genes and environment have a role and that the issue is not to look for a single cause for intelligence but rather to identify the relative influences of both factors. This interaction approach allows for a much more sophisticated model of heredity, as diagrammed in Figure 2.5. Its basic elements are (1) that there are both genetic and nongenetic components to intelligence and (2) that for any two individuals (A and B), the extent to which they share the genetic component of intelligence will be determined by their genetic relation, and the extent to which they share environmental components will be determined by their environmental relation.

Diagrams such as Figure 2.5 can be used to generate expected correlations between test scores. The values of the correlations depend on the strength of the connecting links. Sometimes these strengths can be predicted from genetic theory. For example, in a study of identical twins (monozygotic, MZ), the genetic link would have a strength of 1, indicating the same genetic constitutions (they are clones, genetically the same individual). The link for fraternal twins (dizygotic, DZ) would have a strength of 0.5, because they share, on average, half their genetic inheritance.

For another example, the extreme nature model (variations in intelligence are 100 percent due to genes) can be thought of as a specific case of Figure 2.5 in which the links from environment to intelligence are absent (strength = 0). Or the extreme nurture view (100 percent environment) would have no link to genetics. We could also investigate intermediate models, for example, a model in which the environment–intelligence link is required to have half the strength of the genetics–intelligence link.

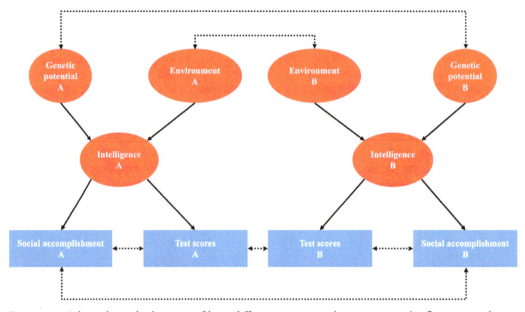

Figure 2.5 A hypothetical schematic of how different genetic and environmental influences might contribute to similar intelligence test scores for two individuals, A and B. We can only observe correlations between manifest variables – test scores and other cognitive variables. Genetic theory can be used to specify the degree of similarity between the genetic potentials of persons A and B (offspring, siblings, twins, unrelated). The other similarities between latent variables must be inferred. Dashed lines suggest noncausal correlations.

In all cases, we would come back to a basic test: which of the many possible models best reproduces the actual correlations between observed variables, that is, the actual link between the four rectangular boxes (observable variables) in Figure 2.5? The important thing to realize now is that research has advanced from evaluating broad assertions about the importance of heredity or environment to a far more sophisticated competition between models that empirically assess parameters to indicate the relative contributions of genetic and environmental causes of intelligence. Chapter 6 discusses this research in detail.

2.2.2 Systems Thinking Complicates the Issue

We have already explained that intelligence is a latent trait that is inferred from observation of various manifest behaviors. These include scores on IQ tests, school grades, and records of performance on the job. Intelligence also is embedded within a complex matrix of other latent variables, including such things as genetic makeup, SES, and educational opportunities. These variables are not independent of each other; they have links between them. For instance, the SES of one's neighborhood is related to the quality of the neighborhood schoolteachers. When variables are linked into complex systems, there are both direct and indirect effects. Untangling them can be difficult. Here is an example regarding verbal intelligence.

Reading is one of the most important skills children acquire in the early school years. Subsequently, reading skills are used to facilitate learning. The sooner and more completely the shift from learning to read evolves into reading to learn, the more the child can get out of the educational system. Learning to read is associated with a child's family background. On average, children from families in the middle and upper SES strata arrive at school better prepared to learn to read than children from a low-SES background. This advantage may continue throughout the school years. Why would this be so?

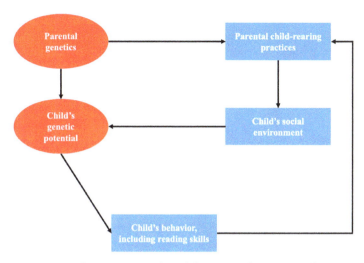

Figure 2.6 Influences on reading skills. Parental genetics will exert a direct effect on a child's genetic potential and also may influence the parent's child-rearing practices (by increased sensitivity to a child's mood). Parental child-rearing practices establish a social environment that, acting through a child's genetic potential, produces a variety of behaviors, including reading skills. These behaviors, in turn, alter the parents' behaviors toward the child.

There is a plausible genetic explanation. To the extent that SES is linked to genetic potential, the children from middle- to high-SES families may be better learners simply because they are biologically smarter. But there is also a plausible environmental explanation. The parents in middle- to high-SES families, on average, invest more time in reading to their preschool children than do parents in low-SES families. Indeed, some of this reading behavior comes very close to teaching reading, as in reading books that tell children that "A is for apple" (in some homes, the technology brand, in other homes, the fruit), and so forth.

We are dealing with a set of skills, embedded in a system. Furthermore, it is a system with feedback loops, for there is little doubt that as children learn to read, their parents respond, and the manner of the response may vary with SES. A simplified version of this forbiddingly complex situation is depicted in Figure 2.6.

Parental genetic effects are of two sorts. Directly, parental genetics determine a child's genetics, and the child's genetic potential interacts with the social environment to influence reading skills. Indirectly, parental genetics will partly produce parental child-rearing behaviors. These behaviors, influenced by the child's genetic potential, will produce the child's behaviors, which may include reading or prereading skills. The child's behavior, in turn, influences future parental behaviors, and the cycle continues.

At this point, you may be wondering if controlled experiments might be informative.

To continue the example of reading skills, there is some evidence that intensive preschool programs, ones that are more prolonged and expensive than the typical Head Start program, can improve children's cognitive skills during the early school years and that the improvements last, in much weakened form, into adulthood (Protzko, 2015). The studies in question involved comparisons between the performances of children who had participated in prolonged programs compared to a matched group who had not, the classic experimental group versus control group study. Such studies might suggest that direct manipulation of a proximal variable, the child's social environment, can have an influence on the development of intelligence.

However, this sort of study does not rule out a role for genetic inheritance, because parental intelligence is not assessed and included in the study design (see Chapter 13 and Section 2.1).

Suppose that we improve the intervention design by measuring parental intelligence and, by appropriate statistical methods, estimating how much the observed improvement in the child's performance can be attributed to the parent's intelligence, how much can be attributed to the intervention, and how much is due to some interaction between the two.

Such studies have been done, especially analyzing adopted children (see Chapter 6 and Section 3.4). The proper design for a study of intelligence depends on what question we are trying to answer. It turns out that the proper theoretical model may also depend on the question at hand.

2.2.3 Intelligence as a Construct in Social Systems

As noted, we need theories and models to help us think about how things work. We use them all the time to understand how things like electrical circuits function. Economists argue about the right model for the world economy, while epidemiologists use models to forecast the seriousness of an outbreak of influenza. Why are models so ubiquitous?

David Geary has argued that humans have evolved a capacity to construct models of the world to satisfy an inherited drive to exert control over their environment (Geary, 2005). To accomplish this goal, the various components of the model have to fit together in a way that we can think about. This means that we may have to conceptualize the same thing in somewhat different ways, depending on the context in which the thing is embedded. That is certainly true of intelligence.

Here is a classic example. During the 1970s, two Stanford University educational psychologists, Lee Cronbach and Richard Snow, studied aptitude by treatment interactions in education (Cronbach and Snow, 1977). They concluded that classroom environments that encourage active exploration and experimentation benefit students with high cognitive ability, while students with lower cognitive ability do better in teacher-structured learning environments. Their findings have been replicated in situations as different as college students learning statistics and elementary school children learning to read.

Conceptually, Cronbach and Snow developed a model of the relationship between psychometrically defined general intelligence and educational practices. The manifest variables they studied, psychometric scores and teaching methods, were measures and actions available to teachers. Statements about, say, the behavioral implications of the level of neural activation of children's brains during functional magnetic resonance imaging (fMRI) would be of little use in the classroom (although this assumption, passionately supported by educational psychologists, can be challenged; Gratton et al., 2018). However, if we were to embed the cognitive abilities that we call intelligence into another system, we might want quite a different conceptualization. Imagine a study in which the investigator is interested in the use of drugs to ameliorate cognitive deterioration during aging. In that situation, relating intelligence to the level of neural activation in a patient's brain might be a reasonable thing to do. If a scientific model is to be used to inform decision makers – that is, to interact with a system – then the theory must be stated in terms of variables that the decision maker can measure and manage.

2.2.4 Reductionism

The crucial role of reductionism in the scientific inquiry of complex phenomena is eloquently expressed by Edward O. Wilson in his book *Consilience: The Unity of Knowledge*: "Complexity is what interests scientists in the end, not simplicity. Reductionism is the way to understand it. The love for complexity without reductionism is art; the love for complexity with reductionism makes science" (Wilson, 1998, p. 59).

Using scientific models to control variables is something of an engineer's view of the scientific enterprise. It is valid, it is practical, and it is not the only use of theory in science. The other use is for pure understanding. Again, the system concept is useful. A system is understood when it is completely closed, so that its behavior can be understood entirely from the actions of its elements, as in Figure 2.6. This produces an ordering of scientific topics: the properties of atoms are derived from the properties of subatomic particles and forces; the properties of chemical compounds are derived from their atomic structure; the properties of biological systems are derived from the chemical and physical properties of their components, and so on. The sequence is called reductionism. It can progress from the bottom up or from the top down.

Reductionist analyses have important implications for the study of intelligence. David Wechsler, the developer of the Wechsler intelligence tests, pointed out that intelligence is revealed by behavior (Wechsler, 1975). Behaviors that we see as exercises of intelligence, or lack thereof, must be under the control of the brain. Therefore, theories of intelligence that relate individual differences in cognitive abilities to individual differences in brain structure and function are important steps in reductionism. They may or may not be useful in understanding how individual differences in intelligence influence academic learning or on-the-job performance. Models of these phenomena might well use concepts such as general intelligence, verbal comprehension skill, and spatial orientation ability without any concern about how these abilities are produced by the brain – just as Newtonian mechanics accepts gravitational force as a basic fact without further explanation. Research into the complexities of understanding intelligence at different levels requires both reductionism and synthesis.

Figure 2.7 diagrams a reductionist view of research on intelligence. All performance depends on the brain. At any moment, the brain is the product of both a person's genetic makeup and their environment.

Environment has to be interpreted in the broadest manner. It does not just refer to obvious physical factors, such as nutrition or injury. It also refers to social factors, including education. Why? Because learning produces physical changes in the brain. This includes the storage of information, but there is more to it than that. The same external task may be processed by the brain in different ways before and after learning. It is often useful to deal with the results of brain action at the functional level rather than at the level of brain action itself.

To illustrate, consider how a person recognizes their country's flag. In principle, it would be possible to describe flag recognition in terms of neural action in the occipital, temporal, parietal, and frontal lobes, but in practice, it is simpler to say that visual pattern–detection and memory-recognition processes are involved. Individual differences in performance on the complex task of flag recognition can then be related to individual differences in vision and memory. Failure to recognize the flag could be due to a distortion of either visual-recognition or memory-retrieval processes. Both processes are theoretical entities. That is why information-processing functions are shown in ellipses in Figure 2.7, while physically observable brain and behavioral processes are shown in rectangles. Nevertheless, there is a place for descriptions of theoretical processes.

The top of Figure 2.7 specifies two types of physically observable behaviors – performance on psychological tests and performance in life activities. In the physical sense, both of these are produced by the brain. Conceptually, though, we can think of both behavior in testing situations and behavior everywhere else as being produced by information-processing capabilities combined with previously acquired knowledge.

The double-headed arrow between behaviors on tests and behaviors elsewhere indicates that the two different types of behavior are correlated but that there is no causal link between them. Psychological test scores do not, in themselves, cause anything. The test scores only indicate possession of

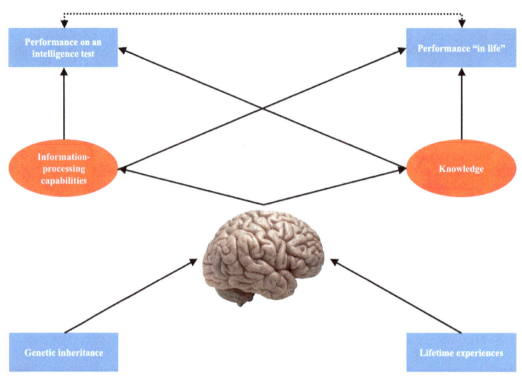

Figure 2.7 Levels of explanation addressed by theories of intelligence. Rectangles indicate measurable manifest variables, and ellipses indicate hypothetical/latent variables.

knowledge and information-processing abilities that control behavior in life outside of the testing session.

The next three sections illustrate these principles by giving brief descriptions of studies that deal with issues at each of three levels of explanation.

2.2.5 *Psychometrics*

Psychometrics refers to the science of test construction, and psychometric theories and models deal with the dimensions that underlie and summarize performance on intelligence tests. One goal is to develop valid and reliable tests. Another goal is to simplify the vast amount of data contained in batteries of cognitive tests by showing that the data from test batteries can be understood in terms of individual variations along a limited number of dimensions of mental ability, called factors. Here we concentrate on an example of factor results from a model that will be fully featured in Chapter 3.

Wendy Johnson and Thomas Bouchard had data from a study in which 436 adult participants had taken forty-two different intelligence tests (Johnson and Bouchard, 2005). There were more than 16,000 numbers in the data set. However, Johnson and Bouchard found that a large part of the individual variation on the forty-two tests could be captured by individual variation on just four factors: general intelligence (g), verbal ability (V), perceptual ability (P), and the ability to rotate images mentally (R), that is, to imagine how something seen from one perspective might look when rotated to a different perspective. They referred to their model as a general intelligence–verbal–perceptual–rotation (g-VPR) model.

Their work is typical of psychometric investigations showing that variation on intelligence tests can be summarized by variation along a small number of dimensions. Psychometric studies also provide measurements that are useful in understanding social systems. Psychometric models are less useful when the task is to explain individual differences on specific cognitive tasks. To illustrate why, we take a look at an act of verbal comprehension.

Here is an excerpt from a newspaper editorial:

> The citizens of Washington and the nation are spending millions and millions of dollars on efforts to protect salmon and save them and the ecosystem they support from the threat of extinction. These goals are being hurt by California sea lions that have outsmarted the system and now wait patiently for the fish to be corralled into the ladders at Columbia River dams. (*Bellingham* [Washington] *Herald*, April 11, 2007)

From the psychometric perspective, to understand the editorial, you have to have a certain level of verbal ability to understand the passage at all and a certain level of general reasoning ability to realize that the problem is that sea lions are eating too many fish to sustain the fisheries. This fact is not stated directly. It has to be inferred from an analysis of the text plus some general knowledge about sea lions, predators, and prey.

Psychometric research could be used to calculate the probability that a person with verbal ability level x and general reasoning ability level y would have probability z of understanding the passage. Something like this sort of analysis is used to design instruction manuals, where the prose has to match the reading and reasoning skills of the people who will read the manual. Psychometric analyses do not tell you what readers have to do to comprehend a manual. For that sort of analysis, we must move to the information-processing level.

2.2.6 *Information Processing*

An examination of what has to be done to understand the salmon editorial reveals that, without effort, we do quite a bit of information processing. Retrieving word meanings is only part of the task. It is complicated enough, for both semantic and syntactical

information must be retrieved and ambiguities must be resolved. Does the word "Washington" in the editorial refer to the city, the state, or the president? Then, even heavier information-processing demands appear during the analysis of sentences and of the text as a whole.

The comprehension process has to operate on words as they appear. However, the meaning of a word, in context, often cannot be determined until following words are received. The initial "the" in the editorial has no meaning until it is attached to "citizens." "The citizens of Washington and the nation" specifies which citizens and resolves the ambiguity of "Washington." Next comes "are." It appears that a descriptor is going to follow, that the collective just described (the citizens) is going to be equated with something, as in "The citizens of Washington and the nation are taxpayers." This interpretation is dashed when "spending" occurs, for now "are" must be interpreted as an auxiliary verb. "Are spending" is not a complete statement; the comprehender must find out what is being spent. The words "millions and millions" cannot be interpreted until "dollars" is encountered.

As this analysis shows, a language comprehender must have a cache for holding information to be integrated with information not yet received. In psychology, this cache is called *working memory*. It is an example of an information-processing concept, in the sense that it refers to an abstract ability to manipulate pieces of information in the mind, without specifying what these pieces of information mean in the world outside the mind.

The result of the construction is a text model summarizing what the text says explicitly, but the text model alone is not enough. To reach full understanding, the comprehender must construct a situation model of what the text means in the context in which it is presented. This means going beyond the text to understanding things that are only referred to obliquely. The editorial is about sea lions eating salmon, but that is not stated explicitly.

Working memory is central to the comprehension process. It follows that individual differences in the capacity of working memory ought to be related to individual differences in verbal comprehension. And they are. Just, Carpenter, and Keller conducted experiments investigating the link (Just et al., 1996). To measure working memory, they used a procedure called the *memory span task*. In a memory span task, an examinee is asked to read a fairly complicated sentence aloud and then remember the last word. Another, unrelated sentence is then presented. After k sentences, the examinee is asked to recall the words. Here is an example with $k = 2$:

> Read: When at last his eyes opened, there was no gleam of triumph, no shade of anger.
> Read: The taxi turned up Michigan Avenue where they had a clear view of the lake.
> Recall: The examinee should then recall the last two words of each sentence: anger, lake.

A person's memory span measure is defined as the highest number of sentences they can read and still recall all the ending words in the proper serial order. The argument is that working memory is taxed by having to hold the ending words of previous sentences while the current sentence is processed. Thus, the memory span is a measure of the capacity of working memory. College students' memory spans on average vary from 2 to 5.5. Just and colleagues found that memory spans are associated with the ability to comprehend text. One of the comprehension examples they used was the ability to detect ambiguities, as in "The experienced soldiers warned about the dangers ... ," which could mean either that the soldiers warned some to-be-specified people or that the soldiers themselves had been warned by some unspecified person, as in "The experienced soldiers warned about the dangers conducted the midnight raid." This contrasts with a phrase like "The experienced soldiers spoke about the dangers ... ," where the soldiers are unambiguously doing the speaking.

These scientists argued that only people with high memory span could afford to keep two interpretations (alternative text models) in mind simultaneously and that this allowed them to achieve a better understanding of complicated texts than could be achieved by people with low memory span and hence low working memory capacities.

These researchers then conducted a new series of experiments that supported their conclusion. For example, high-span people read sentences faster than low-span people. However, high-span people read sentences containing ambiguous verbs, like "warned" in the illustration, more slowly than they read sentences with unambiguous verbs. Low-span people read both types of sentences at about the same rate. Only the high-span individuals noticed the ambiguity and carried both meanings forward until the ambiguity was resolved. This work is an example of the reductionist approach. Individual differences in verbal comprehension, a psychometric construct, were related to working memory capacity, an information-processing concept.

Explanations of behavior that use information-processing concepts such as working memory capacity or retrieval of word meaning are less abstract than explanations based on terms like verbal ability or general reasoning ability, but they are still largely abstract. Biology is more concrete.

2.2.7 Biology

There are two broad categories of biology that contribute to modern intelligence research with their own models of intelligence; both are introduced now and detailed in later chapters.

THE BRAIN
There always has been considerable interest in explaining individual differences in behavior in terms of individual differences in brain structures and functions. Historically, this was done by drawing inferences about normal behavior by analogy to extreme differences in behavior associated with damage to the brain. Today's technology makes it possible to extend the

approach, because we can use neuroimaging to look at and measure the brain in vivo, as people rest comfortably or perform any cognitive task of interest (Barbey et al., 2021; Haier, 2017).

Neuroimaging technologies and intelligence studies are detailed in Chapter 5, but for the introductory purpose of this chapter, they can be divided into three broad categories:

1. Structural magnetic resonance imaging (sMRI) and diffusion tensor imaging (DTI) are used for exploring brain structure (gray matter volume, cortical thickness, cortical surface area, gyrification, and, respectively, white matter integrity and structural connectivity).
2. Functional magnetic resonance imaging (fMRI) provides information on activation or deactivation averaged over one second or so in different regions of the brain while a person is doing a cognitive task or resting.
3. Techniques for recording the electrical signals of neural activity, such as the electroencephalogram (EEG) or magnetoencephalography (MEG), can detect magnetic field fluctuations resulting from aggregations of neurons firing on and off. Both EEG and MEG measure brain activity millisecond by millisecond.

These imaging technologies, along with new genetic approaches based on DNA, have created an exciting new field of neuroscience studies of intelligence that pushes research beyond psychometric and information-processing studies deep into brain neural networks and even to neurons and synapses, as described in Chapters 5 and 6.

GENES AND EXPERIENCE
It could be said that everything of interest for the science of human intelligence begins at the genomic level. Every human is unique from the very beginning, and this fact of nature cannot be ignored. Nevertheless, nongenetic factors are also involved from the outset because they influence many

aspects of intelligence as well as gene expression. For example, life in utero includes many variables that interact with the fetus. The interplay between genetic and nongenetic factors is always present, and this fact complicates the goal of disentangling their relative contributions to the observed differences in intelligence.

The key message is that brain organization is as personal as the genome. How an individual's brain is progressively configured results from the interplay between one's unique genome and how the individual encounters numerous environmental factors during life. From this perspective, it is reasonable to say that the environment itself is uninteresting from a psychological perspective. What is important is how individuals experience their unique environments and how these experiences contribute to configuring their psychological models of the world.

The interplay between the genome and the environment takes place in the brain. How these ingredients are combined should help to explain human behavior, including the most likely response to the question of why some people are smarter than others (Colom, 2016). The remaining chapters explain much more about the answers to this question and discuss new questions based on the results of key studies. Before proceeding, however, let us consider one last question.

2.3 Do We Need All This?

Intelligence tests are necessary because we need efficient quantification of intelligence in applied settings, such as university entrance examinations, industrial employment programs, or for neuropsychological assessments of brain damage. Since the tests are needed, some orderly way of thinking about the results is also required. Theoretical models provide a way of doing this. There is an equally compelling, but quite different, argument for brain-level theories of intelligence. They are an important step in understanding because they link complex behavioral observations, such as verbal comprehension, to events inside the brain. Understanding these links is an essential part of the reductionist program of identifying causal links between individual differences in brain structures/functions and individual differences in intelligence.

Do we also really need a theory of intelligence at the information-processing level, intervening between models of individual differences in the brain and individual differences in test performance and in daily life? We do. An information-processing theory of intelligence provides (1) an understanding of the information-processing mechanisms that underlie intelligence, (2) measurements of individual differences in the application of these mechanisms, and (3) measurements of individual differences in the knowledge bases that are manipulated by the information-processing mechanisms.

This sort of theory is a substantial improvement on a theory that explains the way intelligence is expressed outside the laboratory in terms of correlations between measures of real-world cognition and test scores. A model of intelligence based on the brain alone cannot fulfill this role, because in practice, we are often interested in individual differences in a capacity for processing information, not individual differences in activation levels in this or that brain region or network.

For instance, in assigning air traffic controllers, it is important to know how many different aircraft a given controller can handle at one time, and there are individual differences in this ability. The requirement is stated in terms of information-processing concepts. Psychological models to address this sort of issue must be stated in terms of information-processing abilities. The brain does not have a single dedicated region for a complicated trait like intelligence; rather, it provides a tool kit, the information-processing functions, that, when guided by specific relevant knowledge, can be assembled into intelligent behavior. The way in which the brain is organized for each individual, at the functional level, defines the tool kit.

Concepts like working memory, attention, and speed of information processing are useful ways of talking about the functional underpinnings of intelligence, just as concepts like acceleration, deceleration, gasoline consumption, and turning ratio are useful ways of talking about the functional underpinnings of automobile capabilities. The challenge going directly from brain-based variables to psychometric variables is that the steps are complex. While we now know a great deal about the involvement of different regions and networks of the brain in different types of cognitive activity, we know less about how networks of neurons create memories and conduct inferential reasoning.

Advances in the cognitive neurosciences will tell us about the neural bases of working memory, long-term memory storage, attention, and similar functions (Barbey et al., 2021). But we will still need to understand how these functions are assembled to produce intelligent behavior. And we need to allow for the likely possibility that different people will assemble them in different ways.

These are the basic concepts about intelligence research necessary for learning more about the complex issues faced by researchers. The next chapters tell more of what we know, how we know it, and what we are trying to find out at all levels of explanation.

2.4 Summary

Intelligence has multiple causes and multiple consequences. This poses a challenge because our ability to measure some variables, and hence to study systems involving them, is better than our ability to measure other variables. Test scores and measures of genetic variation are much easier to obtain than measures of success in society or measures of variation in the physical and social environments. Therefore, we have to be vigilant against the error of studying what is easy to analyze, at the expense of missing amorphous but important effects – a tension reflected throughout this book.

Theories are needed for two reasons: (1) to develop models for predicting and understanding systems and (2) to reduce complex phenomena to more basic levels. Scientific theories are further distinguished by a commitment to skepticism, objectivity, and empirical verification.

Theories of intelligence can be stated at the psychometric, information-processing, and biological levels. All are useful for different purposes. Psychometric models provide concise summaries of the dimensions of variation in cognitive competence. They also play an important role in applied settings, such as education, personnel selection, or epidemiology. Information-processing theories provide a functional description of why the variables at the psychometric level correlate in the way they do. This includes coordination with measures of brain structures and functions. Biological theories relate intelligent behavior to individual differences in brain structures, brain functions, genetic inheritance, or any combination of these three. Psychometric-level models and biological-level models relate measurable variables to other measurable variables. Information-processing models are more indirect and abstract. In addition to being the appropriate level at which to predict some aspects of socially relevant behavior, information-processing models provide a vital link between the psychometric and biological levels.

Even the most useful models change as advanced methods generate new and sometimes surprising data. This has been the nature of progress in every science. There is no reason to expect it to be any different in the scientific study of intelligence.

2.5 Questions for Discussion

2.1 How is a latent variable different than a manifest variable?

2.2 Why is a reaction range important?

2.3 What systems are important for understanding intelligence?

2.4 Is reductionism necessary to investigate the nature of intelligence?

2.5 Do you think intelligence is a scientific concept?

References

Ackerman, P. L. 1996. A theory of adult intellectual development: Process, personality, interests, and knowledge. *Intelligence*, 22, 227–257.

Ackerman, P. L., Beier, M. E., & Boyle, M. O. 2005. Working memory and intelligence: The same or different constructs? *Psychological Bulletin*, 131, 30–60.

Barbey, A., Karama, S., & Haier, R. J. 2021. *Cambridge handbook of intelligence and cognitive neuroscience*. Cambridge: Cambridge University Press.

Bouchard, T. J., Jr. 2013. The Wilson effect: The increase in heritability of IQ with age. *Twin Research and Human Genetics*, 16, 923–930.

Collins, F. S. 2010. *The language of life: DNA and the revolution in personalized medicine*. New York: Harper.

Colom, R. 2016. Advances in intelligence research: What should be expected in the XXI century (questions and answers). *Spanish Journal of Psychology*, 19, 1–8.

Colom, R., Bensch, D., Horstmann, K. T., Wehner, C., & Ziegler, M. 2019. The ability–personality integration. Special issue of *Journal of Intelligence*, 7, 13.

Cronbach, L. J., & Snow, R. E. 1977. *Aptitudes and instructional methods: A handbook for research on interactions*. New York: Irvington.

Flynn, J. R. 2018. Reflections about intelligence over 40 years. *Intelligence*, 70, 73–83.

Funder, D. C., & Ozer, D. J. 2019. Evaluating effect size in psychological research: Sense and nonsense. *Advances in Methods and Practices in Psychological Science*, 2, 156–168.

Galton, F. 1869. *Hereditary genius: An inquiry into its laws and consequences*. London: Macmillan.

Geary, D. C. 2005. *The origin of mind: Evolution of brain, cognition, and general intelligence*. Washington, DC: American Psychological Association.

Gignac, G. E., & Szodorai, E. T. 2016. Effect size guidelines for individual differences researchers. *Personality and Individual Differences*, 102, 74–78.

Gratton, C., Laumann, T. O., Nielsen, A. N., et al. 2018. Functional brain networks are dominated by stable group and individual factors, not cognitive or daily variation. *Neuron*, 98, 439–452.

Haier, R. J. 2017. *The neuroscience of intelligence*. Cambridge: Cambridge University Press.

Johnson, W., & Bouchard, T. J., Jr. 2005. The structure of human intelligence: It is verbal, perceptual, and image rotation (VPR), not fluid and crystallized. *Intelligence*, 33, 393–416.

Juan-Espinosa, M., & Palacios, A. 1996. Urban and rural people's conceptions of intelligence in Equatorial Guinea. In En, H., Grad, A. B., & Georgias, J. (eds.), *Key issues in cross-cultural psychology*. Lisse: Sweets & Seitlinger.

Just, M. A., Carpenter, P. A., & Keller, T. A. 1996. The capacity theory of comprehension: New frontiers of evidence and arguments. *Psychological Review*, 103, 773–7–80.

Lynn, R. 2009. What has caused the Flynn effect? Secular increases in the development quotients of infants. *Intelligence*, 37, 16–24.

Mislevy, R. J. 2018. *Sociocognitive foundations of educational measurement*, New York: Routledge, Taylor & Francis.

Pietschnig, J., & Voracek, M. 2015. One century of global IQ gains: A formal meta-analysis of the Flynn effect (1909–2013). *Perspectives on Psychological Science*, 10, 282–306.

Plomin, R. 2018. *Blueprint: How DNA makes us who we are*. Cambridge, MA: MIT Press.

Protzko, J. 2015. The environment in raising early intelligence: A meta-analysis of the fadeout effect. *Intelligence*, 53, 202–210.

Quiroga, M. A., Diaz, A., Román, F. J., Privado, J., & Colom, R. 2019. Intelligence and video games: Beyond "brain-games." *Intelligence*, 75, 85–94.

Quiroga, M. Á., Escorial, S., Román, F. J., et al. 2015. Can we reliably measure the general factor of intelligence (*g*) through commercial video games? Yes, we can! *Intelligence*, 53, 1–7.

Rohrer, J. M. 2018. Thinking clearly about correlations and causation: Graphical causal models for observational data. *Advances in Methods and Practices in Psychological Science*, 1, 27–42.

Ridley, M. 2003. *Nature via nurture: Genes, experience and what makes us human*. London: Fourth Estate.

Sternberg, R. J. 2000. *Practical intelligence in everyday life*. Cambridge: Cambridge University Press.

Warne, R. T., & Burningham, C. 2019. Spearman's *g* found in 31 non-Western nations: Strong evidence that *g* is a universal phenomenon. *Psychological Bulletin*, 145, 237–272.

Wechsler, D. 1975. Intelligence defined and undefined – relativistic appraisal. *American Psychologist*, 30, 135–139.

Wilson, E. O. 1998. *Consilience: The unity of knowledge*. New York: Knopf.

CHAPTER 3

Psychometric Models of Intelligence

3.1 Introduction: ... And Then There Were Tests

For testing the quality of wine produced in Southern France, you do not need to taste every grape in the vineyards of the region. You can taste samples from different regions to obtain a reasonable evaluation.

The same scenario applies to intelligence. Scientists and practitioners do not need to exhaustively follow people's lives to obtain a reasonable estimate of all their cognitive strengths and weaknesses. Facebook, Instagram, and reality shows may give hints about a person's mental abilities, but they are poorly standardized for comparing one person to another (Rozgonjuk et al., 2021; Malanchini et al., 2021). Science needs testing for intelligence under controlled situations, and this is why standardized tests are used.

Intelligence tests sample different mental abilities. Virtually all tests of mental abilities correlate positively with each other. Most likely, individuals with high scores on one intelligence test will not have low scores on another. This was discovered by Charles Spearman at the beginning of the twentieth century (Spearman, 1904), and it is one of the most replicated findings in psychology

(Kovacs and Conway, 2016, 2019). It is known as the positive manifold, and it is the core of the psychometric models that will be described in this chapter. An example of the positive manifold is shown in Table 3.1.

Spearman discovered another interesting principle: the indifference of the indicator. This means that all cognitive tests, *irrespective* of their content, measure or tap intelligence to some degree. Some tests, for example, tap reasoning ability using words or numbers or abstract figures; other tests tap more specific abilities like spelling, knowledge of geography (or any other topic), or aspects of memory. All such tests show the positive manifold. In Table 3.1, for example, there is a correlation between scores on tests of music and French ($r = 0.57$).

The important feature is not the content or form of these problems but rather their complexity level. Cognitive problems vary from quite easy to really difficult. Standardized intelligence tests are designed to determine the level of complexity that a person can manage successfully. This is translated into scores that rank people according to their actual performance. Psychometric models are based on the statistical analysis of these kinds of data. This is represented in

Table 3.1 Correlation coefficients among scholarly subjects reported by Charles Spearman in 1904

Variable	Classics	French	English	Math	Pitch	Music	g
Classics		0.83	0.78	0.70	0.66	0.63	0.96
French			0.67	0.67	0.65	0.57	0.88
English				0.64	0.54	0.51	0.80
Math					0.45	0.51	0.75
Pitch						0.40	0.67
Music							0.65
Average r	0.72	0.68	0.63	0.59	0.54	0.52	

Note. Coefficients are all positive, a pattern called the *positive manifold*. The average correlation (*r*) for each subject with all the other subjects and the loading of these subjects on the general factor (*g*) are also shown (see Figure 3.1).

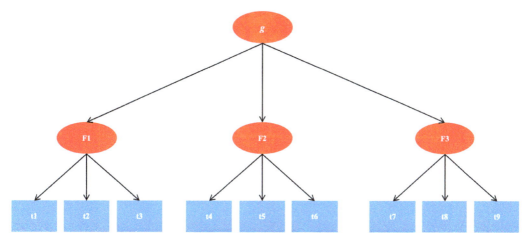

Figure 3.1 The hierarchical structure of mental abilities. The bottom row shows a battery of nine tests of different content and form (t1–t9). As shown in Table 3.1, tests correlate with each other, so it is possible to group tests statistically based on the strength of their correlations (i.e., their similarity to each other). In this example, factors F1–F3 are defined by the tests with the strongest correlations with each other. The factors are also correlated with each other because in this model they share a common general factor (*g*). The arrows indicate a proposed direction of influence (see Box 3.1). This model is one likely possibility for organizing individuals' performance on the administered tests, but there are others.

Figure 3.1, and this hierarchal structure of mental abilities with a general factor at the apex is one of the most important concepts in intelligence research. The *g*-factor represents commonality among all cognitive abilities. It alone usually accounts for about 50 percent of variance among the general population, more than any other single factor. That is why it is a core aspect of intelligence research, although, of course, there is more to the science of human intelligence.

3.2 Sampling Intelligence

Here are some examples of important mental abilities usually tapped by intelligence tests.

3.2.1 *Language*

All intelligence test batteries include evaluation of language skills. There is a good reason for this. Although a person cannot participate fully in human society without language skills, there are substantial individual differences in the degree to which people possess and apply these skills. Everyone learns to produce and comprehend her native language,

but very few people reach the levels of comprehension and expression illustrated by Miguel de Cervantes, William Shakespeare, Virginia Woolf, Alexandre Dumas, Yukio Mishima, or Nadine Gordimer.

In intelligence testing, an important distinction is made between language familiarity and language comprehension. Language familiarity can be assessed by a vocabulary test. Language comprehension is tested by asking about the meanings of sentences or paragraphs. In theory, one might also argue for separate evaluations of comprehension of the written and spoken language. Use of the spoken language is clearly a primary capacity. Normal children learn to speak mostly without explicit instruction simply by being reared by speakers of the local language. Literacy is a secondary skill. Reading and writing are acquired through instruction, and the success of this instruction varies greatly.

3.2.2 *Visuospatial Reasoning*

Intelligence test batteries also include evaluations of some form of *visuospatial reasoning*. The tasks used vary in the extent to which they involve perceptual or reasoning processes. For

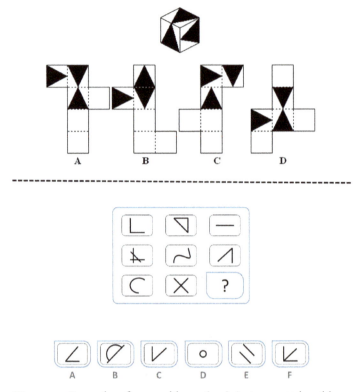

Figure 3.2 Examples of test problems. (top) A visuospatial problem. Which flattened pattern (A–D) corresponds to the three-dimensional cube at the top? The answer is in footnote 1. (bottom) An example of reasoning. Which choice (A–F) completes the 3 × 3 matrix pattern by filling in the missing piece at the lower right? The answer is in footnote 2.
Courtesy of TEA, S.A., Madrid.

evaluating perceptual ability, some tests require the identification of a hidden pattern or of whether two patterns in different orientations are the same or different. An example of a visuospatial problem is shown in Figure 3.2 (top). For evaluating reasoning ability, some tests require identifying a pattern within a visual display, as illustrated in Figure 3.2 (bottom).

3.2.3 *Mathematical Reasoning*

Many tests designed to evaluate intelligence include some form of mathematical reasoning. In the simplest case, this amounts to nothing more than a test of how rapidly one

can do simple arithmetic. More complex problems, for example, may require estimating angles within a geometric figure.

Are tests of mathematical reasoning actually tests of one's academic background, or are they tests of general intellectual ability? The answer is probably a little bit of both. Humans are genetically programmed for some aspects of numerical reasoning, such as the idea of distinguishing specific numbers of objects rather than making a binary distinction between one or many (Tosto et al., 2019). Acquisition of substantial mathematical skill also depends on having an appropriate cultural background. Therefore, mathematical skills beyond a rudimentary level are secondary

---

[1] Answer is A.   [2] Answer is F (hint: it is the only choice made of three lines).

skills. Nevertheless, there is a good reason to assess them as part of an intelligence test battery: skill in mathematics is central to many aspects of functioning in developed societies.

3.2.4 Deductive and Inductive Reasoning

The ability to reason is widely accepted as a sign of intelligence. Two different types of reasoning are recognized: deductive and inductive. In deductive reasoning (the expertise of fictional detectives like Sherlock Holmes, Hercule Poirot, and Anastasiya Kamenskaya) an examinee is told to assume that certain statements are true and is asked to draw conclusions from them. Examples of deductive reasoning follow:

Categorical Syllogism
All the girls in Ms. Sienna's class went on the field trip yesterday.
Ilaria and Adelina are in Ms. Sienna's class.
Where were Ilaria and Adelina yesterday?
Deduction
Helmuth and Klaus drink beer only when they are together.
Yesterday, Helmuth was in town and Klaus was out of town.
Did Helmuth drink beer yesterday?

Inductive reasoning is the process of abstracting general rules from observation of specific cases. Progressive matrix tests (like shown in Figure 3.2, bottom) were developed to evaluate this ability. Inductive reasoning is also evaluated using other formats. Here are some examples:

Similarity
Which of the following cities does not belong in the group:
San Francisco, Las Vegas, San Diego, St. Paul?
Series Completion
Complete the next number in the following series: 3:5:7: _____?
Analogy
Choose the correct answer to complete the analogy:
Black is to white as right is to _____: left, color, up, opposite.

There are logical and empirical arguments for evaluating abstract deductive reasoning.

Syllogisms and categorical reasoning are central to notions of mathematics, law, and rational argument. Parents tell children, "If you don't eat your vegetables, you can't have dessert." Chocolate-loving children are supposed to draw an appropriate conclusion.

Empirically, scores on abstract reasoning tests can be used to predict performance on other applications of intelligence, such as paragraph comprehension, mathematics, and (to a lesser extent) the ability to solve visuospatial reasoning problems. Because abstract reasoning predicts so many other types of thinking, it is reasonable to believe that abstract reasoning is either a central part of intelligence itself or is closely tied to something that is. Many studies indicate that abstract reasoning ability is the single best estimate of the g-factor.

However, there is a case against stressing reasoning too much. Although abstract reasoning is dear to Western academic and scientific circles, outside these circles, some people think of abstract reasoning as a sort of word game, with little intellectual content. Does this view have merit? Here are two relevant examples to consider.

In one of the Sherlock Holmes stories, "Silver Blaze," the detective deduces that an intruder did not break into a stable to steal a horse because a watchdog did not bark in the night. This is an example of the classic syllogism:

Premise: *A implies B. If there is an intruder, the dog will bark.*
Observation: *Not B. The dog did not bark.*
Conclusion: *Therefore, not A. Therefore, there was no intruder.*

This is a fictional story, but in real life, watchdogs do not always bark when strangers appear. The real world is far more complicated than the abstract world of logical reasoning.

In the second example, consider the example of the two men who always drank beer together, given earlier as an example of a syllogism. The example is a paraphrase of an item that was used in an anthropological investigation of reasoning across cultures. Rural

Liberian tribesmen said that the question was not reasonable, on the grounds that they did not know the individuals involved, and one ought not to draw conclusions about things that one has not experienced personally.

Both these arguments can also be used against the inclusion of abstract inductive reasoning items in intelligence testing. And there is another objection: the answer to an inductive reasoning problem is never uniquely determined. This is shown in the example in which an examinee was asked to find the dissimilar city in the set [San Francisco, San Diego, Las Vegas, St. Paul]. One could argue that the first three cities are in the far West of the United States, while St. Paul is in the Midwest; or that the first three cities have Spanish-derived names, while St. Paul does not; or that Las Vegas is the only city not named for a Christian saint. There are different ways in which the four items can be compared; the question does not specify which one is to be used. Actually, this example is so ambiguous that it likely would not be used in a well-constructed test.

Similar arguments can be made against the use of series or analogy problems, or matrix tests. In some cases, the correct answer is decided by consensus, and the consensus opinion may be different in different cultures. Here is another example.

Which two of these animals belong together: fox, cat, dog?

When University of Michigan students were asked this question, most of them said that the fox and the dog belong together because they are both canids. When Central American Indians were asked the same question, their preferred answer was that the fox and the cat belong together because of their similar behavior. The Michigan students preferred a taxonomic grouping, whereas the Central American Indians preferred an ecological one. Who is to say which answer is correct? Usually, however, in the construction of tests, such ambiguous items (e.g., the city example and this example) are eliminated.

Richard Nisbett (2009) has conducted studies of cultural influences on thought and has offered similar examples involving contrasts between the reasoning of North American and Eastern (Asian) styles of thought. He has argued that the American–European emphasis on focusing on only the perceived relevant aspects of a situation, and applying formal logic to those aspects only, is a marked contrast to an Eastern Asian emphasis on being sensitive to the total context of a problem. However, even given cultural influences on reasoning, recall from Chapter 2 that the *g*-factor is found in diverse cultures around the world (Warne and Burningham, 2019).

3.2.5 *Aptitude and Achievement Tests*

A distinction is sometimes made between aptitude and achievement tests. The distinction can be illustrated by comparing three different college admissions tests administered in the United States: the SAT-I, the SAT-II, and the American College Test (ACT). The SAT-I is said to stress aptitude because its content is not tied to specific course curricula. The SAT-II and the ACT contain subtests tied directly to curricula in courses of history, literature, science, and mathematics. What is the difference among these approaches?

Achievement is simple enough to assess. Your success at studying English, mathematics, history, physics, or anything else is demonstrated by answering questions of fact. The more correct answers you give, the greater is your success (at least by this criterion). Achievement implies that a person has done something in a certain field. But what is meant by aptitude? According to Richard Snow (1996), aptitude implies that a person has a talent for doing something – politics, music, athletics, and so forth. A person who has an aptitude for a field likely would do well if they train in that field, but having aptitude does not imply that the training has taken place (and success in any area requires several factors, as discussed in Chapter 8).

You can have an aptitude for playing music without knowing how to play the clarinet. But if you have the aptitude, learning to play should be relatively easy compared to a person with little aptitude. Masterful playing, of course, requires hard work and commitment, attributes not assessed as part of musical aptitude.

The SAT-I was originally called a scholastic aptitude test because it was supposed to identify students who would be successful if they went to college or university, without stressing specific knowledge acquired in particular high school courses. By contrast, the SAT-II and ACT tested knowledge of topics, such as mathematics. However, the SAT-I is not entirely knowledge-free. For instance, it assumes that examinees know the English language but does not assume that they have had a course in English literature.

Progressive matrix tests of reasoning are aptitude tests that require even fewer demands on specific knowledge. However, users of these tests assume that the examinee understands the basic testing situation, and there are various strategies for taking such tests that depend in part on cultural knowledge.

The developer of an achievement test assumes that all examinees will have had certain experiences – for example, a course in human history. The test is intended to determine what the examinee learned from the experience. The argument for using achievement tests as screening devices is that one of the best predictors of how well a person will do at learning something in a new situation is how much the person has learned in previous comparable situations. By this logic, the best predictor of grades in first-year college physics is a test of how much physics the student learned in high school.

In educational settings, it turns out that very much the same admission decisions are made regardless of whether an aptitude or an achievement test is used. While there are cases of people who score highly on the SAT-I and do poorly on the ACT, and vice versa, at the population level, the SAT and the ACT predict success or failure equally well for most people (see Chapter 8).

On the whole, if you have an aptitude for academic studies, then you will have learned a lot from the classes you have already completed. If you do not have the aptitude, you will learn less (Zaboski et al., 2018; Murray,

2008; Pokropek et al., 2021). This is not to deny the fact that late bloomers exist. There are people who have the talent to succeed in academics but who, for a variety of reasons, have not learned very much in high school. Conversely, there are people who do not do well on abstract aptitude tests but are quite good at learning some academic material. These cases are, however, exceptions to the rule. The bottom line is that academic achievement and academic aptitude test scores are highly correlated and predict future academic success equally well (details in Chapter 8).

3.3 Test Design and Test Use

We now move from a consideration of the content of cognitive tests to some additional general issues about how the tests are designed. Recall that *psychometrics* is a term related to the science of test construction.

3.3.1 *Item Selection and Evaluation*

Candidate questions on all intelligence tests initially are designed either by identifying a group of people who represent a range intelligence, and then seeing what sorts of problems they can solve (recall that this is what Alfred Binet did), or by generating questions from a theory of what scientists think the concept of intelligence involves.

The candidate questions are then given to a sample of people selected as typical of the population for whom the test is intended. One way to do this is to insert a candidate question into an existing test. Answers to the trial question are not counted as part of the score on the existing test, but statistics are gathered on how it is answered. If these statistics meet certain psychometric criteria (not too easy or too hard), the candidate question can be incorporated into the next revision of the test. Sometimes a whole section of a test like the SAT has only candidate questions in preparation for the next version of the test to be given in the next year. If a test question is a good one, scores on that question should be positively correlated with scores on the other items on the test. Adding that question may also improve the predictive validity of the total score.

Consider an analogy between taking a test and a jump competition, where competitors have to jump over a bar set at various heights. Suppose a competitor can jump over and clear a 1.75 m bar but cannot jump over a bar set to 1.83 m. We would expect this competitor to be able to jump over a 1.6 m bar but not to be able to jump over a 1.9 m bar. More generally, if a jumper can clear a bar at height x but cannot clear a bar at height y ($y > x$), we expect the jumper to clear all heights lower than x and not to be able to clear a bar at heights higher than y.

Suppose we wanted to construct a ten-item test of the ability to solve word problems. According to psychometrics, we should look for ten problems that

1. could be ordered in terms of difficulty, defined by the percentage of people who can solve each problem;
2. behave like the jump bar – if a person can solve a problem at difficulty level x but cannot solve a problem at difficulty level y, where y is greater than x, then the person should solve all problems with difficulty level less than x and not solve any problems with difficulty level greater than y.

In practice, there will always be some people who fail to solve a problem at one level of difficulty but do occasionally solve problems at a higher level of difficulty. A statistical technique called item response theory (IRT) has been developed to determine the best selection of questions with different degrees of difficulty to measure the same thing (like a latent variable). We discuss more about IRT in Section 3.3.3. The main point to remember is that a great deal of care is taken to select test questions that evaluate the same ability, but at different levels of difficulty.

3.3.2 The Distribution of Test Scores

We need to distinguish between test scores, IQ and similar metrics, and the underlying

concept of intelligence as a property of an individual. This was highlighted in Chapter 2. Again, we draw on an analogy between intelligence testing and jumping.

We can think of a jump competition in the following way. The jumps are ordered by height, from the lowest to the highest. Competitors try to jump at each height and are scored by the number of jumps they clear. A competitor's score would not depend on the order in which jumps were attempted.

To see this, consider a person who has the ability to jump 1.5 m high and a contest in which the jumps are, in order of height, 1 m, 1.5 m, and 2 m. That person will succeed on two jumps, regardless of the order in which they are presented. An Olympic-level athlete would almost certainly succeed on all three. In a cognitive test, each question is associated with a level of difficulty, just as the height of a bar is associated with difficulty in jumping. The raw score on a test is the number of questions the person can answer correctly.

What would be the distribution of scores for the population of people who either compete in jumping or take the test in intelligence research? That depends on two things: how high we set the bar (or how difficult we make the items) and how skilled the competitors (test takers) are.

If we were testing high school–level men competitors using the 1, 1.5, and 2 m bars, virtually everyone would clear the lowest bar, most would clear the middle bar, and a few would clear the highest bar. The distribution of scores would contain very few 0s, some 1s, mostly 2s, and a few 3s.

We could change the distribution of scores by changing competitors. The score for men college jumpers would be mostly 3s, for women jumpers mostly 2s. Or we could change the distribution of scores by changing the height of the bars. Suppose the bars were set at 1.4, 1.6, and 1.8 m. The scores for men college jumpers would pile up in the 3s, and we would begin to see more 3s in the women's competition.

We could do exactly the same thing in intelligence testing, except that we would manipulate item difficulty instead of the height of the jump bar. For a given population (high school students, military recruits, etc.), the distribution of raw scores (number of questions answered correctly) depends on how many items the test contains at different levels of difficulty. For a given test, the distribution of scores depends on the distribution of cognitive abilities in the population.

Many cognitive tests are constructed so that they yield a normal distribution of test scores in their intended population. For instance, SAT scores are approximately normally distributed over the population of people who apply to US colleges and universities. This is a selected group of individuals of higher-than-average intelligence. By contrast, scores on the Wechsler Adult Intelligence Scale (WAIS) are intended to reflect a person's intelligence relative to all people in the population.

A student who receives a near-average score (somewhere around 1100 for the combined math and verbal scales) on the SAT-I would probably have an IQ score of over 100, simply because the mean intelligence level of the high school students who apply to college is higher than the mean intelligence level of all people in that age group.

As a result, raw scores on intelligence tests are approximately normally distributed because the questions on IQ tests have been selected to produce a normal distribution.

Contrast this to height, which (within sexes) also is approximately normally distributed. Since the measurement procedure for height is dictated by a standard way to assess all kinds of length (measuring tapes with no appreciable error), the fact that height is normally distributed is a discovery about nature. The fact that IQ and similar cognitive scores are distributed normally is a consequence of the way the tests are constructed.

There are marked advantages in requiring that test scores be normally distributed over an appropriate population. The statistical procedures for dealing with normally distributed scores are well known. If scores are normally distributed, the standard score metric provides a convenient way of comparing individuals to each other in terms of

different cognitive abilities, measured on the same population. For example, raw scores can be converted to standard scores based on the average and the variance of the distribution. A raw score 15 points above the mean (100) on an IQ test is 1 standard deviation and converts to a standard score of +1, 30 points above the mean converts to +2, and so on (negative standard scores denote raw scores below the average). This was illustrated in Figure 1.1.

When scores are normally distributed for any variable, there is also a useful translation from standard scores to percentiles. For instance, for a normally distributed set of test scores, a person with a standard score of +1 will have a score that is above the scores of approximately 85 percent of the population (Figure 1.1). Percentiles are the most common way IQ scores are interpreted for individuals.

3.3.3 Item Response Theory (IRT): Beyond Raw Scores

Raw scores are easy to compute and understand in simple situations. IRT provides an alternative method of scoring that has the advantage of depending less on an arbitrary selection of item difficulties to approximate a normal distribution of scores but has the disadvantage of being harder to understand. We want to describe it because it is widely used in modern psychometric test construction, including intelligence tests. The analogy between intelligence testing and jump competition is again useful. This time, suppose that the judges have lost their measuring devices, so they don't know how high the bars are. They can still conduct the meet and assign sensible scores to the competitors. Here is how this would work.

We can assume that every competitor has a trait, jumping ability. Every bar, although of unknown height, has a quality we will call jumping difficulty. We can measure jumping difficulty directly by seeing what percentage of people can clear a bar, even if we do not know how high that bar is.

The insight of IRT is that jumping ability and jumping difficulty must have the same scale. We now assume that in some reference population, jumping ability is distributed normally, with a standard score mean of 0 and a standard deviation of 1. Arbitrarily, let us decide that the population of men high school students will be the reference population. We find the bar (of unknown height, it doesn't matter) such that half of all high school students can clear this bar. From our assumption of the normal distribution, half of all high school students have a jumping ability above the mean, and half below, that is, a standard score of 0. Therefore, the bar that just half the students can clear must have a jumping difficulty score of 0. By the same token, if approximately 16 percent of the students clear a second, higher bar, then the second bar must have a jumping difficulty of about 1, because, by the properties of the normal distribution, 16 percent of the population has a standard score above 1.

If we carry out this norming procedure for, say, thirty bars of different heights, we will have a test of thirty items, each of which has a jumping ability defined by the percentage of people in the reference population who cleared each bar. Note that the raw scores (numbers of bars cleared) would not necessarily follow a normal distribution.

The analogy to cognitive testing is exact:

1. Take a reference population.
2. Assume that the cognitive ability of interest is normally distributed in the reference population.
3. Infer difficulty levels for questions by observing the percentages of people in the reference population who can answer each question.

The resulting test can then be used to measure ability levels in populations other than the reference population, using the scale derived from norming in the reference population.

In the jumping example, we could compare the scores of college-level women competitors to those of high school–level women competitors, using the scale established by the high school men. In an intelligence application, we could compare the intelligence level of, say, Cambridge students to Oxford students using a scale established at University College London.

The only arbitrary assumption is that the underlying ability is normally distributed in the reference population. If the reference population represents a wide range of ability, this assumption can be defended. If intelligence is produced by the cumulative effects of many different causes – ranging from inheriting certain genes to going to good schools, and so on – and if these causes are independent of each other and no one of them has a strong effect, then intelligence will be distributed normally in a large population.

3.3.4 *The Importance of Norming*

Choosing a reference population is an essential step in test construction, regardless of whether raw or IRT scores are used. The appropriate reference population depends on the purpose of the test. Probability samples (also called random samples) of entire populations (all Spaniards, Brazilians, Chinese, or Germans for the appropriate version of the test) are required for norming purposes. Because tests are intended to be applicable to all ages, the norming sample ought to contain a large number of people of different ages.

The problem of norming is much easier when a test is intended for a clearly defined subset of a population. The PISA (Program for International Student Assessment) tests, for example, are intended primarily for fifteen-year-old high school students worldwide, so norming is carried out only in samples from this demographic. When we compare studies using different tests, we have to keep in mind the effect of different tests having been normed in different populations.

Test designers try to make a test maximally sensitive to changes in the underlying trait in the middle ranges of the reference population. This means that if a test is used to study a population that is very different from the reference population, the test may not do a good job of distinguishing between people in the new population. To see this, imagine that for some perverse reason it was decided that test X and test Y should switch roles. Colleges and universities use X, while the military uses Y. It would be hard to distinguish between the better and the best students applying to universities, because both groups would be getting very high scores on X, which is markedly easier than Y. This is called a ceiling effect. For the converse, the military would find it hard to distinguish between applicants who were marginal but acceptable and ones who were unacceptable because both groups would be getting very low scores on the harder test Y. This is called a floor effect.

Of course, no one is going to switch test X and test Y. However, changes in reference populations do occur and can have important practical consequences. For example, one version of a matrices reasoning test may be sufficiently difficult to yield a normal distribution of scores in the general population, but the same test would, on average, be easier for university graduate students, so there would be mostly high scores and less variation among people. A harder version of the test would be more appropriate.

3.4 Intelligence Measurement: Summary

Although different tests have been designed based on different theoretical rationales, there is a surprising commonality of content across all tests. Similar methods for evaluating verbal intelligence, quantitative skills, and abstract reasoning appear over and over again.

Some tests are given individually; others are given to groups of examinees. Beginning around 1995, there was a movement toward presenting questions by computer. While this has advantages in terms of test administration, simply putting items on a computer does not change the nature of the psychological traits being evaluated. A vocabulary test is a vocabulary test. The underlying psychometrics of a test are the same.

Reliance on snapshot testing ("drop in from the sky") limits what can be evaluated.

With conventional testing, it is difficult to evaluate capacity for reflection, or creativity, that could be considered part of the intelligence concept.

When all is said and done, however, conventional intelligence tests have good predictive validity for success in varied life settings. Validity coefficients range from 0.30 to 0.85, as we will see in Chapter 8. These values represent substantial effect sizes in real life (Funder and Ozer, 2019). Moreover, perfection is not a reasonable goal. Life success is not solely influenced by personal characteristics, including intelligence. Nor can a test possibly evaluate all the personal traits that might be important in every situation. The fact that there are limits on predictive validity does not imply that tests should be disregarded. In fact, the evidence is overwhelming to support the conclusion that intelligence test scores are meaningful indexes of intelligence as commonly defined.

3.5 Psychometric Models in Brief

Psychometric models try to explain how variations in mental abilities and cognitive performance across many different situations can be summarized by individual differences in a smaller number of basic factors, such as general reasoning, verbal ability, and visuospatial reasoning.

3.5.1 *Essence of Psychometric Models*

When a battery of different mental tests is given to a large number of people, representative of the general population, we can find out how the tests relate to each other. Figure 3.1 illustrates one simple model for organizing the data. But how can we determine the best ways to group the different tests? Statistical analyses are used to identify the latent traits underlying performance on different tests. The next section is a brief description of factor analysis, a key statistical tool for building and refining these psychometric models.

3.5.2 *Factor Analysis*

A psychometric model takes the correlations among a set of measures as the data to be organized. For example, a group of people representative of the population complete nine intelligence tests: Information, Vocabulary, Similarities, Matrices, Blocks, Puzzles, Coding, Symbol Search, and Cancellation. For now, the actual content of each test is not important. Individuals who do well on one test will also tend to do well on the other tests, and individuals who do poorly on one test tend to do poorly on the others. We described these relationships earlier in this chapter as the positive manifold and noted that it is one of the best replicated findings in psychology.

This manifold, however, is a matter of degree. Correlations among tests are not perfect. The best individual on the Vocabulary test might not be the best on the Matrices or the Symbol Search test, but they would probably be toward the top. The same thing would be true of the worst individual on the Vocabulary test – this person would probably be toward the bottom on the Matrices and Symbol Search tests, but not necessarily right at the bottom. Performance on one test would be a useful predictor of performance on the remaining tests, but the prediction would not be perfect.

This reflects the reality of intelligence testing. All tests of intelligence performance exhibit positive manifold to some degree, and they often do so quite strongly. Therefore, it may seem reasonable to suppose that there are three underlying abilities involved in the nine tests we used in the example: (1) verbal, (2) visuospatial, and (3) processing speed. If the three abilities are themselves correlated, it might be appropriate to identify a general ability common to all three. But how are we to test this supposition and find out empirically whether we can summarize performance differences assessed by the nine intelligence tests to three common dimensions based on underlying traits?

The statistical techniques known as factor analyses help answer this question. Factor analysis comes in two major varieties – exploratory factor analysis (EFA) and confirmatory factor analysis (CFA). Understanding the basic workings of factor analyses is required for understanding the logic behind psychometric models. First, we consider EFA.

3.5.3 *Exploratory Factor Analysis (EFA)*

Table 3.2 shows examples of scores obtained by twenty individuals on the nine intelligence tests. The actual data set includes 1,002 individuals, representative of the population in terms of age, sex, education, residence, and so forth. The table shows the raw scores on the nine tests obtained by each examinee. Looking at every score of each of the 1,002 individuals across the nine intelligence tests would be interesting, but certainly boring and not illuminating from a theoretical standpoint of trying to identify patterns in the data.

Knowing if these individuals can be ordered according to their performance levels across tests requires computing the correlations among their scores as a first step. We can do that for every pair of tests. Table 3.3 shows the computed correlations for all these intelligence tests according to the scores obtained by the complete group of individuals. For example, the correlation (r) between Information and Vocabulary is 0.63, whereas the correlation between Information and Cancellation is 0.24. Both values are far from perfect ($r = 1$).

Looking carefully at the values in Table 3.3, we can see the positive manifold. All these correlations are positive, and therefore, individuals with better performance in one test are also those with greater scores on the remaining tests. This happens for all tests, but the values change, and these variations tell a story to scientists. Thus, for instance, Vocabulary and Similarities show a correlation of 0.68, whereas the correlation between Vocabulary and Cancellation is 0.21. Cancellation and Symbol Search show a correlation of 0.51, whereas the correlation between Cancellation and Puzzles is 0.29.

Table 3.2 Examples of scores obtained by twenty individuals after completing nine intelligence tests

	Information	Vocabulary	Similarities	Matrices	Blocks	Puzzles	Coding	Symbol Search	Cancellation
1	16	28	18	24	50	19	54	23	40
2	19	46	30	23	56	18	87	45	59
3	18	40	20	13	49	15	108	38	37
4	6	24	21	19	58	18	47	22	33
5	9	36	29	21	50	13	69	30	32
6	24	33	25	21	54	21	84	40	53
7	15	28	23	22	30	17	72	40	49
8	24	37	23	26	54	20	67	41	51
9	20	35	23	23	53	18	78	25	38
10	16	43	28	21	56	24	71	30	45
11	16	20	28	23	43	17	81	34	55
12	12	28	23	14	36	7	48	22	59
13	17	33	24	23	52	22	90	36	39
14	12	32	27	20	34	18	77	33	41
15	23	37	26	24	53	23	69	36	41
16	25	48	28	25	60	22	70	24	40
17	21	36	24	20	47	15	68	31	42
18	7	21	14	23	41	10	37	24	34
19	6	27	10	11	28	7	20	12	23
20	17	29	14	21	45	18	37	28	39

Note. Each row represents an individual, whereas each column represents the scores of the examinees in a given test.

Table 3.3 Correlation values for nine intelligence tests completed by 1,002 individuals representative of their population

Test	I	V	SIM	M	B	P	COD	SS	CAN
Information		0.63	0.67	0.59	0.54	0.50	0.56	0.52	0.24
Vocabulary			0.68	0.60	0.53	0.50	0.59	0.55	0.21
Similarities				0.65	0.59	0.57	0.63	0.60	0.28
Matrices					0.76	0.71	0.76	0.72	0.38
Blocks						0.74	0.74	0.72	0.33
Puzzles							0.67	0.66	0.29
Coding								0.86	0.49
Symbol Search									0.51
Cancellation									

EFA analyzes all these correlation values simultaneously and then groups those tests with the most similar correlations. The technical details are not relevant here; there is nothing magical about them. We only need to know that these groupings are based on the average correlation between each test and the remaining tests in a given intelligence battery. Tests with greater average correlation values share more common cognitive requirements. Tests with smaller average correlation values share fewer cognitive requirements. Generating a correlation matrix like that in Table 3.3 is the first step for EFA.

If we submit the correlation matrix shown in Table 3.3 to an EFA – which can be easily done using a computer program – the result would be the numbers shown in Table 3.4. This is the second step in EFA. In this example, three groupings (or factors) of tests are identified.

Table 3.4 demonstrates that all tests show relatively high values on the three obtained factors or underlying dimensions (Factor I, Factor II, and Factor III). However, within each factor, some tests show higher values than others. These values are called factor loadings and are derived, as noted, from the

Table 3.4 Factor matrix derived from the correlation matrix shown in Table 3.3

Test	Factor I	Factor II	Factor III
Information	0.33	0.70	0.41
Vocabulary	0.39	0.70	0.41
Similarities	0.41	0.71	0.46
Matrices	0.59	0.57	0.72
Block	0.57	0.43	0.80
Puzzles	0.48	0.43	0.75
Coding	0.84	0.45	0.59
Symbol Search	0.86	0.40	0.57
Cancellation	0.59	0.34	0.42

Note. The values in columns I, II, and III are called factor loadings.

average correlation between each test and the remaining tests in the battery. The optimal number of factors that can be derived from any matrix depends on several technical issues, but the details do not concern us here.

In this example, the highest loadings in Factor III are for Matrices (0.72), Block (0.80), and Puzzles (0.75). The loadings on Factor III are smaller for the remaining six tests – smaller, but still substantial. It is of paramount importance to remember that all these computations are based on individuals' scores calculated from their performance on the nine intelligence tests.

We can inspect Factors I and II and see that some tests show higher loadings than others. Those tests showing greater loadings delineate the psychological interpretation of the findings derived from the statistical computations. Factor II appears to be mostly a verbal factor given the highest loadings are on Vocabulary, Information, and Similarities. Factor III appears to be mostly a visuospatial factor with highest loadings on Matrices, Block, and Puzzles. Factor I appears to be mostly a speed of processing factor with highest loadings on Coding, Symbol Search, and Cancellation. The descriptive labels we assign to factors are somewhat arbitrary and not that important, as noted humorously by Douglas Detterman (1989, p. 169): "It is impossible to name factors and still have friends. Any attempt at naming factors produces instant hostility."

In the final step, we can ask whether the three factors summarizing people's performance are correlated. Looking at the loadings of Table 3.4, we see the answer: yes, they are correlated, because each factor has loadings on every test. The actual correlations among the three factors are shown in Table 3.5.

The psychological meaning of these correlations is straightforward: people showing better performance on the tests summarized by Factor I will also show better performance on the tests summarized by Factors II and III, and vice versa. This indicates that there might be a common factor underlying the three factors derived from the nine tests. The factor common to all three factors is the g-factor, shown in Figure 3.1. From an applied perspective, people can be ordered according to their objective performance on the tests, and we can obtain a general score (g) along with scores related with the verbal, visuospatial, and processing speed factors.

Table 3.5 Correlations among the factors shown in Table 3.4

	Factor I	Factor II	Factor III
Factor I		0.51	0.69
Factor II			0.59
Factor III			

Another noteworthy issue refers to the fact that the scores obtained by any given examinee on a given test can be attributed to three sources of variance: the g-factor, the group factor the test taps, and the specific skills required by the test. Higher or lower scores on, say, the Vocabulary test can be obtained because of g, verbal ability (the group factor mainly tapped by the test), and the specific skills related to the acquisition and use of the vocabulary present in a given cultural context. For example, Vladimir and Xi could obtain identical scores on a vocabulary test, but for different reasons. It would be incorrect to conclude that scores on this test imply the same level of verbal ability. The performance assessed by the vocabulary test is influenced by general cognitive ability (g) along with specific abilities and skills required by the test. The score obtained by Vladimir might be attributed mainly to g, whereas the same score obtained by Xi might attributed mainly to his specific verbal abilities and skills.

From a theoretical perspective, we can build models for representing the structure of the intelligence construct from the evidence described in the example of the nine tests completed by the 1,002 individuals who represent their population, from the empirical fact of the positive manifold. This procedure has been applied for decades, and the effort made by scientists has led to proposing different psychometric models of intelligence. Some models fit the actual data better than other models. CFA provides a way of deciding how well a set of competing theoretical models matches the observations in a psychometric study. Here is how CFA works.

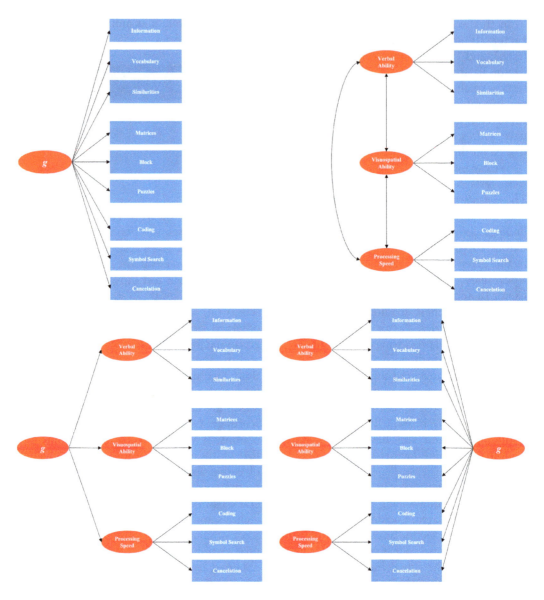

Figure 3.3 Four models of how people's test performance might be organized based on completing nine intelligence tests tapping verbal (Information, Vocabulary, and Similarities tests), visuospatial (Matrices, Block, and Puzzles tests), and processing speed (Coding, Symbol Search, and Cancellation tests) abilities. The model at the top left represents a key role for a single factor (*g*); the top right represents organization of correlated factors (without *g*); the bottom left shows a hierarchical organization (*g* at apex); and the bottom right shows a bifactor organization (both *g* and factors contribute independently). CFA computations help to choose the best model for describing the actual pattern of correlations.

3.5.4 *Confirmatory Factor Analysis (CFA)*

Scientists can decide, in advance, the relationships they think link factors and tests, and factors to other factors. This advances

factor analysis from a method for identifying relationships from data to a method for testing hypotheses about data.

We can use the same data set from the explanation of EFA to build several models of how people's performance might be organized according to their scores on the nine tests. Figure 3.3 shows four candidate models with different assumptions. The boxes, ellipses, and arrows establish the structure of a factor analytic model. These are the models' assumptions: (upper left) people's performance can be summarized by one single dimension representing general intelligence (g); (upper right) people's performance can be summarized by three correlated dimensions representing verbal, visuospatial, and processing speed abilities; (lower left) people's performance can be summarized by the latter dimensions, but these three can be further summarized by a general higher-order dimension (g) (hierarchical model); and (lower right) people's performance can be summarized by a general dimension (g) and three dimensions representing verbal, visuospatial, and processing speed unrelated with the general factor (bifactor model).

We can submit the data set to a CFA computer program and see how well the results fit the models. The fit between the relationships represented by the models and the data can be excellent, appropriate, or poor. The better the fit, the more we can trust the model. The details of these criteria are not important here. The point is that there is an objective way to quantify which models are best for fitting the actual data.

3.5.5 *Limits of Factor Analysis*

The development of factor analysis has enhanced the precision of psychometric research. Factor analytic studies are used to identify underlying, unobservable traits (latent variables). Tests with factor loadings close to 1 are important pieces of evidence, because a loading of 1 indicates that variation in observable test performance is perfectly linked to variation in the trait. For instance, the argument that the ability to detect patterns is an important part of general intelligence is supported by the fact that progressive matrix tests often have high loadings on the g-factor, and that pattern detection is an important step in solving

progressive matrix problems. In this case, we say that progressive matrix test scores are good indicators for general intelligence.

Unfortunately, it is easy to misinterpret this statement. A high loading between a trait and a test does not mean that the quality indicated by the trait is required for performance on the test in the way that, say, strength is required to push a piano across your living room. It means that variation *in test performance* across people is related to variation *in the trait* across people.

Factor loadings are statements about individual differences as observed in a particular group or population. A person's standing on both traits and tests is defined relative to the performance of other individuals in the group, not by the individual's absolute performance. The mathematics whiz in high school may not be that outstanding at MIT. This has an important implication for factor analysis. The test loadings derived from factor analysis result from an interaction between the properties of the test itself, the other tests in the battery, and the distribution of traits in the population being tested. Some thought experiments illustrate what this means.

Here is a new list of nine tests tapping various mental abilities:

1. a vocabulary test;
2. a test of paragraph comprehension based on newspaper editorials;
3. a test requiring judgments of the syntax of German sentences;
4. a test requiring people to evaluate the logical argument contained in a political speech of Angela Merkel;
5. a test in which people are asked to judge whether meaningless patterns presented at different orientations are identical or whether one is the mirror image of another (this means that the examinee has to rotate figures in the mind's eye – such tests are called rotation tests);
6. a test in which people view a room and then imagine that they are at one location, facing in a particular direction, and point to a target object (e.g., "Imagine

that you are at the desk, facing the file cabinet. Point to the flowerpot");

7. a test in which the task is to find a picture hidden in another (e.g., finding a triangle in a Star of David figure);

8. a map-reading test;

9. a word problem test in which people read passages about an explorer going through a jungle and then answer questions about the spatial layout of objects (e.g., "As Friedrich proceeded southward, he realized that the python was keeping up with him on the right and the tiger on the left." Question: Was the python west of the tiger?).

The first four tests evaluate language skill; the next four evaluate visuospatial reasoning; and the last test evaluates both language skill and visuospatial reasoning. Suppose that these tests have been constructed so that the scores are normally distributed in the high school population. Because visuospatial and language skills are correlated in the high school population, it would be possible to extract a general factor along with visuospatial and verbal factors independent of the general factor.

Imagine now two further studies. In the first we give the test battery to a representative sample of lawyers. In the second it is given to a representative sample of architects. What would happen to the factor loadings? Stop and think for a moment.

Lawyers are highly selected for language-related skills. Therefore, the four tests of language skills, although appropriate for a high school sample, would be quite easy for the lawyers. As a result, there would be little variation in scores on the verbal tests. Lawyers are not selected for their visuospatial skills, so we might find considerable variation on the visuospatial tests. An analysis for a single general factor would result in a factor with high loadings on visuospatial tests, because that is where the variance would be. Test 9, which draws on both verbal and visuospatial skills, would have a high loading on the visuospatial factor and a low loading on the verbal factor.

On the other hand, the visuospatial skills tests would be easy for the architects, while tests of verbal skills should show more variation. Accordingly, it is likely that the general factor, as defined for architects, would have high loadings on verbal tests. Test 9 would now have a high verbal loading and a low visuospatial loading. These changes in loading have nothing to do with the absolute abilities of the individuals involved. They are driven by the amount of variation in the population, not the ability level.

These examples represent actual situations that may occur. It is common practice to conduct psychological studies using college students as participants. College students are selected largely on verbal skills and on general intelligence. As a result, the variation among college students on these traits is less than the variation in the population. Therefore, a study of the importance of intelligence in college students is likely to underestimate the importance of intelligence in the population at large.

The next thought experiment deals with another problem that limits the generality of factor analytic findings. The loading of a test on a trait can vary depending on what other tests are in the battery, even though the same population is studied. Suppose that instead of using all nine of the tests in the example, we compose a smaller battery consisting of the first five tests – four tests of language skills and the rotation test. The test battery is given to a representative set of adults. We then extract a single general factor from the data. It would be defined by the traits required by the four verbal tests. The single spatial test (test 5) would have a low loading on the general factor and a large specific term. The vocabulary test (test 1) would have a high loading on the general factor, for it has repeatedly been found that people with high general verbal skills tend to have large vocabularies.

Now construct a new battery consisting of the vocabulary test and the four visuospatial tasks, tests 5–8. The general factor will be determined largely by the visuospatial tasks. The rotation test would have a high loading on this factor, while the vocabulary test would have a low loading and a high specific term.

These thought experiments show what might happen, not necessarily what does happen. In practice, scientists use balanced and similar test batteries, which leads to replicability of results but at the same time produces a restricted definition of intelligence. This is so because the same mental skills are being evaluated from study to study in only slightly different ways. Factor structures for a given battery are usually replicated across populations (college students, military recruits, population samples of adults). The restrictions on generality described here are warnings to keep in mind when interpreting results, not compelling arguments for regarding all factor analysis results as meaningless artifacts to discard. The next sections disprove the myth that the *g*-factor and related cognitive abilities are merely artifacts of the method.

3.6 To *g* or Not to *g*

Now you know what *g* is and how it is derived. Scientists who advocate the *g* model acknowledge that there are other factors of mental performance, such as verbal, visuospatial, or speed ability, but they focus on the weight of evidence that shows that each of these factors accounts for much less of the variation in cognitive performance than does *g*. It is safe to say that *g* is at once the best-known, most praised, and most vilified psychometric concept of intelligence. Here is some additional history.

3.6.1 *General Intelligence (g)*

As noted in Chapter 1, the *g* construct was first proposed by Charles Spearman (1904, 1923, 1927), a Briton who is recognized as the first modern psychometrician. After serving for some time as an army officer, Spearman entered academia, receiving a PhD from Wilhelm Wundt's experimental psychology laboratory in Germany. He was attracted to Francis Galton's concept of intelligence as generalized mental fitness,

although he did not share Galton's enthusiasm for what we would today call information-processing measures of cognition. Instead, he thought that one should find a general factor through the analysis of complex measures of thinking, such as school grades. To investigate this conception, he developed both the basis for the rank order correlation coefficient and early factor analysis, thus making substantial contributions to the budding science of statistics. David Wechsler, the developer of the famous Wechsler scales, was a student of Spearman's, as was Raymond B. Cattell. Spearman's work also influenced the ideas of Philip E. Vernon, Hans Eysenck, and Arthur Jensen.

Although Spearman emphasized the importance of *g*, he recognized the presence of group factors (*s*), intellectual skills, such as facility with language or facility in dealing with visuospatial patterns, that are less general than *g*, but more general than an ability or skill to take a specific test. All intelligence tests would be expected to load on the *g*-factor, but the verbal tests would also be expected to load on a language factor, and the visuospatial tests would be expected to load on a visuospatial factor.

Spearman's assumption that the ability to detect patterns is central to intelligence led his student John Raven to develop the Raven's Progressive Matrices (RPM) test. The effort was successful, for this test provides a good estimate of *g*. Other tests that appear to call on different sorts of problem solving than what are required on the RPM tests also are good estimates of *g*.

Spearman's model has lasted longer than most psychological theories do. Arthur Jensen (1998), a prominent advocate of *g* theory, has claimed that after literally a century of exploration, the *g* model provides a simple, accurate summary of a massive number of studies of intelligence. Jensen built his case on three lines of evidence: (1) the positive manifold is widely observed (all tests of mental abilities are positively correlated), (2) measurements of general intelligence are among the best predictors of performance both in school and in the workplace, and (3) measures of *g* are related to a relatively small set of information-processing functions and to certain brain-based and genetic measures.

3.6.2 *The Positive Manifold and* g

In 1993, John B. Carroll published an extensive survey of the results from studies of intelligence test batteries. He found a positive manifold and evidence for a g-factor in all 461 studies ($N > 130,000$) done in nineteen countries. There appears to be little doubt that analysis of any battery of mental tests, in any population within an industrial society, will identify a g-factor.

Nevertheless, demonstrations of the positive manifold in different populations and with different test batteries does not prove that all studies have found the same g. Carroll observed that to reach such a conclusion, one would have to administer several test batteries to the same individuals, extract the g-factor from each battery, and show that they were highly correlated.

Wendy Johnson and colleagues (2004) tested this. They administered three different test batteries to a group of nearly 500 adults. The obtained correlations among g-factors extracted from different batteries were a perfect 1.0. These results strongly supported the contention that g is the same trait regardless of the test battery used. Because the Johnson et al.'s study provided such strong evidence for a general reasoning factor, as both a statistical and a psychological phenomenon, we need to take a detailed look at the study.

The batteries they used were the WAIS and two batteries that had been designed to sample a comprehensive range of verbal, quantitative, and nonverbal reasoning abilities. Their sample consisted of adults from their twenties to their seventies who were participating in an extensive behavior genetic study of adoptees. The participants were generally of European ancestry and of middle or higher socioeconomic class. Such a sample is certainly not statistically representative of the world population, although it is fairly representative of an important segment of the population within the industrial/postindustrial nations.

This massive amount of data shows a simple, clear-cut, and important fact. Within Western culture, people can be classified reliably by the extent to which they display general intelligence (g). Anyone who asserts that a meaningful assessment of an individual' mental ability must test several independent cognitive traits is simply wrong. A person who is very good at verbal reasoning may not be the best at quantitative reasoning or reasoning about visuospatial displays, but they are unlikely to be poor at these other tasks. The same argument applies to people who are very good at quantitative or visuospatial reasoning. Their verbal reasoning may not be as good as their quantitative reasoning, but it is unlikely to be bad (assuming they speak the language of the test).

3.6.3 *The Nature of* g

The positive manifold has a strong empirical basis. There is some disagreement, however, about why evaluations of intellectual performance are always positively correlated. The unitary g hypothesis is the most common explanation (Rindermann et al., 2020). It regards the g-factor as reflecting a mental trait of cognitive strength, much like physical strength reflects a general factor based on various muscle groups. But this is not conclusive. If A (unitary g) implies B (positive manifold) and B is observed, then A might be the case, but B might have arisen for other reasons (Savi et al., 2019).

What hypotheses other than the unitary g hypothesis imply a positive manifold? A positive manifold arises if people possess distinct, specialized mental traits, such as verbal and visuospatial reasoning, and these traits are correlated in the population for either environmental or genetic reasons. This is the correlated traits hypothesis. Correlated traits would be analogous to the statistical association between blond hair and blue eyes, neither of which causes the other.

A third possibility is that there are separate cognitive traits but that, over the life span, there are positive interactions among these traits, so that high or low performance in one trait affects the development of other traits. The interaction could be due to biology, the environment, or some combination of the two. For example, children who appear to be highly verbal may be singled out for special instruction, which improves their cognitive capacity in other areas, opening further opportunities. The opposite may also happen – children who are perceived as being slow speakers or readers could be treated as if they were not too bright, leading to poor performance in other areas.

The first position, that *g* is a unitary trait, is the one held by advocates of the general intelligence model. But what is the psychological trait behind the statistical abstraction? The question can be answered at different levels of explanation. At the psychometric level, general intelligence could be defined by examining the marker tests that have high loadings on the general factor and attempting to identify the common cognitive challenges they present. At the information-processing level, general intelligence could be defined by showing that the general factor is highly correlated with particular information-processing functions, such as working memory, the ability to control attention, and general information-processing speed. At the biological level, general intelligence could be associated with individual differences in brain structures and functions and with variations in the genome. All three levels of explanation have been investigated; the present chapter focuses on psychometric evidence. (Chapter 4 details the information-processing level, and Chapters 5 and 6 detail the brain and genetic levels.)

After reviewing findings from a great number of studies, John Carroll (1993) found that four different classes of tests have substantial factor loadings on *g*. These were tests of inductive reasoning, visualization (the ability to imagine movements of relatively complex forms), quantitative reasoning, and verbal ability: "the eventual interpretation [of *g*] must resort to analysis of what processes are common to the tasks used in the measurement [of the abilities just listed] and to the analysis of what attributes of such tasks are associated with their difficulties" (p. 597).

It is unclear if a single cognitive process underlies all four of these categories.

Accordingly, let us look at the correlated traits hypothesis, the idea that there are some cognitive functions that, although not what we would normally think of as intelligence in themselves, are required for a large variety of cognitive tasks. Therefore, individual differences in these functions would produce *g* as a statistical phenomenon, even though no such thing as *g* exists as an underlying feature of the mind.

How would this work? Consider an analogy to carpentry. Carpenters use the same tools to make a great many things, ranging from tables and kitchen cabinets to fences. Suppose that there are individual differences in the quality of tools available to different carpenters. Instead of thinking of a modern carpenter, imagine a bit of time traveling, where we ask a modern carpenter, a medieval carpenter, a Bronze Age carpenter, and a Stone Age carpenter to make us some furniture, each using the tools appropriate to the historic era. We would not think of a saw, hammer, or adze as indicating carpentry skill in itself. Nevertheless, across the ages, we could extract a positive manifold in making furniture solely because of the quality of the tools.

Unlike carpentry tools (but like carpentry skills), psychological processes are dynamic. In an absolute sense, cognitive competence increases through adolescence and declines in old age. Short-term memory processes and the speed of making simple decisions show a similar rise and fall with age. The same principle applies to knowledge. Knowledge acquired in mathematics can be used while studying physics. One way to find support for the correlated traits hypothesis is to show that practice in solving one task involving general intelligence will produce improvement on another, seemingly very different task.

For example, a study involved training children to use the abacus (Irwing et al., 2008). This device requires concentration of attention and recognition of visual patterns. Sudanese children were given an age-appropriate progressive matrices test and then divided into experimental groups that received substantial training in the use of the abacus and control groups that did not. Prior to training, both groups had equivalent test scores. Following training, the experimental group outperformed the control group. This is an interesting example, but such change scores are difficult to interpret (Haier, 2014), as discussed in Chapter 13.

Certain cognitive skills are useful in a variety of contexts. These include verbalization, which makes different aspects of the problem open to conscious inspection, and concentration of attention, which is a more basic operation. No one of these skills alone is general intelligence, but collectively, they are relevant. If the possession of one of these skills is statistically associated with the possession of others, the positive manifold and its correlate, g, will result, even though no single cognitive skill can be pointed to as intelligence.

Why should the cognitive skills that contribute to general intelligence be correlated? One possibility is that all these skills draw on a common biological capacity, such as efficiency of neural processing, and that there are substantial individual differences in that capacity. This argument makes the positive manifold and g manifestations of a biological phenomenon.

Another possibility is that positive manifold emerges from interactions in which the development of one cognitive skill facilitates another. This is called *mutualism*. For instance, practicing verbalization during problem solving might facilitate the ability to control attention. If this is the case, the possession of key cognitive skills would become correlated over time, even though they were initially uncorrelated (Kan et al., 2019; van der Maas et al., 2006).

To summarize, the statistical evidence for g and the positive manifold could be produced by a pervasive general intelligence factor, or it could be produced by correlations between more specialized abilities. This leaves open the question of what these abilities are. Box 3.1 discusses an example of how we can use findings observed from *training-to-think* research (sometimes called *experimental psychometrics*).

3.6.4 *Reservations about* g

The extent to which a given study supports the g model depends on the extent to which the tests administered show a positive manifold. If the tests are highly correlated, the g model is supported; if they are not, it is not. It cannot be stressed too strongly that the weight of evidence from many well-designed studies is consistent with the g model.

Nevertheless, two reservations are in order. The first is that psychologists who have studied intelligence have been conservative about what a test of cognition is. The bulk of the evidence for g has been obtained by analyzing conventional tests that focus on problem solving restricted to the testing situation and by excluding harder-to-measure constructs that may be relevant for intelligence (see "drop in from the sky" testing limitations discussed in Chapter 2). If we expand the definition of intelligence to include problem solving in situations not amenable to the conventional testing session, then other factors might be found, and g might or might not hold up as well. But what might happen is not what would happen. Whether studies of cognition in expanded situations would provide additional evidence for g should be decided by empirical research, not by the intuitions of people who support or oppose the model. For now, there is a paucity of data on this topic, but there is some progress (Chuderski and Jastrzebski, 2018).

The second reservation is that studies of specialized populations often do not show strong evidence for g. In part, this is simply a statistical issue of restricted range in which the sample does not represent the full range of scores in the population. Take the case of college students at a highly selective institution, such as Stanford or Cambridge. These students have been selected by a process that, to a considerable extent, evaluates their general reasoning ability (in a similar way that professional basketball players have been selected by height). Therefore, within the selected group, there will be a restricted range of g, and other factors will determine individual differences in test scores. The situation is analogous to the fact

---

Box 3.1 Experimental Psychometrics

John Protzko (2016) discussed the failure of targeted cognitive training for increasing intelligence. He suggested that the situation is of great theoretical relevance regarding the psychometric models of the concept of intelligence. If the most likely structure of intelligence is hierarchical, then the observed failure is inevitable. If the structure of intelligence is a causal measurement model, then working memory and intelligence are not casually connected. This might explain why training working memory and finding improvements fails to enhance intelligence: "A top-down causal structure makes upward causation from subfactors (like working memory) to general intelligence impossible. ... Failures to find transfer may be indicative of the correct causal direction between intelligence and its subprocesses" (p. 1022).

The arrows shown in Figure 3.1 might mean something important. These arrows go downward, not upward. Furthermore, there are no direct arrows between latent factors. Therefore, improvements in, say, F_1 will only impact their respective markers/tests.

The nature of targeted training and the absence of transfer among cognitive domains are ideal for studying the causal connections that psychometric models of intelligence presumably represent. If a hierarchical structure is more than a simple summary, then targeted cognitive training aimed at increasing verbal ability will fail to enhance intelligence because the latter causes the former (not the other way around).

Experimental psychometrics may help in understanding the causal structure of human intelligence, identifying the cognitive process that we may want to improve. Afterward, we must find the way for enhancing just this process and show that there is an increment on this process at the latent level.

Protzko summarizes his perspective using this analogy: "much as we cannot move the hand on the barometer in the hopes it will change the weather, the structure of intelligence may make it impermeable to changing subfactors in the hopes of upward effects" (p. 1030). Jonathan Haidt (2012, p. 57) wrote something close when discussing morality: "you can't make a dog happy by forcibly wagging its tail." Protzko and Colom (2021) applied this experimental psychometrics approach to use the impact of focal brain lesions for testing the likelihood of a number of psychometric models beyond statistics, as we will discuss in Chapter 5.

---

that height is not closely related to the ability to score points in professional basketball because almost all the players are already very tall.

This poses a practical problem for intelligence research because, in general, the populations that are easiest to study are the ones where such restrictions of the range of reasoning ability may occur. We do not have to go to highly selective colleges to see this restricted range effect. All college and university students have been subject to some selection on general reasoning ability.

The same situation occurs when people try to obtain voluntary samples from the general population. People who volunteer for research may not be representative of the population. A truly random sample is hard to obtain. For example, it is easier to recruit people from the middle and upper socioeconomic classes (SES) than to recruit people with low SES. This bias, and many other recruitment biases, operates to produce samples with restricted ranges on *g*, thus often underestimating the importance of the trait in the population. Restricted range of *g* is only one problem. There are some systematic changes in the ubiquity of *g* that are not solely due to statistical issues, as described next.

3.6.5 *Spearman versus Thurstone*

In the 1930s, Louis Leon Thurstone, a professor at the University of Chicago (and former assistant of Thomas Alva Edison), challenged Spearman's *g* model (Thurstone, 1938). Thurstone thought that intelligence is based on several distinct primary abilities, rather than on a single general reasoning factor. The primary abilities were as follows:

Inductive reasoning: The ability to see patterns.

Spatial reasoning: The ability to reason about figural representations.

Number facility: The ability to do relatively simple numerical computations quickly.

Verbal relations: The ability to comprehend verbal statements.

Word fluency: The ability to produce simple words and statements rapidly.

Memory: The ability to recall information.

Perceptual speed: The ability to detect simple figures in a display.

Based on test data he collected, Thurstone claimed that these abilities are essentially statistically independent. Being good or poor on one of them does not predict whether a person is good, poor, or average on another. This conclusion is diametrically opposed to Spearman's claim that these abilities would be correlated because intelligence is mainly produced by a single general reasoning factor.

In the 1940s and 1950s, there was debate over whether the discrepancy between Thurstone's and Spearman's results might have been due to different groups having used different factor analytic methods. The development of modern computerized techniques has essentially ended that discussion. Different factor analytic methods were not the cause.

The discrepancy was probably due in part to restriction of range effects. In general, Spearman and other British psychologists analyzed data from the testing of schoolchildren. While Thurstone did similar studies, he relied on studies of University of Chicago students. Chicago was a highly selective institution, so his college sample undoubtedly had a highly restricted range of scores on *g*.

Another likely reason for the discrepancy could be that the structure of intelligence is more differentiated at high levels than at lower levels. In concrete terms, unusually high scores on a test of verbal reasoning might be only moderate predictors of unusually high scores on a test of mathematical reasoning, while unusually low scores on the verbal test could be good predictors of unusually low scores on the mathematical test. To the extent that this is true, factor analysis would reveal small correlations among factors in a high-ability group, while revealing a strong *g*-factor in a low-ability group because of large correlations among specific factors. To examine this possibility, Douglas Detterman and Mark Daniel (1989) divided the WAIS standardization sample into five ability groups. The strength of the *g*-factor was highest in the low-ability group and declined as group IQ score increased.

Similar results have been obtained by other investigators, using other tests in both national and international settings. For instance, Francisco J. Abad and colleagues (2003) analyzed two samples, one of high school graduates ($N = 3,430$) and the other obtained from the standardization of the WAIS-III for Spain ($N = 823$). The first sample is clearly unrepresentative of the general population. The second represented the population in Spain in terms of sex, age, education, and so forth. Sophisticated procedures were applied for selecting high- and low-ability individuals from these samples. The researchers computed the amount of variance explained by *g* in high- and low-ability individuals. Figure 3.4 depicts the results for the first and second samples. The percentage is higher in low-ability subjects in both samples but stronger for the sample representative of the population (WAIS-III).

This is an important result because it is relevant to a social issue regarding the distribution of intelligence at high levels of cognitive functioning. High levels of cognitive ability appear to be relatively specialized

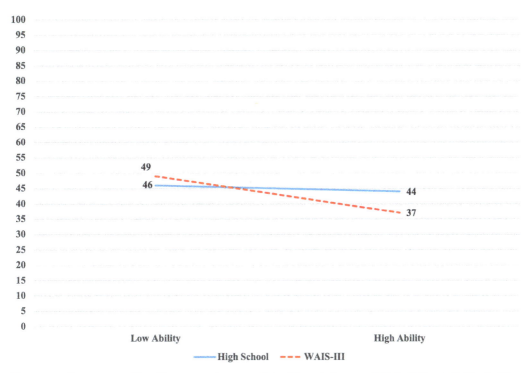

Figure 3.4 Percentage of intelligence variance explained by *g* in low- and high-ability individuals. The solid line depicts results for a sample unrepresentative of the general population, whereas the dotted lines show results for a sample representative of the population. It can be seen that the relevance of the general factor is much smaller for high-ability individuals taken from the representative sample (Abad et al., 2003).

(i.e., less *g*), while lower cognitive ability seems to have greater general effects. Why might this be the case?

In part, it may be due to experience and formal education. Modern society encourages specialization to a much greater extent than did past societies. Studies of expertise in a variety of fields, ranging from athletics to chess, have shown that acquiring a high level of expertise takes a great deal of time and effort. At high levels of talent, therefore, social pressures lead to a differentiation of cognitive competences due to specialized training. But this cannot be the whole picture, because the differentiation of abilities at high levels (or, conversely, generalization at low levels) occurs in children, as evidenced by studies on differentiation involving the WISC (Juan-Espinosa et al., 2000). Research is continuing on this issue (Breit et al., 2020); see Box 3.2 for a related issue highly relevant in applied settings.

3.7 The Cattell–Horn–Carroll (CHC) Model

R. B. Cattell studied with Spearman in England. Afterward, he moved to the University of Illinois and finally to the University of Hawai'i. He and his student at Illinois John Horn (subsequently a professor at the Universities of Denver and Southern California) believed that Spearman's theory did not give sufficient weight to group or specific factors. They were also skeptical of the idea that *g* is a psychological trait. Instead, they believed that the positive manifold is due to individual tests drawing on several broad factors. To follow their argument, consider this mathematical word problem:

A train leaves station A and proceeds to station B, traveling at 60 km per hour. At the same time that this train leaves A, another train leaves B, bound for A,

---

Box 3.2 Beyond *g*?

Psychometric models comprise several specific and broad cognitive abilities. And psychometric batteries designed for measuring some of these abilities usually provide separate scores. Researchers, however, wonder if these specific scores do meet the required standards for use in applied settings.

Marley Watkins and Gary Canivez (2021) addressed this issue considering the latest version of the Wechsler Intelligence Scale for Children (WISC-V). Although the interpretation of separate scores is widely used in clinical practice, basic researchers have raised doubts regarding inadequate reliability, validity, and diagnostic usefulness of these specific scores.

Watkins and Canivez applied an evidence-based approach for testing the psychometric utility of IQ scores. The factor structure of the WISC-V includes five broad factors (verbal comprehension, visuospatial, fluid reasoning, working memory, and processing speed) along with a general intelligence factor (*g*).

After computing their statistical analyses, the results indicated that the general factor accounted for 68 percent of the common variance.

These were the key conclusions from this research:

1. Because similar results are found for other intelligence and educational measurement batteries, the conclusion that there is little practical relevance for other factors beyond *g* might be considered universal.
2. There is meager evidence supporting the diagnostic utility of specific factor scores beyond *g*.
3. There are two questions requiring answers before using specific ability scores in applied setting: (1) Is the specific score a reliable measure of the relevant specific construct? and (2) Is the specific score different from other scores tapping distinguishable constructs?

In the authors' own words, "it is unlikely that cognitive profiles will exhibit sufficient reliability for clinical decisions" (p. 629).

---

traveling at 30 km per hour. The distance from A to B is 270 km. How long will it be before the trains meet?[3]

A student who tries to solve this problem must know the meaning of words, the syntax of English, and sentence comprehension. Several facts also must be kept in mind as others are received, calling on short-term memory ability. Numerical facility also is required.

These narrowly defined abilities are examples of primary or first-stratum abilities. These first-stratum abilities are themselves grouped into broader, second-stratum abilities. According to Cattell and Horn, the most important of these are fluid intelligence (Gf)

and crystallized intelligence (Gc). They are defined as the ability to deal with new and unusual problems (Gf) and the ability to apply previously acquired knowledge to the current problem (Gc). In many contexts, visuospatial ability (Gv), the ability to deal mentally with spatial and visual images, is also important.

The primary abilities of inductive and deductive reasoning are grouped under fluid intelligence. General cultural knowledge and lexical knowledge abilities are grouped under crystallized intelligence. The abilities to compare visuospatial forms and to manipulate visuospatial forms in the mind's eye are grouped under visuospatial ability.

As the theory has evolved with new studies, additional second-stratum abilities have

---

[3] Answer is 3 hours.

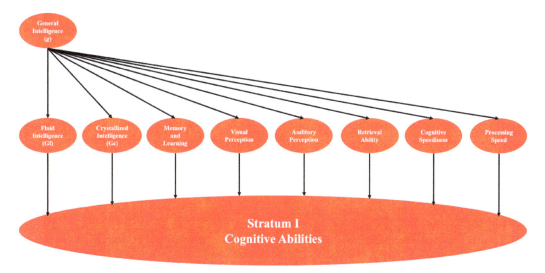

Figure 3.5 Carroll's original three-stratum model (Carroll, 1993). Stratum I includes more than sixty narrow cognitive abilities (bottom row). These abilities can be grouped into eight broad abilities (stratum II; middle row), and they are ordered according to their relevance to *g* (from left to right). The closer to the third stratum a single ability is (*g*), the greater is the relevance for intelligent behavior. Therefore, fluid intelligence is more relevant than crystallized intelligence. Processing speed is the least relevant second stratum ability.

been identified. These include factors for retrieval from short- and long-term memory, the ability to deal with auditory as well as visual stimuli, quantitative ability, and two factors reflecting processing speed: cognition in general (cognitive processing speed, Gs) and another dealing with the speed with which very simple decisions are made (decision reaction time, Gt).

Cattell and Horn had long and active careers, during which they had colleagues who conducted research investigating such things as the processing of auditory and tactile stimuli. In the typical extension of the theory, a study would be conducted with a battery that included some tests already identified as a marker of abilities found by previous research and some new tests that explored different primary abilities and skills. This process inevitably resulted in the definition of still more second-stratum factors.

3.7.1 *Extensions and Applications of the Three-Stratum Model*

On the basis of his quantitative review of the literature, J. B. Carroll (1993) concluded that the Gf–Gc model provided a good fit to most of the more than 460 data sets that he reanalyzed. However, in almost all cases, abilities at the second stratum were themselves correlated. He took this as evidence for a single third-stratum factor, general intelligence (*g*), as shown in Figure 3.5. Conceptually, this is the *g*-factor advocated by Spearman (1904, 1927) and, almost a century later, by A. R. Jensen (1998). The resulting three-stratum theory has been used as a basis for building psychometric batteries such as the Woodcock–Johnson Test Battery. However, the heart of the theory lies in the broader, second-stratum abilities.

W. Joel Schneider and Kevin S. McGrew (2018,) developed the latest version of the CHC model. They define this model as "a comprehensive taxonomy of abilities embedded in multiple overlapping theories of cognition. … It provides a common framework and nomenclature for intelligence researchers to communicate their findings without getting bogged in endless debates about whose version of this or that construct is better" (p. 73). Intelligence is considered multidimensional, but integrated. The identified factors can be related

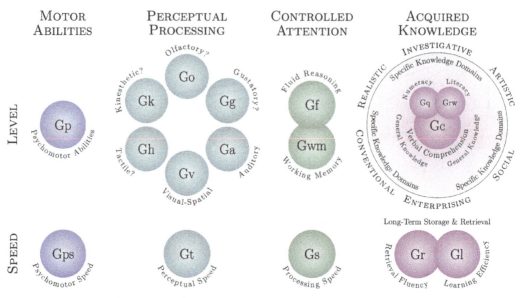

Figure 3.6 Conceptual groupings of CHC broad abilities divided into speed and level categories. Speed is based mainly on speeded performance, whereas level is based on the achievement scores people attain regardless of time constraints (Schneider and McGrew, 2018).
Courtesy of W. Joel Schneider.

to knowledge from other sources, such as cognitive, biological, or developmental psychology. Carroll (1993) identified eight broad abilities, and this latest version of the model suggests that there might be twenty such abilities. Figure 3.6 depicts a comprehensive summary of the latest CHC model. The families of cognitive abilities are clustered and defined in Table 3.6 (along with the corresponding narrow abilities). Carroll and Cattell disagreed with respect to the *g*-factor; that is why *g* is not included in Figure 3.6.

Schneider and McGrew accept the relevance of *g*, but they have reservations about its nature. Is it an ability? They properly note that there are many things that can influence brain structure and function and that "the overall level of a person's intelligence can be assessed without necessarily referring to a causal force called general intelligence" (p. 89). However, they subscribe to the conclusion that we still lack definitive answers in this regard (Mansolf and Reise, 2017).

Gf (fluid intelligence), Gwm (working memory), Gl (learning efficiency), Gr

(retrieval fluency), Gs (processing speed), Gt (reaction and decision speed), and Gps (psychomotor speed) are domain-free general capacities defined by their lack of association with sensory systems. Moreover, individuals can recover relevant information when required (intelligence as knowledge). The factors included here are Gc (comprehension knowledge), Gkn (domain-specific knowledge), Grw (reading and writing), and Gq (quantitative knowledge). The broad abilities identified within "sensory and motor linked abilities" are supported by well-defined brain regions and networks. The factors considered here are Gv (visual processing), auditory processing (Ga), Go (olfactory abilities), Gg (tactile abilities), Gk (kinesthetic abilities), and Gp (psychomotor abilities).

Finally, Schneider and McGrew state that the perceptual processing factors (Gv, Ga, Go, Gh, and Gk) underlie the level of complexity that the individual can manage. Attention is involved in moving percepts to consciousness, and it is required for managing information in the short term, storing and retrieving the relevant information in

Table 3.6 Broad and narrow cognitive abilities included in the latest version of the CHC model

Broad abilities	Narrow abilities
Fluid intelligence: The use of deliberate and controlled procedures (often requiring focused attention) to solve novel, "on-the-spot" problems that cannot be solved by using previously learned habits, schemas, and scripts.	Induction, general sequential reasoning, quantitative reasoning, reasoning speed, Piagetian reasoning
General working memory: The ability to maintain and manipulate information in active attention.	Auditory short-term storage, visuospatial short-term storage, attentional control, working memory capacity
Learning efficiency: The ability to learn, store, and consolidate new information over periods of time measured in minutes, hours, days, and years.	Associative memory, meaningful memory, free-recall memory
Retrieval fluency: The rate and fluency at which individuals can produce and selectively and strategically retrieve verbal and nonverbal information or ideas stored in long-term memory.	Ideational fluency, expressional fluency, associational fluency, sensitivity to problems, originality, speed of lexical access, naming facility, word fluency, figural fluency, figural flexibility
Processing speed: The ability to control attention to automatically, quickly, and fluently perform relatively simple repetitive cognitive tasks. Gs may also be described as attentional fluency or attentional speediness.	Perceptual speed, perceptual speed-search, perceptual speed-compare, number facility, reading speed, writing speed
Reaction and decision speed: The speed of making very simple decisions or judgments when items are presented one at a time.	Simple RT, choice RT, inspection time, semantic processing speed, mental comparison speed
Psychomotor speed: The ability to perform skilled physical body motor movements (e.g., movement of fingers, hands, legs) with precision, coordination, fluidity, or strength.	Speed of limb movement, writing speed, speed of articulation, movement time
Crystallized intelligence: The ability to comprehend and communicate culturally valued knowledge. Gc includes the depth and breadth of both declarative and procedural knowledge and skills such as language, words, and general knowledge developed through experience, learning, and acculturation.	Language development, lexical knowledge, general knowledge, listening ability, communication ability, grammatical sensitivity
Domain-specific knowledge: The depth, breadth, and mastery of specialized declarative, and procedural knowledge (knowledge not all members of a society are expected to have).	General science information, knowledge of culture, mechanical knowledge, foreign language proficiency, knowledge of signing, skill in lipreading
Reading and writing: The depth and breadth of declarative and procedural knowledge and skills related to written language.	Reading comprehension, reading decoding, reading speed, writing speed, language usage
Quantitative knowledge: The depth and breadth of declarative and procedural knowledge related to mathematics.	Mathematical knowledge, mathematical achievement

Table 3.6 (cont.)

Broad abilities	Narrow abilities
Visual processing: The ability to make use of simulated mental imagery to solve problems – perceiving, discriminating, manipulating, and recalling nonlinguistic images in the mind's eye.	Visualization, speeded rotation, imagery, flexibility of closure, closure speed, visual memory, spatial scanning, serial perceptual integration, length estimation, perceptual illusions, perceptual alternations, perceptual speed
Auditory processing: The ability to discriminate, remember, reason, and work creatively (on) auditory stimuli, which may consist of tones, environmental sounds, and speech units.	Phonetic coding, speech sound discrimination, resistance to auditory stimulus distortion, maintaining and judging rhythm, memory for sound patterns, musical discrimination and judgment, absolute pitch, sound localization
Psychomotor ability: The ability to perform physical body motor movements (e.g., movement of fingers, hands, legs) with precision, coordination, or strength.	Manual dexterity, finger dexterity, static strength, gross body equilibrium, multilimb coordination, arm–hand steadiness, control precision, aiming
Emotional intelligence: The ability to perceive emotions expressions, understand emotional behavior, and solve problems using emotions.	Emotion perception, emotion knowledge, emotion management, emotion utilization
Olfactory ability: The ability to detect and process meaningful information in odors.	
Tactile ability: The ability to detect and process meaningful information in haptic (touch) sensations. This domain includes perceiving, discriminating, and manipulating touch stimuli.	
Kinesthetic ability: The ability to detect and process meaningful information in proprioceptive sensations.	

Note. See Box 3.3 for the application of the CHC model to a wide set of intelligence batteries. RT = reaction time.

long-term memory, reasoning, or acting. They underscore the importance of the high correlation between fluid intelligence and working memory capacity (which will be discussed extensively in the next chapter).

3.7.2 What Is a Natural Kind of Ability: g Gf, or Gc?

A natural kind of ability is a phenomenon that exists in nature and is to be discovered. This contrasts with an artifactual classification, which is constructed by human thought. The distinction between men and women is a natural kind. The distinction between legal and illegal residents of the European Union is an artifactual one. Artifactual classifications can be useful in some settings. Nevertheless, they are the result of human categorization, not of a law of nature.

In studies of intelligence, the broad sensory modalities and memory factors are natural kinds of ability, for they are defined as individual differences in human biological capacities. The distinctions among g, Gf, and Gc, however, are debatable as to whether they are natural kinds. The Gf–Gc distinction is based in part on an individual's social and cultural history. Consider how the following problem might be solved:

Box 3.3 CHC and Intelligence Measurement Batteries

Caemmerer and colleagues (2020) applied a sophisticated psychometric approach (cross-battery confirmatory factor analyses) for testing the likelihood of the CHC model regardless of the administered intelligence measurement battery.

Data from 3,927 children and adolescents (age range six to eighteen years) were obtained for studying sixty-six tests from six different intelligence batteries (Woodcock–Johnson Test of Cognitive Abilities/WJ-III, Kaufman Assessment Battery for Children/KABC-II, Kaufman Test of Educational Achievement/KTEA-II, Wechsler Intelligence Test for Children/WISC-V, Wechsler Individual Achievement Test/WIAT-III, Differential Abilities Scale/DAS-II), and these were the main conclusions after the computed statistical analyses:

1. The factor loadings of the six identified CHC broad abilities (Gf/fluid reasoning, Gc/crystallized ability, Gv/visuospatial ability, Gwm/working memory, Gl/long-term memory, and Gs/processing speed) on the general factor (g) were quite substantial.
2. Gf and g were statistically indistinguishable (0.99). Gv had the next strongest loading on g (0.87), followed by Gc (0.81), Gl (0.78), Gwm (0.68), and Gs (0.61).
3. Therefore, "novel problem solving, visuospatial problem solving, and the depth and breath of general knowledge and the ability to retrieve that knowledge efficiently are strong indicators of overall intelligence (g)" (p. 9).

This comprehensive research again demonstrated the inevitability of the positive manifold when cognitive demands are made. Look at the correlation matrix computed from the broad ability factors identified here.

	Gf	Gv	Gc	Gl	Gwm
Gv	0.90				
Gc	0.78	0.67			
Gl	0.75	0.64	0.68		
Gwm	0.65	0.55	0.60	0.54	
Gs	0.57	0.57	0.42	0.59	0.51

Two freight trains approach each other on a single track. Between them there is a short side route capable of holding only one engine or box car. How can the two trains pass each other?

Most readers will see this as a novel problem and solve it using a mixture of reasoning about new problems (Gf) and manipulation of visuospatial images (Gv). In fact, it is a problem that occurs often in railroad switching yards and has a standard solution. The railroad-experienced reader will treat the problem as an exercise in Gc.

Test developers get around such issues by restricting tests of Gc to questions that draw on generally culturally accepted knowledge. This automatically restricts most Gc tests to

the industrially developed countries, although the concept of cultural knowledge applies equally well to a nomadic or hunter-gatherer culture. However, it would have to be tested using questions appropriate to the hunter-gatherer culture. A test of Gc in one culture could be a test of Gf in another. This is an important distinction, for the two types of intelligence make different demands on our information-processing capacities. To solve a novel problem, you must develop a way of representing it mentally. This can be a difficult task, involving the development of information structures to be held in short-term memory. To solve a problem by applying previously acquired knowledge, you must have had appropriate experiences and coded them in long-term memory in such a way that they are accessible in the present context. Working and long-term memory draw on different brain structures and functions. This argument suggests that Gc and Gf are two different natural kinds of ability.

The argument is not resolved by demonstrations that *g* exists as a statistical phenomenon. Cattell argued that we acquire Gc largely by using Gf to discover appropriate problem-solving procedures, a process that he referred to as investing Gf in the acquisition of Gc. If two people enter into some experience that is unfamiliar to both of them, the one with the higher Gf will learn more from the experience and end up with a higher Gc. Repeated throughout life, this would produce a correlation between Gf and Gc, and hence statistical evidence for *g*, even though there is no natural ability that corresponds to the higher-order factor of general intelligence.

The contrary argument, made most notably by Jensen, is that pervasive individual differences in brain processes cause general mental competence, and these differences are causes of differences for both Gc and Gf tasks (Jensen, 2006). There is no way to decide between the Cattell and Jensen positions from an analysis of the psychometric data alone. There might be, however, a nonpsychometric argument in favor of Cattell's proposal, as we see next. But there is also nonpsychometric evidence in favor of Jensen's position (Protzko and Colom, 2021), as we will show in Chapter 5.

Experimental psychologists show that two underlying human capacities are different by showing that they respond to a change of conditions in different ways. For instance, one of the strongest pieces of evidence for a distinction between conscious and unconscious memory systems is that conscious memory retrieval falters when a person is distracted, while unconscious retrieval does not (Jacoby et al., 1993).

This sort of argument can be used to distinguish between Gc and Gf, using a universally occurring experimental condition: aging. Performance on tests that tap Gf (matrix problems) declines with age over the adult life span. Performance on tests that tap Gc (vocabulary tests and tests of cultural knowledge) does not. In fact, measures of Gc may increase until advanced old age. This provides compelling evidence that Gf and Gc are indeed two separate, albeit correlated, abilities (Schaie, 2013). Nevertheless, the aging situation is complex and intricate, as we discuss in Chapter 11.

3.8 The *g*-VPR Model

By the year 2000, the evidence favored the CHC model over a simple *g* model. However, in 2005, Wendy Johnson and Thomas Bouchard published new analyses that questioned the three-stratum model and offered an alternative (Johnson and Bouchard, 2005b). Johnson and Bouchard began with a nearly forgotten model of intelligence proposed by the Canadian psychometrician Philip E. Vernon in the 1960s. Vernon himself barely mentioned this model in his 1979 book on heredity, environment, and intelligence.

Vernon's structure of intelligence model contained three factors: a general factor (*g*) and two factors statistically independent of each other with no overlap with *g*. One he identified as a verbal/educational factor, reflecting the emphasis on verbal skills in the educational system, and the other as a perceptual/motor factor representing skill in identifying and manipulating objects. Vernon also suggested the presence of a third special factor, mathematical skills, but felt that this was closely related to the perceptual/motor factor. Johnson and Bouchard proposed a four-stratum model with the structure shown in Figure 3.7. Only three strata are shown in this figure because the fourth comprises the individual tests from which the other three strata are derived.

The first level consists of the primary traits evaluated by individual tests, such as a test of the ability to do simple computations or to solve anagrams (again, this level is not shown in Figure 3.7). The second stratum consists of eight broader but still fairly narrow abilities. For instance, at this

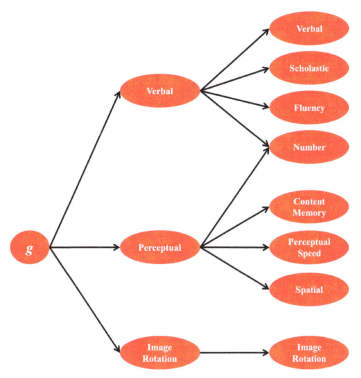

Figure 3.7 *g*-VPR model. The figure shows three strata of latent factors and not the fourth level of individual tests from which the eight second-stratum latent factors are identified (Johnson and Bouchard, 2005b).

level, there is a distinction between word fluency, which is essentially a measure of speed of producing verbal associations, and verbal comprehension (their term, verbal), which is characterized by vocabulary and the understanding of proverbs. A similar distinction was made between memory for meaningful material and memory for arbitrary, experimenter-presented associations, such as arbitrary lists of number–noun pairs.

In all the data sets analyzed, they found substantial correlations among second-level factors, which indicated a need for a third stratum in which the number of factors would be reduced and where a second-level factor could have loadings on more than one third-order factor.

The third stratum, which is the heart of the model, contains three factors – Vernon's verbal and perceptual skills factors and a third perceptual ability, the ability to envision motion of a static figure, most clearly seen in tasks that require mental rotation of

a visual figure. As was the case for Vernon's model, the VPR model does not contain a memory factor. This is consistent with research in cognitive psychology, which has identified different types of memory, as discussed in Chapter 4.

Johnson and Bouchard's third-level factors were correlated to each other, indicating a need for a fourth stratum. They found that only one factor was required at this level, which they identified as general intelligence, *g*. Developing an acronym from the names of the third-level factors, Johnson and Bouchard refer to their model as the *g-VPR* model.

3.8.1 *Psychometric Evidence for the g-VPR Model*

Johnson, Bouchard, and their colleagues have presented two arguments for preferring the *g*-VPR model to either the *g* model (which it improves, rather than replaces) or the CHC model (Johnson and Bouchard, 2005a, 2005b, 2007). The first argument is based on psychometric evidence, whereas the second argument is based on biological plausibility. We look first at the psychometric evidence.

They computed a comparative analysis of three diverse data sets, in which they compared the *g*-VPR model to the Gf–Gc model and to Vernon's model of independent (orthogonal) verbal and spatial abilities. The first data set came from 400 adults who participated in the Minnesota Study of Twins Raised Apart. This is a large database in which the same adults completed three different test batteries. The second data set was one Thurstone had used to justify his primary mental abilities model. Thurstone's data set contained scores from a study done in Chicago in the 1930s. Some of the participants had taken sixty different tests. The third data set was based on forty-six different tests given to more than 500 seamen in the Dutch Navy in the 1960s.

In all three data sets, the *g*-VPR model showed better statistical fit indices than any of the other models. A comparison to the three-stratum model is particularly informative. Johnson and her colleagues subsumed Gc into a verbal factor (V), identified the Gf factor with *g*, and split the Gv factor into two perceptual factors: the analysis of static visual images (P) and the ability to conduct mental manipulations of visual objects (R). Because none of the batteries they considered contained tests involving auditory presentations, they had no opportunity to uncover an auditory (Ga) factor, if there is one.

The V, P, and R factors were not independent, but, as would be expected, the correlation between the P and R factors was higher than the correlation between the V factor and the other two. While none of these samples are representative of a particular population (as compared to the standardization samples for batteries such as the WAIS), one has to be impressed by the uniformity of the results obtained over a wide variety of tests and using markedly different samples tested in different decades.

Johnson and Bouchard did not find a need to identify broad memory factors, although they did identify some specific memory factors in the second stratum. This does not mean that memory is unimportant to cognition; obviously it is. What it means is either that memory factors are subsumed by one of the broad reasoning factors or that memory abilities are highly specific to the type of material being tested. There is evidence for subsuming the ability to use short-term memories under *g* and to view long-term memory abilities as being more specific.

Johnson and Bouchard regard their psychometric evidence as a disconfirmation of the CHC model, but this is perhaps too strong a statement. Whether you accept it depends on your approach to statistical hypothesis testing. If one takes the classic approach to hypothesis testing, none of the models, including the *g*-VPR model, accounted for the data perfectly. In every comparison, there was a statistically significant deviation of the data from that predicted by the Gf–Gc, CHC, original Vernon, or *g*-VPR models. Johnson and Bouchard took the relativistic approach of comparing the models to each other, using a sophisticated statistical approach. This analysis identifies the best model within a set of models to be compared, rather than testing to see if there are significant deviations from a particular (true) model. The *g*-VPR model was the winner, although it did not account for all the data. All these models are versions of the model shown in the bottom left of Figure 3.3; they are hierarchical but organize test relationships in different ways.

3.8.2 *Logical Arguments for the g-VPR Model*

Johnson and Bouchard were also critical of the Gf–Gc distinction on logical grounds. Gc is supposed to represent the use of previously acquired knowledge to solve the current problem. But how can you construct a test for this ability? It makes little sense to test a person's level of a skill unless the examinee has had a chance to acquire the skill. Therefore, tests of Gc have to be tests based on information that is widely available in the examinee's culture. Indeed, most Gc questions are of this nature. The vocabulary tests used to tap Gc are roughly at the level of vocabulary used in television dramas.

Suppose that tests have been constructed such that we can be certain that every examinee has been exposed to the information needed to do well on the test. Individual differences in test scores will then be produced by differences in examinees' ability to extract this information from their common experiences. A great deal of cultural knowledge is based on induction from experience, rather than explicit instruction. This is especially true of our ability to understand the meaning of words in different contexts. "How are you feeling?" spoken by a waitress at a restaurant is a different question and requires a different response than "How are you feeling?" spoken by a physician in an emergency room. To understand such distinctions, a person has to recognize patterns of usage. Pattern recognition is by definition Gf and Spearman's g.

This muddles the Gf–Gc distinction. Either a test of Gc is not fair, because it evaluates information to which the individual has not been exposed, or it is actually a disguised test of Gf.

Johnson and Bouchard also point out another problem with the Gf–Gc distinction, one that had actually been raised by Horn (1998). If Gf is an initial ability that is invested to produce Gc, then Gf measures should be more responsive to individual biological variables than Gc measures. This is simply not what happens. The heritability coefficients for Gc and Gf are approximately the same, in violation of the argument that Gc reflects cultural experiences, while Gf does not. By contrast, heritability analyses show a coherent pattern of genetic association for the variables in the g-VPR model (Johnson et al., 2007).

The g-VPR model aligns with some neuroscientific findings. The ubiquity of g suggests that there are individual differences in some pervasive brain processes. At least three candidate processes have been suggested: individual differences in the ability to control attention, individual differences

in the speed and accuracy of neural conduction, and individual differences in the plasticity of neural connections. All would affect the ability to acquire and retrieve information. Language processing and perceptual processing are carried out by mostly separate brain systems. At least one biological distinction mirrors the distinction between mental rotation and the analysis of static figures. There are sex differences, in favor of men on average, in the ability to conduct rotation-like tasks (Halpern et al., 2007). Sex differences in other perceptual tasks are smaller and, in some cases, are in favor of women again on average (see Chapter 10).

All in all, Johnson and Bouchard make a persuasive case for their model. Psychometric models of intelligence alone, however, are not sufficient to explain intelligence even if they meet strong statistical criteria. Cognitive and biological models of intelligence also must be considered. That is why we answered yes to the question "Do we need all this?" raised at the end of Chapter 2.

3.9 Summary

All tests of mental ability are positively correlated with each other; this is the positive manifold. Any psychometric model of intelligence must account for the degree to which the positive manifold is identified. The manifold is stronger at the bottom of the population ability distribution than at the top. The positive manifold could be produced by a unitary g or the development of separate modules of cognition that have positive influences on each other's development.

Correlations among tests indicate that a theory of intelligence has to include a general factor common to all tests (g). But g alone is not enough to explain all the relationships among diverse mental abilities. The debate is about the appropriate structure of the broad abilities that lie below g in the structure of intelligence.

The CHC model makes a distinction between problem solving based on the manipulation of working memory (Gf) and problem solving based on retrieval of previously acquired information (Gc). The Gc component of the three-stratum model is very closely associated with verbal reasoning. The *g*-VPR model makes the distinction between brain structures involved in working memory and the control of attention. The VPR components of the *g*-VPR model are closely tied to pathways of sensory information processing. The *g*, CHC, and *g*-VPR models can all account for the psychometric data in a general way. If we look at relative statistical accuracy, the data indicate that the *g*-VPR model fits the data best, but more research is needed before any model is discarded. All of these psychometric models are attempts to find structure in the data from conventional intelligence testing, and each model has different theoretical implications about the nature of intelligence.

3.10 Questions for Discussion

3.1 What are some different kinds of reasoning that researchers include when studying intelligence?

3.2 What advantage does item response theory have for constructing mental tests?

3.3 What does factor analysis show?

3.4 Do you think the way the *g*-factor is defined makes sense?

3.5 How are different psychometric models of intelligence evaluated?

References

Abad, F. J., Colom, R., Juan-Espinosa, M., & Garcia, L. F. 2003. Intelligence differentiation in adult samples. *Intelligence*, 31, 157–166.

Breit, M., Brunner, M., & Preckel, F. 2020. General intelligence and specific cognitive abilities in adolescence: Tests of age differentiation, ability differentiation, and their interaction in two large samples. *Developmental Psychology*, 56, 364–384.

Caemmerer, J. M., Keith, T. Z., & Reynolds, M. R. 2020. Beyond individual intelligence tests: Application of Cattell–Horn–Carroll theory. *Intelligence*, 79, 101433.

Carroll, J. B. 1993. *Human cognitive abilities: A survey of factor-analytic studies*. Cambridge: Cambridge University Press.

Chuderski, A., & Jastrzebski, J. 2018. Much ado about aha! Insight problem solving is strongly related to working memory capacity and reasoning ability. *Journal of Experimental Psychology: General*, 147, 257–281.

Detterman, D. 1989. Detterman's laws of individual differences research. In Sternberg, R. J., & Detterman, D. K. (eds.), *Current topics in human intelligence*, vol. 1. New York: Ablex.

Detterman, D. K., & Daniel, M. H. 1989. Correlations of mental tests with each other and with cognitive variables are highest for low-IQ groups. *Intelligence*, 13, 349–359.

Funder, D. C., & Ozer, D. J. 2019. Evaluating effect size in psychological research: Sense and nonsense. *Advances in Methods and Practices in Psychological Science*, 2, 156–168.

Haidt, J. 2012. *The righteous mind: Why good people are divided by politics and religion*. New York: Pantheon Books.

Haier, R. J. 2014. Increased intelligence is a myth (so far). *Frontiers in Systems Neuroscience*, 8, 34.

Halpern, D. F., Benbow, C. P., Geary, D. C., et al. 2007. The science of sex differences in science and mathematics. *Psychological Science in the Public Interest*, 8, 1–51.

Horn, J. 1998. A basis for research on age differences in cognitive capabilities. In McArdle, J. J., & Woodcock, R. W. (eds.), *Human cognitive abilities in theory and practice*. Mahwah, NJ: Erlbaum.

Irwing, P., Hamza, A., Khaleefa, O., & Lynn, R. 2008. Effects of abacus training on the intelligence of Sudanese children. *Personality and Individual Differences*, 45, 694–696.

Jacoby, L. L., Toth, J. P., & Yonelinas, A. P. 1993. Separating conscious and unconscious influences of memory – measuring recollection. *Journal of Experimental Psychology: General*, 122, 139–154.

Jensen, A. R. 1998. *The g factor: The science of mental ability*. Westport, CT: Praeger.

Jensen, A. R. 2006. *Clocking the mind: Mental chronometry and individual differences*. New York: Elsevier.

Johnson, W., & Bouchard, T. J., Jr. 2005a. Constructive replication of the visual–perceptual–image rotation model in Thurstone's (1941) battery of 60 tests of mental ability. *Intelligence*, 33, 417–430.

Johnson, W., & Bouchard, T. J., Jr. 2005b. The structure of human intelligence: It is verbal, perceptual, and image rotation (VPR), not fluid and crystallized. *Intelligence*, 33, 393–416.

Johnson, W., & Bouchard, T. J., Jr. 2007. Sex differences in mental ability: A proposed means to link them to brain structure and function. *Intelligence*, 35, 197–209.

Johnson, W., Bouchard, T. J., Jr., Krueger, R. F., McGue, M., & Gottesman, I. I. 2004. Just one *g*: Consistent results from three test batteries. *Intelligence*, 32, 95–107.

Johnson, W., Bouchard, T. J., Jr., McGue, M., et al. 2007. Genetic and environmental influences on the verbal–perceptual–image rotation (VPR) model of the structure of mental abilities in the Minnesota Study of Twins Reared Apart. *Intelligence*, 35, 542–562.

Juan-Espinosa, M., Garcia, L. F., Colom, R., & Abad, F. J. 2000. Testing the age related differentiation hypothesis through the Wechsler's scales. *Personality and Individual Differences*, 29, 1069–1075.

Kan, K. J., van Der Maas, H. L. J., & Levine, S. Z. 2019. Extending psychometric network analysis: Empirical evidence against *g* in favor of mutualism? *Intelligence*, 73, 52–62.

Kovacs, K., & Conway, A. R. A. 2016. Process overlap theory: A unified account of the general factor of intelligence. *Psychological Inquiry*, 27, 151–177.

Kovacs, K., & Conway, A. R. A. 2019. A unified cognitive/differential approach to human intelligence: Implications for IQ testing. *Journal of Applied Research in Memory and Cognition*, 8, 255–272.

Malanchini, M., Rimfeld, K., Gidziela, A., et al. 2021. Pathfinder: A gamified measure to integrate general cognitive ability into the biological, medical, and behavioural sciences. *Molecular Psychiatry*, 26, 7823–7837.

Mansolf, M., & Reise, S. P. 2017. When and why the second-order and bifactor models are distinguishable. *Intelligence*, 61, 120–129.

Murray, C. 2008. *Real education: Four simple truths for bringing America's schools back to reality*. New York: Crown Forum.

Nisbett, R. E. 2009. *Intelligence and how to get it: Why schools and cultures count*. New York: W. W. Norton.

Pokropek, A., Marks, G. N., & Borgonovi, F. 2021. How much do students' scores in PISA reflect general intelligence and how much do they reflect specific abilities? *Journal of Educational Psychology*, 114, 1121–1125.

Protzko, J. 2016. Does the raising IQ–raising *g* distinction explain the fadeout effect? *Intelligence*, 56, 65–71.

Protzko, J., & Colom, R. 2021. A new beginning of intelligence research: Designing the playground. *Intelligence*, 87, 101559.

Rindermann, H., Becker, D., & Coyle, T. R. 2020. Survey of expert opinion on intelligence: Intelligence research, experts' background, controversial issues, and the media. *Intelligence*, 78, 101406.

Rozgonjuk, D., Schmitz, F., Kannen, C., & Montag, C. 2021. Cognitive ability and personality: Testing broad to nuanced associations with a smartphone app. *Intelligence*, 88, 101578.

Savi, A. O., Marsman, M., van Der Maas, H. L. J., & Maris, G. K. J. 2019. The wiring of intelligence. *Perspectives on Psychological Science*, 14, 1034–1061.

Schaie, K. W. 2013. *Developmental influences on adult intelligence: The Seattle longitudinal study*. New York: Oxford University Press.

Schneider, W. J., & McGrew, K. S. 2018. The Cattell–Horn–Carroll theory of cognitive abilities. In Flanagan, D. P., & McDonough, E. M. (eds.), *Contemporary intellectual assessment: Theories, tests, and issues*. New York: Guilford Press.

Snow, R. E. 1996. Aptitude development and education. *Psychology Public Policy and Law*, 2, 536–560.

Spearman, C. 1904. General intelligence objectively determined and measured. *American Journal of Psychology*, 15, 201–293.

Spearman, C. 1923. *The nature of "intelligence" and the principles of cognition*. London: Macmillan.

Spearman, C. 1927. *The abilities of man; their nature and measurement*. New York: Macmillan.

Thurstone, L. L. 1938. *Primary mental abilities*. Chicago: University of Chicago Press.

Tosto, M. G., Garon-Carrier, G., Gross, S., et al. 2019. The nature of the association between number line and mathematical performance: An international twin study. *British Journal of Educational Psychology*, 89, 787–803.

van der Maas, H. L. J., Dolan, C. V., Grasman, R., et al. 2006. A dynamical model of general intelligence: The positive manifold of intelligence by mutualism. *Psychological Review*, 113, 842–861.

Vernon, P. E. 1979. *Intelligence, heredity and environment*. San Francisco: W. H. Freeman.

Warne, R. T., & Burningham, C. 2019. Spearman's *g* found in 31 non-Western nations: Strong evidence that *g* is a universal phenomenon. *Psychological Bulletin*, 145, 237–272.

Watkins, M. W., & Canivez, G. L. 2021. Assessing the psychometric utility of IQ scores: A tutorial using the Wechsler Intelligence Scale for Children–Fifth Edition. *School Psychology Review*, 51, 619–633.

Zaboski, B. A., Kranzler, J. H., & Gage, N. A. 2018. Meta-analysis of the relationship between academic achievement and broad abilities of the Cattell–Horn–Carroll theory. *Journal of School Psychology*, 71, 42–56.

CHAPTER 4

Cognitive Models of Intelligence and Information Processing

4.1 Introduction

Psychometric models do not explain the processes that underlie thinking. They are not intended to do so, but they nevertheless contribute to understanding intelligence. This has been the case since at least 1923, when Charles Spearman wrote *The Nature of Intelligence and the Principles of Cognition*. As Sternberg (2016, p. 236) highlighted, "Spearman believed that apprehension of experience, education of relations, and education of correlates are the basic overlapping information processes of intelligence. ... The great psychometricians of all time – Spearman and Carroll – were also astute cognitive psychologists."

John B. Carroll (1993) devoted part III of his encyclopedic treatise to discussing the role of psychometric models within the broader psychological realm. He stated that the psychological interpretation of the factors included in psychometric models "often imply hypotheses as to the processes that underlie performances in tests of ability. . . . Factor analysis can help in differentiating processes" (pp. 644–645). In other words, psychometric and cognitive models are complementary. Both help to better trace the big picture of the science of human intelligence.

Imagine two individuals, Rebeca and Yukio. We can determine their psychometric intelligence and then ask them to solve two problems. The first problem makes use of the English language rules that permit center embedding, putting one relative clause inside another. Rebeca and Yukio are presented with sentences of the following form:

The rat ate the cheese.
The rat the cat chased ate the cheese.
The rat the cat the dog scared chased ate the cheese.
The rat the cat the dog the man owned scared chased ate the cheese.

and so on. We can determine at what point each person finds the sentence incomprehensible.

This is the second problem:

The distance between Berlin and Athens is approximately 2,600 km by air. If an aircraft leaves Berlin at 9 AM and travels toward Athens at 900 km per hour at an altitude of 9,000 m, while at the same time an aircraft leaves Athens and travels to Berlin at 850 km per hour at an altitude of 8,500 m, when will they pass by each other, and how far is this point from Athens?

Now assume that we had conducted an experiment not involving Rebeca and Yukio in which problems like these were included in a large psychometric study using the tests considered by the CHC or g-VPR models described in Chapter 3. If we knew Rebeca's and Yukio's scores on the psychometric dimensions/factors, we could estimate the probability that each of them

would correctly solve the two problems. However, we could not go beyond this point because the psychometric approach does not provide detailed models of the problem-solving mental processes. Psychometrics does not tell us what the elementary cognitive steps are in the problem-solving process or how individual differences in the ability to execute these steps translate into individual differences in problem-solving ability. This includes the ability to take intelligence tests, and this is why cognitive psychology models are complementary to psychometric models.

4.2 Cognitive Psychology

Suppose that we were to ask cognitive psychology professors to explain Rebeca's and Yukio's problem-solving behavior for the sentence problems. They would want to look at the requirements of the problem-solving task. They would first observe that Rebeca and Yukio would need to retrieve word meanings from memory. This in itself is not a trivial task. But there is more. To untangle center-embedded sentences, a person has to store information in memory about a noun phrase until it can be connected to its verb. To understand *The rat the cat chased ate the cheese*, Rebeca and Yukio would need to temporarily store the phrase "the rat," then process "the cat chased," then attach the result of the processing to "the rat" (the particular rat that the cat chased), and then connect the modified idea to the verb phrase "ate the cheese." The professors would want to know what temporary memory storage capacities Rebeca and Yukio had for holding unresolved noun phrases and whether they could organize that storage so that the noun phrases would be available when their verbs were encountered.

It is not easy to be a cognitive psychologist.

For the mathematics word problem, the professors would first observe that Rebeca and Yukio retrieve word meanings from memory and have sufficient temporary storage space in memory so that they can analyze the meanings of the sentences.

However, the passage does not contain deeply center-embedded sentences, so sentence comprehension itself would not present a challenge to their capacity for temporary memory storage. There would be a different challenge.

Using information extracted from the sentences, Rebeca and Yukio would have to develop an internal representation of the problem. This could either be in the form of equations or in the form of a mental picture. The professors would probably remark that both representations require internal memory storage of information but differ in the kinds of information stored. A representation in terms of equations is a symbolic representation that at least approximates syntactical analysis. A map-like mental representation requires visuospatial memory storage. The professors would also note that the problem requires answering two questions, the time the two aircraft meet and the distance from Athens. The problem solver has to set up a goal–subgoal structure to decide which problem to attack first, work on that problem, and then switch to the second problem. The professors would like to have an estimate of how many goals and subgoals Rebeca and Yukio can track and how long it takes them to switch from working on one goal to working on another.

Stepping back from the particulars of the two problems, cognitive psychologists offer explanations of behavior in terms of elementary cognitive tasks (ECTs), such as retrieving the meaning of a letter or word from memory, holding a piece of information in temporary memory storage, and switching attention from one cognitive task to another. Information-processing theories of intelligence attempt to relate individual differences in intelligence to individual differences in the execution of elementary cognitive tasks that tap specific cognitive processes.

Cognitive psychologists stress that such explanations are bound to be incomplete, because problem-solving ability will be influenced both by the information the person has (knowledge) and how well the person can manipulate information in the abstract. An information-processing model addresses only the latter. Nonetheless, information-processing models stand somewhere between psychometric and brain-based models of intelligence (Figure 4.1).

To clarify, we return to the example of the center-embedded sentence. Knowing where Rebeca ranks defined by the CHC or g-VPR models tells us the probability that she will be able to understand a sentence with, say, three levels of embedding. We might observe that when Rebeca tries to understand a center-embedded sentence, the frontal, anterior parietal, and left temporal parts of her brain show heightened activity.

Yukio might obtain a different rank in the CHC or g-VPR models and have a different probability of understanding the sentence, and while trying to understand the sentence, Yukio might show different levels of brain activation than Rebeca.

It is useful to have a level of explanation that stands between the psychometric and brain-based models by explaining in functional cognitive terms how Rebeca and Yukio are both enabled and limited by their information-processing abilities. We do not explain what a computer does because it is turned on. We do it looking at the programs and data over which it operates.

To understand attempts to connect information-processing models of thought to the concept of intelligence, we must understand theories of cognition as information processing. Generally, these theories are based on concepts developed for designing digital computing systems. The contention is that any computation, including thought, has to rely on elementary actions and that studying performance of individual actions is useful when studying cognition. Any computing device has to be able to do three things:

1. Perception: Sense the environment.
2. Categorization: Classify the environment into states relevant to the device.
3. Memory retrieval: Relate these classifications to previously stored information.

In people, the result of these computations is an internal representation of the current

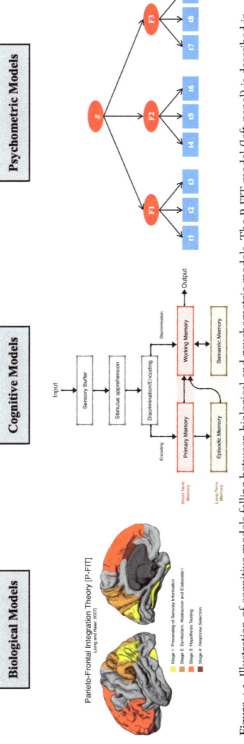

Biological Models

Parieto-Frontal Integration Theory [P-FIT]
(Jung and Haier, 2007)

Stage 1: Processing of Sensory Information
Stage 2: Symbolism, Abstraction and Elaboration
Stage 3: Hypothesis Testing
Stage 4: Response Selection

Cognitive Models

Input

Sensory Buffer

Stimulus apprehension

Discrimination/Encoding

Discrimination

Working Memory

Output

Semantic Memory

Encoding

Primary Memory

Episodic Memory

Short-Term Memory

Long-Term Memory

Psychometric Models

g

F1   F2   F3

t1 t2 t3   t4 t5 t6   t7 t8 t9

Figure 4.1 Illustration of cognitive models falling between biological and psychometric models. The P-FIT model (left panel) is described in Chapter 5.
Courtesy of Kenia Martínez.

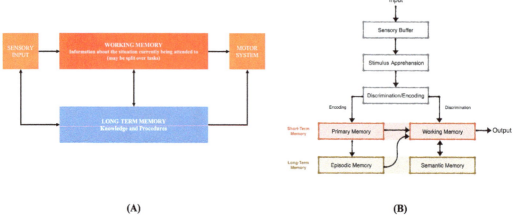

(A)   (B)

Figure 4.2 (A) The simple blackboard model of cognition. Information about the current situation is held in a temporary working memory store. Sensory input enters working memory if attention is being directed to the appropriate input channel. Information on the blackboard and from the sensory input channel activates problem-solving procedures that have been stored in long-term memory. These place further information in working memory. Sensory information may also enter long-term memory directly, thus lowering the threshold required to activate some problem-solving procedures or, in some cases, initiating an action without placing information on the blackboard. (B) The detailed blackboard model of cognition. Each box represents a cognitive process, and the arrows represent the information flow within the system. The sensory buffer filters the incoming information by distinguishing the relevant from the irrelevant. Only the former becomes the focus of attention. Apprehension requires detecting changes in the information, discrimination implies distinguishable responses depending on the stimuli, and encoding involves identification. Short-term memory is capacity limited, and it may be passive (primary memory) or active (working memory). Long-term memory can be episodic (contextualized information) or semantic (abstract and symbolic information).
Adapted from Jensen (1998).

situation as interpreted by memory of past situations. The internal representation may be used to select a response (decision-making). If the computing device operates continuously in a dynamic environment, it must maintain a record of the situations it has encountered, the responses that have been made, and the results of the responses. People also have to be concerned with the storage and retrieval of information in long-term memory in a way that ensures its accessibility when needed.

Although one can conceive of a robot mind that executed each of the tasks – perception through decision-making – in serial order, in the human brain, they are interleaved with a great deal of feedback between them. A model of how this exchange of information takes place is known as a cognitive architecture. Psychologists who study human information processing have

converged on some version of the "blackboard" model of cognitive architecture (Anderson, 1996; Hunt and Lansman, 1986; Meyer and Kieras, 1997a, 1997b; Newell, 1990). A simplified version of this model is shown in Figure 4.2A. Figure 4.2B depicts a more complete model. We detail it further because it demonstrates seven key cognitive concepts relevant for intelligence research:

1.   Information from the environment is perceived and then classified into progressively higher-order categories by activation of related information in long-term memory. Long-term memory is not thought of as a static storage process, as is the case in computer systems. Items in long-term memory exist in various states of activation, depending on how frequently and how recently they and related pieces of information have

been attended to. Therefore, the interpretation of a percept will be influenced by the context in which it occurs.

2. The information from the environment, together with relevant information retrieved from long-term memory, is placed in working memory. There, further processing results in an internal representation of the current situation. The internal representation may include interpretations in contexts and the identification of goals and subgoals in problem solving. The processes acting on the internal representation are referred to as executive processes, whose role is to update the internal representation and establish priorities for action. One of the most important subsets of the executive processes are the attentional processes that highlight some pieces of information and suppress others.

3. Organizing information in this way enables the ability to deal with two or more tasks simultaneously – for example, talking while driving an automobile. The impression of simultaneity is something of an illusion, as both tasks will compete for working memory resources and for attention. When two tasks are done together, therefore, close examination almost always shows that one or both are performed less well than they would be if performed alone.

4. Working memory acts as a blackboard that broadcasts information into long-term memory. Alternatively, you can think of long-term memory processes as actively watching working memory to see if their cue has been called, rather like actors at a stage production.

5. The storage section of working memory is thought of as being divided into modality-specific sections for linguistic, auditory-nonlinguistic, and visuospatial information. There is no implication that working memory is located at a single place in the brain. It is a system resulting from the integration of processing at several locations.

6. Information in working memory is consolidated in long-term memory. This takes time. Therefore, the probability that a piece of information in working memory will be stored in long-term memory is partially determined by the time it remains in working memory. Items briefly attended to are not likely to be remembered.

7. Sometimes the information in working memory will initiate a process that calls for a motor response – speaking or making a physical movement. While cognitive information-processing psychologists are concerned with response production, psychologists interested in intelligence have done rather little work on this aspect of cognition.

Cognitive psychologists are concerned with the typical characteristics of each of the processes in the cognitive architecture. Psychologists interested in intelligence want to know how individual differences in these characteristics are related to individual differences in intelligence, as measured by psychometric tests. These two perspectives are detached historically, but cognitive researchers have increasingly focused more on the individual difference approach (Box 4.1).

Using the psychometric models from Chapter 3, we now look at two aspects of cognitive architecture: mental processing speed and working memory. Both are related to individual differences in general reasoning ability. Then we will examine cognitive processes related to verbal and visuospatial abilities. At the end of this chapter, we will describe a comprehensive model aimed at integrating psychometric, cognitive, and biological models of intelligence (process overlap theory). The discussion of this integrative model will introduce Chapters 5 and 6, which are devoted to the biological basis of intelligence.

4.3 Mental Processing Speed

The blackboard model (Figure 4.2) depicts thinking as the shuttling of attention from one piece of information to another. To

Box 4.1 Can One Size Fit All? No, It Cannot

Thomas Grandy and colleagues (2017) addressed this question, and although the resulting report was unpublished, it is a good example of applying an individual difference approach to basic cognitive research. Their key finding was that there is no one general memory search process. People can have the same performance by using distinguishable cognitive processes and strategies. This type of result usually has been ignored in experimental psychology and cognitive neuroscience.

In their experiment, thirty-two individuals completed 1,488 trials of the classic Saul Sternberg short-term memory task. In this well-known task, the individual must remember various stimuli; a memory set of various numbers of stimuli is presented, and after a brief blank period, a target stimulus is presented. The individual must decide if the target stimulus was or was not within the memory set. For deciding, the individual must compare the target stimulus with the stimuli already memorized. The general prediction is that the larger the number of stimuli in the memory set is, the more time will be required for deciding.

Grandy et al. observed that thirteen individuals applied self-terminating memory search (the search stops when the target is identified), whereas thirteen other individuals applied exhaustive memory search (the search involves scanning the whole memory set): "The majority of participants provided data consistent with one of the two major competing memory search models which prohibits the adequate description of memory search with a single model. ... When individuals differ among each other in the cognitive processes they use to solve a task, or when a given individual solves a task in more than one way from trial to trial, then the aggregate neural response, either across or within persons, will represent a mixture of neural processing configurations. The appreciation of

heterogeneity and variation is a necessary challenge for cognitive psychology and cognitive neuroscience" (p. 48). In other words, one size cannot fit all, and this has real consequences for understanding human behavior from a cognitive and neuroscientific perspective.

The same approach was used by Juan Botella and colleagues (2019), considering a conflict-monitoring attention task. Here, the authors analyzed a large group of 1,159 individuals for addressing the question of whether seeking a unique cognitive model is a reasonable and fruitful scientific strategy. These researchers compared the conventional approach in experimental psychology with analyses at the individual level. Key findings revealed that at least four models are required to account for the performance of the 1,159 individuals.

These results are shown in Figure 4.3. Four out of ten participants failed to show stimulus–response congruency effects in the experimental task, whereas the remaining 60 percent followed distinguishable theoretical models (consistent with conflict-monitoring theory and/or priming and episodic memory effects). The authors concluded that individuals' cognitive mechanisms might help explain some of the reproducibility issues that are currently of great concern in psychology (Nosek et al., 2022).

The main conclusion was that general/experimental and differential/correlational psychological approaches work better together for addressing relevant theoretical issues in psychological research. This is not a new perspective, however. This same message was delivered to the members of the American Psychological Association in the presidential address of Lee J. Cronbach (1957, p. 673) six decades ago: "it is shortsighted to argue for one science to discover the general laws of mind or behavior and for a separate enterprise concerned with individual minds."

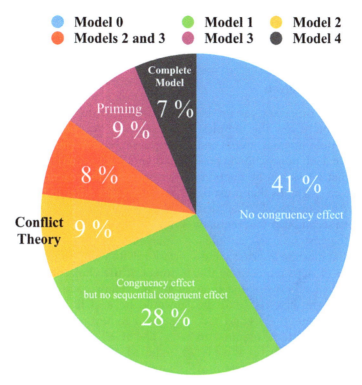

Figure 4.3 Percentage of sample ($N = 1,159$) using different cognitive models for resolving the same conflict-monitoring experimental task (Botella et al., 2019).

illustrate this, consider the common task of determining whether a view of a scene matches a description. The task could be as simple as showing a person a scene and asking a question like "Are the car keys on the dining room table?" The listener must form an internal representation of the linguistic statement, form an internal representation of the visual scene, and compare the two. Different people use different strategies for comparing sentences to pictures, but no matter which strategy they use, they must shift back and forth from processing one piece of information to processing another. All this has to happen quickly enough that the thinker can keep up with the environment. If a speaker is speaking at a normal conversational tempo, listeners are expected to keep up.

There is also an internal demand for speed in mental processing. Competing lines of thought vie for the limited storage resources in working memory, and the

mental competition is fierce. Have you ever been introduced to someone, immediately embarked on a conversation, and then realized that you did not know the person's name, even though the two of you had just been introduced? If information in working memory is usurped by another task, the original contents of working memory may be lost before it can be moved to long-term memory and consolidated (Figure 4.2).

These considerations illustrate why processing speed is important in thought. But is there a single speed of processing or different speeds for different elementary cognitive tasks (ECTs)? Processing speed is a pervasive factor because all mental actions depend on neural processing. If people differ in the effectiveness of processing speed at the neural level, then there should be generalized individual differences in processing speed no matter what the task (Jensen, 2006). Accordingly, speed of processing should be positively correlated with standard measures of intelligence. To test this hypothesis, reliable measures of mental processing speed are required.

4.3.1 *Experimental Paradigms*

Elementary cognitive processes cannot be measured in isolation. Even the simplest tasks require multiple mental steps and may mix cognitive and noncognitive processes. Suppose that a task can be fractionated into a sequence of k steps that must be conducted in order. Suppose further that you can construct a second task that requires all the steps of the first task and an additional $k + 1$ step that must be inserted somewhere in the series of steps required by the first task. The difference between the time required to execute the first task and the time required to execute the second task will be a measure of the time required to complete the inserted step. That is, if we let $R(x)$ be the time required to complete process x, then R(inserted process) = R(task with insertion) − R(original task).

The approach is based on three assumptions: (1) the processes involved must be executed in sequence; (2) for any pair of adjacent processes in the sequence, the first process is not initiated until the second is completed; and (3) the speed with which a process is completed is independent of the speed with which any other process is completed. Collectively, these conditions are referred to as the independent serial processing assumptions. The logic applies only to situations in which the entire task is completed successfully. Researchers interested in intelligence have not always paid as much attention as they should either to the need to prove that the independent serial processing assumptions apply to their tasks or to restricting their analyses to trials with correct responses. Despite these limitations, this experimental paradigm has provided useful data.

Experimental paradigms have also been designed to measure the speed with which a person can make an elementary decision, without reference to memory. Such studies are called choice reaction time (CRT) tasks. At the beginning of the experiment, the participant is told that one of a small number of directly perceivable stimuli may occur. For instance, the participant might be told that either a red or a green light could be flashed on a screen. The participant's task is to identify as quickly as possible which of the possible stimuli has been presented by pressing the respective response button (for this instance, Q for red and P for green). The time it takes for a response is the reaction time (RT). Choice reaction tasks differ in the number of possible stimuli that might be presented. Discriminating between red and green lights is an example of a two-choice reaction time task (2CRT). In a more complicated experiment, if the choice were to be among red, green, yellow, and blue lights, the task would be a 4CRT task. Real experiments seldom go beyond the 8CRT task.

The 2CRT condition represents the time required to make a binary decision. Logically, this should be central to cognition. However, the response time in the 2CRT condition will also include the time required to press the button once the cognitive decision is made (this is called movement time, MT). This contaminates cognitive and noncognitive processes. To avoid such problem, we can measure the manner in which choice reaction times increase with the number of choices. An analysis of how reaction time changes as the number of choices is increased can provide an estimate of cognitive processing speed, independent of motor processes.

The basic CRT approach depends on assumptions of strict serial processing and independence of processes from each other. An alternative approach is to construct a task that is thought to be primarily responsive to a particular cognitive process and use it as a measure of the time required for that process, ignoring the fact that other processes may also influence the measure. One such paradigm is the lexical identification task, which measures the time required to retrieve information from long-term memory. The subject must decide that a string of letters either does or does not constitute a common word, for example, MALEC (no) or CAMEL (yes).

A generalization of this task is the semantic identification task. In semantic identification, an object name and a category name are presented, for example, CAMEL ANIMAL. The subject indicates whether the object named is a member of the category. In such tasks, RT reflects both retrieval of information from long-term memory and the motor processes involved in making a response.

In cognitive psychology, without regard to individual differences, the typical experiment focuses on how experimental conditions change average reaction times in an identification task. For instance, one of the major uses of the lexical identification task is to study how observation of semantically related items changes the speed of identification of a target item. To take a frequently cited example, the word NURSE will be recognized faster if the immediately preceding word was DOCTOR than it will if it was BUTTER.

A similar logic applies to studies of individual differences. When we study the relation of elementary cognitive processes to intelligence, we are interested in the differences in reaction times among people indexed by their scores on standard intelligence tests. The correlation coefficient quantifies the strength of both differences (at the psychometric and cognitive levels). If people who have high intelligence scores are consistently faster in lexical identification tasks than people who have low scores, then this is evidence that at least one of the cognitive processes required for lexical or semantic identification is associated with intelligence. Many studies show that this is the case. In the population of college students, lexical identification RT has a correlation of $r = -0.40$ with scores on tests of verbal intelligence (e.g., Palmer et al., 1985). The negative direction is expected because shorter (faster) reaction times are associated with higher intelligence test scores. The magnitude of the correlation in a more heterogeneous population than college students would be higher because of the lack of restriction of range in the population sample.

It would be helpful to obtain as direct a measure as possible of cognitive processing speed. To see this, consider some facts obtained from general experimental psychology. We can think of any cognitive process as consisting of the following steps:

Perceive the stimulus $\rightarrow$ Select a response $\rightarrow$ Make a response

The middle step is what we normally think of as cognition. The first step reflects sensory-perceptual processes, and the third step reflects motor processes (Figure 4.2B).

In information-processing studies, the individual reacts to information presented on a computer screen. Studies in perception suggest that it takes from twenty to fifty milliseconds to detect a simple visual figure, depending on the illumination and the complexity of the surrounding visual field. It takes a healthy college student about 250 ms to press a button indicating detection of a visual signal like a light flash (no decision about the flash is required – just acknowledging the flash). This kind of task is called single (or simple) reaction time (SRT), and it minimizes the cognitive step in the sequence.

If a cognitive task is introduced, such as asking the student to recognize whether CAMEL is a word, reaction time rises to 500–600 ms. This means that of the roughly 550 ms taken for the decision task, somewhere between 250–300 ms, about half the response time, will be taken up by perceptual and motor processes. The other half is taken up by cognitive processes. Therefore, the reaction time provides a measure of cognitive processing speed.

However, this analysis ignores an important issue. In general, the longer a person waits to make a response in a choice task, the more likely the person is to make the correct choice. This is called the speed–accuracy trade-off (SATO). Different people adopt different criteria for accuracy. Rebeca may be satisfied if her responses are correct on 90 percent of the trials, Yukio may prefer 99 percent, and Hans may accept 95 percent. Studies of SATO have shown that above accuracy rates of 90 percent, small differences in accuracy may be associated with substantial differences in RT (Lohman, 2000). Therefore, if different individuals adopt different criteria for acceptable accuracy, another source of variation has been introduced, further confusing the interpretation of correlations between CRT and intelligence scores.

Some of the problems with CRT tasks are avoided by diffusion models (Box 4.2) and

inspection tasks. In the typical visual inspection time (IT) task, two vertical lines are presented simultaneously for a brief time, followed by a visual mask to avoid afterimage postprocessing. The individual indicates which line (right or left) is longer (here is an animation: https://youtu.be/PQmnB9_1x38). The amount of time the two lines are visible in a presentation of the lines is

Box 4.2 Diffusion Models, Cognitive Measurement Models, and a Reversed Train of Thought

Diffusion models allow the separation of distinguishable elementary cognitive processes. The so-called drift rate provides appropriate estimates of mental processing speed regardless of motor speed and speed–accuracy trade-offs (SATOs). Interestingly, drift estimates combine reaction time and accuracy. Nevertheless, this model is limited to decision tasks with two response options (binary tasks). When the stimuli are shown, the examinee accumulates information in a continuous way before reaching a critical threshold regarding (1) response options and (2) response accuracy. Knowing the distance between these thresholds provides an estimate of how much information is required before deciding. Uninteresting nondecision time is controlled for applying mathematical models.

This approach might be interesting to test if the inverted-U-shaped relationship between processing speed and intelligence (Figure 4.5), that is, increased correlations between ECT performance and intelligence with increased task complexity (up to 1,200 mc) but decreased correlation afterward (for longer RT tasks), is a genuine phenomenon or results from how mental speed is usually estimated. This was tested in one study by administering eighteen binary RT tasks that included long RT tasks (up to 3,000 mc) (Lerche et al., 2020). Findings showed a remarkable correlation $(r = 0.68)$

between a general drift rate factor and general (psychometric) intelligence (g). This relationship was valid for simple and complex binary RT tasks. However, the authors did not conclude that mental processing speed and intelligence are causally related: "rather, structural properties of the brain may give rise to the association between mental speed and intelligence" (p. 2226).

In this later regard, Anna-Lena Schubert and Gidon Frischkorn (2020, pp. 142–143) discussed the neurocognitive psychometrics of intelligence, relying on the finding that speeded information processing across elementary cognitive tasks correlates with psychometric intelligence: "More intelligent individuals benefit from a greater velocity of evidence accumulation, from both sensory input and memory, but do not show greater encoding or motor response speed. ... Greater white matter tract integrity enhances the speed and capacity of information processing, which in concert positively affect reasoning ability." These researchers favor the integration of experimental and neurobiological approaches for refining our understanding of the relationships between elementary cognitive processes and higher-order cognition. Taking a different view, a study by Willoughby and Lee (2021) fails to support a key role for axonal conduction velocity. Their findings favor the potential relevance of properties of brain networks such as connectivity patterns. Of course, these are not mutually exclusive approaches.

Box 4.2 *(continued)*

The fact is that the search for the elementary cognitive processes associated with intelligence is far from easy and highly elusive. A comprehensive review of the evidence addressed usual suspects, such as processing speed, working memory capacity, and executive processes, reaching the tentative conclusion that single cognitive processes might not be key (Frischkorn et al., 2022). Cognitive measurement models are strongly required, but, as acknowledged by these researchers, "the application of these models for understanding the cognitive basis of intelligence is still in its infancy" (p. 8). These measurement models are, moreover, circumscribed to specific domains. The diffusion model, for two-choice decisions, described earlier, is a good example.

Given this uncertain state of affairs, Adam Chuderski (2022, p. 5) suggested a reversed train of thought: "instead of elementary cognitive processes underpinning intelligence, it is equally plausible that it is intelligence, understood as a set of neurocognitive parameters validly captured by intelligence tests, which translates into scores on elementary cognitive processes." In fact, addressing the relationship between psychometric general intelligence (*g*), working memory capacity, mental processing speed, and several broad psychometric cognitive abilities, Colom and colleagues (2004, p. 288) proposed structural models, across three related research studies, on which *g* was the predictor of a wide array of cognitive factors instead of the criterion: "within any task domain, individual differences will emerge when the task demands consume sufficient capacity to exhaust some person's resources. These individual differences have much to do with individual differences in the overall capacity and efficiency of mental processes."

---

called the inspection time (IT). The experimenter adjusts the IT until it is so short that the observer can no longer make the distinction beyond a given level of accuracy. This is usually set at 75 percent correct responding. Some people perceive the difference in lines faster than others; their IT would be shorter.

The IT paradigm avoids mixing cognitive processing and motor responding. However, it does so at the expense of minimizing the cognitive processes involved because the task is mostly perceptual. Therefore, it is not surprising to find that when IT measures are included in an intelligence test battery, the IT measures load on a cognitive speediness factor. This is a broad second-order factor in the CHC model discussed in Chapter 3. It might also indicate an ability to control attention (Nettelbeck, 2001). When discussing all cognitive tasks, always remember that the same experimental task may not measure exactly the same thing in different individuals (Putnick and Bornstein, 2016).

4.3.2 *Experimental Results for Processing Speed and Intelligence*

The correlations between intelligence test scores and SRT typically are weak or nonexistent. But intelligence test correlations with CRT and IT experimental paradigms are typically in the −0.6 to −0.7 range once corrected for restriction of range (many studies are done with college students, who do not represent the general population, e.g., Jensen, 2006). These correlations indicate there is some common core to the psychometric test scores and elementary information-processing measures. We know from factor analysis studies that processing speed is one component of intelligence, so

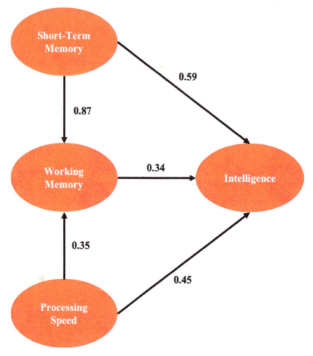

Figure 4.4 Relationships of short-term memory, working memory, and processing speed to intelligence in a sample representative of the population (Colom et al., 2008).

these correlations make sense. In the early days of the modern study of individual differences on information-processing tasks, Earl Hunt and colleagues (1973) pointed out that the study of individual differences in information processing is related to, but different from, the study of psychometrically defined intelligence. As we see, combining the approaches is informative and increases our scientific knowledge.

The testing procedures required to evaluate elementary cognitive tasks are time consuming. While there are laboratories where banks of computer-controlled test stations are available, for the most part, this research takes place using small participant groups. Partly for this reason, most of the research on individual differences in elementary cognitive tasks has been conducted using university undergraduates with a restricted range of intelligence. Generalizing to other samples can produce inconsistent and contradictory results.

For instance, in a study of undergraduates done in Spain, Roberto Colom and colleagues

(2008) found no reliable relationship, at the latent variable level, between information-processing speed and intelligence. However, when they extended their study using exactly the same procedures, to include high school students, a reliable and substantial relationship appeared (0.45), as shown in Figure 4.4.

In another study, researchers contrasted information processing in university undergraduates and people in the low-IQ range. Both the evidence for a pervasive speed factor and the relation of that factor to intelligence test scores were much stronger in the mentally challenged group than in the university undergraduates (Detterman and Daniel, 1989). Deary and Der (2005) considered a Scottish representative sample of people from their late teenage years to their sixties, finding correlations between reaction times and performance on a complex arithmetic task in the −0.3 to −0.4 range.

The potential size of the effects achieved by expanding the study of the information processing–intelligence relationship beyond the college population should not be underestimated. If we compare fast populations, such as college students, to slower populations, such as the elderly or mildly mentally disabled individuals, estimates of processing speed may vary by as much as five times or more, depending on the apparatus and parameters involved (Hunt, 1980, 1987; Salthouse, 1996).

What are we to make of the relation between mental processing speed and psychometric measures of intelligence? Two extreme positions could be taken:

1. The various elementary cognitive tasks evaluate the speeds of different mental processes, each of which contributes to cognition. In this view, for instance, the lexical identification task and the inspection time task are measuring different things. Therefore, each of the processes evaluated should make its own contribution to the prediction of performance on intelligence tests.
2. At the other extreme, it might be that all of the different information-processing measures reflect a fundamental speediness property of a person's nervous system. If this is the case, all the different information-processing measures would be regarded as measures of the same generalized property and would have similar correlations with test scores.

As usual, the reality seems to lie somewhere in between. Measures of performance, such as the mean CRT, tend to have higher correlations with scores on intelligence tests than do measures of basic processing like SRT. In theory, the latter measures should evaluate a specific cognitive process. In fact, the most important variable seems to be the total time required to complete the information-processing task. The longer the time required is, the higher is the correlation between performance on the task and performance on an intelligence test (Jensen, 2006). This is what would be expected on the assumption that all measures tap a single mental speed trait, because the longer (slower) tasks would provide a better

measure of individual differences in processing speed. However, it seems that beyond a given point (RTs > 1.5 s), the correlations between RTs and intelligence begin to decline, and accuracy measures begin to be more sensitive to individual differences in information processing. This is shown in Figure 4.5; Box 4.3 describes an example of how processing speed can be linked to brain processes. In addition, different tasks have some specificity. For instance, lexical decision times correlate most highly with verbal intelligence measures, while inspection time measures correlate most highly with nonverbal tests. Nevertheless, the general trend is clear: general cognitive speediness is a reliable trait and a component of intelligence.

Some authors have argued that some societies distrust rapid responders, believing that the more intelligent individual is the one who stops to weigh alternatives before speaking. The objection might, however, miss the point of the aims of basic research. Studies of individual differences in processing speed are concerned with how rapidly a person can grasp a situation, given a fixed amount of information. Deciding to make an overt response is a separate action. There is a difference between noticing the logical flaw in your father-in-law's political views and deciding to point that flaw out to him. These two separate acts of cognition both benefit from rapid thinking. Indeed, rapid thinking will probably help you suppress inappropriate or impolitic responses in many social settings, including on social media.

4.4 Working Memory Capacity

Scientists distinguish between two types of memory: short-term (immediate) memory, which lasts a few seconds at most, and long-term memory, lasting the individual's lifetime (although retrieval becomes difficult as time passes). The blackboard model (Figure 4.2) uses this distinction. The blackboard is itself an immediate memory device and is thought of as a limited resource (Cowan, 2001, 2017), while

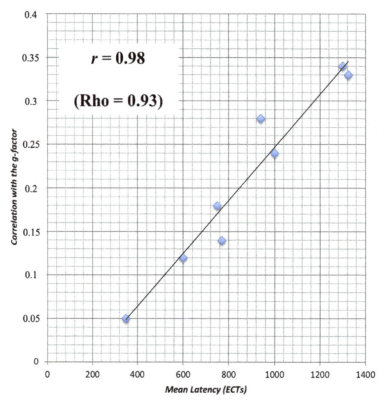

Figure 4.5 (*y*-axis) *g*-Factor scores obtained from eight tests from the Armed Services Vocational Aptitude Battery (ASVAB). (*x*-axis) ECT task complexity (estimated by mean response latency on each task). The correlation between these two variables is 0.98, showing that *g*-loadings increase linearly with increased mean latencies (task complexity). Rho is a correlation based on rank orders of both variables.
Adapted from Jensen (1998).

long-term memory is presumed to be limitless. We now take an even closer look at the structure of the blackboard model and the role of working memory capacity (also see Box 4.4).

Intelligence test batteries, such as the Wechsler batteries, contain separate subtests for short-term memory and long-term memory. Short-term memory can be evaluated by digit span tests. In a forward digit span test, the examiner presents a randomly chosen series of three to eight digits, for example, 3, 9, 7, 4. The examinee's task is to repeat the digits immediately after they have been presented. In the more challenging backward digit span task, the examinee must repeat the digits in reverse order, 4, 7, 9, 3 in this example.

The forward digit span task is regarded as a test of short-term storage capacity. Digit span tests typically have a loading of about 0.5 on the *g*-factor extracted from an intelligence test battery. Backward digit span introduces a processing as well as a memory component; instead of reading from the memory space as it was inputted, the examinee has to make a transformation of the memory space by reversing the order of the digits once they are all presented. The backward digit span shows a higher loading on the *g*-factor than the forward span. This suggests that we cannot regard the blackboard/immediate memory system as just a passive storage system (Colom et al., 2016; Martínez et al., 2011; Unsworth et al., 2014). It also means that digit span, especially

Box 4.3 Looking Down on Human Intelligence

Information processing tells a high-level story occurring within a prosaic brain. Therefore, it would be illustrative to connect both levels of analysis (Figure 4.1) for answering the question about the relationship between mental processing speed and intelligence.

A group of German researchers, led by Anna-Lena Schubert, did interesting research testing the most likely relationship between processing speed and intelligence (Schubert et al., 2017). The title of the article summarizes the key finding: "Is General Intelligence Little More than the Speed of Higher-Order Processing?" The take-home message was that "more intelligent individuals benefit from a more efficient transmission of information from frontal attention and working memory processes to temporal-parietal processes of memory storage" (p. 1498).

In their study, 122 participants completed a set of experimental tasks (single and choice RT tasks, the Sternberg memory scanning task, and the Posner letter-matching task), along with the Raven's Advanced Progressive Matrices (RAPM) and the Berlin Intelligence Structure Test. Electroencephalogram (EEG) was also recorded during the completion of the experimental tasks, and several event-related potentials (ERPs) were obtained for separating early and late information-processing times.

The correlation between general intelligence and general behavioral processing speed revealed a value of $r = 0.43$. However, the findings were sharply different when early and late EEG timing components were considered separately. The values now were 0.33 and 0.89, respectively: "the result suggests that smarter individuals do not have a general, but a very specific advantage in the speed of higher-order information processing" (Schubert et al., 2017, p. 1508).

These researchers related the finding to the parieto-frontal integration theory of intelligence (P-FIT) (Jung and Haier, 2007), which we will discuss in the next chapter, and concluded that greater speed of higher-order information processing enhances both working memory and intelligence. This goal is presumably achieved by increasing the efficiency of selective attention and short-term memory updating. Moreover, the powerful prediction obtained from later ERPs (associated with higher-order cognition) may involve executive processes, which may be consistent with the main assumption of the process overlap theory proposed by Kovacs and Conway (2016), which we discuss in the final section of this chapter.

backward, is a good single-test estimate of g, although single-test estimates are less appropriate than a g latent variable extracted from a test battery.

When the British psychologist Alan Baddeley (1986; Baddeley and Hitch, 1974) introduced the idea of working memory, he pointed out that when people solve problems, they have to hold key pieces of information readily at hand, even though they are not the immediate focus of consciousness. In addition, information must be transformed, mental representations have to be built, and attention has to be focused on information relevant to the task at hand, while irrelevant information must be suppressed. Baddeley's view of working memory, which is incorporated into the blackboard model, is that it is a mental workspace containing (1) modality-specific memories for small amounts of auditory and spatial-visual information and (2) a central executive that supervises the flow in and out of the storage systems.

The central executive was not well specified. At minimum, there must be provision

Box 4.4 The Many Faces of Working Memory

Nelson Cowan (2017) published a thoughtful article regarding the working memory concept. He noted nine different definitions of working memory:

1. *Computer working memory*: A holding place of information to be used temporarily, with the possibility of many working memories being held concurrently.
2. *Life-planning working memory*: A part of the mind that saves information about goals and subgoals needed to carry out ecologically useful actions.
3. *Multicomponent working memory*: A multicomponent that holds information temporarily and mediates its use in ongoing mental activities.
4. *Recent-event working memory*: A part of the mind that can be used to keep track of recent actions and their consequences in order to allow sequences of behavior to remain effective over time.
5. *Storage and processing working memory*: A combination of temporary storage and the processing that acts on it, with a limited capacity for the sum of storage and processing activities. There is not a clear commitment to multiple storage components, only a separation between storage and processing.
6. *Generic working memory*: The ensemble of components of the mind that holds a limited amount of information temporarily in a heightened state of availability for use in ongoing information processing.
7. *Long-term working memory*: The use of cue and data-structure formation in long-term memory that allows the information related to an activity to be retrieved relatively easily after a delay.
8. *Attention-control working memory*: The use of attention to preserve information about goals and subgoals for ongoing processing and to inhibit distractions from those goals. It operates in conjunction with short-term storage mechanisms that hold task-relevant information in a manner that does not require attention.
9. *Inclusive working memory*: The mental mechanisms that are needed to carry out a complex task. It can include both temporary storage and long-term memory, insofar as both of them require attention for the mediation of performance.

This list demonstrates the disparate perspectives in experimental psychology regarding this core cognitive system. Since our focus is on human intelligence, we underscore that Cowan notes that generic working memory may be especially relevant for researchers with a psychometric, individual differences approach, for which core aspects of working memory (1) are strongly correlated across storage and processing and (2) account well for general intelligence (even for experimental tasks emphasizing storage without additional processing if, and only if, storage is properly measured; Unsworth and Engle, 2007; Colom, 2016).

for a process that focuses attention on some information in working memory to make it available for processing, while at the same time preventing loss of information that is not currently the focus of attention but that has been accessed in the recent past and may be needed in the immediate future.

This idea may sound complicated, but it is encompassed by the term *situation awareness*, a phrase popular in engineering psychology and in the military. The term refers to the need to be aware of what is going on about you, even if some aspects of the situation are not of immediate concern. For instance, an automobile driver must keep aware of the positions of nearby cars, including those behind the driver's vehicle and hence not in immediate view.

Similar requirements occur in many situations quite far removed from driving. When attorneys question witnesses during criminal trials, they have to be aware of jurors' reactions as they are interrogating the witness. When people attack problems with logic, they have to be aware of different interpretations of the premises. When chess players consider a move, they must evaluate different countermoves that the opponent might make (Blanch, 2021).

From an information-processing view, controlled manipulation of information in working memory is an extremely important part of cognition. Therefore, we would expect individual differences in the effectiveness of the working memory system to be related to general intelligence – and many research studies show that they are.

4.4.1 *The Measurement of Working Memory Capacity*

The memory span task is a popular technique for evaluating working memory capacity. A person is asked to perform a series of simple actions, like reading a sentence, while holding increased amounts of information in memory as each sentence is read. The simple task will be called the primary task, while the act of holding information in memory is the secondary task. The idea is that the processing requirements of the primary task interfere with holding the information needed for the secondary task. Therefore, the amount of information that can be held in memory in the face of the periodic interruptions by the primary task provides a measure of the temporary storage capacity of working memory.

When the primary task is reading, the span measure has a 0.5 correlation with measures of verbal comprehension, including the paragraph comprehension tasks that are often found on tests of verbal intelligence (Daneman and Merikle, 1996). When different types of span tasks are included in a comprehensive battery of reasoning and information-processing tasks, (1) all the span tasks reflect a common underlying factor, referred to as working memory capacity (WMC), and (2) WMC has a high correlation with the *g*-factor extracted from conventional intelligence tests (Colom et al., 2006a).

In dual tasks, people are asked to attend to at least two streams of information at once. A person who listens to the radio while driving a car is performing a dual task. Many laboratory tasks are formalizations of this everyday situation. In a dual task, attention must be switched from one stream of information to another, and then back again, with minimum loss of information during the switch. Dual tasks call on both the storage and attentional control aspects of working memory, but research evidence shows that there is no need to administer dual tasks for measuring working memory capacity: simple and complex memory span tasks rely on the same shared general cognitive resources (Colom et al., 2006a; Cowan, 2017; Unsworth and Engle, 2007).

4.4.2 *The Relation between Working Memory and Intelligence*

In a landmark series of studies by Patrick Kyllonen and his colleagues at the US Air Force's Armstrong Laboratory (San Antonio, Texas), recruits performed a number of working memory tasks, including span tasks, and also took highly *g*-loaded tests of reasoning. The recruits' scores on the army's intelligence test battery (the Armed Services Vocational Aptitude Battery; ASVAB) were also available. A common factor extracted from the working memory tasks showed an almost perfect correlation to the general factor extracted from the ASVAB (Kyllonen and Christal, 1990). Similar findings have been reported by other investigators (Colom et al., 2004; Oberauer et al., 2005) even in neuroscience research (Bowren et al., 2020), as we will see in Chapter 5. This raises an interesting question: is intelligence just working memory?

This work is important for two primary reasons. First, the populations studied included USAF enlistees, thus broadening the scope beyond the usual range of college students. Second, the number of different working memory tasks used ensured that the result did not depend on the unique information-processing requirements of particular tasks.

The finding poses something of an intellectual challenge. As Roberto Colom and colleagues (2004, p. 227) put it, "working memory comprises the functions of focusing attention, conscious rehearsal, and transformation and mental manipulation of information." Kyllonen's work and the studies that followed it show that working memory, a complex information-processing concept, is strongly related to general intelligence (g). But what part of working memory is crucial? Ian Deary (2000) has pointed out that we do not advance understanding by showing that one mysterious concept is linked to another. However, there are two ways to address this objection when considering memory–intelligence relationships:

1. One, two, or all of the information-processing functions that comprise working memory make separate (albeit related) contributions to the relationship to g. We can reduce the intelligence–working memory link to links between intelligence and the component parts of working memory.
2. We can accept the linkage at the intelligence–working memory level and challenge whether there is any need for further reduction.

To discriminate between these possibilities, we need to have separate measures of four aspects of working memory proposed by Colom and colleagues (2008): (1) memory span tests, to evaluate the ability to store information while processing another, concurrent task; (2) tests that evaluate short-term storage without processing, for instance, recall of a string of digits without any intervening or concurrent processing (remember Cowan's caution in this regard

noted in Box 4.4); (3) tests that evaluate the ability to control attention without any memory component; and (4) tests that evaluate processing speed, such as a choice reaction time (CRT) task. This is considered necessary because speed of stimulus identification, decision-making, and rehearsal of information are all involved in the working memory tasks.

A statistical method called structural equation modeling (SEM) can be used to identify latent traits underlying the four aspects of working memory and the measures of intelligence, to identify the relationships among the defined latent traits. Several such studies have been done. Within each of the studies, a coherent picture emerges, often identifying a key component determining the relation between working memory and intelligence. Across studies, however, there is very little agreement over what the component is.

Interestingly, the array of these studies has an international flavor. Studies of North American university students produce results indicating that the key component is the ability to control attention (Engle et al., 1999; Kane et al., 2004; Kane et al., 2001; Heitz et al., 2005). Studies of European and Asian school and university students indicate that the storage process (including executive updating) is the key component (Chuderski, 2014; Chuderski et al., 2012; Colom et al., 2008; Hornung et al., 2011; Martínez et al., 2011; Oberauer, 2003; Shahabi et al., 2014).

It is unlikely that working memory would work differently in Europe, Asia, and the United States – unlikely, but not impossible. A study that explicitly compared the architecture of information processing in Greece and China showed that the overall relations between structures were the same in each (Demetriou et al., 2005). Notably, however, Chinese children and adolescents were better than Greeks at tasks involving visuospatial reasoning. But there was no difference in variables specifying the relation between working memory and reasoning. These are fascinating results, but remember, no one

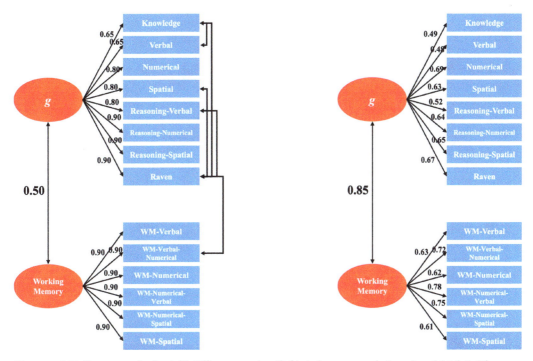

Figure 4.6 Different methods yield different results. (left) Ackerman et al. (2005) and (right) Oberauer et al. (2005) analyzed the same data set with different methods and found different results regarding the pattern of correlations between general intelligence (*g*) and working memory.

study is definitive. Further cross-cultural research is required to establish a strong weight of evidence.

There is little doubt that working memory is an important component of general intelligence, although the two are not identical, either in the sense of being perfectly correlated with each other statistically or in the sense of being conceptually identical. This was the main conclusion drawn by Phillip Ackerman and colleagues (2005). However, Klaus Oberauer and colleagues (2005) came to a different conclusion after their analysis of the same data set based on a different method of factor analysis. Instead of fixing the factor loadings of the measures using the values obtained on a previous model, Oberauer and colleagues fitted the model without any restriction. The resulting model delivered a remarkably higher correlation between *g* and working memory (0.85 compared to 0.50, as shown in Figure 4.6).

It is not surprising to find that the constraining features of working memory may

differ in different populations. For instance, as noted, studies in the ubiquitous university student population show relatively small relationships between processing speed and intelligence test performance. If we extend the studies over the full adult range, processing speed becomes more prominent. Finally, confusion may result because studies that try to isolate the most important aspect of the working memory–intelligence relationship may be trying to carry the analysis to a finer level of detail than is possible given the complexity of both variables (Box 4.5).

Roberto Colom and colleagues (2016) reviewed the evidence regarding the general factor of intelligence and the components relevant for the working memory concept. Their main finding was that inhibition, interference resolution, and attention control are not related to fluid intelligence once short-term memory storage is controlled. The implication is that complex (dual) span tasks are not stronger predictors of intelligence than simple span tasks. The crucial

Box 4.5 Brunswickian Symmetry

When we compare two sets of behaviors, it is important to maintain the same level of complexity for each behavior. The German psychologist Werner Wittmann referred to this balance as *Brunswickian symmetry*, based on ideas from Egon Brunswick, a Hungarian American psychologist. The symmetry argument can be applied to studies of the relationships between working memory and intelligence.

The terms *intelligence* and *working memory* both refer to systems of psychological functions, rather than individual functions. Consider the case of possibly the best single measure of general intelligence, the progressive matrix test (Figures 4.7 and 3.2, bottom).

What does it take to solve this sort of problem? You need to pick apart the individual elements of the figures, recognizing that each figure has certain attributes, rather than reacting to the figure as a whole. You have to be able to recognize that each attribute of a component of the figure appears once in each row and that each attribute of another component appears once in each column. Because visual attention is limited, you must be able to hold in memory information about recognizing an item in a row as you look down the rows and across the columns. Then you have to infer the missing pattern. Finally, you have to hold the inferred pattern in short-term memory as you search the alternative answers provided, until you find one that matches the pattern. There is no one sufficient working memory function. All of them are required, albeit to different degrees.

Working memory and intelligence are both concepts that refer to the ability to coordinate a system of elementary functions into a coherent whole. There are relations at the system level, but trying to establish relations between the elementary functions, below the level of overall system functioning, is not likely to be useful. The quest for a single information-processing function that explains intelligence is no more likely to succeed than the quest for the Holy Grail (Hunt, 1987), but it cannot yet be ruled out entirely (Chuderski, 2019).

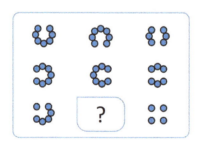

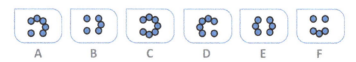

Figure 4.7 A sample progressive matrix problem. The task is to decide which of the figures A–F replaces the blank item (?) in the 3 × 3 matrix. The answer is in footnote 1.
Courtesy of TEA, S. A. (Madrid)

[1] Answer is D.

feature may be the temporary and reliable storage of the pertinent information. You cannot use cognitive processing on information that is no longer there (because it is no longer stored in the short-term memory) or that is low-quality (unreliable). Their key conclusion was that executive control alone is almost irrelevant for explaining individual differences in intelligence using working memory capacity components. Instead, the strong overlap between intelligence and working memory may be explained by a common capacity to build and keep relevant connections in the short-term memory space (Chuderski, 2019). Evidence from lesion patients also indicates that "reasoning is dependent on the ability to first hold task-relevant information in working memory" (Jolly et al., 2020, p. 1158).

4.5 Verbal Comprehension and Intelligence

Verbal comprehension is the process of understanding written or spoken communication. Comprehension is somewhat different from production, although the abilities to understand and to generate linguistic messages are closely related. Tests of verbal comprehension are included in almost all intelligence test batteries. For instance, people who take the SAT or the language test from the PISA survey must show that they can extract meaning from passages that are roughly the length of a newspaper editorial and deal with serious topics. The 2018 PISA report revealed the worrisome finding that nine out ten high school students (fifteen years old) in the Organisation for Economic Co-operation and Development (OECD) countries cannot distinguish facts and opinions implicit in a written document (Schleicher, 2018).

Over the general range of ability, intelligence and understanding language are closely intertwined but not perfectly correlated. In cases from the extreme tails of the normal IQ distribution, the processes can become distinct. Poets are not necessarily good at mathematical reasoning, nor are mathematicians necessarily literary virtuosos. However, it is important to keep such differences in perspective. People who have an elegant command of the language often do not display incompetence in logical and mathematical reasoning, nor are logicians and mathematicians likely to be inarticulate. At the other end of the intelligence scale, low general intelligence is typically associated with a pervasive low level of verbal comprehension. However, dissociations sometimes occur between general reasoning and aspects of verbal skill. An example is Williams syndrome.

Williams syndrome is a genetically determined type of mild to moderate mental disability. Patients with this syndrome display language abilities that are considerably higher than their general reasoning abilities. At one time it was thought that this was evidence for a dissociation between linguistic skills and general cognition, but further research has shown that that is not accurate. Williams syndrome patients show surprising verbal skill compared to other people suffering from the same level of general mental disability. However, their use of language lags behind that of age-matched children with average range-mental abilities. One of the most interesting features is their tendency to react to the literal meaning of statements, thus failing to grasp the meaning of metaphorical statements or irony (Mervis et al., 2003). A Williams syndrome patient might profoundly misinterpret statements like *He's full of baloney* or *I suppose he'll relax by going skydiving*.

So just what is language comprehension? Walter Kintsch (1998) has developed a classic useful framework for understanding the comprehension process. He distinguished three stages in comprehension. First, there are the purely linguistic processes of retrieving word meaning (*lexical retrieval*) and applying syntactic rules to extract meaning from phrases and sentences. These tasks use low-level comprehension. Low-level comprehension is followed by the second stage of high-level comprehension, in which sentence and phrase meanings are incorporated into a text model, essentially an understanding of what a text says. As the text model is built, some information is dropped out, and other information is highlighted. The text model is then incorporated into a situation model, the third stage, which represents the meaning of the text in light of other information that the comprehender has and sees as relevant.

Properly, Kintsch presents text comprehension as a highly interactive process. Sentence comprehension, building the text model, and building the situation model do not take place sequentially. There is simultaneous processing of each stage and many feedback loops. Consider the following statement by Senator John McCain about Senator Hillary Clinton, at a time when both were candidates for their respective party's presidential nominations (October 29, 2007, at a Republican debate in Orlando, Florida):

> In case you missed it, a few days ago, Senator Clinton tried to spend $1 million on the Woodstock Concert Museum. Now, my friends, I wasn't there. I'm sure it was a cultural and pharmaceutical event. ... I was tied up at the time.

The following propositions make up one possible text model:

> Senator Clinton tried to spend a million dollars.
> The money was for the Woodstock Concert Museum.
> McCain was not at the Woodstock concert.
> McCain believes the concert was a cultural and pharmaceutical event.
> McCain was occupied with other matters at the time.

Certain statements in the text are unimportant – for example, the qualifier "in case you missed it." They drop out of the text model. Strictly speaking, the statement "I wasn't there" is ambiguous; it could refer either to the concert or to the time when Senator Clinton tried to acquire money for the museum. The situation model puts this in

the context of unstated but (we hope) widely known information about the Woodstock concert, and about McCain himself. Of course, such a model depends on the comprehender's knowledge. Here is what the situation model would be (information not specifically stated in the text base is shown in parentheses):

> Clinton wants to appropriate $1 million for a museum commemorating the Woodstock concert (government money, not her own).
> The Woodstock concert was a cultural event (it was the icon of the radical youth culture of the late 1960s and early 1970s).
> There was a great deal of (illegal) drug use at the concert.
> At the time of the concert, McCain was not present (he was a prisoner of war in Vietnam, where he was badly treated – he was literally tied up).

An even wider situation model assumes that McCain was contrasting his patriotic record during the Vietnam War to Clinton's attempt to memorialize an event that many in his Republican audience considered an example of moral decay during the hippy era. Other readers might develop other models. In each case, the situation model will depend heavily on knowledge that the comprehender brings to bear. This depends on both what the comprehender knows and what is seen as relevant to understanding the current statement.

Our interest is in the role of individual differences in information processing during both the low-level stage of understanding what the text literally says and the high-level stage of understanding what it means. Let us take each in turn.

4.5.1 Low-Level Linguistic Skills

With few exceptions, every human being acquires a complex set of rules for speaking and listening up to the level of analysis of sentences and phrases in a stunning display of tacit learning. But we are not all equally adept linguists. There are differences among individuals in their ability to understand language.

People know different words, and a score on a brief vocabulary test can be a guide to a person's general intelligence (Carroll, 1993). For instance, the vocabulary test included in the WAIS has a loading of 0.80 on the general factor (Gignac, 2006). Having a small vocabulary does not necessarily reflect a deficiency in information processing, for differences in vocabulary may simply reflect differences in a person's social and educational environments. Word knowledge itself is a component of intelligence; your vocabulary contributes, in part, to how well you function in society. Just knowing words, however, is not an information-processing component of intelligence.

People also differ in the speed with which they can retrieve well-known words, which is an information-processing ability. People who have high scores on tests of verbal comprehension recognize common words more rapidly than people with low test scores (Palmer et al., 1985). They also recognize well-known semantic relationships, answering questions like "Is a deer an animal?" more rapidly than their low-scoring counterparts (Goldberg et al., 1977). This appears to be partly a general processing speed effect and partly a more restricted verbal skill.

There is one aspect of word recognition that is not related to verbal comprehension test scores. In general, people are quicker to recognize a word when it is presented shortly after a related word. As noted earlier, the word NURSE will be recognized more quickly if a person is first asked to recognize a related word, like DOCTOR, than if an unrelated word, like BUTTER, is presented first. This phenomenon is called *semantic priming*. While priming is useful in verbal comprehension, there are only small individual differences in priming, and they are not related to individual differences in verbal comprehension (Palmer et al., 1985).

Moving from word recognition to sentence recognition, we find both effects of accuracy of lexical retrieval and also effects of working memory. Marcel Just and Patricia Carpenter (1992) measured working memory using a reading span task (Macdonald et al., 1992). They determined the time that the participants took to analyze complex sentences and how accurately they were able to analyze them. Here are two illustrations from their work. Consider the following two sentences:

> The evidence examined by the lawyer shocked the jury.
> The defendant examined by the lawyer shocked the jury.

In the first sentence, "examined" must introduce a relative clause modifying "evidence," because "evidence" is inanimate and cannot examine anything. In the second case, "examined" might refer to an activity by the defendant, or it might introduce a relative clause. The ambiguity cannot be resolved until the phrase "by the lawyer" is encountered. At an absolute level, people with higher working memory capacity read both sentences faster than people with lower working memory capacity. However, people with higher capacity read the first sentence more rapidly than the second; apparently, they take advantage of the ambiguity resolution afforded by the inanimate quality of "evidence." People with lower memory spans read both sentences at about the same speed.

Now consider this kind of processing for the following two sentences that we discussed briefly in Chapter 2:

> The experienced soldiers warned about the raid before the midnight attack.
> The experienced soldiers warned about the raid conducted the midnight attack.

The phrase "The experienced soldiers warned about the raid" is ambiguous; it could refer to a warning either given by or given to the soldiers. The ambiguity is resolved when the word "before" or "conducted" is read. Readers with higher working memory capacities slowed down when they encountered the ambiguity-resolving word. People with lower working memory capacities did not. When asked questions about the interpretation of the second question, such as "Did anyone warn the soldiers?" people with higher spans were more likely to answer the question correctly.

Just and Carpenter interpreted these results as showing that the high-span (and, by inference, high-working-memory-capacity) readers carried forward both meanings of the ambiguous phrase until the ambiguity was resolved, while the low-span readers carried forward only the preferred alternative, that the soldiers did the warning.

Just and Carpenter also considered the effects of an extrinsic task on sentence comprehension. In the memory span task, a person reads a sentence, stores the last word, then reads another sentence, stores the last word, and so forth. Memory span is assessed by the number of words that can be recalled in the proper sequence as more and more sentences are read. Just and Carpenter reversed this procedure. First, they established people's memory spans using a fixed set of sentences. In a separate procedure, they presented material to be retained in memory and then presented sentences of varying complexity and determined whether the readers understood them. This gave them a measure of sentence-processing capability in the presence of a concurrent short-term memory load. Understanding decreased as sentence complexity increased, but people with higher spans were less bothered by the concurrent memory load than were people with lower spans.

Just and Carpenter concluded that working memory should be considered a capacity for processing, rather than a reflection of either accurate lexical retrieval or short-term storage capacity. However, working memory is a system of storage, retrieval, processing, and attention control functions. The important thing is how well these functions work together to create an integrated system for managing relevant information. Integration is key.

When dealing with verbal comprehension, we regard working memory as a component of verbal ability. When we deal with reasoning more generally, we talk about the effect of working memory on *g*. In either case, individual differences in the effectiveness of the working memory system are central to individual differences in higher-order cognition.

4.5.2 Higher-Order Comprehension

It is difficult to distinguish higher-order verbal comprehension from intelligence. Some of the most complicated items on cognitive tests are verbal comprehension items. Doing well on the PISA tests, for example, requires comprehension of increasingly complex statements. It is informative to present an analysis of how texts are comprehended, continuing with the use of Kintsch's model. The analysis will indicate the points at which information-processing constraints become important.

The following passage appeared in the *Yale Alumni Magazine*:

> Whatever fails to accord with the values of political liberalism fits uncomfortably within the range of possibilities that the prevailing conception of diversity permits students to acknowledge as serious contenders in the search for an answer to the first-person question of what living is for. (Kronman, 2007, p. 26)

With a substantial cognitive effort, this forty-five-word sentence makes sense. You just have to take it stage by stage.

Whatever fails to accord with political liberalism → nonliberal ideas

Parse phrase, refer to long-term memory for semantic references, and store in memory.

fits uncomfortably within the range → is outside of, not permitted by

Parse phrase, refer to long-term memory for semantic references, store result in memory. Working memory now holds *nonliberal ideas not permitted by.*

of possibilities that the prevailing conception of diversity permits → ideas permitted by the current concept of diversity

Parse phrase, refer to long-term memory, identify *possibilities* with *ideas*. Working memory now holds *nonliberal ideas not permitted by the current concept of diversity.*

Next, we encounter *students*. This changes the presumed grammatical structure of the sentence. It turns out that "concept of diversity" is the subject of the sentence, not the object of the preposition *by*. We have an example of updating, one of the functions of working memory. Appropriate rearrangement of the contents of working memory produces

The current concept of diversity does not permit students

and *nonliberal ideas* is free floating until the next word occurs:

to acknowledge

This anchors the term *nonliberal ideas.* Working memory now holds

The current concept of diversity does not permit students to acknowledge nonliberal ideas

We next encounter a mouthful (memory-ful?) of relative clauses:

as serious contenders in the search for an answer to the first-person question of what living is for

These require similar parsing and analysis. The result is fairly simple:

Current concepts of diversity do not let students consider nonliberal views.

We have gone from forty-five words in the statement to eleven in the text base. We can move to the situational model by recalling that Kronman was writing in 2007 and that

he was addressing Yale alumni. A completed part of the situation model is

There has been concern over academic biases against conservative thought.
Professor Kronman is trying to explain the current situation to the Yale alumni.
Kronman says that views in favor of cultural diversity prevail on campus.
These views do not permit students to consider competing ideas.

Perhaps Kronman could have written more simply (but it was for Yale).

The purpose here is not to edit his writing. It is to show that comprehension of language, a very important part of human intelligence, requires a complex interplay between the working memory and long-term memory systems. Long-term memory does not just act as a provider of information and syntactic rules. It also holds information needed to understand what the text means in the context in which it is stated. Neither McCain's nor Kronman's statements can be understood unless the comprehender knows, and sees as relevant, a great deal of cultural information about US political and social issues in the early twenty-first century.

The ability to bring such information to bear is an important part of intelligence. It has to be evaluated if one wants to claim that a test measures verbal intelligence in any meaningful way. This poses something of a problem to the test maker. There may be situations in which the test maker does not want to evaluate possession of knowledge, because possession of knowledge depends on a particular cultural background. PISA tests are crystal-clear examples because of the requirement of testing heterogeneous populations across the globe. At the same time, the test maker does want to evaluate the ability to use knowledge to build a situation model.

What to do? The answer is to be clever. Here is an example of evaluating an elementary school student's ability to build a situation model. It was a test item in an Australian examination intended to assess a student's ability to understand irony.

Text: *Lovely mosquito, sitting on my arm stay right where you are, I mean you no harm.*
Still as a statue, stand right where you're at. I only mean to give you a pat.
Question: *Does the writer like the mosquito?*

Computer programs for automatic language comprehension would have a hard time answering this question. A reasonably intelligent fourth grader replied with an evil grin: *He wants to kill it.*

We now turn to other relevant factors of intelligence.

4.6 Visuospatial Ability and Intelligence

Visuospatial ability is included in psychometric models as the Gv (visual intelligence) second-order factor in the CHC model and by the perceptual and rotational dimensions in the *g*-VPR model (Carroll, 1993; Johnson and Bouchard, 2005; Schneider and McGrew, 2019). The visual term is justified because the ability refers to detecting, recognizing, and analyzing objects in the visual field, and, in visual imagery, it is the ability to manipulate objects mentally. The two are closely related (Burton and Fogarty, 2003; Kosslyn, 1980; Poltrock and Brown, 1984). The spatial term refers to the ability to reason about real or imagined movement in space, including one's own position and movement relative to other objects (orientation). Orientation ability typically is not measured in the major intelligence test batteries, but it is evaluated in some personnel-selection situations.

A classic study by two Australian researchers, Lorell Burton and Gerald Fogarty (2003), provides a good idea of the nature of visuospatial ability. Previous research on the CHC model had identified the following five specialized factors for visuospatial abilities:

1. *Closure of forms (CF)*: The ability to detect a hidden form in a larger display, particularly if the detection requires that the examinee break down a figure into component parts. As an example, count the number of vertical lines in this sentence. To do so, you have to detect features within familiar letters.

2. *Speed of closure (CS)*: The speed with which you can recognize easily detectable forms, such as a triangle or circle. This is closely related to the more general concept of processing speed.

3. *Speed of rotation (SR)*: The speed with which one can recognize a simple figure (a letter) rotated out of its usual orientation.

4. *Visualization (Vz)*: The ability to imagine what a visual figure will look like if its orientation is changed. An example is the paper folding test, in which the examinee is shown a piece of paper with dotted lines on it to indicate a fold. One or more holes are shown punched into the paper. The examinee is to indicate what the paper looks like (where the holes are) if the paper is unfolded.

5. *Memory for shapes (MS)*: The ability to hold a shape in short-term memory for a brief while.

In the terms of the *g*-VPR model, these abilities would be second-stratum factors – more general than specific tasks, but still specific abilities. They would be subsumed by the more general perceptual and rotation factors at the third stratum of the model (Figure 3.7).

Tests of the abilities just listed, and a number of others, were given to more than 200 undergraduate students (Burton and Fogarty, 2003). Table 4.1 shows the resulting correlation matrix. The correlations are substantial and positive, an indication of a single underlying or common visuospatial ability factor.

Similar support for a general visuospatial factor was identified by a genetically informative study by Kali Rimfeld and colleagues (2017). In it, 1,367 twin pairs completed a comprehensive battery of ten visuospatial tests. Findings revealed an underlying common single factor, both at the phenotypic and genotypic levels. Moreover, this general visuospatial factor

Table 4.1 Correlations between different visuospatial abilities

Narrow factor	CF	CS	SR	VZ
CS	0.65			
SR	0.60	0.32		
VZ	0.58	0.65	0.77	
MS	0.52	0.43	0.43	0.46

Note. CF = closure of forms. CS = speed of closure. SR = speed of rotation. VZ = visualization. Data are excerpted from Burton and Fogarty (2003).

was clearly distinguishable from the general factor of intelligence (*g*). Nevertheless, this research failed to include relevant everyday life measures of navigation abilities, such as wayfinding or map-reading (see the next section).

4.6.1 *Imagery*

Visuospatial tasks require examinees to think about something they can see. Studies of imagery require thinking about things that are imagined. For instance, you could ask a person to imagine a letter *E* in its normal form and then ask to describe how it would look if it were to be rotated ninety degrees clockwise (Kosslyn, 1980). If operations on perceived and imagined figures use the same mental operations, individual differences in visuospatial ability should be related to individual differences in imagery.

Burton and Fogarty (2003) included a number of objectively measured tasks that required mental imagery in their test battery. Performance on these tasks reduced to two factors: image quality (the accuracy of information contained in a person's image) and imaging speed (how fast a person can construct an image).

Table 4.2 shows the loadings of both perceptual and imagery scores on a single visuospatial reasoning factor. Imagery and visuospatial reasoning are closely related. Interestingly, people's reports of their

imagery have a lower relation to the visuospatial factor than do the behavioral measures of imagery. This is noteworthy because many studies of imagery have relied on self-report, which may not be accurate.

4.6.2 *Spatial Orientation*

Spatial orientation is the ability to develop an internal representation of an exterior space, including one's own position in that space. There are large individual differences in the ability to do so. Some individuals have a keen awareness of their spatial orientation, both with respect to objects in their immediate environment and with respect to larger spatial layouts. Think about the importance of this ability for basketball players and other sports participants. Who is behind you on your right? Can you draw the layout of the arena? Other people have a great deal of trouble answering such questions. Poor spatial orientation does not mean that people go around getting lost. Instead, they often can rely on memories for routes between key locations.

A number of laboratory tasks have been designed to evaluate different aspects of orientation ability. This research generally has been conducted outside of research on intelligence, even though maintaining orientation can reasonably be considered part of intelligence. The reason for this situation is understandable. To evaluate someone's orientation ability, you have to assess how well they can explore an unfamiliar environment. This is an expensive, time-consuming process that does not fit into the conventional testing paradigm. Virtual environment technologies provide a way of getting around the logistical problems inherent in evaluating spatial orientation (Waller et al., 2004). Using this technology, the examinee moves through computer-generated artificial worlds to solve orientation problems. Ability to do so differs among people, and these differences are related to both conventional tests of visuospatial ability and skill at orientation in real-world environments (Waller et al., 2001; Hegarty and Waller, 2005).

Table 4.2 Loadings of visuospatial and imagery first-order factors on a general visuospatial reasoning factor

	CF	CS	SR	VZ	Image quality	Image speed	Image self-report
Loading	0.74	0.69	0.80	0.87	0.83	0.59	0.46

Note. CF = closure of forms. CS = speed of closure. SR = speed of rotation. VZ = visualization. Data are excerpted from Burton and Fogarty (2003).

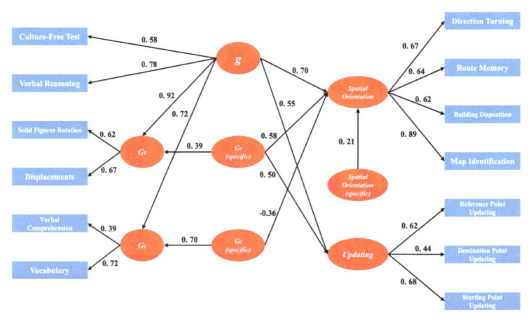

Figure 4.8 Wayfinding and intelligence. The main finding is that general intelligence (*g*) predicts spatial orientation (0.70) and updating one's position in space (0.55). A visuospatial general factor independent of *g* (Gv-specific) also contributed to the predictions (0.58 and 0.50, respectively). The complex model depicts all the measures (rectangles) that defined the latent factors (*g*; Gv, general visuospatial ability; Gc, crystallized intelligence; spatial orientation; and updating). Components of Gv were also considered (Gv-specific, Gc-specific, and spatial orientation). Numbers in the figure represent the loadings from structural equation modeling (SEM) analyses of the intelligence latent factors (*g*, Gv, and Gc) and the spatial orientation latent factors (spatial orientation and updating).
From Juan-Espinosa et al. (2000).

In a comprehensive study (originally designed by Earl Hunt), Manuel Juan-Espinosa and colleagues (2000) tested 111 psychology undergraduates with a battery of tests and tasks tapping the general factor of intelligence (*g*), visuospatial ability (Gv), crystallized ability (Gc), and several orientation processes for wayfinding (updating one's position, route representation, and survey representation). Wayfinding was measured by seven tasks (two were field tasks and five were computerized). Figure 4.8 depicts the main complex result.

Wayfinding processes represent cognitive action for solving orientation problems. This comprehensive study concluded that *g* predicts wayfinding to a substantial degree. In a separate analysis (not shown in Figure 4.8), the authors computed an exploratory hierarchical factor analysis including all the variables measured. The factor loadings on the resulting general factor were all substantial: culture fair intelligence test (0.53), verbal comprehension (0.36), vocabulary (0.56), verbal reasoning (0.71), displacements (0.69), solid figures rotation (0.61), reference point updating (0.57), starting point updating (0.62), destination point updating (0.53), directional turning (0.44), route memory (0.61), map identification (0.73), and building disposition (0.51). This analysis reinforced the key conclusion that spatial orientation is strongly related to the general factor of intelligence (*g*). Remember that this research studied college undergraduates and therefore these factor loadings might be distorted to some degree.

4.6.3 Visuospatial Ability and Information Processing

How do information-processing models enhance our understanding of visuospatial ability? The answer is very little, because they are incorporated into several of the psychometric tests used to identify visuospatial processing. This is seen by contrasting two workhorse tasks used to define verbal intelligence, vocabulary tests and paragraph comprehension tests, to two workhorse tasks used to define visuospatial ability, closure and rotation tasks.

Closure tasks require the detection of lines, assignment of boundaries to objects, detection of surfaces, and development of representations of objects from representations of surfaces. There is something of an analogy to building a text model or even a situation model. Accurate perception requires that you (1) have some clue about what you might see and (2) are given a reference point. In a rotation task, a person is shown a relatively simple picture and asked what it would look like if components were moved to different orientations or positions. Figure 4.9 presents an example.

The two verbal reasoning tasks are complex. Recognizing the meaning of a string of letters requires identifying the string as a word, retrieving its meanings, and determining the correct meaning in context. The

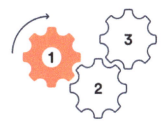

Which gears are rotating clockwise?

Figure 4.9 A rotation problem. If gear 1 is moved in the direction indicated by the arrow, which way will gear 3 move?

selection process is not trivial. In English, which can be a particularly ambiguous language, the typical word has 2.5 meanings (Hunt and Agnoli, 1991). Paragraph comprehension requires recognition of words, retrieval and selection of word meanings, plus analysis of sentences, and then the construction of text and situation models. Neither psychometric task rises to the level of detail in examining verbal processing that do the information-processing measures.

There are substantial individual differences in the ability to perform tasks like this (Hegarty et al., 1988). Rotation tasks appeared on psychometric tests of visuospatial orientation before they were studied in experimental laboratories. The only difference between the psychometric and information-processing techniques is that the psychometric tests assess how many rotation problems a person can solve in a fixed period of time, while the laboratory procedures determine how long a person takes to solve individual rotation problems.

Visuospatial factors generally have high loadings on *g* (Figure 4.8) (Juan-Espinosa et al., 2000). From an information-processing view, this is not surprising. Visuospatial tasks often require the development and manipulation of internal representations of visuospatial information. Baddeley's model of working memory contained a visuospatial component, and subsequent research shows that measures of this component correlate with measures of verbal working memory (Ackerman et al., 2002; Colom et al., 2006b).

As was the case in discussing the working memory relationship to *g*, trying to pick apart visuospatial working memory into separate tasks does not reveal any single, sufficient process that explains visuospatial ability. The visuospatial working memory system is just that – a system. Reasoning about percepts, images, and space is an emergent property produced by the integration of storage, attention to, and processing of percepts and images (see Box 4.6).

Box 4.6 Fluid Intelligence, Processing Speed, Working Memory Capacity, and the Ability to Build Complex Cognitive Structures from Simple Parts

The Polish psychologist Adam Chuderski published a thought-provoking study showing that the latent relationship between fluid intelligence and working memory is progressively reduced when time restrictions for completing the fluid tests are removed (Chuderski, 2013).

The correlation was perfect ($r = 1$) when the fluid tests were administered under strict time restrictions. However, when these restrictions were eliminated (untimed administration), the correlation with working memory decreased to $r = 0.62$. Low-working memory individuals are thought to compensate their capacity limitations in unspeeded conditions. If this argument is correct, then low–working memory individuals must show fluid scores more similar to high–working memory individuals with increased administration times.

Roberto Colom and colleagues tested this later prediction. The Raven's Advanced Progressive Matrices (RAPM) was given to three independent groups using three administration times: forty, thirty, and twenty minutes (Colom et al., 2015). In addition, all participants completed a battery of six verbal, numerical, and visuospatial working memory tasks, and they were classified according to their general working memory capacity. The key finding was that individuals with higher working memory capacity and the shortest time for solving the RAPM (twenty minutes) have higher intelligence scores than individuals with lower working memory capacity and the longest time for solving the RAPM (forty minutes). Therefore, time alone cannot explain their performance differences. But, as usual, there is more to the story.

John Duncan and colleagues (2017) proposed that the critical function in fluid reasoning is splitting a complex whole into simple, separately attended parts (a process they called *segmentation*). They argued that fluid reasoning differences are not fully explained by working memory capacity and processing speed. They manipulated a set of problems for minimizing load on integration, working memory, and processing speed. However, this experiment found that performance was poor in participants with lower fluid intelligence scores. This indicated that the role of working memory capacity and processing speed could not be minimized. Nonetheless, Duncan et al. maintained, "the universal requirement for building a complex whole from focused parts may be at least one major explanation for the finding of universal positive correlations between fluid intelligence and even simple tasks. As tasks become more complex, however, it is increasingly challenging to separate them into clearly focused parts" (p. 5298). This fact may help explain why having more time for completing the RAPM is insufficient for achieving higher scores if you have lower working memory capacity.

4.7 Process Overlap Theory: Connecting Psychometric, Cognitive, and Biological Models of Intelligence

Kristof Kovacs and Andrew Conway (2016) proposed the process overlap theory (POT) in a brave attempt to combine psychometric, cognitive, and biological models of intelligence. According to this model, the positive manifold emerges from a battery of tasks tapping domain-general executive processes. The prefrontal and parietal cortices are considered key brain regions for supporting these executive processes, which show greater overlap with domain-specific processes in mental test performance. Indeed, individual differences in executive processes might be a bottleneck for application of specific cognitive processes. Fluid intelligence and working memory may be strongly correlated because of their overlapping executive requirements. The overlap of executive and specific processes may account for the positive manifold, but also for

the hierarchical structure of cognitive abilities. This perspective is not new (see (Bartholomew et al., 2013; Thomson, 1916; Detterman, 1994, 2000). The key idea in the model is that specific cognitive abilities are related because they tap overlapping psychological processes. In other words, the general factor of intelligence (*g*) is the common consequence and not the common cause of the correlations among specific abilities. Figure 4.10 shows an illustration of how this might work.

Figure 4.10 depicts a verbal (Gc), a visuospatial (Gv), and a fluid (Gf) ability factor. Verbal (v), visuospatial (s), and fluid executive (black dots) processes are also shown. Process overlap is not the exclusive source of the positive correlations usually observed in research. Cognitive complexity refers to the extent to which a given problem taps executive control processes and this might explain why working memory is strongly related to general intelligence (*g*), at least when latent modeling is computed.

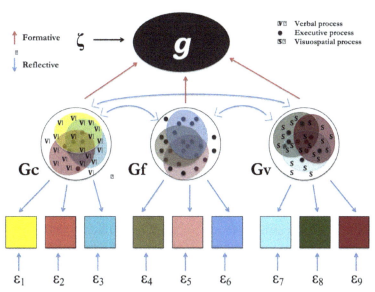

Figure 4.10 Illustration of process overlap theory as a latent variable model where (*g*) is the common consequence (arrows point upward), not the common cause of the correlations among specific abilities, in this example, crystallized intelligence (Gc), fluid intelligence (Gf), and visuospatial intelligence (Gv). This is why *g* is considered a "formative" factor. On the contrary, Gc, Gf, and Gv are considered reflective factors (arrows point downward). V = verbal processes. S = spatial processes. Black dots represent executive processes. The Greek letters represent unexplained or error variance of no interest for understanding the essence of the model (Kovacs and Conway, 2016).

Furthermore, the model predicts the processes determining whether individual differences are domain-general and unrelated to the cognitive domain. Abstract cognitive problems and problems based on numbers, for instance, show substantial correlations because of the overlap between the processes tapped by both types of problems (inductive reasoning and inhibition). As noted by Kovacs and Conway (2016, p. 166), this is a nice example of the principle of the indifference of the indicator (Spearman, 1904) described in Chapter 3: "categorization of tasks according to the kind of material, by domain or content, is not necessarily instrumental in understanding the determinants of individual differences."

Of course, the story is not that simple or straightforward. In one study, for example, basic executive processing components (shifting, inhibition, and updating) showed correlations among themselves (from 0.42 to 0.63) (Miyake et al., 2000). However, a replication study published twenty years later failed to identify executive control (defined by inhibition tasks) as a latent variable (Rey-Mermet et al., 2019). Furthermore, the executive control measures were not related to fluid intelligence or working memory capacity. On the other hand, Naomi Friedman and colleagues (2006) have shown that only executive updating shows substantial correlations with intelligence (both fluid and crystallized, approx. $r = 0.75$). Similarly, Kenia Martínez and colleagues (2011) reported that executive updating and working memory show a near-perfect correlation with intelligence.

The latter study is especially relevant within this context because one simple short-term memory span latent factor showed near-perfect correlations with both working memory and executive updating latent factors. Furthermore, these three latent factors showed near-perfect correlations with fluid reasoning ability. The correlation between these latent factors and executive inhibition/mental processing speed was substantially smaller.

The main implication derived from these findings is this: there are cognitive systems more prone to coping successfully with several diverse challenges. Concurrent processing requirements leave less capacity for temporary storage of information, which diminishes the reliability of the stored information, which in turn is responsible for the behavioral effects observed in memory span tasks. Shared capacity limitations might derive from the number of items that can be kept active in the short term or the number of relationships between elements that can be kept active during the reasoning process necessary for solving problems like those on standard intelligence tests (Halford et al., 2007). The limitations shared by intelligence and memory span factors might be based on the ability to build and keep relevant and reliable connections in the short term (Chuderski, 2019).

Regarding the brain, Kovacs and Conway (2016) argue that so far, the search for a neuro-g has failed (Haier et al., 2009). We will discuss this further in the next chapter, but here we note a study by Román and colleagues (2014). They looked at brain correlates of latent variables at different levels of the intelligence hierarchy (as described in Chapter 3). Their key finding was that as one moves upward in the hierarchy (from tests to specific cognitive abilities to g), there are fewer brain correlates, and they are increasingly concentrated in the frontal lobes. They note, "Factors capturing the variance common to both specific measures and group factors partial out the specificity present at the measurement [test] level. Removing specific variance reveals that frontal regions in the brain are crucial for supporting human intelligence" (p. 3816). This is exactly what the process overlap theory predicts: "the results of Román et al. can be interpreted as the neural equivalent of the psychological explanation proposed by the process overlap theory" (Kovacs and Conway, 2016, p. 168).

The POT proposed by Kovacs and Conway (2016) received comments from researchers of varied psychological disciplines (Colom et al., 2016; Oberauer, 2016; Haier and Jung, 2016; Gottfredson, 2016; Sternberg, 2016; Deary et al., 2016;

Ackerman, 2016; Cowan, 2016; Kan et al., 2016; Kaufman, 2016; Detterman et al., 2016; Flynn, 2016). Their evaluations ranged from very positive to highly negative, but the important thing to underscore is this: "the process overlap theory is a thoughtful consideration of current *g*-related issues and a road map for neuroimaging studies that might succeed in testing the respective validities of competing psychometric models" (Haier and Jung, 2016, p. 218). Independent research groups are testing POT assumptions, but there is not yet a weight of evidence (Wang et al., 2021).

Conceptual models are essential for making sense of the vast amounts of newly available data and the novel findings from neuroscience and molecular genetic approaches to intelligence research (Haier, 2017; Colom, 2016). We explore and discuss developments based on neuroimaging and genetic/DNA research in the next two chapters.

4.8 Summary

Individual differences in information processing are related to individual differences in intelligence. However, the two are not identical. Information-processing capacity constrains intelligence and intelligence, conceived as a set of neurocognitive factors tapped by standardized tests, might translate into the dependent variables considered by measures of elementary cognitive processes (Chuderski, 2022). The important question to ask about the information-processing contribution to intelligence is not whether this or that isolated process is important; rather, it is whether and how the whole system is functioning. Remember: integration is key.

Human thought results from the process of building internal representations of external problems. This means that there must be a way to store the representation of percepts and images. There must also be processes for fetching information from storage and using it. There must also be a way of prioritizing the processing of crucial information. Everything has to work together, in an integrated way. And, finally, all this happens quickly, but more quickly in some people than in others.

The nature of information-processing constraints varies between different groups. For example, there are age effects in studies of the role of attention, short-term memory storage, and information-processing measures in general. Attentional control is weak in young children and the elderly. Processing speed decreases markedly with age, even during an adult's working lifetime (see Chapter 11). Information-processing constraints appear to be stronger in people of lower intelligence than in those with average or higher intelligence.

Most cognitive psychology research is conducted with samples of college students. Generalizing results from these studies (with restricted ranges) has limitations. Future research on the relation between intelligence and information processing should ask how information-processing capacities constrain intelligence in different populations, as well as how information processing constrains thought in people in general.

Cognitive models like the process overlap theory integrate large amounts of data and inconsistent findings. Such attempts help drive scientific knowledge forward.

4.9 Questions for Discussion

4.1 How are cognitive and biological models of intelligence related?

4.2 Why is reaction time useful for studying intelligence?

4.3 What do elemental cognitive tasks have to do with intelligence research?

4.4 How does working memory relate to intelligence?

4.5 Can you describe the core concept in the process overlap theory?

References

Ackerman, P. L. 2016. Process overlap and *g* do not adequately account for a general factor of intelligence. *Psychological Inquiry*, 27, 178–180.

Ackerman, P. L., Beier, M. E., & Boyle, M. O. 2002. Individual differences in working memory within a nomological network of cognitive and perceptual speed abilities. *Journal of Experimental Psychology: General*, 131, 567–589.

Ackerman, P. L., Beier, M. E., & Boyle, M. O. 2005. Working memory and intelligence: The same or different constructs? *Psychological Bulletin*, 131, 30–60.

Anderson, N. H. 1996. Cognitive algebra versus representativeness heuristic. *Behavioral and Brain Sciences*, 19, 17.

Baddeley, A. D. 1986. *Working memory*. New York: Oxford University Press.

Baddeley, A. D., & Hitch, G. 1974. Working memory. In Bower, G. (ed.), *Recent advances in learning and motivation*. New York: Academic Press.

Bartholomew, D. J., Allerhand, M., & Deary, I. J. 2013. Measuring mental capacity: Thomson's Bonds model and Spearman's *g*-model compared. *Intelligence*, 41, 222–233.

Blanch, A. 2021. *Chess and individual differences*. New York: Cambridge University Press.

Botella, J., Privado, J., Suero, M., Colom, R., & Juola, J. F. 2019. Group analyses can hide heterogeneity effects when searching for a general model: Evidence based on a conflict monitoring task. *Acta Psychologica*, 193, 171–179.

Bowren, M., Adolphs, R., Bruss, J., et al. 2020. Multivariate lesion-behavior mapping of general cognitive ability and its psychometric constituents. *Journal of Neuroscience*, 40, 8924–8937.

Burton, L. J., & Fogarty, G. J. 2003. The factor structure of visual imagery and spatial abilities. *Intelligence*, 31, 289–318.

Carroll, J. B. 1993. *Human cognitive abilities: A survey of factor-analytic studies*. New York: Cambridge University Press.

Chuderski, A. 2013. When are fluid intelligence and working memory isomorphic and when are they not? *Intelligence*, 41, 244–262.

Chuderski, A. 2014. The relational integration task explains fluid reasoning above and beyond other working memory tasks. *Memory and Cognition*, 42, 448–63.

Chuderski, A. 2019. Even a single trivial binding of information is critical for fluid intelligence. *Intelligence*, 77, 101396.

Chuderski, A. 2022. Fluid intelligence emerges from representing relations. *Journal of Intelligence*, 10, 51.

Chuderski, A., Taraday, M., Necka, E., & Smolen, T. 2012. Storage capacity explains fluid intelligence but executive control does not. *Intelligence*, 40, 278–295.

Colom, R. 2016. Advances in intelligence research: What should be expected in the XXI century (questions and answers). *Spanish Journal of Psychology*, 19, 1–8.

Colom, R., Abad, F. J., Quiroga, M. Á., Shih, P. C., & Flores-Mendoza, C. 2008. Working memory and intelligence are highly related constructs, but why? *Intelligence*, 36, 584–606.

Colom, R., Chuderski, A., & Santarnecchi, E. 2016. Bridge over troubled water: Commenting on Kovacs and Conway's process overlap theory. *Psychological Inquiry*, 27, 181–189.

Colom, R., Privado, J., Garcia, L. F., et al. 2015. Fluid intelligence and working memory capacity: Is the time for working on intelligence problems relevant for explaining their large relationship? *Personality and Individual Differences*, 79, 75–80.

Colom, R., Rebollo, I., Abad, F. J., & Shih, P. C. 2006a. Complex span tasks, simple span tasks, and cognitive abilities: A reanalysis of key studies. *Memory and Cognition*, 34, 158–171.

Colom, R., Rebollo, I., Palacios, A., Juan-Espinosa, M., & Kyllonen, P. C. 2004. Working memory is (almost) perfectly predicted by *g*. *Intelligence*, 32, 277–296.

Colom, R., Shih, P. C., Flores-Mendoza, C., & Quiroga, M. Á. 2006b. The real relationship between short-term memory and working memory. *Memory*, 14, 804–813.

Cowan, N. 2001. The magical number 4 in short-term memory: A reconsideration of mental storage capacity. *Behavioral and Brain Sciences*, 24, 87–114; discussion 114–185.

Cowan, N. 2016. Process overlap theory and first principles of intelligence testing. *Psychological Inquiry*, 27, 190–191.

Cowan, N. 2017. The many faces of working memory and short-term storage. *Psychonomic Bulletin and Review*, 24, 1158–1170.

Cronbach, L. J. 1957. The 2 disciplines of scientific psychology. *American Psychologist*, 12, 671–684.

Daneman, M., & Merikle, P. M. 1996. Working memory and language comprehension: A meta-analysis. *Psychonomic Bulletin & Review*, 3, 422–433.

Deary, I. J. 2000. *Looking down on human intelligence: From psychometrics to the brain*. Oxford: Oxford University Press.

Deary, I. J., Cox, S. R., & Ritchie, S. J. 2016. Getting Spearman off the skyhook: One more in a century (since Thomson, 1916) of attempts to vanquish *g*. *Psychological Inquiry*, 27, 192–199.

Deary, I. J., & Der, G. 2005. Reaction time, age, and cognitive ability: Longitudinal findings from age 16 to 63 years in representative population samples. *Aging Neuropsychology and Cognition*, 12, 187–215.

Demetriou, A., Kui, Z. X., Spanoudis, G., et al. 2005. The architecture, dynamics, and development of mental processing: Greek, Chinese, or universal? *Intelligence*, 33, 109–141.

Detterman, D. K. 1994. Theoretical possibilities: The relation of human intelligence to basic cognitive abilities. In Detterman, D. K. (ed.), *Current topics in human intelligence: Vol. 4. Theories of intelligence*. Norwood, NJ: Ablex.

Detterman, D. K. 2000. General intelligence and the definition of phenotypes. In Bock, G. R., Good, J. A., & Webb, K. (eds.), *The nature of intelligence*. New York: John Wiley.

Detterman, D. K., & Daniel, M. H. 1989. Correlations of mental tests with each other and with cognitive variables are highest for low IQ groups. *Intelligence*, 13, 349–359.

Detterman, D. K., Petersen, E., & Frey, M. C. 2016. Process overlap and system theory: A simulation of, comment on, and integration of Kovacs and Conway. *Psychological Inquiry*, 27, 200–204.

Duncan, J., Chylinski, D., Mitchell, D. J., & Bhandari, A. 2017. Complexity and compositionality in fluid intelligence. *Proceedings of the National Academy of Sciences of the United States of America*, 114, 5295–5299.

Engle, R. W., Tuholski, S. W., Laughlin, J. E., & Conway, A. R. 1999. Working memory, short-term memory, and general fluid intelligence: A latent-variable approach. *Journal of Experimental Psychology: General*, 128, 309–331.

Flynn, J. R. 2016. No population is frozen in time: The sociology of intelligence. *Psychological Inquiry*, 27, 205–209.

Friedman, N. P., Miyake, A., Corley, R. P., et al. 2006. Not all executive functions are related to intelligence. *Psychological Science*, 17, 172–179.

Frischkorn, G. T., Wilhelm, O., & Oberauer, K. 2022. Process-oriented intelligence research: A review from the cognitive perspective. *Intelligence*, 94, 101681.

Gignac, G. E. 2006. Evaluating subtest 'g' saturation levels via the single trait-correlated uniqueness (STCU) SEM approach: Evidence in favor of crystallized subtests as the best indicators of 'g.' *Intelligence*, 34, 29–46.

Goldberg, R. A., Schwartz, S., & Stewart, M. 1977. Individual differences in cognitive processes. *Journal of Educational Psychology*, 69, 9–14.

Gottfredson, L. S. 2016. A *g* theorist on why Kovacs and Conway's process overlap theory amplifies, not opposes, *g* theory. *Psychological Inquiry*, 27, 210–217.

Grandy, T. H., Lindenberger, U., & Werkle-Bergner, M. 2017. When group means fail: Can one size fit all? *bioRxiv*, 126490; doi: https://doi.org/10.1101/126490.

Haier, R. J. 2017. *The neuroscience of intelligence*. New York: Cambridge University Press.

Haier, R. J., Colom, R., Schroeder, D. H., et al. 2009. Gray matter and intelligence factors: Is there a neuro-*g*? *Intelligence*, 37, 136–144.

Haier, R. J., & Jung, R. E. 2016. The psychometric brain. *Psychological Inquiry*, 27, 218–219.

Halford, G. S., Cowan, N., & Andrews, G. 2007. Separating cognitive capacity from knowledge: A new hypothesis. *Trends in Cognitive Sciences*, 11, 236–242.

Hegarty, M., Just, M. A., & Morrison, I. R. 1988. Mental models of mechanical systems: Individual-differences in qualitative and quantitative reasoning. *Cognitive Psychology*, 20, 191–236.

Hegarty, M., & Waller, D. A. 2005. Individual differences in spatial abilities. In Miyake, P. S. A. (ed.), *The Cambridge handbook of visuospatial thinking*. New York: Cambridge University Press.

Heitz, R. P., Unsworth, N., & Engle, R. W. 2005. Working memory capacity, attention control, and fluid intelligence. In Engle, O. W. R. W. (ed.), *Handbook of understanding and measuring intelligence*. Thousand Oaks, CA: SAGE.

Hornung, C., Brunner, M., Reuter, R. A. P., & Martin, R. 2011. Children's working memory: Its structure and relationship to fluid intelligence. *Intelligence*, 39, 210–221.

Hunt, E. 1980. Intelligence as an information-processing concept. *British Journal of Psychology*, 71, 449–74.

Hunt, E. 1987. The next word on verbal ability. In Vernon, P. A. (ed.), *Speed of information processing and intelligence*. New York: Ablex.

Hunt, E., & Agnoli, F. 1991. The Whorfian hypothesis – a cognitive-psychology perspective. *Psychological Review*, 98, 377–389.

Hunt, E., Frost, N., & Lunneborg, C. E. 1973. Individual differences in cognition: A new approach to intelligence. In Bower, G. S. (ed.), *Advances in learning and motivation*. New York: Academic Press.

Hunt, E., & Lansman, M. 1986. Unified model of attention and problem-solving. *Psychological Review*, 93, 446–461.

Jensen, A. R. 1998. *The g factor: The science of mental ability*. Westport, CT: Praeger.

Jensen, A. R. 2006. *Clocking the mind: Mental chronometry and individual differences*. New York: Elsevier.

Johnson, W., & Bouchard, T. J., Jr. 2005. The structure of human intelligence: It is verbal, perceptual, and image rotation (VPR), not fluid and crystallized. *Intelligence*, 33, 393–416.

Jolly, A. E., Scott, G. T., Sharp, D. J., & Hampshire, A. H. 2020. Distinct patterns of structural damage underlie working memory and reasoning deficits after traumatic brain injury. *Brain*, 143, 1158–1176.

Juan-Espinosa, M., Abad, F. J., Colom, R., & Fernandez-Truchaud, M. 2000. Individual differences in large-spaces orientation: g and beyond? *Personality and Individual Differences*, 29, 85–98.

Jung, R. E., & Haier, R. J. 2007. The parieto-frontal integration theory (P-FIT) of intelligence: Converging neuroimaging evidence. *Behavioral and Brain Science*, 30, 135–154; discussion 154–187.

Just, M. A., & Carpenter, P. A. 1992. A capacity theory of comprehension: Individual-differences in working memory. *Psychological Review*, 99, 122–149.

Kan, K.-J., van der Maas, H. L. J., & Kievit, R. A. 2016. Process overlap theory: Strengths, limitations, and challenges. *Psychological Inquiry*, 27, 220–228.

Kane, M. J., Hambrick, D. Z., Conway, A. R. A., & Engle, R. W. 2001. The generality of working memory capacity: A latent variable approach. *Journal of Experimental Psychology: General*, 130, 169–183.

Kane, M. J., Hambrick, D. Z., Tuholski, S. W., et al. 2004. The generality of working memory capacity: A latent-variable approach to verbal and visuospatial memory span and reasoning. *Journal of Experimental Psychology: General*, 133, 189–217.

Kaufman, S. B. 2016. Commentary on Kovacs and Conway, process overlap theory: A unified account of the general factor of intelligence. *Psychological Inquiry*, 27, 229–230.

Kintsch, W. 1998. *Comprehension: A paradigm for cognition*. New York: Cambridge University Press.

Kosslyn, S. M. 1980. Mental images. *Recherche*, 11, 156–163.

Kovacs, K., & Conway, A. R. A. 2016. Process overlap theory: A unified account of the general factor of intelligence. *Psychological Inquiry*, 27, 151–177.

Kronman, A. T. 2007. Against political correctness: A liberal's cri de coeur. *Yale Alumni Magazine*.

Kyllonen, P. C., & Christal, R. E. 1990. Reasoning ability is (little more than) working-memory capacity. *Intelligence*, 14, 389–433.

Lerche, V., Von Krause, M., Voss, A., et al. 2020. Diffusion modeling and intelligence: Drift rates show both domain-general and domain-specific relations with intelligence. *Journal of Experimental Psychology: General*, 149, 2207–2249.

Lohman, D. 2000. Complex information processing and intelligence. In Sternberg, R. J. (ed.), *Handbook of intelligence*. New York: Cambridge University Press.

Macdonald, M. C., Just, M. A., & Carpenter, P. A. 1992. Working memory constraints on the processing of syntactic ambiguity. *Cognitive Psychology*, 24, 56–98.

Martínez, K., Burgaleta, M., Román, F. J., et al. 2011. Can fluid intelligence be reduced to "simple" short-term storage? *Intelligence*, 39, 473–480.

Mervis, C. B., Robinson, B. F., Rowe, M. L., Becerra, A. M., & Klein-Tasman, B. R. 2003. Language abilities of individuals with Williams syndrome. *International Review of Research in Mental Retardation*, 27, 35–81.

Meyer, D. E., & Kieras, D. E. 1997a. A computational theory of executive cognitive processes and multiple-task performance. 1. Basic mechanisms. *Psychological Review*, 104, 3–65.

Meyer, D. E., & Kieras, D. E. 1997b. A computational theory of executive cognitive processes and multiple-task performance. 2. Accounts of psychological refractory-period phenomena. *Psychological Review*, 104, 749–791.

Miyake, A., Friedman, N. P., Emerson, M. J., et al. 2000. The unity and diversity of executive functions and their contributions to complex "frontal lobe" tasks: A latent variable analysis. *Cognitive Psychology*, 41, 49–100.

Nettelbeck, T. 2001. Correlation between inspection time and psychometric abilities: A personal interpretation. *Intelligence*, 29, 459–474.

Newell, A. 1990. Better models of the cognitive agent: Some prospects for management and organizational science. *Mathematical Social Sciences*, 20, 309–309.

Nosek, B. A., Hardwicke, T. E., Moshontz, H., et al. 2022. Replicability, robustness, and reproducibility in psychological science. *Annual Review of Psychology*, 73, 719–748.

Oberauer, K. 2003. Selective attention to elements in working memory. *Experimental Psychology*, 50, 257–269.

Oberauer, K. 2016. Parameters, not processes, explain general intelligence. *Psychological Inquiry*, 27, 231–235.

Oberauer, K., Schulze, R., Wilhelm, O., & Suss, H. M. 2005. Working memory and intelligence – their correlation and their relation: Comment on Ackerman, Beier, and Boyle (2005). *Psychological Bulletin*, 131, 61–65.

Palmer, J., Macleod, C. M., Hunt, E., & Davidson, J. E. 1985. Information-processing correlates of reading. *Journal of Memory and Language*, 24, 59–88.

Poltrock, S. E., & Brown, P. 1984. Individual-differences in visual-imagery and spatial ability. *Intelligence*, 8, 93–138.

Putnick, D. L., & Bornstein, M. H. 2016. Measurement invariance conventions and reporting: The state of the art and future directions for psychological research. *Developmental Review*, 41, 71–90.

Rey-Mermet, A., Gade, M., Souza, A. S., Von Bastian, C. C., & Oberauer, K. 2019. Is executive control related to working memory capacity and fluid intelligence? *Journal of Experimental Psychology: General*, 148, 1335–1372.

Rimfeld, K., Shakeshaft, N. G., Malanchini, M., et al. 2017. Phenotypic and genetic evidence for a unifactorial structure of spatial abilities. *Proceedings of the National Academy of Sciences of the United States of America*, 114, 2777–2782.

Román, F. J., Abad, F. J., Escorial, S., et al. 2014. Reversed hierarchy in the brain for general and specific cognitive abilities: A morphometric analysis. *Human Brain Mapping*, 35, 3805–3818.

Salthouse, T. A. 1996. The processing-speed theory of adult age differences in cognition. *Psychological Review*, 103, 403–428.

Schleicher, A. 2018. PISA 2018. www.oecd.org/pisa/ PISA%202018%20Insights%20and% 20Interpretations%20FINAL%20PDF.pdf.

Schneider, W. J., & McGrew, K. S. 2019. Process overlap theory is a milestone achievement among intelligence theories. *Journal of Applied Research in Memory and Cognition*, 8, 273–276.

Schubert, A.-L., & Frischkorn, G. T. 2020. Neurocognitive psychometrics of intelligence: How measurement advancements unveiled the role of mental speed in intelligence differences. *Current Directions in Psychological Science*, 29, 140–146.

Schubert, A.-L., Hagemann, D., & Frischkorn, G. T. 2017. Is general intelligence little more than the speed of higher-order processing? *Journal of Experimental Psychology: General*, 146, 1498–1512.

Shahabi, S. R., Abad, F. J., & Colom, R. 2014. Short-term storage is a stable predictor of fluid intelligence whereas working memory capacity and executive function are not: A comprehensive study with Iranian schoolchildren. *Intelligence*, 44, 134–141.

Spearman, C. 1904. General intelligence objectively determined and measured. *American Journal of Psychology*, 15, 201–293.

Spearman, C. 1923. *The nature of "intelligence" and the principles of cognition*. London: Macmillan.

Sternberg, R. J. 2016. Groundhog Day: Is the field of human intelligence caught in a time warp? A comment on Kovacs and Conway. *Psychological Inquiry*, 27, 236–240.

Thomson, G. H. 1916. A hierarchy without a general factor. *British Journal of Psychology*, 8, 271–281.

Unsworth, N., & Engle, R. W. 2007. On the division of short-term and working memory: An examination of simple and complex span and their relation to higher order abilities. *Psychological Bulletin*, 133, 1038–1066.

Unsworth, N., Fukuda, K., Awh, E., & Vogel, E. K. 2014. Working memory and fluid intelligence: Capacity, attention control, and secondary memory retrieval. *Cognitive Psychology*, 71, 1–26.

Waller, D., Beall, A. C., & Loomis, J. M. 2004. Using virtual environments to assess directional knowledge. *Journal of Environmental Psychology*, 24, 105–116.

Waller, D., Knapp, D., & Hunt, E. 2001. Spatial representations of virtual mazes: the role of visual fidelity and individual differences. *Human Factors*, 43, 147–58.

Wang, T., Li, C., Ren, X., & Schweizer, K. 2021. How executive processes explain the overlap between working memory capacity and fluid intelligence: A test of process overlap theory. *Journal of Intelligence*, 9, 21.

Willoughby, E. A., & Lee, J. J. 2021. Parsing information flow in speeded cognitive tasks: The role of *g* in perception and decision time. *Journal of Experimental Psychology: Learning, Memory, and Cognition*, 47, 1792–1809.

CHAPTER 5

Intelligence and the Brain

5.1 Introduction

The brain is the mediator of every aspect of intelligence. Of the many mysteries locked inside human brains, solving how intelligence works may have the most far-reaching consequences. In the short term, knowing how the brain creates intelligence from genetic and nongenetic influences may redefine intelligence in terms of quantifiable brain characteristics and provide brain-based ways to assess individual differences in intelligence. In the longer term, if we learn how to tinker with brain mechanisms to increase reasoning ability, we might enter a new phase of personal achievement and societal well-being. Such knowledge might even

create more geniuses on the level of Einstein, Newton, Cervantes, or Da Vinci. Increasing intelligence could even raise the bar for artificial intelligence to catch up to humans (Hawkins, 2021).

This chapter focuses on what we know about intelligence and the brain. There is considerable research to present, and the pace of discovery is increasing rapidly, mostly driven by advances in neuroimaging technology. Box 5.1 is a refresher of basic brain anatomy, and Box 5.2 describes frequently used neuroimaging methods to study intelligence.

In the first edition of this book, a literature search for papers on "brain and intelligence" retrieved 5,648 citations using

Box 5.1 Basic Brain Anatomy

The human brain is a 1.5 kg (three-pound) bundle of tissue that sits at the upper end of the spinal cord. Structurally, there are two sides of the brain, called the right and left *hemispheres*. There are four major *lobes*: the frontal, temporal, parietal, and occipital. Each lobe is split into both hemispheres. It is too simplistic to say that each hemisphere or lobe has its own set of mental functions. Instead, based on the neuroimaging research discussed in this chapter, we now have a more nuanced picture of brain structure–function relationships that goes into much finer resolution than lobes. The connectivity among discrete brain areas within lobes has replaced the general preimaging characterizations of frontal lobes doing the thinking, temporal lobes (sides of brain) doing memory, parietal lobes (upper back of the brain) doing integration of the senses, and the occipital lobe (back of the brain) doing visual perception. Together, the lobes are the *cerebral cortex*. There are different systems for labeling regions within the cortex (e.g., numerical Brodmann areas, or stereotaxic coordinates from any of several brain atlases). The *cerebellum* sits below and to the rear of the cerebral cortex and is generally involved with automatic motor coordination, although it is also involved in some higher-order cognition.

If you were to view the brain from above, you would see a deep fissure that divides the right and left hemispheres and the bundle of neural connections that bridge the fissure. The largest of these bundles is the *corpus callosum*, which provides the main communication link between the two hemispheres. In general, somatosensory and motor information is represented contralaterally; the right side of the brain controls the left side of the body, and vice versa. Certain functions that are not directly tied to a side of the body may also be differentially localized across the hemispheres. Language, which primarily depends on structures in the left hemisphere, is the best known of the lateralized functions.

There are remarkable individual differences in brain structure. Some left-handed people have their language centers located in the right hemisphere, and there are differences between men and women. These will be discussed in Chapter 10, which describes sex similarities and differences in intelligence. Individual differences in brain structure are key to understanding individual differences in intelligence.

Several subcortical structures are important in cognition. The *cingulate gyrus* functions as a communication hub between various areas of the cerebral cortex. Below the cingulate gyrus is the *limbic system*, which includes the *hippocampus* and the *amygdala*. The hippocampus plays an important role both in memory and in visuospatial reasoning. The amygdala, along with a midbrain structure called the *hypothalamus*, is involved with emotions.

Six terms are used to define locations in the brain:

Frontal or anterior: The forward part of the brain in general or the forward part of any brain area (e.g., the anterior hippocampus).
Posterior: The opposite of frontal.
Ventral: Toward the bottom of the brain or an area.
Dorsal: Toward the top of the brain or an area.
Medial: Toward the middle of the brain or an area.
Lateral: Toward the side of the brain or an area.

These terms may be combined. For instance, the *dorso-lateral prefrontal cortex* (DLPFC) is an area at the extreme front of the brain (prefrontal), on the upper (dorso) outside (lateral) surface.

Box 5.1 *(continued)*

All information held in the brain (memories) and all processing of that information (thoughts, percepts, images) are determined by the physical form of networks of nerve cells, or *neurons*. The term *physical* is important; ultimately, all thinking comes down to the manipulation of neurons. So is the term *network*: we store and process information by changing the configurations of networks of connected neurons, not by changing the states of individual neurons. Our memories of our grandmother, or of yellow Volkswagens, are coded as configurations of neural elements. Within limits, two different individuals may code the same information in somewhat different neural configurations.

Neurons are classified anatomically as either *gray matter* or *white matter*. *White matter* refers to neurons whose axons are coated with a fatty substance called *myelin*. Gray matter consists of masses of neurons that are involved in computations within a local area of the brain. The white matter provides long-distance connections between brain regions. There is a loose analogy here to distributed computing, where networks of gray matter play the roles of local computing centers and the white matter provides the cabling that connects the local centers to each other.

Introductory textbooks often contain maps showing where different functions lie in the brain. The most famous examples involve speech. *Broca's area*, in the left posterior frontal region, is associated with speech expression, and *Wernicke's area*, in the left temporal lobe, is associated with retrieval and understanding of semantic and syntactic relationships. A great deal has also been written about how the left brain is specialized for analytic, sequential reasoning, while the right brain conducts intuitive, parallel reasoning. This is an oversimplification, to say the least. Information processing is well integrated between the hemispheres, among the lobes, and with numerous subcortical areas.

The brain does contain centers that are specialized for certain kinds of processing, but it does not contain regions dedicated to broad cognitive or emotional functions. *The brain functions as a system.* For understanding human intelligence, we need to consider what cognitive functions are performed in different regions of the brain, but understanding the complexities of intelligence requires understanding both where different specialized processes are and how they mesh to produce thinking. This is recognized in the parieto-frontal integration theory (P-FIT) of intelligence, detailed in Section 5.3.1.

PsycINFO (May 30, 2009). Using the same terms and database at this writing (May 2, 2022), that number is 38,970 citations. Neither we nor anyone else has read them all, but we will present representative studies that contribute to the emerging weight of evidence on key topics, including the identification of brain areas and networks related to intelligence, the assessment of efficient information processing related to intelligence, and, perhaps most thought-

provoking, predicting intelligence from brain images based on patterns of connectivity unique to individuals like fingerprints. Many analogies have been used to help conceptualize the various aspects of brain–intelligence relationships. Two of the most useful are described in Box 5.3.

Before getting to these topics, a brief note of caution about perspective is in order. We have colleagues who believe that virtually all psychology, including the study of

Box 5.2 Technologies for Seeing and Quantifying the Brain

This box presents a brief view of the major neuroimaging technologies used to relate intelligence differences to brain structural and functional differences. Each technology was developed from basic discoveries in physics that made it possible to detect weak chemical, electrical, and magnetic signals emanating from the brain. Making sense of the signals required the development of complicated computer algorithms and rapid computer processing, whose development also required basic advances in physics. The imaging technologies represent a striking example of how basic research in one field can have important practical implications in other fields.

Technologies for Imaging Brain Structures

Computerized Axial Tomography (CAT or CT Scanning). CT scanning derives from X-ray medical imaging. Marie Curie received the Nobel Prize for her discovery of X-rays in 1911. This started a string of Nobel awards building on Curie's discovery. The first CT scanning devices were announced in 1972, and the inventors were awarded Nobel Prizes in 1979.

In CT scanning, a large series of low-intensity X-rays are passed through the body from multiple positions on a circular ring surrounding the body. After the X-rays pass through the body, they are detected by sensitive electronic receivers. Low-intensity X-rays are impeded to some degree by different densities of soft tissues as they pass through the body. Each of many two-dimensional images, taken from multiple different angles, are combined mathematically by a computer program that reconstructs a detailed image. The method is applicable to all parts of the body, not just to the brain. The spatial resolution for brain tissue, however, is not as good as it is for MRI.

Magnetic Resonance Imaging (MRI), Also Called Structural MRI (sMRI). Magnetic resonance imaging (MRI) derived from work in the 1960s and 1970s, based on discoveries about electromagnetism that date from the 1930s and 1940s, including Nobel Prize–winning research by Isidor Rabi. The first medical scanners based on MRI were introduced in the early 1980s (MRI originally was called nuclear magnetic resonance, NMR). The developers of modern sMRI, Paul Laterbauer of the University of Illinois and Peter Mansfield of the University of Nottingham, were awarded the Nobel Prize in 2003.

In MRI, the person being scanned is placed in a tube that is surrounded by a large magnetic field. A radio pulse is directed toward the part of the body being scanned. The pulse frequency is chosen to resonate with hydrogen atoms. This causes hydrogen atoms in the body to move out of alignment with the magnetic field. Between each rapid pulse, the atoms briefly return to alignment and, as they do so, emit a weak but detectable electromagnetic signal. Variations in this signal are used mathematically to construct an image of the body. MRI images have higher spatial resolution than CAT scans, so considerably more detail is visible. These images also can be quantified to assess gray and white matter volumes and cortical thickness and surface area. Spatial resolution is excellent in the millimeter range (depending on the strength of the magnetic field), so many anatomical details are visible. Whereas CAT scans expose the person to X-ray radiation, there is no radiation exposure at all for MRI, a major advantage making it practical to repeat scans on the same person. Russell Poldrack (2021) describes an interesting demonstration based on himself having more than 100 MRI scans during eighteen months. A good summary of MRI and how it is used in studies of intelligence can be found in Drakulich and Karama (2021).

Box 5.2 *(continued)*

Diffusion Tensor Imaging (DTI). Diffusion tensor imaging (DTI) is a form of magnetic resonance imaging based on signals that are sensitive to the diffusion of water molecules along the myelin sheaths that coat neurons (white matter). Signals are transmitted along myelinated neurons considerably faster than they are along unmyelinated neurons (gray matter). Columns of myelinated neurons are involved in transmitting signals from one region of the brain to another, while unmyelinated neurons are involved in local computations within a region. As a loose analogy, DTI provides a way of imaging the cabling connections among the computing centers of the brain. Measurements of white matter integrity and connectivity can be generated from DTI data. Since DTI is a form of MRI, there is no radiation exposure, and repeat scans are practical. A good summary of DTI and how it is used in studies of intelligence can be found in Genc and Fraenz (2021).

Technologies for Imaging Brain Activity

Positron Emission Tomography (PET) Scanning. The basic ideas behind positron emission tomography were developed in the 1950s. The first use in humans was in the 1970s, due in part to the need for ultra-high-speed computing to support the detector technology.

The PET procedure begins with an injection of a radioisotope into a person's bloodstream. As the isotope decays, positrons are emitted, and when a positron encounters an electron, both are annihilated, emitting a gamma ray. Rapidly decaying isotopes are required to avoid damage to tissue as the gamma rays pass through the body. For use in brain imaging, the isotope is chemically created so that it can be absorbed into the brain (typically using glucose or oxygen or a neurotransmitter label as the carrier for the positron emitter). The more active any brain area is, the more isotope it will absorb (neurons require glucose to function, and glucose is delivered by blood flow). Hence more gamma rays will be emitted from the active areas. After a short period of absorption, sensors in the PET scanner, arranged in rings around the head, detect the gamma rays as they exit the body and record whenever two are detected at a 180 angle. Computer programs are then used to calculate the source of each gamma ray based on millions of detections, and this information is mathematically transformed into an image that maps the amount of activity throughout the brain. PET data allow quantification of the glucose metabolic rate, which changes within a person as they perform different cognitive tasks during the isotope uptake. The time resolution of a PET scan is limited to several seconds to thirty-two minutes, depending on the half-life of the isotope used. This means that brain activity is averaged over this period, which is long for brain functions that happen in seconds or in milliseconds. The isotopes used in PET are short-lived and therefore must be created with a cyclotron, an expensive piece of equipment to obtain and manage, nearby the PET scanner. Even though radiation exposure is within the limits for medical imaging, repeat scans in the same person are limited. The need for a nearby cyclotron and the cost of creating isotopes make PET a difficult and costly technology to use compared to fMRI.

Functional Magnetic Resonance Imaging (fMRI). fMRI is a version of sMRI. Some radio wave pulse sequences are so fast that changes in blood hemoglobin can be detected and turned into images of blood flow. This is possible because neurons take up oxygen from

Box 5.2 (continued)

the blood as they fire, and the more neural activity there is, the more oxygen is consumed, and the more hemoglobin is taken to the active areas. This creates a signal called the blood oxygen level dependent (BOLD) response. The BOLD response can be mapped across the brain to show what areas are most and least active while a person performs different cognitive tasks. fMRI has a time resolution of about a second and a spatial resolution of millimeters. Although scanners are expensive, overall, it is now brain researchers' most widely used imaging technology, and many psychology departments own and operate their own fMRI scanners.

Electroencephalograms (EEG) and Event-Related Potentials (ERP). Neural events generate electrical signals that can be detected by electrodes placed on the scalp. This was first demonstrated in 1929. It is the basis for the modern electroencephalogram (EEG). EEG recording is important in medicine because physiological states have characteristic EEG signatures. These include characteristic patterns for epilepsy and for different stages of sleep. The event-related potential (ERP) is an EEG signal in response to a specific external event, such as a flashing light. The ERP has proven to be a useful tool in cognitive psychology, as different mental events have characteristic ERP signals. For example, when people hear semantically meaningless sentences, such as THE COOK ROASTED THE

CEMENT, they display a characteristic ERP. Syntactically anomalous sentences, such as WOMAN THE LOVE CATS, have a different ERP signature.

Modern EEG signals are recorded from as many as 100 sites on the scalp. Computer-based analysis is required to identify signature waveforms and to infer their location in the brain. A major advantage of EEG and ERP methods is that they measure changes in electrical activity (neuron firing) millisecond by millisecond. This is the fastest time resolution now possible and better than the time resolution of seconds or minutes from fMRI or PET. The spatial resolution is based on interpolation of signals among the electrode sites and is not as good as MRI of PET.

Magnetoencephalography (MEG). MEG measures fluctuations in the weak magnetic fields generated when neurons fire. MEG has the advantage of assessing these fluctuations deep inside the brain with little distortion, whereas EEG signals are best assessed from the cortical surface. Like EEG, MEG has a time resolution of milliseconds but limited spatial resolution. MEG is a difficult technology, and its potential is still emerging for intelligence research. The same is true for magnetic resonance spectroscopy (MRS), which can assess biochemical signatures in each voxel and correlate them to variables like IQ scores. Good summaries of this technique and its uses in intelligence research can be found in Jung and Chohan (2021) and Raz (2021).

intelligence, eventually will be reduced to studies of the brain. We are sympathetic to this view, but we recognize that there is more to the story. We know, for instance, that raising your finger from a closed hand involves neurons firing in your motor cortex. But when you raise just your middle finger, there is more involved. How you came to

know this and when you decide to use this knowledge may be explainable at the level of neurons, but there is a relevant sociocultural aspect associated to this particular finger movement that enriches any explanation of the neural events.

One more caution is, when reading about intelligence and the brain, it is important to

Box 5.3 Is Intelligence More Like an Orchestra or an Automobile?

As you read this chapter, either of these analogies might be helpful.

Orchestras vary in their ability to play music. The same piece played by your local high school orchestra and by the New York Philharmonic may be recognizably the same piece, but the two orchestras do not sound the same. There is a positive correlation between the number of musicians and musical quality. High school orchestras tend to be smaller than citywide amateur orchestras, which are smaller than the major philharmonics. There even is variation in the quality of music performed by orchestras of the same size. Adding performers to your high school orchestra will not make it sound like a philharmonic.

As we shall see in Section 5.2, there is also a correlation between intelligence and brain size, but the same cautions apply. By analogy to studies of brain injury, imagine studying the essence of orchestral quality by removing one player at a time. To start, remove the conductor. The orchestra begins to play in a flat, hesitant fashion and can play only those pieces for which it has a lot of practice. But it can play, and there are interorchestra differences even when the conductor is there. In addition, there is the puzzling problem that the conductor clearly influences the music being played, despite that the conductor does not make any noise.

Now try removing all or part of the string section, the brass, the woodwinds, or the percussion instruments. Each removal would affect performance, but the effects would depend on the musical work played. Removing the strings would affect most pieces, but not quite all. Playing "76 Trombones" does not require a violin; percussion instruments are not needed to play chamber music. And there is still the problem that there is

considerable variation in the performance of intact orchestras. Now try an analogy to modern studies of how brain metabolism varies with mental activity. The orchestral equivalent would be measuring the sound level in different parts of the orchestra. You would quickly find that in all but the simplest pieces, the sound comes from all over, and that the pattern of sound varies far more with the piece being played than with the quality of the orchestra. The same thing might be true in cognition: the pattern of activation might change more with the mental activities than with the individual performing them. However, research fails to support this. As demonstrated by Caterina Gratton and colleagues (2018, p. 447), functional network organization arises from stable factors (genetics, structural connections, and long-term histories of coactivation among regions); day-to-day cognitive variations are less relevant: "the large subject-level effects in functional networks highlights the importance of individualized approaches for studying properties of brain organization" (see also Kraus et al., 2021).

Instead of assessing activity all over the orchestra, suppose we arrange a "laboratory study" that isolates the performance of individual players, so that we can rate their performance. This is somewhat like what the information-processing psychologist does, by designing situations that isolate working memory or visual perception, instead of having them work together, as they do in everyday problem solving. There would be a substantial correlation between the quality of individual performers and the quality of the orchestra. Musicians in philharmonics are much better than musicians in high school orchestras. However, if you were to look at a narrower range – say, between the members of the Royal Concertgebouw, Berlin, Vienna, London, and Chicago Philharmonics – you would find that the differences were small. Then

Box 5.3 *(continued)*

there is the pesky problem of the conductor. You would find it difficult to evaluate a conductor without an orchestra. They would look more like a person suffering from a minor seizure than a musician and would not make any sound at all. How can you reconcile this with the fact that orchestras play considerably better with a conductor than without?

Suppose instead you conduct a "metabolic" study, by looking at how much orchestra musicians are paid. This is an indicator. Musicians in major orchestras earn considerably more than musicians in minor ones, and amateurs are not paid at all. But is this because being paid more produces better music, or possibly the other way around? The silent conductor is getting paid more than anyone.

The brain is more complex than an orchestra, and cognition is more complicated than musical performance. Cognition and music are emergent properties. They depend partly on measurable qualities of their component parts and partly on the interactions among those parts. We can make progress in understanding cognition (and music) by making measurements on parts, but these measurements alone will not provide a full explanation. You should keep this in

mind as we discuss brain research that tells us key aspects of intelligence, but not the whole story.

There is no duality of mind and brain. Theoretically, if we knew the nature of every connection among the approximately 5 billion neurons in a person's brain, and if we knew how the brain activates and alters these connections, we would know everything there is to know about that person's cognition. Until we have that knowledge, there is a place for nonbiological models of intelligence.

Another simple analogy for understanding the brain and intelligence relationships is automotive. There are many kinds of cars. Some are more powerful than others and get you to a destination faster. Others get you to a destination more efficiently. They all need fuel, they all have propulsion systems, steering systems, tires, and stopping systems. How all the parts are arranged into the systems determines how well your car functions compared to other cars. There are also many different ways to measure how parts and systems work. Research on any one part or system is only part of the story for understanding automobile function. It is the integrative nature that is the essence of human intelligence.

keep three things in mind: (1) no story about the brain is simple, (2) no one study is definitive, and (3) it takes many years to sort through inconsistent results and establish a compelling weight of evidence (Haier, 2017). Let's start with an old question about brain size.

5.2 Are Bigger Brains Smarter?

General intelligence has evolved in humans and nonhuman mammals (Burkart et al., 2017). Cross-species and evolutionary

records indicate a reliable relationship between brain size and cognitive power (Roth and Dicke, 2005). Compared to the brains of other mammals, human brains are relatively large for their body size. If you accept the idea that the progression from throwing sticks and making stone tools to throwing bombs and making metal tools represents an increase in intelligence, then there is clear evidence that across species, increases in brain size parallel increases in intelligence. But does this relationship apply to individuals within the same species, like us?

The first empirical evidence for a within-species brain size–intelligence relationship was published by Karl Pearson (1906), the professor who developed the correlation coefficient. A number of data sets were analyzed with his new method, including Galton's measurements of skull sizes and grades of Cambridge University students. The result was $r = 0.11$. This relationship was positive but weak. At the time, the only ways to estimate cranial capacity (and, by inference, brain size) in a living person were indirect and subject to error, like measuring the exterior circumference of the head. Another approach was to fill skulls with bird seed, sand, or shotgun pellets to estimate cranial volume and brain size (this method was not used with Cambridge or other living students).

These were indirect, unreliable measurements, but the basic finding of a relationship between such measures and intelligence has been verified beyond doubt. Lee Willerman of the University of Texas at Austin and his colleagues (1991) reported the first reliable correlation between intelligence and whole-brain volume measured using magnetic resonance imaging (MRI; see Box 5.2). As MRI machines became more accessible, other researchers reported similar results, and in 2005, a meta-analytic review of thirty-seven samples across 1,530 individuals (including men and women of diverse ages) concluded that the whole-brain size–IQ correlation is about 0.33 on average (McDaniel, 2005).

More recent analyses concur (Gignac and Bates, 2017; Nave et al., 2019; Pietschnig et al., 2015). One of these analyses classified studies according to their quality of intelligence assessment (fair, good, and excellent). The r values increased with this quality: 0.23, 0.32, and 0.39, respectively (Gignac and Bates, 2017). The weight of evidence created by these findings helps confirm that IQ scores are related to quantifiable brain characteristics, a finding that would not be possible were IQ scores meaningless. Even earlier studies with positron emission tomography (PET) also made the same point, as discussed later in this chapter.

So, the answer to the opening question is, yes, bigger brains tend to be smarter. On the one hand, more gray and white matter in bigger brains could result in better problem solving associated with higher test scores. On the other hand, people with higher scores might seek experiences and environments that lead to bigger brains. There is now good data that the former is more likely. A genetically informative study (with replication) using Genome-Wide Association Study (GWAS) data (see Chapter 6) indicated that brain size had a causal effect on intelligence (Lee et al., 2019; see also Jansen et al., 2020; Grasby et al., 2020).

Fortunately, in the last two decades, research has vaulted light-years beyond this simple correlation, and we are now finding many more detailed brain–intelligence associations. As the rest of this chapter presents, recent studies use advanced neuroimaging techniques (Box 5.2), including functional MRI (fMRI), diffusion tensor imaging (DTI), and advanced image analysis methods that quantify cortical thickness (CT), cortical surface area (CSA), and various gray and white matter characteristics. There are also advances in measures of brain connectivity that correlate with intelligence test scores. The next section begins our discussion of these advances with another deceptively simple question.

5.3 Where in the Brain Is Intelligence?

It would be convenient if there were an intelligence center in the brain, that is, one critical area that was the source of all reasoning ability. If people with higher intelligence test scores had a bigger intelligence center than people with lower scores, researchers could focus on how this area develops and what features within the area are most relevant to the differences among people.

Early research suggested that it was reasonable to assume that the frontal lobes might contain an intelligence center because lesion studies, EEG studies, and neuropsychology studies implied that the frontal lobes were involved in complex reasoning and planning. One problem with this idea came from research on people who had frontal lobotomies (an early surgical "treatment" for schizophrenia and other mental disorders that cut the connections between the frontal lobes and the rest of the brain – no longer practiced because it didn't work). The problem was that IQ scores did not decrease much after the procedure compared to premorbid testing (Duncan et al., 1995).

While it might be naive to think that intelligence is located in any one place in the brain, it is reasonable to believe that different parts of the brain are involved to different degrees. The advent of modern neuroimaging provided tools for direct quantitation of regional brain structure and function, assessed simultaneously throughout the entire brain. Moreover, the spatial resolution of these measures was in millimeters, allowing for much finer localization of key areas than that inferred from neuropsychological tests, EEG, and small-scale lesion studies.

The first intelligence study specifically addressing the "where" question used PET, while volunteers solved problems from the Raven's Advanced Progressive Matrices (RAPM) (Haier et al., 1988). There were two main findings. First, compared to individuals who were scanned performing attention and control tasks, the group that was scanned while solving RAPM problems showed elevated glucose metabolic rates in several brain areas that were not limited to the frontal lobes. This suggested that complex reasoning involved multiple brain regions. This was not entirely surprising, but showing it with functional imaging was exciting. Second, and completely unexpected, within the areas activated during the RAPM, the more glucose that was consumed while taking the test, the *lower* was the person's score. This suggested that intelligence might not be related to how hard the brain is working but rather to how efficiently it is working.

When this study was published in 1988, there were still critics who believed that intelligence tests were meaningless and that

the g-factor was a statistical artifact. Finding correlations between glucose metabolic rate in the brain and RAPM scores undercut both beliefs, and today neither one is taken seriously by anyone familiar with the empirical data from studies with thousands of people.

5.3.1 *The P-FIT Model of Distributed Networks*

Because of the expensive nature of neuroimaging, another eighteen years passed before additional PET and MRI intelligence–imaging studies were reported. These studies, thirty-seven in all, were the basis of a systematic review with the goal of identifying common brain areas related to intelligence (Jung and Haier, 2007). This was a formidable challenge because the thirty-seven studies used different structural and functional imaging methods (sMRI, fMRI, PET), different image analysis/quantification procedures, different systems for anatomical localization of brain areas, different intelligence tests, and different inclusion criteria for defining their samples.

An earlier review attempted the same goal for imaging studies of cognition, which were far more numerous (Naghavi and Nyberg, 2005). This review of cognitive research was based on more than 200 studies and tabulated how many times each brain area was found to be related to cognitive variables across all studies. Brain areas that appeared in 50 percent or more of the studies were deemed noteworthy, irrespective of different imaging methods, different cognitive assessments, and other differences.

Jung and Haier (2007) used the same procedure. Figure 5.1 shows their summary analysis. Combining all thirty-seven studies, fourteen brain areas related to intelligence test scores were found in more than 50 percent of the thirty-seven studies (either structural or functional). These fourteen areas were mostly in the left hemisphere or bilateral. The areas are identified using a standard nomenclature developed by Brodmann (1909). Since the boundaries of each Brodmann area (BA) vary among

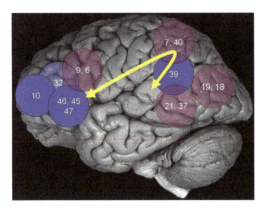

Figure 5.1 Parieto-frontal integration theory of intelligence (P-FIT) (Jung and Haier, 2007). Numbers refer to Brodmann areas of the cortex. Dark blue circles are left hemisphere intelligence score/brain findings. Light red and light blue circles are bilateral findings. The arrow represents a white matter tract connecting temporal, parietal, and frontal regions (the arcuate fasciculus).

individuals, they are depicted as standard circles to represent the general location of the findings that met the greater than 50 percent criterion.

Most of the fourteen areas are in the parietal and frontal lobes, so the pattern was named the parieto-frontal integration theory (P-FIT) of intelligence. The name emphasizes that the salient areas are distributed across the brain, not focused solely in the frontal lobes, and that communication (integration) among the salient areas is a crucial consideration.

The P-FIT proposed that the ability to do the sort of thinking captured by intelligence measures is supported by a network of brain regions involving the dorsolateral frontal cortex, the parietal lobe, the anterior cingulate gyrus, and parts of the temporal and occipital cortex. Each of these regions performs some of the functions needed for abstract reasoning and problem solving; no one of them is sufficient alone.

The P-FIT hypothesized that individuals would differ in subnetworks defined by different combinations of the salient areas and/or the sequence of information processing among the subnetworks. In other words, intelligence differences among people could

be related to different aspects of the P-FIT networks. This could be the case for the *g*-factor and for other specific abilities/factors, such as verbal or visuospatial. The P-FIT also specifically proposed that efficient communication through white matter tracts that connect brain areas would be related to higher intelligence test scores (data on brain efficiency are presented in Section 5.4).

The P-FIT implies a number of testable hypotheses. For example, a sequence of information processing necessary for reasoning/intelligence is proposed, as illustrated in Figure 5.2 (Colom et al., 2010). Four broad stages are proposed: (1) the *input processing stage*, when occipital and temporal areas process sensory information using the extrastriate cortex (BAs 18 and 19) and the fusiform gyrus (BA 37) involved with recognition of imagery and elaboration of visual inputs, as well as Wernicke's area (BA 22) for analysis and elaboration of syntax of auditory information; (2) the *integration and abstraction stage*, when parietal areas process the sensory information using BAs 39 (angular gyrus), 40 (supramarginal gyrus), and 7 (superior parietal lobule); (3) the *problem solving, evaluation, and hypothesis testing stage*, when the parietal areas interact with the frontal lobes using frontal BAs 6, 9, 10, 45, 46, and 47; and (4) the *response selection*, when the anterior cingulate (BA 32) is implicated to either select or inhibit responses once the best solution is determined in the previous stage. These stages are not necessarily linear, and information may flow back and forth among relevant areas many times within a few seconds.

At the time of the P-FIT review, involvement of the frontal cortex was not unexpected given the considerable research that ties it to reasoning, planning, and organization. The importance of the parietal lobe, however, was less expected, even though the parietal cortex is responsible for integrating information from various sensory modalities and for providing temporary storage of information. Parietal involvement with intelligence would be consistent with its established role in controlling attention to particular parts and objects in the sensory fields (Posner et al., 2006), as well as with

Parieto-Frontal Integration Theory [P-FIT]

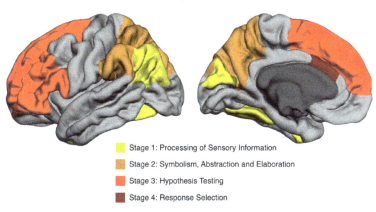

Stage 1: Processing of Sensory Information
Stage 2: Symbolism, Abstraction and Elaboration
Stage 3: Hypothesis Testing
Stage 4: Response Selection

Figure 5.2 Information-processing stages suggested by the P-FIT framework.
Courtesy of Kenia Martínez.

mental speed and its role in human evolution (Bruner et al., 2014; Bruner et al., 2011). The important point is that intelligence appears to involve not only higher-order cognition in the frontal lobe but also individual differences in early, more integrative sensory processing in parietal areas, and possibly in occipital areas. The integration aspect of the P-FIT refers to variability in communication in the networks that link all these areas (Pineda-Pardo et al., 2016).

The P-FIT proposes that the anterior cingulate gyrus is involved for response selection. This includes directing decisions with substantial input from the frontal lobes. The anterior cingulate gyrus also seems to evaluate likely consequences of taking an action. An important part of the frontal cortex–cingulate gyrus interaction may be modulation of the emotional and nonemotional interactions of decision-making (Blair, 2006).

The P-FIT networks are also implicated in many studies of fundamental cognitive abilities, especially aspects of working memory, attention, and language, suggesting that the complexities of intelligence evolved from more basic mental abilities. For example, gyrification of the cortex in parieto-frontal areas is positively correlated to working memory, a key component of intelligence (Green et al., 2018). Gyrification (i.e., more folding)

increases surface area and shortens the distance of white matter tracts, thereby increasing speed of processing information among adjacent areas, possibly improving efficiency of working memory (data on brain efficiency are presented in Section 5.4). A better working memory may play a role in learning complex material. P-FIT language areas include parts of BA 45 (Broca's area) and BAs 39/40 (Wernicke's area), and several of the P-FIT areas are also associated with components of attention, including parts of the default mode network (DMN). Importantly, individual brain areas can be involved in more than a single function, an observation with a long history in neuropsychology (Basso et al., 1973; Duncan, 2010; Seghier and Price, 2018).

Individual differences are central for interpretations of the P-FIT. If an intelligence test score is obtained from a test battery like the WAIS, an individual's score is calculated by a weighted composite of scores on the subtests. Therefore, two individuals can obtain the same *g* or IQ score if they have a different combination of subtest scores. For example, one person could have a high IQ score on the WAIS by obtaining a high verbal IQ and moderate performance IQ, while another person could obtain the same IQ score with a moderate verbal score and a high performance IQ.

Before Controlling for *g*

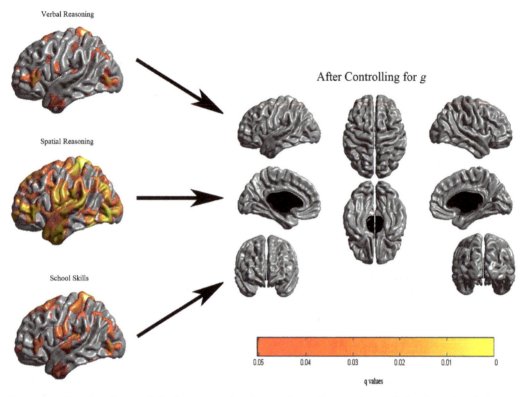

Figure 5.3 Results of cortical thickness correlated to each intelligence factor (left column) and the result after controlling for the general factor of intelligence (*g*). The color bar scale represents statistical significance; *q*-values are analogs of standard *p*-values (Karama et al., 2011).

More generally, when we ask someone to demonstrate "intelligence," they execute a medley of information-processing acts, using different parts of the brain for the different acts. If a general measure is obtained from a single marker, such as a progressive matrix test, the questions on the test will be fairly complex and will yield to different strategies. The same principle applies in the case of a test battery: two people may obtain the same score using different elementary processing actions. When this is the case, they will not display the same pattern of brain activation (and the correlates of the activation), even though they obtain equivalent test scores (Johnson et al., 2008).

These issues may confound the search for a neural basis for the *g*-factor (Haier et al., 2009). There is not yet a definitive weight of evidence about specific neural correlates of *g*

or tests of alternative theories like the process overlap theory (described in Chapter 4). Nonetheless, to assess brain activity related to other specific cognitive abilities and factors, the large overall effect of *g* on all abilities must be considered. For example, Figure 5.3 illustrates the ubiquity of *g* in a developmental neuroimaging study using a sample of children representative of the population (*N* = 207; mean age 11.8 years, range 6 to 18.3). Cortical thickness was assessed with MRI and correlated to different cognitive tests (Karama et al., 2011). The resulting statistical maps in Figure 5.3 indicate different patterns of regional correlates for each test. However, Figure 5.3 also shows that all these patterns effectively disappear when *g* is removed from each test. This illustrates that neuroimaging studies of any cognitive task can give misleading results unless there is adjustment for *g*. A similar caution was discussed in Box 3.2 regarding psychometric models and measurements. The results further validate the importance of the *g*-factor but, of course, do not explicate anything about its neural basis.

Individual difference considerations also are demonstrated in neuroimaging studies of patients with brain damage that examine not only *g* but other intelligence factors as well. For example, Gläscher and colleagues (Gläscher et al., 2010; Gläscher et al., 2009) studied 241 neurological patients using voxel-based lesion symptom mapping (VBLSM). They found that lesions in parietal and frontal areas were associated with deficits in the *g*-factor, consistent with the P-FIT as shown in Figure 5.4 (top). Furthermore, deficits in other intelligence factors (perceptual organization, verbal comprehension, working memory, and processing speed) were associated with different locations of lesions, as shown in Figure 5.4 (bottom).

Barbey and colleagues (2012a) studied 182 Vietnam War veterans with brain damage using a similar approach (VBLSM). Their data extended findings to measures of executive functioning. Figure 5.5 depicts the findings for the general factor of intelligence (*g*) and for executive function (EF). The overlap between both psychological factors is also shown (see Box 5.4).

Since the P-FIT was published in 2007, a number of increasingly sophisticated neuroimaging/intelligence studies have reported findings that are consistent with the importance of the parietal/frontal network (Deary et al., 2010; Santarnecchi et al., 2017a). Ulrike Basten and colleagues (2015) computed a meta-analysis that distinguished studies showing an average difference in a brain characteristic between two groups defined by a difference in intelligence (task approach) from studies that showed individual differences within a group as correlations between a brain characteristic and intelligence test scores (individual differences approach). These two approaches were combined by Jung and Haier (2007).

The Basten group considered only studies using the individual differences approach,

and they analyzed structural and functional correlates separately. Their method provided a quantitative assessment of brain/intelligence findings common across twenty-eight studies that included more than 1,000 participants. They found substantial overlap with the original fourteen P-FIT areas, especially a parietal-frontal network (Figure 5.6). They also suggested that three additional areas could be added to the P-FIT pending replication: the posterior cingulate/precuneus, the caudate, and the midbrain. In an updated review of neuroimaging results relating specific brain area function to intelligence, Basten and Fiebach (2021) concluded that key areas were distributed across the brain but mainly in the prefrontal and parietal cortices and, to a lesser extent, temporal and occipital cortices along with subcortical structures, such as the thalamus and the putamen.

As noted, the identification of specific brain areas related to intelligence raises the question of whether variability in the characteristics of those areas among people might predict individual differences in intelligence test scores. An example of this effort comes from Ritchie and colleagues (2015), who studied 672 older adults in their seventies. They used structural MRI to assess six different quantified brain characteristics and then used a multivariate statistical method called structural equation modeling to assess how much *g*-factor variance among the individuals could be predicted by combinations of the six variables. Only four of the six were predictive: total brain volume, total cortical thickness, white matter structure, and white matter hyperintensity load. Together these variables predicted up to 23 percent of the variance in scores. The single best predictor was total brain volume (about 12 percent), and the next best was total cortical thickness (about 5 percent). The authors suggested that even more *g*-factor variance could be predicted if functional measures were included and the age range broadened. Brain volume and cortical thickness within specific areas like the ones in the P-FIT might also increase the prediction.

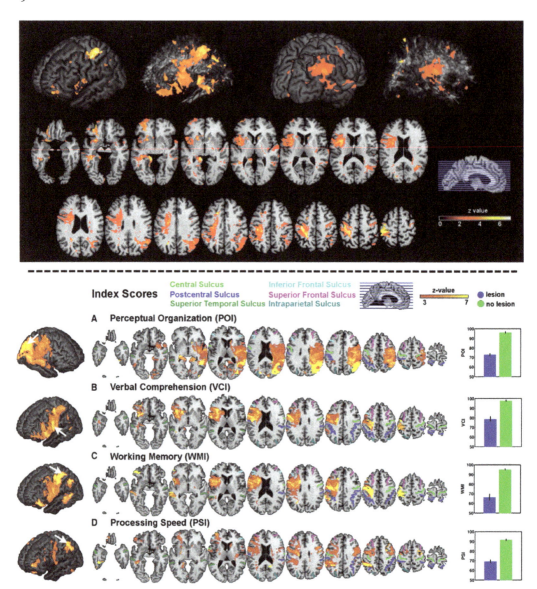

Figure 5.4 (top) Lesion mapping of the general factor of intelligence (*g*) (Gläscher et al., 2010). (bottom) Lesion mapping of specific cognitive abilities measured by the Wechsler Battery: perceptual organization, verbal comprehension, working memory, and processing speed. The graphs on the right show the mean difference on each ability score between those patients whose lesions included the voxel showing the maximum effect (white arrows on the 3-D display at the left) and those whose lesions did not include it (Gläscher et al., 2009).

This study illustrated that specific aspects of basic brain anatomy contribute unique variance to individual differences in the *g*-factor. This basic finding is also reported in a comprehensive study of structural MRIs in more than 3,000 individuals (Reardon et al., 2018). This report showed complex normative relationships among brain size, shape, specific networks, and intelligence along with neurobiological markers that provide clues about how the salient brain features function. As we see in the next sections, research into the "how" question becomes central.

One word of caution is in order. Given the results of imaging studies, by now you might

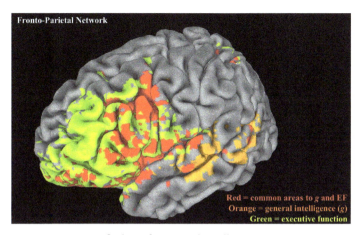

Figure 5.5 Lesion findings for general intelligence (*g*), executive function (EF), and their overlap, suggesting that EF (green) is more localized in and involves more of the frontal lobe than *g* (orange); areas in common are red (Barbey et al., 2012b).

Box 5.4 Brain Lesion Mapping and Intelligence: Moving toward Causation Inferences

Bowren and colleagues (2020) applied a lesion mapping approach to 402 lesion patients roughly similar to those analyzed by Gläscher and colleagues (2010). In their introduction section, they note that "*g* has become central to cognitive science" (p. 8924). This is an emerging recognition that the scientific interest devoted to the general factor of intelligence (*g*) is not limited to old-fashioned and venerable psychometricians. Instead, it is a psychological construct of current multidisciplinary interest.

The lesion approach allows for causal inferences by determining if *g* relates to specific neuroanatomical structures and knowing if these structures are common to specific cognitive abilities (crystallized ability, visuospatial ability, long-term memory, working memory, and processing speed). These were the main findings:

1. There was a perfect correlation between *g* and working memory at the behavioral level.

2. Scores for *g*/working memory were related to the integrity of left temporo-parietal white matter, left parieto-occipital white matter, left frontal lobe white matter, right subinsula/clastrum, left posterior middle/inferior temporal gyrus, and right posterior middle temporal gyrus. Nevertheless, the strongest finding highlighted the left temporo-parietal white matter. The arcuate fasciculus was especially prominent, a specific result of connectivity hypothesized by the P-FIT model (Jung and Haier, 2007).

3. Removing *g* from the remaining specific cognitive abilities resulted in null findings, consistent with what was reported by Karama et al. (2011) of a representative sample of developing children and adolescents.

4. The neuroscience evidence found here led the investigators to conclude that "working memory is constitutive of *g* and we can build on our understanding of the mechanisms of domain-general cognition by reframing individual differences in *g* as being largely driven by individual differences in working

Box 5.4 *(continued)*

memory" (p. 8934). This general conclusion appeals to the integrative nature of both psychological constructs.

In short, parieto-frontal connectivity was underscored in this large-scale research with lesion patients. Moreover, white matter was found as more relevant than gray matter for understanding the brain basis of intelligence differences. This study also raises doubts regarding the analysis of separate cognitive abilities. We might think that the focus of our research is language ability, but cognitive differences in this ability may be mainly explained by integrating a host of mental abilities, namely, general intelligence (*g*). In the authors' own words, "research that aims to focus on a specific cognitive process may be unintentionally studying the confluence of *g*/working memory and the intended domain of study, rather than the unique aspect of the targeted process" (p. 8935).

This and other cautions were raised by Haier et al. (2009) for improving the quality of the neuroscientific research of the intelligence construct. They are enumerated elsewhere in this chapter.

The study of brain lesion patients is of paramount relevance for overcoming the limitations of correlational studies. In this regard, Protzko and Colom (2021) applied this lesion framework for helping to choose among candidate models intended to account for the positive manifold (the positive correlation among cognitive tests found across the globe, as discussed in previous chapters). The authors tested three classes of models (common-cause, sampling/network, and interconnected models). They began by highlighting the fact that statistical criteria are insufficient and limited to choose among these competing models. Instead, they analyzed the cognitive impact of focal cortical lesions for testing the causal connections predicted by these three models. The key of their rationale was that focal cortical lesions lead to local instead of global deficits: "only models that can accommodate a deficit in a given ability without effect on other covarying abilities can accommodate focal lesion evidence" (p. 1).

After extensive testing of the competing models, the authors concluded that the bifactor model (a specific instance of the common-cause models; see Figure 3.3, bottom right) is the best candidate to date. Nevertheless, finding the best model was not their main aim. Instead, the aim was to demonstrate a connection between psychometrics and neuroscience to advance our understanding of the intelligence construct because "measurement models are more than ways of picturing an idea. They are causal models, scientific theories, with necessarily causal connections and testable predictions" (p. 11).

---

be wondering if there are any brain areas not involved with intelligence. It is a good question. The P-FIT identified fourteen areas of the cortex using Brodmann nomenclature, and subsequent research indicates that a few more are relevant. For perspective, there are fifty-two basic Brodmann areas in the cortex plus some (but not all) subcortical areas (and divisions within them). The networks linking the areas salient for intelligence are distributed throughout the brain, but they represent only a portion of the brain, and most known networks have not been related to intelligence.

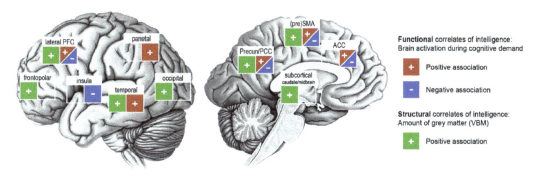

Figure 5.6 Brain areas related to intelligence based on a meta-analysis of imaging studies. (left) Lateral and (right) medial surfaces of the brain. ACC = anterior cingulate cortex. PCC = posterior cingulate cortex. PFC = prefrontal cortex. (pre)SMA = (pre-)supplementary motor area. Precun = Precuneus. VBM = voxel-based morphometry (Basten et al., 2015).

As neuroimaging analyses become increasingly sophisticated, more support is apparent for expanding the P-FIT framework, especially with respect to how the individual salient brain areas communicate with each other. A key methodological advance is the development of statistical ways to quantify connections among brain areas. An advanced method is called graph analysis and variations of it have explicated new findings about the P-FIT, predicting intelligence and brain efficiency, as reviewed in the next two sections.

5.3.2 *Patterns of Brain Connectivity*

Identifying a set of brain areas related to intelligence is just a beginning. How these areas communicate with one another is key. When one brain area is active, it may excite or inhibit activity in other areas simultaneously or in a rapid sequence measured in milliseconds, and the sequence may differ over time or among individuals. For example, do higher scorers on *g* need fewer or more brain areas involved and/or do they have faster information processing among the key areas?

Brain areas can be connected to many other areas structurally or functionally. For example, brain white matter fibers are bundled into tracts that physically connect local and distant brain areas. These structural pathways can be visualized and assessed with DTI (see Box 5.2). Functional connectivity can be inferred when activity (increased or decreased) in one area is time linked to activity in other areas (either simultaneously or in reliable sequences).

The application of a mathematical method called graph analysis has advanced empirical studies of brain connectivity. Here is the basic way it works, although there are many statistical variations. Every neuroimage irrespective of how it is generated is constructed from millions of voxels. A voxel is a three-dimensional pixel because brain images have depth as well as height and width. Each voxel contains information about that specific location in the image. For PET scans, the information within the voxel may be the amount of glucose consumed; for MRI, the information may be the amount of gray or white matter; for fMRI, the information may be a value for the BOLD response/blood flow (Box 5.2).

Now imagine taking the information in one voxel and correlating it to the information in all the other millions of voxels in the brain image. You get a map of how the information in that voxel relates to all other voxels. Now repeat this process, correlating the information in each of the millions of voxels to every other voxel. From this giant matrix of correlations, patterns can be identified and displayed as a map of relationships (networks) that shows that some voxels are related to a few other voxels and some voxels are related to many other voxels. An example of a resulting map of connections is given as Figure 5.7.

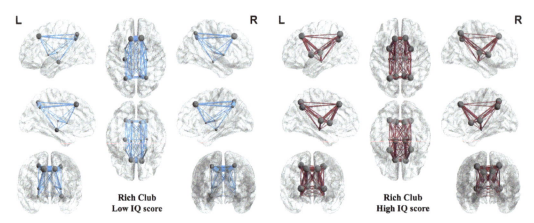

Figure 5.7 Example of graph analysis for a low-IQ person and a high-IQ person. Nodes are represented by spheres and are located at anatomical locations; the larger the sphere is, the more connections there are to other brain areas. Straight lines connecting nodes are called *edges*; thicker lines represent more connections. Rich clubs are nodes with many connections. Note that in this comparison, the node patterns are similar, but the high-IQ person has more edges (connections) among the nodes. Such connectivity patterns can be based on structural or functional data. Patterns tend to be unique to individuals and also can be averaged to compare groups.
Courtesy of Javier Santonja.

In graph analysis, voxels are called "nodes," and nodes of adjacent voxels can be large or small, depending on the strength of relationships among them. Larger spheres in the illustration in Figure 5.7 show larger nodes. The connections among nodes are called *edges*. Edges can be strong or weak, depending on the strength of relationships between connecting nodes. Thicker lines in the illustration in Figure 5.7 show stronger edges. Nodes with many edge connections are called *rich clubs* (van den Heuvel and Sporns, 2011). Nodes connected to other nearby nodes are called *small world connections*. Nodes with more distant connections are called *global connections*.

Nodes and edges can be calculated to define connectivity patterns for an individual, or average connectivity patterns can be compared among groups (or for the same person performing different tasks). Graph analysis is a way to test hypotheses like whether high-IQ individuals have more connections among P-FIT areas or whether they may have more efficient connections. The sizes of nodes and strengths of edges also can be correlated to variables like IQ scores.

For example, early studies suggested that connections among P-FIT areas correlated to IQ scores (Song et al., 2008; Shehzad et al., 2014) and that another set of connections called the default mode network (DMN) also were related to IQ (Hearne et al., 2016). The DMN is a set of brain areas active in nonrandom ways when the brain is in a resting state, thought necessary to maintain basic neural functioning. Other studies report similar results, adding to the weight of evidence supporting the P-FIT (Santarnecchi et al., 2017a; Santarnecchi et al., 2017b) and explicating brain–intelligence relationships, even in children (Kim et al., 2016).

Connectivity analyses are exciting because the results tie intelligence test scores directly to complex brain characteristics rather than to a list of brain areas. This is bad news for those who still might argue that intelligence is too amorphous or meaningless a concept for scientific study or who find psychometric methods unconvincing. It is good news for advancing research to the next level of questions, such as, can intelligence be predicted from brain images?

5.3.3 *Predicting Intelligence from Brain Connectivity*

Predicting intelligence from brain images has been a lofty goal since the first PET study in 1988. Why is this important? Generally, in science, identifying variables that predict a phenomenon gives important clues about the nature of the phenomenon (Yarkoni and Westfall, 2017). If we know what aspects of the brain predict intelligence scores, we can focus on understanding more about those aspects, including how they develop and what else might influence them.

Early prediction studies were limited by small samples and rudimentary methods; none were compelling (Haier, 2017). Later studies are more informative. For example, Penke and colleagues (2012) used DTI in 420 older adults to assess multiple white matter tracts throughout the brain. They correlated the integrity of each tract with intelligence scores, looking for the connections most related to intelligence. These correlations were not significant for any one tract, but when all white matter was combined, global white matter was correlated with the information-processing speed subtest score and accounted for 10 percent of the variance. This was a small result but encouraging. A preliminary DTI study of more than 2,000 research participants reports robust relationships between *g* scores and white matter in specific tracts (Stammen et al., 2022).

An early study based on graph analysis invigorated the pursuit of IQ prediction with an incredible claim (Finn et al., 2015). This study was done on an fMRI database collected from a sample (*N* = 126) obtained by the Human Connectome Project (HCP). These researchers took the individual differences approach seriously. Instead of comparing the average group functional connectivity patterns for each of the six task conditions (including working memory, language, emotion, motor, and two resting conditions) performed during the image acquisition, they computed functional connectivity patterns among 268 preselected nodes for each person for each task condition. The comparisons of the six functional patterns for each individual

revealed that the patterns were stable across tasks. This was surprising, since function changes depending on task. The apparent stability indicates that the pattern of relationships among areas may not change much even if the amount of activity within areas does. Because of the stability, each person could be characterized by their average connectivity pattern, which was unique enough to identify that person, similarly to a fingerprint. For example, one person might have the strongest functional connectivity between frontal and parietal areas, irrespective of task, and another person's pattern might show the strongest functional pattern between frontal and temporal areas.

More surprisingly, and to our main interest, each person's functional connectivity pattern was related to their fluid intelligence score. Connectivity of areas consistent with the P-FIT and the DMN were among the most predictive of fluid intelligence. The correlation between predicted and actual scores was 0.50 for the frontal–parietal network connectivity patterns, and 0.35 for the DMN. Interestingly, these values are greater than those reported by combining structural brain features (total brain volume and cortical thickness) as reported by Ritchie and colleagues (2015).

The analyses in the Finn et al. (2015) report are complex, but there are independent replications and extensions with similar results from other research groups. For example, Dubois and colleagues (2018) computed a correlation of 0.46 between the *g*-factor and resting-state functional connectivity among areas widely distributed across the brain; Figure 5.8 illustrates how they computed *g* (left) and the correlation between this *g* and the *g* predicted from connectivity (right). Their study was based on 884 individuals carefully selected from the HCP.

Neuroimaging prediction research is at an early stage and is influenced by at least three general problems that often create inconsistencies among individual studies:

1. *Statistical*: Correlations can be calculated among many brain features in many brain areas and among different intelligence

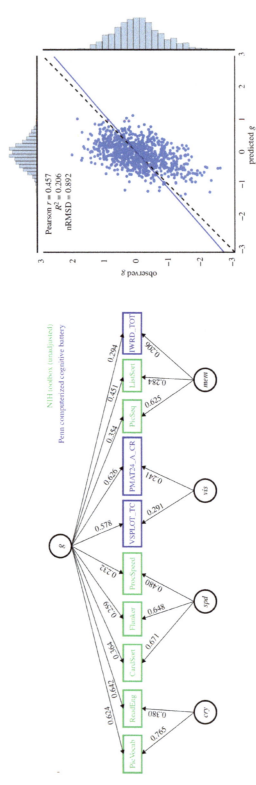

Figure 5.8 Predicting *g* from brain imaging. (left) Bifactor model (see Figure 3.3, bottom right) representing how ten tests are organized to obtain general (*g*) and specific ability scores for crystallized ability (cry), processing speed (spd), visuospatial ability (vis), and memory (mem). (right) Scatterplot of predicted values of the general factor of intelligence (*g*) based on functional connectivity (*g*) versus actual (observed) values. From Dubois et al. (2018).

test scores. Given the large number of possible comparisons, undoubtedly some findings would be due to chance, and others would be missed if the sample size was too small to detect an actual result. Recent studies with large samples minimize false-positive and false-negative results, especially when replication samples are reported.

2. *Individual differences*: Not all brains work the same way, so computing a group average always results in loss of information. New studies that focus on analyses of individuals are just emerging (Dubois and Adolphs, 2016).

3. *Measures*: Although estimates of *g* from different test batteries are virtually identical, many studies still are limited to a single measure. Two people with the same test score, or even the same estimate of *g*, may have different underlying neuro-structures (Haier, 2009). This is why adding multiple measures of intelligence to extract a stable *g* is invaluable, especially when combined with large and well-defined samples.

Here are five recommendations for optimally investigating intelligence and neuroimaging analyses (Haier et al., 2009):

1. Use diverse measures tapping abstract, verbal, numerical, and spatial content domains.

2. Use three or more measures to define each group factor. These group factors should fit the main factors in models such as the CHC or the *g*-VPR model.

3. Measures for each group factor should include both speeded and nonspeeded tests.

4. Use three or more group factors to define the *g*-factor.

5. Separate sources of variance contributing to participants' performance on the measures. Scores result from *g*, cognitive abilities (group factors), and cognitive skills (test specificities). Separate brain correlation analyses for each source would be most informative.

Predicting individual differences in intelligence from neuroimages is now possible. We think this is an amazing advance that would awe the early generation of intelligence researchers working only with psychometrics. Whether the prediction will ever be strong enough to replace psychometric assessments is an empirical question (Gabrieli et al., 2015). Most probably, the upper limit of predicting intelligence from neuroimages likely has not yet been achieved as we await studies that combine structural and functional brain features and connectivity patterns with DNA, as we will see in Chapter 6.

Potentially, imaging research may provide new ways to assess intelligence with greater reliability and validity than current methods. Imagine a brain scan that can predict important real-world outcomes better than IQ tests, the SAT or PISA, or vocational testing. Scans that take a few minutes already are less expensive than test preparation courses that require active participation for weeks. Furthermore, unlike currently used tests, cheating or faking a brain scan would be difficult (as long as the identity of the person in the scanner can be verified, perhaps with DNA). Think about that.

5.4 Brain Efficiency

We have already mentioned brain efficiency as a possible explanation for the finding of lower brain activity in people who perform better on a cognitively complex or demanding ability test (i.e., inverse correlations between cerebral glucose metabolic rate and RAPM scores) (Haier et al., 1988). At the time, this inverse relationship was surprising, since no previous intelligence research had used direct measures of brain energy and previous PET studies did not examine individual differences in performance. The inverse correlations were interpreted as evidence that it was not how hard the brain worked but rather how efficiently it worked that was related to intelligence.

The rudiments of the idea that smarter brains are more efficient can be traced back to Francis Galton's studies of reaction time. Brighter people responded to stimuli faster and more accurately. This general finding has been replicated many times with increasingly sophisticated measures of reaction time to a variety of decision-making tasks (Jensen, 1998, 2006). Charles Spearman (1923) thought about intelligence metaphorically as mental energy (see also Lykken, 2005). You can imagine that intelligence might be related to having more mental energy or to using mental energy more efficiently. Reaction time can be thought of as an indirect measure of invested mental energy. A direct measure of brain energy is quantified by glucose metabolic rate assessed with PET.

How a brain might be efficient is still an open question. There are a number of possibilities, and more than a single explanation is likely. For example, more intelligence is associated with bigger brains and more gray matter. This might mean that an intelligent person has more cognitive resources to apply to a problem and hence can solve it faster with less activity overall. Or, an intelligent person might have more white matter connectivity among salient areas, resulting in faster information processing. Perhaps intelligence is related to characteristics of neurons or synapses. There are empirical data that address these possibilities, as we now present.

The 1988 PET results became the modern version of the brain efficiency hypothesis (BEH; sometimes called the neural efficiency hypothesis, or NEH). The individual differences approach of correlating brain activity with task performance was adopted by many subsequent imaging studies. Haier's group also followed up with a prediction that learning a complex task would result in lower glucose metabolic rate (GMR) if the brain, somehow, became more efficient by learning what networks were *not* necessary for good performance. They tested this hypothesis using the famous video game *Tetris*. As predicted, GMR decreased after students became expert at *Tetris*, even though after fifty days of practice their scores were higher and the

game play more difficult (Haier et al., 1992). Haier's group also predicted that people with low IQ would show higher GMR, suggesting inefficient brains. This was the case in a PET study that compared low-IQ individuals, people with Down's syndrome, and matched controls (Haier et al., 1995).

The BEH attracted a number of scientists. For example, studies led by the Austrian differential psychologist Aljoscha Neubauer and colleagues (2005) used EEG methods and research designs that incorporated task content, task complexity, and sex. They found that the amplitudes of neural responses to verbal or visuospatial problems were negatively correlated with measures of verbal or visuospatial reasoning, respectively, consistent with the BEH. There was also an interesting interaction with the sex variable. The relations between EEG amplitude and test scores were strongest for men solving visuospatial problems and for women solving verbal problems, mirroring average sex differences in verbal and visuospatial abilities (see Chapter 10).

Anna-Lena Schubert and colleagues (2017) also used EEG to study the relationship between individual differences in processing speed and intelligence. They found evidence that "more intelligent individuals benefit from a more efficient transmission of information from frontal attention and working memory processes to temporal-parietal processes of memory storage" (p. 1498). Their results showed that speed of higher-order information processing explained an estimated 80 percent of the variance in general intelligence (Box 4.2).

The BEH generally finds support in some studies that used graph analyses and examined P-FIT areas (van den Heuvel et al., 2009; Cole et al., 2012; Penke et al., 2012; Ryman et al., 2016; Santarnecchi and Rossi, 2016; Santarnecchi et al., 2015; Vakhtin et al., 2014; Hilger et al., 2017). For example, the Basten group showed that a distinction between activated and deactivated brain networks during task performance was helpful for understanding individual differences in brain efficiency (Basten et al., 2013).

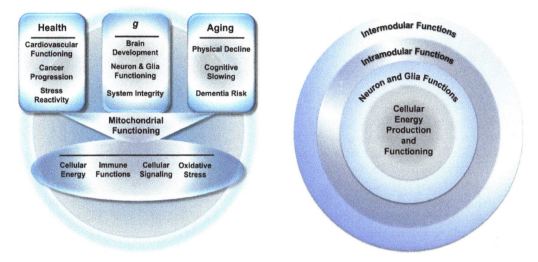

Figure 5.9 A role for mitochondria? (left) The multiple functions supported by mitochondria, including cellular energy production, are critical to general health, brain development and functioning, cognition, and various age-related changes in health and cognition. (right) Individual differences in *g* could be influenced by performance at one or several levels of brain functioning. Deficits at lower levels will compromise functioning at all higher levels, but compromises at higher levels do not necessarily imply deficits at lower levels (Geary, 2018).

However, not all studies are consistent with the BEH. For example, a large comprehensive study (*N* = 1,096 with resting-state fMRI from the HCP) has not replicated the relationship between efficient connectivity (assessed with multiple methods) and scores for either fluid or crystallized intelligence (Kruschwitz et al., 2018). Meta-analyses and insightful reviews of the BEH found only limited support (Basten et al., 2015; Basten and Fiebach, 2021). One comprehensive review of the relevant literature concludes that "while efficiency does characterize many brain-ability relationships, it does not rise to the level of a general functional principle that uniformly applies across networks and situations" (Euler and McKinney, 2021, p. 88).

But there are newer data from studies using advanced connectivity analysis techniques. One study with 812 volunteers (mean age = 28.6 years), for example, indicates that multiple functional systems in the brain are related to the *g*-factor and that functioning among these systems is reconfigured according to the task (Thiele et al., 2022). Individual differences among people in reconfiguration are related to intelligence differences in ways that suggest that higher

intelligence is related to less reconfiguration. The authors conclude, "Our results support neural efficiency theories of cognitive ability and reveal insights into human intelligence as an emergent property from a distributed multitask brain network" (p. 1).

Other new approaches continue to test the BEH, although the weight of evidence for these is not yet established. For example, Haier and colleagues (1988) speculated that brain efficiency may have its origins at the level of the synapse or neuron. Inside neurons, for instance, mitochondria provide cells with energy. Might there be differences in neuronal mitochondria between better and worse problem solvers (Haier, 2003)?

David Geary (2018, 2019) has proposed a comprehensive framework for examining whether mitochondria functioning is a relevant biological mechanism underlying intelligence. The model (1) incorporates empirical findings relating the *g*-factor, health, and aging; (2) has levels of analysis from cellular energy production up to brain function within and across regions (Figure 5.9); and (3) includes a detailed list of evidence that would refute the specific hypotheses. And just as science is supposed to work, there is evidence that refutes key aspects of the mitochondria model (Matzel et al., 2020). Matzel and colleagues wrote, "Geary's hypothesis should serve as a role-model on how to create and polish new theories of intelligence. His theory makes precise and testable predictions and explores ideas outside the beaten track. However, as was the case for previous attempts to describe a unitary mechanism behind intelligence, tests of Geary's theory might not prove fruitful as the theory does not match well and sometimes is even incompatible with what is known. Déjà vu all over again" (p. 7). Nonetheless, this model and the subsequent testing of its hypotheses, even with negative results, illustrate a key development in the history of psychology: intelligence research is moving into mainstream neuroscience with testable hypotheses (Burgoyne and Engle, 2020).

A provocative example of intelligence research moving deeper into the brain is provided by Genc and colleagues (2018). They used an advanced MRI technique called neurite orientation dispersion and density imaging. This technique allows quantitative inferences about neuron structure. Using multiple samples, they found that *less* density and *less* arborization of dendrites (the branching ends of neurons) are related to higher intelligence scores. The authors interpret the inverse correlations as suggesting that "the neuronal circuitry associated with higher intelligence is organized in a sparse and efficient manner, fostering more directed information processing and less cortical activity during reasoning. ... These results offer a neuroanatomical explanation underlying the neural efficiency hypothesis of intelligence" (p. 1). Key findings are illustrated and explained in Figure 5.10.

Another research group (Goriounova et al., 2018), moving deeper into the brain, found a *positive* correlation between intelligence and complexity of dendrites in temporal lobe pyramidal neurons : "Larger dendritic trees enable pyramidal neurons to track activity of synaptic inputs with higher temporal precision, due to fast action potential kinetics. Indeed, we find that human pyramidal neurons of individuals with higher IQ scores sustain fast action potential kinetics during repeated firing. These findings provide ... evidence that human intelligence is associated with neuronal complexity, action potential kinetics and efficient information transfer from inputs to output within cortical neurons" (p. 1). The key findings are shown in Figure 5.11. This finding was based on the temporal lobe (the only area studied) and at first glance seems to be in the opposite direction of what Genc et al. found throughout the cerebral cortex.

However, in a follow-up study, the authors provided a more detailed microstructural analysis of cortical layers in temporal lobe. They showed that thicker cortex in subjects with higher intelligence did not contain more neurons but rather similar numbers of larger cells at lower neuronal densities (Heyer et al., 2021). Thus, lower neurite density in the Genc et al. study could manifest at a cellular level as a lower density of neurons with larger dendrites.

How do these features of individual neurons, such as long dendrites and fast signals, relate to the functional network of the whole brain as can be measured by fMRI or MEG? This question was put to the test in neurosurgery patients by measuring their resting-state network activity before the surgical procedure and quantifying the cellular features in resected cortical tissue after surgery. Using this method, Douw et al. (2021) found that a measure of local network integration – a greater network centrality – correlated with more integrative properties at the cellular level in the area of interest. In patients who had neurons with larger dendrites and faster signals in the temporal cortical area, this area played a more important role as a hub in the global brain network. This finding emphasizes that cellular properties associated with intelligence do not only exert their effect locally but lead to an increased functional integration within the whole brain.

So, it appears that positive and negative correlations between intelligence and

low-IQ individuals

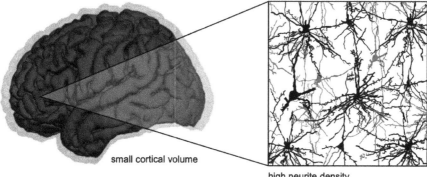

small cortical volume

high neurite density
high neurite orientation dispersion

high-IQ individuals

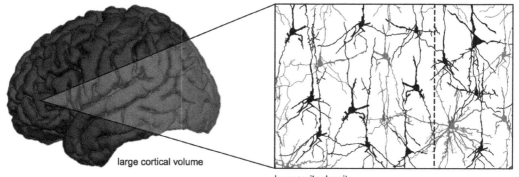

large cortical volume

low neurite density
low neurite orientation dispersion

Figure 5.10 Neurons and human intelligence. The depiction of differences between low-IQ and high-IQ individuals for brain volume, neurite density, and arborization of dendritic trees within the cortex. High-IQ individuals are likely to possess more cortical volume than low-IQ individuals, which is indicated by differently sized brains (left side) and differently sized panels showing exemplary magnifications of neuron and neurite microstructure (right side). The difference in cortical volume is highlighted by the shadow around the upper brain. Owing to their larger cortices, it is conceivable that high-IQ individuals benefit from the processing power of additional neurons, which are marked by the dotted line in the lower panel. The cerebral cortex of high-IQ individuals is characterized by a low degree of neurite density and orientation dispersion, which is indicated by smaller and less ramified dendritic trees in the respective panel. Intellectual performance is likely to benefit from this kind of microstructural architecture because restricting synaptic connections to an efficient minimum facilitates the differentiation of signals from noise while saving network and energy resources. Neurons and neurites are depicted in black and gray to create a sense of depth. Please note that this depiction does not correspond to the actual magnitude of effect sizes reported in the study. For the purpose of an easier visual understanding, differences in both macrostructural and microstructural brain properties are highly accentuated (Genc et al., 2018).
Courtesy of Erhan Genc.

neuron features both can be interpreted as consistent with the BEH. There are important methodological differences between these studies and other nuances that could be resolved with additional research. But can opposite findings ever be interpreted to support the same efficiency hypothesis?

Consider that efficiency of brain networks is a dynamic concept. During problem solving, there is a sequence of activation and

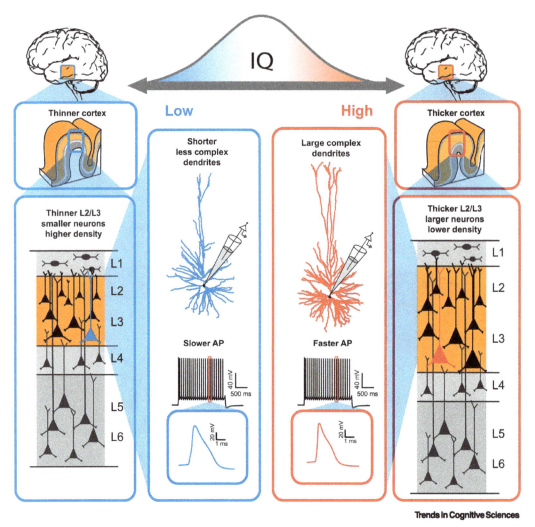

Figure 5.11 A cellular basis of human intelligence. Higher IQ scores associate with larger dendrites, faster action potentials during neuronal activity, and more efficient information tracking in pyramidal neurons of the temporal cortex (Goriounova et al., 2018).
Courtesy of Natalia Goriounova.

deactivation among and within brain areas (Haier, 2017). For example, in an fMRI study, brain areas in the DMN and the dorsal attention network showed that activation during complex problem solving shifted back and forth, and this switching was modulated by fluid intelligence (Ramchandran et al., 2019). The authors termed this *distributed neural efficiency*. The concept of distributed neural efficiency is similar to the adaptive functional connectivity proposed by Aron Barbey (2018) and to dynamic modularity observed in the fMRI study by Hilger et al. (2020; see also the review by Cohen and D'Esposito, 2021). These models help explain individual differences because not all brains will respond to a problem with precisely the same pattern of activations and deactivations in salient networks with the same sequencing or timing. This could even be the case for individuals matched on intelligence test scores like IQ, since the same score can be obtained with different combinations of subtest scores. This underscores the crucial relevance of going from the group to the individual level (Colom and Román, 2018).

Box 5.5 Is There a *g*-Neuron?

Based on evidence from neuroscience, psychiatry, neurology, gerontology, and comparative psychology, Oliver Bruton (2021) proposed a reductionistic hypothesis regarding the neurological basis of general intelligence (*g*): although there are more than 86 billion neurons in the human brain, the von Economo neuron (VEN, for short) represents a crucial biological component of *g*. But why VENs? Because they can work by "rapidly inducing the coherence of neural oscillations within a functionally invariant parieto-frontal network underlying higher cognition, thereby facilitating mental efficiency" (p. 1). VENs have been identified in the anterior insula, the DLPFC, the ACC, and the medial frontopolar cortex of the human brain.

As the author acknowledges, empirical testing of this hypothesis is still required, and the very idea that "only a single subclass of neurons should determine a trait as complex as *g* appears rather fantastic" (p. 12). Nonetheless, this is the kind of novel and testable hypothesis that can drive further advances for understanding intelligence.

All these studies, along with Geary's framework for mitochondria, highlight advances of intelligence research delving into the brain based on empirical findings and conceptual refinements (see Box 5.5 and Goriounova and Mansvelder, 2019). Progress continues, even down to the synapses that link neurons to each other. There is now methodology to map billions of synapses of different neurotransmitter systems throughout the mouse brain (Zhu et al., 2018). When similar methodology becomes available for humans, synaptome maps may provide insights into individual differences in intelligence at the level of synaptic activity, including testing whether some synapses regulated by specific neurotransmitters might be more efficient than others.

So, what is the weight of evidence regarding the BEH? The emerging weight of evidence appears to favor the general idea. The original concept was simple, but the brain is complex. During problem solving or at rest, the brain is likely to dynamically engage both more activity and more efficient activity, perhaps simultaneously in different networks or in flexible sequences as information is processed throughout the brain. The three cautions noted earlier (the so-called Haier's laws) are applicable: no story about the brain is simple, no one study is definitive, and it takes time to establish a compelling weight of evidence.

5.5 Linking Intelligence, Brain Connectivity, and Neurons

The P-FIT and the BEH have stimulated research studies and inspired other network frameworks to help understand intelligence. The process overlap theory (Kovacs and Conway, 2016), discussed in Chapter 4, focuses on working memory and executive control as the crux of *g*. The multiple demand (MD) theory, like the P-FIT, emphasizes parieto-frontal areas, but the focus is on specific brain areas that are commonly activated across diverse tasks, and any one area can contribute to different tasks (Duncan, 2010). Aron Barbey (2018, 2021) has proposed the network neuroscience theory (NNT) to tie brain networks to intelligence. This framework is based on connectivity studies that find that fluid intelligence is related to *weaker* brain connections, especially parieto-frontal networks, perhaps permitting more flexible reasoning. Crystallized intelligence is more related to the small world local connection hubs, especially in the DMN, necessary to retrieve semantic and episodic information.

The NNT emphasizes the dynamic functional organization of the brain.

Euler and McKinney (2021) have reviewed all these models and a few others. All of them propose somewhat different explanations of how brain connectivity might explain *g* and other intelligence factors. The good news is that they all agree with the distributed nature of brain areas related to intelligence. They also link psychometric observations of the hierarchical nature of *g* and intelligence factors to possible neuroimaging findings that suggest hierarchies in the brain (Román et al., 2014). We also note that brain network studies are not unique to intelligence. Personality and creativity have also been studied and may even share some networks with intelligence (Jung and Chohan, 2019), but that is beyond the scope of this chapter.

Hilger and Sporns (2021) also reviewed connectivity models related to intelligence. They further examined connectivity within modular subnetworks. Figure 5.12 illustrates their summary of graph theory applied to brain efficiency and modularity. Figure 5.13 illustrates their summary of multiple studies that define structural and functional connectivity models of intelligence. Figure 5.13 is the culmination of more than thirty years of neuroimaging studies of intelligence and nicely demonstrates how clarity among inconsistent findings evolves over time (Haier's third law). There are also new projects that aim to combine the results of all fMRI studies into one database (e.g., Neurosynth.org) so we are likely to see figures like Figure 5.13 develop in greater detail. Moreover, a number of multiple international consortia are now pooling resources into mega data sets that include neuroimaging, genetic, and, in early stages, cognitive test batteries (e.g., ENIGMA, Connectome, the Brain Initiative, and the Human Brain Project; see Khundrakpam et al., 2021). The creation of these collaborative projects is a major advance in science and bodes well for intelligence research.

The quantification of complex patterns of brain connectivity is a major scientific advance for intelligence research. Nonetheless, we are still mostly in a descriptive stage. Recall the orchestra analogy (Box 5.3). It is one thing to say that the conductor is the person standing on the podium at the front of the orchestra. It is another to say that the conductor maintains the tempo and signals emphasis for the various orchestral sections. It is still another matter to say how the conductor maintains tempo, and so forth. We cannot stop by saying that the P-FIT model acts like a conductor. To understand intelligence, we must learn how brain networks function at a deeper level. Studies of neurons, synapses, and intelligence are just beginning. At some point, the next stage may be the experimental manipulation of brain features to test hypotheses about how they work with respect to intelligence. One outcome could be finding ways to enhance intelligence, as discussed in Chapter 13. But, before we go there, we need to consider genetics. Genes play a role in understanding the "how" of intelligence. Probably no topic in intelligence research is more misunderstood and misrepresented than the genetic story discussed next, in Chapter 6.

5.6 Summary

Psychometric intelligence test scores are correlated to individual differences in brain structure and function. Early neuroimaging studies identified specific brain areas associated with individual differences in scores. Newer studies show that connectivity among brain areas is related to intelligence and that brain connectivity can be used to predict intelligence scores.

Neuroimaging results are fascinating and have advanced our knowledge. Nevertheless, we are still mostly in a descriptive stage despite the increasing sophistication of brain–intelligence models. We have to explain how brain structures and networks function. Connectivity and efficiency investigations ever deeper into the brain are promising, but the "how" of intelligence remains a mystery

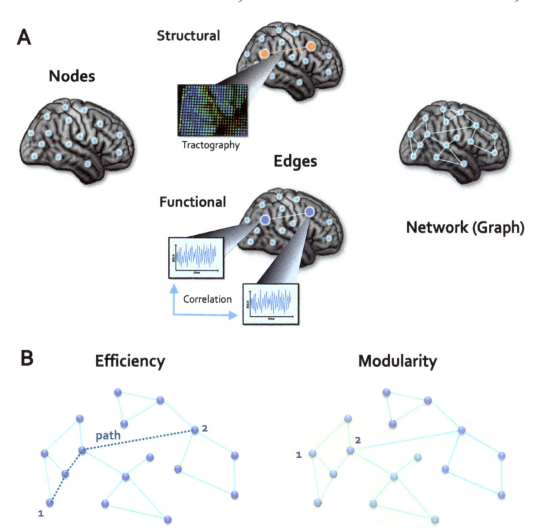

Figure 5.12 Schematic of structural and functional brain network construction and key network metrics. (A) Network construction. First (left), network nodes are defined based on, for example, an anatomical brain atlas. Second (middle), edges are defined between pairs of nodes by measuring white matter fiber tracts (structural network, e.g., measured with DTI) or by estimating temporal relationships between time series of BOLD signals (functional network, e.g., measured with resting-state fMRI). Third (right), nodes and edges together define a graph (network) whose topological properties can be studies with global (whole-brain) and nodal (region-specific) graph-theoretical measures. (B) Key network metrics. Network efficiency (left) is derived from the lengths of shortest paths between node pairs. In this example, the path between nodes 1 and 2 has a length of three steps. Network modularity (right) partitions the network into communities or modules that are internally densely connected, with sparse connections between them. In this example, the network consists of four modules illustrated in different colors. Individual nodes differ in the way they connect to other nodes within their own module (within-module connectivity) and to nodes in other modules (diversity of between-module connectivity, nodal participation). Here node 1 has low participation, while node 2 has high participation (Hilger and Sporns, 2021).
Courtesy of Kirsten Hilger and Olaf Sporns.

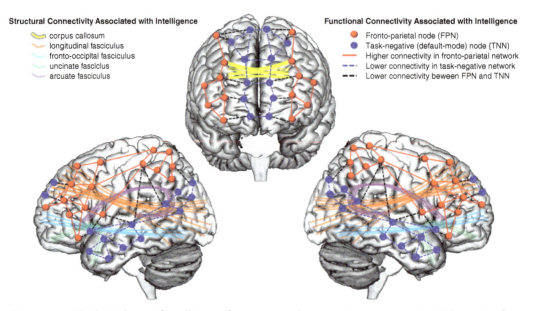

Figure 5.13 The brain bases of intelligence from a network neuroscience perspective. Schematic of selected structural and functional brain connections associated with intelligence across different studies (Hilger and Sporns, 2021).
Courtesy of Kirsten Hilger and Olaf Sporns.

that likely requires advanced genetic research. Solving this mystery is the ultimate aim of brain–intelligence theories and research.

5.7 Questions for Discussion

5.1   Do you find that correlations between intelligence test scores and neuroimaging measurements increase your confidence that test scores are assessing real differences among people?

5.2   Why do you think bigger brains might be associated with higher intelligence?

5.3   How would you explain the main difference between the P-FIT framework and newer connectivity theories of intelligence?

5.4   If it becomes possible to predict academic performance from brain images at least as good as current tests, would you prefer a brain image for college entrance instead of standardized tests?

5.5   Do you have any personal experiences that support or do not support the brain efficiency hypothesis?

References

Barbey, A. 2018. Network neuroscience theory of human intelligence. *Trends in Cognitive Sciences*, 22, 8–20.

Barbey, A. 2021. Human intelligence and network neuroscience. In Barbey, A., Karama, S., & Haier, R. J. (eds.), *Cambridge handbook of intelligence and cognitive neuroscience*. New York: Cambridge University Press.

Barbey, A. K., Colom, R., & Grafman, J. 2012a. Distributed neural system for emotional intelligence revealed by lesion mapping. *Social Cognitive and Affective Neuroscience*, 9, 265–272.

Barbey, A. K., Colom, R., Solomon, J., et al. 2012b. An integrative architecture for general intelligence and executive function revealed by lesion mapping. *Brain*, 135, 1154–1164.

Basso, A., De Renzi, E., Faglioni, P., Scotti, G., & Spinnler, H. 1973. Neuropsychological evidence for the existence of cerebral areas critical to the performance of intelligence tasks. *Brain*, 96, 715–728.

Basten, U., & Fiebach, C. J. 2021. Functional brain imaging of intelligence. In Barbey, A., Karama, S., & Haier, R. J. (eds.), *The Cambridge handbook of intelligence and cognitive neuroscience*. New York: Cambridge University Press.

Basten, U., Hilger, K., & Fiebach, C. J. 2015. Where smart brains are different: A quantitative meta-analysis of functional and structural brain imaging studies on intelligence. *Intelligence*, 51, 10–27.

Basten, U., Stelzel, C., & Fiebach, C. J. 2013. Intelligence is differentially related to neural effort in the task-positive and the task-negative brain network. *Intelligence*, 41, 517–528.

Blair, C. 2006. How similar are fluid cognition and general intelligence? A developmental neuroscience perspective on fluid cognition as an aspect of human cognitive ability. *Behavioral and Brain Sciences*, 29, 109.

Bowren, M., Adolphs, R., Bruss, J., et al. 2020. Multivariate lesion-behavior mapping of general cognitive ability and its psychometric constituents. *Journal of Neuroscience*, 40, 8924–8937.

Brodmann, K. 1909. *Vergleichende Lokalisationslehre der Grosshirnrinde in ihren Prinzipien dargestellt auf Grund des Zellenbaues*. Leipzig: Barth.

Bruner, E., De Lazaro, G. R., De La Cuetara, J. M., et al. 2014. Midsagittal brain variation and MRI shape analysis of the precuneus in adult individuals. *Journal of Anatomy*, 224, 367–376.

Bruner, E., Martin-Loeches, M., Burgaleta, M., & Colom, R. 2011. Midsagittal brain shape correlation with intelligence and cognitive performance. *Intelligence*, 39, 141–147.

Bruton, O. J. 2021. Is there a "g-neuron"? Establishing a systematic link between general intelligence (*g*) and the von Economo neuron. *Intelligence*, 86, 101540.

Burgoyne, A. P., & Engle, R. W. 2020. Mitochondrial functioning and its relation to higher-order cognitive processes. *Journal of Intelligence*, 8, 14.

Burkart, J. M., Schubiger, M. N., & Van Schaik, C. P. 2017. The evolution of general intelligence. *Behavioral and Brain Sciences*, 40, e195.

Cohen, J. R., & D'Esposito, M. 2021. An integrated, dynamic functional connectome underlies intelligence. In Barbey, A., Karama, S., & Haier, R. J. (eds.), *Cambridge handbook of intelligence and cognitive neuroscience*. New York: Cambridge University Press.

Cole, M. W., Yarkoni, T., Repovs, G., Anticevic, A., & Braver, T. S. 2012. Global connectivity of prefrontal cortex predicts cognitive control and intelligence. *Journal of Neuroscience*, 32, 8988–8999.

Colom, R., Karama, S., Jung, R. E., & Haier, R. J. 2010. Human intelligence and brain networks. *Dialogues in Clinical Neuroscience*, 12, 489–501.

Colom, R., & Román, F. J. 2018. Enhancing intelligence: From the group to the individual. *Journal of Intelligence*, 6, 11.

Deary, I. J., Penke, L., & Johnson, W. 2010. The neuroscience of human intelligence differences. *Nature Reviews Neuroscience*, 11, 201–211.

Douw, L., Nissen, I. A., Fitzsimmons, S. M. D. D., et al. 2021. Cellular substrates of functional network integration and memory in temporal lobe epilepsy. *Cerebral Cortex*, 32, 2424–2436.

Drakulich, S., & Karama, S. 2021. Structural brain imaging of intelligence. In Barbey, A., Karama, S., & Haier, R. J. (eds.), *Cambridge handbook of intelligence and cognitive neuroscience*. New York: Cambridge University Press.

Dubois, J., & Adolphs, R. 2016. Building a science of individual differences from fMRI. *Trends in Cognitive Sciences*, 20, 425–443.

Dubois, J., Galdi, P., Paul, L. K., & Adolphs, R. 2018. A distributed brain network predicts general intelligence from resting-state human neuroimaging data. *Philosophical Transactions of the Royal Society of London, Series B*, 373, 1–13.

Duncan, J. 2010. The multiple-demand (MD) system of the primate brain: mental programs for intelligent behaviour. *Trends in Cognitive Sciences*, 14, 172–179.

Duncan, J., Burgess, P., & Emslie, H. 1995. Fluid intelligence after frontal lobe lesions. *Neuropsychologia*, 33, 261–268.

Euler, M. J., & McKinney, T. L. 2021. Evaluating the weight of the evidence: Cognitive neuroscience theories of intelligence. In Barbey, A., Karama, S., & Haier, R. J. (eds.), *The Cambridge handbook of intelligence and cognitive neuroscience*. New York: Cambridge University Press.

Finn, E. S., Shen, X., Scheinost, D., et al. 2015. Functional connectome fingerprinting: Identifying individuals using patterns of brain connectivity. *Nature Neuroscience*, 18, 1664–1671.

Gabrieli, J. D. E., Ghosh, S. S., & Whitfield-Gabrieli, S. 2015. Prediction as a humanitarian and pragmatic contribution from human cognitive neuroscience. *Neuron*, 85, 11–26.

Geary, D. C. 2018. Efficiency of mitochondrial functioning as the fundamental biological mechanism of general intelligence (*g*). *Psychological Review*, 125, 1028–1050.

Geary, D. C. 2019. Mitochondria as the linchpin of general intelligence and the link between *g*, health, and aging. *Journal of Intelligence*, 7, 25.

Genc, E., & Fraenz, C. 2021. Diffusion-weighted imaging of intelligence. In Barbey, A., Karama, S., & Haier, R. J. (eds.), *The Cambridge handbook of intelligence and cognitive neuroscience*. New York: Cambridge University Press.

Genc, E., Fraenz, C., Schluter, C., et al. 2018. Diffusion markers of dendritic density and arborization in gray matter predict differences in intelligence. *Nature Communications*, 9, 1905.

Gignac, G. E., & Bates, T. C. 2017. Brain volume and intelligence: The moderating role of intelligence measurement quality. *Intelligence*, 64, 18–29.

Gläscher, J., Rudrauf, D., Colom, R., et al. 2010. Distributed neural system for general intelligence revealed by lesion mapping. *Proceedings of the National Academy of Sciences of the United States of America*, 107, 4705–4709.

Gläscher, J., Tranel, D., Paul, L. K., et al. 2009. Lesion mapping of cognitive abilities linked to intelligence. *Neuron*, 61, 681–691.

Goriounova, N. A., Heyer, D. B., Wilbers, R., et al. 2018. Large and fast human pyramidal neurons associate with intelligence. *Elife*, 7, e41714.

Goriounova, N. A., & Mansvelder, H. D. 2019. Genes, cells and brain areas of intelligence. *Frontiers in Human Neurosciences*, 13, 44.

Grasby, K. L., Jahanshad, N., Painter, J. N., et al. 2020. The genetic architecture of the human cerebral cortex. *Science*, 367, 6690.

Gratton, C., Laumann, T. O., Nielsen, A. N., et al. 2018. Functional brain networks are dominated by stable group and individual factors, not cognitive or daily variation. *Neuron*, 98, 439–452.

Green, S., Blackmon, K., Thesen, T., et al. 2018. Parieto-frontal gyrification and working memory in healthy adults. *Brain Imaging and Behavior*, 12, 303–308.

Haier, R. J. 2003. Positron emission tomography studies of intelligence: From psychometrics to neurobiology. In Nyborg, H. (ed.), *The scientific study of general intelligence*. New York: Elsevier Science.

Haier, R. J. 2009. Neuro-intelligence, neuro-metrics and the next phase of brain imaging studies. *Intelligence*, 37, 121–123.

Haier, R. J. 2017. *The neuroscience of intelligence*. Cambridge: Cambridge University Press.

Haier, R. J., Chueh, D., Touchette, P., et al. 1995. Brain size and cerebral glucose metabolic rate in nonspecific mental retardation and Down syndrome. *Intelligence*, 20, 191–210.

Haier, R. J., Colom, R., Schroeder, D. H., et al. 2009. Gray matter and intelligence factors: Is there a neuro-*g*? *Intelligence*, 37, 136–144.

Haier, R. J., Siegel, B. V., Jr., Maclachlan, A., et al. 1992. Regional glucose metabolic changes after learning a complex visuospatial/motor task: A positron emission tomographic study. *Brain Research*, 570, 134–143.

Haier, R. J., Siegel, B. V., Jr., Nuechterlein, K. H., et al. 1988. Cortical glucose metabolic-rate correlates of abstract reasoning and attention studied with positron emission tomography. *Intelligence*, 12, 199–217.

Hawkins, J. 2021. *A thousand brains: A new theory of intelligence*. New York: Basic Books.

Hearne, L. J., Mattingley, J. B., & Cocchi, L. 2016. Functional brain networks related to individual differences in human intelligence at rest. *Scientific Reports*, 6, 32328.

Heyer, D. B., Wilbers, R., Galakhova, A. A., et al. 2021. Verbal and general IQ associate with supragranular layer thickness and cell properties of the left temporal cortex. *Cerebral Cortex*, 32, 2343–2357.

Hilger, K., Ekman, M., Fiebach, C. J., & Basten, U. 2017. Efficient hubs in the intelligent brain: Nodal efficiency of hub regions in the salience network is associated with general intelligence. *Intelligence*, 60, 10–25.

Hilger, K., Fukushima, M., Sporns, O., & Fiebach, C. J. 2020. Temporal stability of functional brain modules associated with human intelligence. *Human Brain Mapping*, 41, 362–372.

Hilger, K., & Sporns, A. 2021. Network neuroscience methods for studying intelligence. In Barbey, A., Karama, S., & Haier, R. J. (eds.), *The Cambridge handbook of intelligence and cognitive neuroscience*. New York: Cambridge University Press.

Jansen, P. R., Nagel, M., Watanabe, K., et al. 2020. Genome-wide meta-analysis of brain volume identifies genomic loci and genes shared with intelligence. *Nature Communications*, 11, 5606.

Jensen, A. R. 1998. *The g factor: The science of mental ability*. Westport, CT: Praeger.

Jensen, A. R. 2006. *Clocking the mind: Mental chronometry and individual differences*. New York: Elsevier.

Johnson, W., Te Nijenhuis, J., & Bouchard, T. J., Jr. 2008. Still just 1 *g*: Consistent results from five test batteries. *Intelligence*, 36, 81–95.

Jung, R. E., & Chohan, M. O. 2019. Three individual difference constructs, one converging concept: Adaptive problem solving in the human brain. *Current Opinion in Behavioral Sciences*, 27, 163–168.

Jung, R. E., & Chohan, M. O. 2021. Biochemical correlates of intelligence. In Barbey, A., Karama, S., & Haier, R. J. (eds.), *The Cambridge handbook of intelligence and cognitive neuroscience*. New York: Cambridge University Press.

Jung, R. E., & Haier, R. J. 2007. The parieto-frontal integration theory (P-FIT) of intelligence: Converging neuroimaging evidence. *Behavioral and Brain Sciences*, 30, 135–154; discussion 154–187.

Karama, S., Colom, R., Johnson, W., et al. 2011. Cortical thickness correlates of specific cognitive performance accounted for by the general factor of intelligence in healthy children aged 6 to 18. *Neuroimage*, 55, 1443–1453.

Khundrakpam, B. S., Poline, J., & Evans, A. 2021. Research consortia and large-scale data repositories for studying intelligence. In Barbey, A., Karama, S., & Haier, R. J. (eds.), *The Cambridge handbook of intelligence and cognitive neuroscience*. New York: Cambridge University Press.

Kim, D.-J., Davis, E. P., Sandman, C. A., et al. 2016. Children's intellectual ability is associated with structural network integrity. *Neuroimage*, 124, Part A, 550–556.

Kovacs, K., & Conway, A. R. A. 2016. Process overlap theory: A unified account of the general factor of intelligence. *Psychological Inquiry*, 27, 151–177.

Kraus, B. T., Perez, D., Ladwig, Z., et al. 2021. Network variants are similar between task and rest states. *Neuroimage*, 229, 117743.

Kruschwitz, J. D., Waller, L., Daedelow, L. S., Walter, H., & Veer, I. M. 2018. General, crystallized and fluid intelligence are not associated with functional global network efficiency: A replication study with the human connectome project 1200 data set. *Neuroimage*, 171, 323–331.

Lee, J. J., McGue, M., Iacono, W. G., Michael, A. M., & Chabris, C. F. 2019. The causal influence of brain size on human intelligence: Evidence from within-family phenotypic associations and GWAS modeling. *Intelligence*, 75, 48–58.

Lykken, D. T. 2005. Mental energy. *Intelligence*, 33, 331–335.

Matzel, L. D., Crawford, D. W., & Sauce, B. 2020. Déjà vu all over again: A unitary biological mechanism for intelligence is (probably) untenable. *Journal of Intelligence*, 8, 24.

McDaniel, M. A. 2005. Big-brained people are smarter: A meta-analysis of the relationship between in vivo brain volume and intelligence. *Intelligence*, 33, 337–346.

Naghavi, H. R., & Nyberg, L. 2005. Common fronto-parietal activity in attention, memory, and consciousness: Shared demands on integration? *Consciousness and Cognition*, 14, 390–425.

Nave, G., Jung, W. H., Karlsson Linner, R., Kable, J. W., & Koellinger, P. D. 2019. Are bigger brains smarter? Evidence from a large-scale preregistered study. *Psychological Sciences*, 30, 43–54.

Neubauer, A. C., Grabner, R. H., Fink, A., & Neuper, C. 2005. Intelligence and neural efficiency: Further evidence of the influence of task content and sex on the brain–IQ relationship. *Cognitive Brain Research*, 25, 217–225.

Pearson, K. 1906. On the relationship of intelligence to size and shape of head, and to other physical and mental characteristics. *Biometrika*, 5, 105–146.

Penke, L., Maniega, S. M., Bastin, M. E., et al. 2012. Brain white matter tract integrity as a neural foundation for general intelligence. *Molecular Psychiatry*, 17, 1026–1030.

Pietschnig, J., Penke, L., Wicherts, J. M., Zeiler, M., & Voracek, M. 2015. Meta-analysis of associations between human brain volume and intelligence differences: How strong are they and what do they mean? *Neuroscience and Biobehavioral Reviews*, 57, 411–432.

Pineda-Pardo, J. A., Martínez, K., Román, F. J., & Colom, R. 2016. Structural efficiency within a parieto-frontal network and cognitive differences. *Intelligence*, 54, 105–116.

Poldrack, R. A. 2021. Diving into the deep end: A personal reflection on the MyConnectome study. *Current Opinion in Behavioral Sciences*, 40, 1–4.

Posner, M. I., Sheese, B. E., Odludas, Y., & Tang, Y. Y. 2006. Analyzing and shaping human attentional networks. *Neural Networks*, 19, 1422–1429.

Protzko, J., & Colom, R. 2021. Testing the structure of human cognitive ability using evidence obtained from the impact of brain lesions over abilities. *Intelligence*, 89, 101581.

Ramchandran, K., Zeien, E., & Andreasen, N. C. 2019. Distributed neural efficiency: Intelligence and age modulate adaptive allocation of resources in the brain. *Trends in Neuroscience and Education*, 15, 48–61.

Raz, N. 2021. Good sense and good chemistry: Neurochemical correlates of cognitive performance assessed in vivo through magnetic resonance spectroscopy. In Barbey, A., Karama, S., & Haier, R. J. (eds.), *The Cambridge handbook of intelligence and cognitive neuroscience*. New York: Cambridge University Press.

Reardon, P. K., Seidlitz, J., Vandekar, S., et al. 2018. Normative brain size variation and brain shape diversity in humans. *Science*, 360, 1222–1227.

Ritchie, S. J., Booth, T., Hernandez, M., et al. 2015. Beyond a bigger brain: Multivariable structural brain imaging and intelligence. *Intelligence*, 51, 47–56.

Román, F. J., Abad, F. J., Escorial, S., et al. 2014. Reversed hierarchy in the brain for general and specific cognitive abilities: A morphometric analysis. *Human Brain Mapping*, 35, 3805–3818.

Roth, G., & Dicke, U. 2005. Evolution of the brain and intelligence. *Trends in Cognitive Sciences*, 9, 250–257.

Ryman, S. G., Yeo, R. A., Witkiewitz, K., et al. 2016. Fronto-parietal gray matter and white matter efficiency differentially predict intelligence in males and females. *Human Brain Mapping*, 37, 4006–4016.

Santarnecchi, E., Emmendorfer, A., Pascual-Leone, A., et al. 2017a. Dissecting the parieto-frontal correlates of fluid intelligence: A comprehensive ALE meta-analysis study. *Intelligence*, 63, 9–28.

Santarnecchi, E., Emmendorfer, A., Tadayon, S., et al. 2017b. Network connectivity correlates of variability in fluid intelligence performance. *Intelligence*, 65, 35–47.

Santarnecchi, E., & Rossi, S. 2016. Advances in the neuroscience of intelligence: From brain connectivity to brain perturbation. *Spanish Journal of Psychology*, 19, 1–7.

Santarnecchi, E., Tatti, E., Rossi, S., Serino, V., & Rossi, A. 2015. Intelligence-related differences in the asymmetry of spontaneous cerebral activity. *Human Brain Mapping*, 36, 3586–3602.

Schubert, A. L., Hagemann, D., & Frischkorn, G. T. 2017. Is general intelligence little more than the speed of higher-order processing? *Journal of Experimental Psychology: General*, 146, 1498–1512.

Seghier, M. L., & Price, C. J. 2018. Interpreting and utilising intersubject variability in brain function. *Trends in Cognitive Sciences*, 22, 517–530.

Shehzad, Z., Kelly, C., Reiss, P. T., et al. 2014. A multivariate distance-based analytic framework for connectome-wide association studies. *Neuroimage*, 93(Pt 1), 74–94.

Song, M., Zhou, Y., Li, J., et al. 2008. Brain spontaneous functional connectivity and intelligence. *Neuroimage*, 41, 1168–1176.

Spearman, C. 1923. *The nature of "intelligence" and the principles of cognition*. London: Macmillan.

Stammen, C., Fraenz, C., Grazioplene, R. G., et al. 2022. Robust associations between white matter microstructure and general intelligence. *bioRxiv*, 2022.05.02.490274.

Thiele, J. A., Faskowitz, J., Sporns, O., & Hilger, K. 2022. Multitask brain network reconfiguration is inversely associated with human intelligence. *Cerebral Cortex*, 19, 4172–4182.

Vakhtin, A. A., Ryman, S. G., Flores, R. A., & Jung, R. E. 2014. Functional brain networks contributing to the parieto-frontal integration theory of intelligence. *Neuroimage*, 103, 349–354.

van den Heuvel, M. P., & Sporns, O. 2011. Rich-club organization of the human connectome. *Journal of Neuroscience*, 31, 15775–15786.

van den Heuvel, M. P., Stam, C. J., Kahn, R. S., & Pol, H. E. H. 2009. Efficiency of functional brain networks and intellectual performance. *Journal of Neuroscience*, 29, 7619–7624.

Willerman, L., Schultz, R., Rutledge, J. N., & Bigler, E. D. 1991. Invivo brain size and intelligence. *Intelligence*, 15, 223–228.

Yarkoni, T., & Westfall, J. 2017. Choosing prediction over explanation in psychology: Lessons from machine learning. *Perspectives on Psychological Science*, 12, 1100–1122.

Zhu, F., Cizeron, M., Qiu, Z., et al. 2018. Architecture of the mouse brain synaptome. *Neuron*, 99, 781–799.

CHAPTER 6

The Genetic Basis of Intelligence

6.1 Introduction

In the last chapter we presented evidence tying intelligence differences among individuals to quantified details of brain structure and function. There were never doubts that intelligence is a function of the brain, so modern neuroimaging findings were not controversial in principle.

Genetics is a different story. In the 1960s and 1970s, there was a debate about whether genetic variation was related to intelligence differences in any way. The stakes were high because of the widespread, but incorrect, assumption that genes were always deterministic. Thus, if intelligence had a genetic basis, enriching environments, including education, would have little if any impact on improving reasoning ability. This would not be good news for parents, teachers, and social scientists in general.

We now know two things. (1) The earlier debate is resolved. Genes influence intelligence differences among individuals.

(2) Genetic influences on complex traits like intelligence are more probabilistic than deterministic because they affect the brain and intelligence in complex combinations with multiple factors, including nongenetic ones and random events, during brain development. Untangling these interactions to understand how they work is among the most challenging problems at the interface of biology and psychology.

How do we know that genes play a role in the intelligence differences among people? This chapter tells the story of genes and intelligence as we know it today. We start with basic human evolution and then we delve into one of the most misunderstood, or misrepresented, central concepts in this story: heritability.

On one hand, a basic question is why something like intelligence appears to run in families from one generation to the next. On the other hand, it is not so easy to answer this simple question. Later in this chapter, we discuss compelling evidence from classic and modern quantitative behavioral genetic studies of adoptees and twins. Then we will present the newest evidence from molecular genetic studies of intelligence looking for specific genes and how they might function to influence the brain. One potentially powerful statistical method is based on sampling thousands or millions of DNA sites looking for small differences related to intelligence (or other complex traits or conditions). These DNA sites, whose content varies among individuals, are called single-nucleotide polymorphisms (SNPs). With this information, it is possible to compute a "polygenic" score for each person and use it to predict individual differences in intelligence (or other complex traits/conditions). It is a conceptually simple and potentially informative method, although not without controversy, as we will discuss.

Knowing about genetics is a basic education requirement for being well informed in the twenty-first century, no matter what your educational background or personal politics. Whether you find genetic research fascinating or scary, please read on with an open mind. By the end of this chapter, you will find that the genetics of intelligence is a highly positive and exciting story. If you need an introduction or a refresher on basic genetics, we recommend chapter 10 of Robert Plomin's (2018) book and chapter 3 of Kevin Mitchell's (2018) book.

6.2 Evolution and Genes

First came the monkeys. Then the apes. Then us. We and our other ape cousins lost those elegant monkey tails, developed a different limb structure, and have bigger brains. The monkey–ape genetic changes took place more than 20 million years ago; fewer than 10 million years ago, more genetic changes produced the hominids.

Somewhere around 100,000 years ago or so (some estimates are even earlier), our own species, *Homo sapiens*, appeared. We humans are a big-brained species with a limb structure that is fine for walking but terrible for climbing trees. Humans began spreading out into the world from our place of origin in Africa. Somewhat less than 15,000 years ago, humans had reached the southern tip of South America, thus populating all the continents except Antarctica.

One of the reasons that humans were able to settle all over the globe, in very different ecological systems, is that our big brains and enhanced learning ability allowed us to develop different cultures to adapt to and, crucial for intelligence, to shape different environments for overcoming our limitations when adaptation is not enough. Then these cultures took on a life of their own, as they were socially transmitted and modified from generation to generation. So, there may be differences in intelligence influenced by the sociocultural milieu, as we will discuss in Chapter 7.

Although it is beyond the scope of this book, the genetic story of ancient human migration patterns is now under study with advanced DNA testing that can identify geographical origins and subsequent routes out of Africa. You may already have learned your own DNA ancestry using one of the commercial companies that provide an analysis from your saliva. These analyses often include probability statements of your chance to have certain physical traits and medical conditions. Someday they may routinely provide estimates of your intelligence. As we will see, Section 6.3.3 suggests that this may be closer than you think, although in fairness, there are researchers who believe this goal is impossible.

Genetic changes due to natural selection, the engine of evolution, are slow and incremental. But they did not stop as humans wandered across the globe for these many thousands of years. Almost all of the DNA content of the human genome, perhaps close to 99 percent, is shared by all people, but there are significant genetic differences in the remainder between people and populations. This is because the human genome includes about 20,000 protein-coding genes and, in total, more than 3 billion base pairs (see Box 6.6). It is differences in these base pairs that underlie most individual differences. One percent of 3 billion allows for considerable variation among people and populations defined by a genetic ancestry. For example, Swedes, Tanzanians, Chinese, and the Quechua Amerindians (who live in the Andes) differ in their appearance, within the range of human differences, and they differ in other ways as well. These include sensitivity to sunlight, the ability to metabolize milk and milk products, and susceptibility to a number of diseases. Whether psychological differences are included among these features is an empirical question. Chapter 12 addresses world differences in cognitive abilities.

The bad news about genetics and intelligence is clear: more than 300 known genetic conditions virtually guarantee mental disability; in these cases, genes are essentially deterministic. People with Down's syndrome, Fragile X, and Williams syndrome are examples where low general intelligence is part of the condition. No one doubts the importance of genetic research on these conditions with the hope of finding treatments, cures, and prevention.

The good news about genetics and intelligence is more sensitive, but a new perspective is evolving. As we obtain knowledge about specific genes involved, there is optimism for changing their negative effects, even if hundreds or thousands of genes may contribute to a complex condition like schizophrenia or traits like intelligence. Genetic risk for heart disease or high blood pressure, for example, can be ameliorated by lifestyle changes, including exercise and diet. Whether genes that influence intelligence can also be modified to increase intelligence is an open question. But it is a question that undermines the old worry that genetic influences are deterministic and always constrain or prevent change. For this reason, the more a complex trait like intelligence has genetic involvement, the more optimistic we might be about discovering ways to modify the genetic influences.

Reaction to modern genetic research on intelligence often faces an old presumption that any differences among individuals are all or mostly due to environmental factors. It certainly seems like common sense that a child's early environment, upbringing, and schooling would be strong influences, if not determinants, for a complex trait like intelligence. The extreme environment-only position originated in the blank slate view of human nature. This view of easily malleable human traits like intelligence dominated social science, including psychology, for most of the twentieth century. It still biases discussions against the role for genetics. However, as Steven Pinker (2002, p. 421) has written, the blank slate should be considered a hypothesis, and when tested, it fails. We discuss evidence supporting environmental contributions to cognitive differences in Chapter 7. Modern behavioral genetics can estimate heritability as a first step toward identifying molecular genetic mechanisms that affect the brain and influence intelligence differences among people. This is an exciting time in this field, so let us see how it began.

6.3 Behavioral Genetics

Behavior genetics attempts to identify the extent to which variations in human traits are associated with genetic *and* environmental factors. There are different statistical methods for making these estimates. All are based on studies of families (kinship studies), especially twin and adoption studies that can separate these domains to some extent. This is necessary because it is impossible to test causal hypotheses by controlling either genetic makeup or environmental conditions in humans in the way that might be done in genetic studies of plants or nonhuman animals.

If identical twins are more similar to each other on intelligence test scores than fraternal twins, we can infer that genes play a role in the similarity. Comparing the strengths of identical and fraternal twin pair correlations of intelligence test scores allows us a quantified estimate of how much genes are involved. Doubling the difference is one estimate of how much genetic and nongenetic factors are involved. But twins typically grow up in the same family environment, confounding the genetic contribution to any psychological differences. Comparing identical to fraternal twins may not completely deal with this difficulty. Adoption is a natural experiment that helps separate genetic and nongenetic variance. Adoption is an even more powerful research design if the adoptees are twins separated at a young age and each one is raised in a different family or cultural environment. What do twin and adoption studies of intelligence tell us about heritability and the genetics of intelligence?

6.3.1 *Heritability*

Robert Plomin (1999, p. 27) has written, "The most far-reaching implications for science, and perhaps for society, will come from identifying genes responsible for the heritability of *g*."

When we say something is heritable, does this mean it is genetic? This simple question illustrates a common confusion about what heritability means. *Heritable* and *genetic* are not synonyms. Heritability is a measure that estimates the contribution of genetics (DNA) to the variance of a particular condition or trait like intelligence. Variance is key.

Variance among individuals on any variable (e.g., height, weight, eye color, extraversion) can be wide or narrow, depending on the environment in which it is measured. Intelligence in the population is normally distributed and has wide variance around the mean (see Figure 1.1). The variance is measured with a statistic called the standard deviation. In a sample of students at highly selected universities, the mean intelligence test score will be higher and variance will be narrower than in the general population. Heritability estimates of intelligence likely would differ if computed separately in these two groups. Similarly, if you try to estimate the heritability of height in a sample of people selected for degrees of malnutrition, most of the variance will be due to the environment (availability of food), whereas the heritability of height in a random, normally nourished sample of the population will be due mostly to genetics. This is why heritability estimates are useful only for describing the populations from which they are calculated and not individuals.

There are several common misunderstandings about heritability regarding intelligence. Here is an example of the main one. In this chapter, we shall review the data that indicate that the heritability of intelligence ranges between 40 percent and 80 percent, depending on age, whereas the remaining percentage is due to nongenetic factors. This, however, does not mean that 40 percent or 80 percent of any one person's intelligence is genetic. For any individual, nongenetic factors may be more important than for another person. The estimates apply only to a population within broadly homogeneous environments.

Robert Plomin (2018, pp. 190–195) has detailed additional common misunderstandings around heritability. Another main one is that anything with strong genetic contributions to heritability must be immutable and, therefore, limiting. We know, however, that many genes may increase the likelihood of having a particular trait or condition and that the probability can be influenced by external factors like diet and exercise in the case of genetic risk for heart disease. For intelligence, external influences are not yet established with the same certainty or specificity as for some medical conditions.

There are different ways to estimate heritability. A simple way is to obtain the correlations between twin pairs for a variable like a cognitive test score. Doubling the difference between the monozygotic (MZ) correlation and the dizygotic (DZ) correlation provides a rough estimate. The more similar the MZ twins are compared to the DZ twins, the higher the heritability.

Genetic and environment are two broad sources of variance to include in heritability estimates. A more elaborate and commonly used method is called the ACE model. This model assumes three sources of variation in test scores or similar indices:

A – *Additive genetic effects*: These are the sum of the effects of different genes, without considering any modifier effects such as genetic dominance.

C – *Shared environmental effects*: These refer to variations in the environment that are common to related individuals. The household where siblings grow up would be an example of shared environment, as would their community and school if they attended the same one.

E – *Nonshared environmental effects*: These are effects of the environment that differ for related individuals. Siblings can have different teachers and friends even if they share the same household, school, and neighborhood.

Each source of variance, A, C, and E, can be quantified based on how scores (or other variables) are correlated between related individuals, where MZ and DZ twins are most informative.

The ACE model is used for many behavioral genetic studies of cognitive and personality traits. It has been criticized for failing to allow for gene-by-environment interactions and correlations among different influences. If these are present, they will be erroneously assigned either to environmental or genetic main effects. Alternative models that do permit measurement of interactions have been developed (Sauce and Matzel, 2018), but they require more observed correlations than the ACE model requires.

With this simplified understanding of heritability, the rationale for twin and adoption studies, and the example of the ACE model, we now turn to the data about intelligence.

6.3.2 *Adoption Studies*

Adoption studies have a long history. Box 6.1 is a verbatim news account about one study from 1931 (Lawrence, 1931). Can you tell what other details you would like to see in this news account?

Adoption studies are not easy to do. In the ideal adoption study, (1) the children will have been adopted very early in life, (2) data will be available on the intelligence of both biological parents and both adoptive parents, and (3) the adoptees will be followed until adulthood to allow for dissipation of environmental effects and the effect of genetic traits that become apparent during adolescence, adulthood, and possibly old age. Such an ideal study has not been done, but there are good ones.

Here are two classic adoption studies carried out in the United States. They are somewhat different, but the findings are compatible. Taken together, they provide valuable information about the relative influences of genetic and nongenetic factors on intelligence.

THE TEXAS ADOPTION PROJECT
The Texas Adoption Project (TAP) began with children adopted between 1963 and 1971 (Loehlin et al., 2007; Horn, 1983; Loehlin et al., 1989, 1997). It focused on 300 adoptees from a church-affiliated adoption home for unwed mothers. Both the biological mothers and the adoptive parents were mostly of European ancestry and middle class. This probably reflects the social mores of the day. A woman who was a single parent faced greater social disapproval in the 1960s than she would today. There was more social pressure to give up a baby for

Box 6.1 From 1931: The Inheritance of Intelligence – a Verbatim News Account

Dr. Evelyn Lawrence has made a new contribution to the old controversy on nature versus nurture, her special problem being the relative importance of heredity and environment in determining the level of intelligence.

For this purpose, she has applied intelligence tests to a considerable number of children drawn from various institutions and schools, relating the results to the social class of the parents – a criterion of intelligence whose inadequacy she recognizes. As regards the distribution of intelligence her observations are in keeping with those of previous investigators. As we go down the social scale from the higher to the lowest classes the intelligent quotient gradually shifts over from above the 100 level to considerably below it.

There has, however, been little clear evidence hitherto to explain how these differences of level arise, how much they are due to inherited differences, and how much to the immense environmental differences.

Dr. Lawrence finds a correlation between the intelligence of children and the social class of their parents when they have never seen those parents, but have been brought up in institutional life. This is "fairly conclusive evidence," she says, "that the correlation so generally found for children in their own homes is not mainly due to the direct social influence of the home, but is a genuinely biological fact."

On the other hand, the correlation she finds is by no means a high one, it is of the order of 0.2 to 0.3, so that inheritance is by no means necessarily the only, or even the dominating, factor. She herself concludes that the association is rather smaller in the case of institution children – the figures in

support are not very convincing – and that "there is little doubt that environmental conditions have some weight in influencing the response to tests."

A further test was made by the comparison of children taken away from home at an early age with those in the same institutions who were removed from their parents later in life.

The figures suggest that the correlation between the child's intelligence and the social status of the parents is somewhat less among those who entered the institution earliest in life, but it is very doubtful if these differences are significant. They are not supported by the fact that the children taken from bad homes to institutions showed almost no increase of intelligence with the improvement of their surroundings. There was also little difference in variability between the children in the uniform environment of the institution and those in their own diverse homes, though if environment can affect intelligence scores, we should certainly expect children reared in an institution to be more alike than those brought up in the widely varying surroundings outside.

Various health-ratings, such as height-weight index and nutrition, were correlated with intelligence without significant result. With such small correlations between intelligence of child and status of parents any generalization about social classes is a hazardous business. Dr. Lawrence concludes that "for any definite plans for social reform based upon the differential inheritance of intelligence, social class is not a satisfactory grouping. Either some basis of classification resulting in more homogeneous and more widely differentiated groups needs to be found, or we must realise that the only safe unit by which to assess intelligence is the individual, or at most the family."

adoption. Therefore, neither the biological nor the adoptive parents likely are a representative sample. The adoptees, their biological mothers, adopting parents, and the biological children of adopting parents completed intelligence testing. There was some restricted range for IQ in both parent groups, which would reduce correlations between parent and child test scores compared to the equivalent correlations in a representative population. On the plus side, in this study, the biological mothers and the adoptive parents had roughly equivalent SES. This contrasts with other studies where adoptive parents tend to be both better educated and economically better off (and hence better able to afford to raise a child) than the biological mothers.

There were two waves of testing: an initial wave when the adopted children were eight years old on average (with a considerable range), and a follow-up of about 50 percent of the original sample ten years later. A variety of cognitive, personality, and health measures were taken. Twenty years after the initiation of the study, a questionnaire was sent to all adoptees and biological children of participants and to the adoptive parents. The questionnaire covered various aspects of life adjustment, such as occupational and marital status.

There are three main comparisons of interest for this chapter:

1. *Parents and children in standard families (no adoption).* Here, any similarity on intelligence (or other variables) cannot be attributed to genetic or nongenetic factors, because these factors are confounded; similarity is assessed by correlation coefficients.
2. *Adoptive parents and their adopted children.* Here, any similarity can be attributed to their shared environment because their genetic similarity is no greater than if chosen at random from the population.
3. *Biological parents and their adopted-away children.* Here, any similarity can be attributed to their shared genomes because they have not shared their family environment.

Figure 6.1 shows the results for these three group comparisons (Loehlin et al., 1997). In both early childhood (age eight) and adolescence (age eighteen), the strongest parent–child IQ correlations were for biological parents with their adopted-away children (0.3 and 0.45, respectively), indicating a genetic effect. By comparison, adoptive parent IQ correlations with the IQ of the adoptee was near zero by adolescence (0.02) and only slightly higher at age eight (0.12), indicating little if any effect of parenting on IQ similarity. All the correlations, including the low correlations within the standard family comparisons, may be artifacts of the restricted ranges found in these samples.

THE COLORADO ADOPTION STUDY
The Colorado Adoption Project (CAP) began in the 1970s and was a twenty-year study of children recruited from two adoption agencies in Denver, Colorado (Petrill and Deater-Deckard, 2004; Defries et al., 1981; Petrill et al., 2004). As was the case for the TAP, the participants were largely of European ancestry and middle class. In addition to studying the development of adopted children, the project included a comparison group of more than 200 families of parents and their biological children. Intelligence tests were conducted when the children were one, two, three, four, seven, twelve, and sixteen years of age. This made it possible to compare stability of intelligence, in the sense of the relative standing of a person compared to others of the same age, from infancy to adolescence.

Figure 6.2 shows the main findings:

1. The parent–child IQ correlation for the standard families was $r = 0.19$ when children were three years old and $r = 0.31$ when they were sixteen years old. Therefore, parent–child similarity increased as children aged. This increase in similarity could be attributed to more genetic influence as more genes are expressed with age (the number of genes inherited does not change with age, but their functions might) or greater influence of their shared family

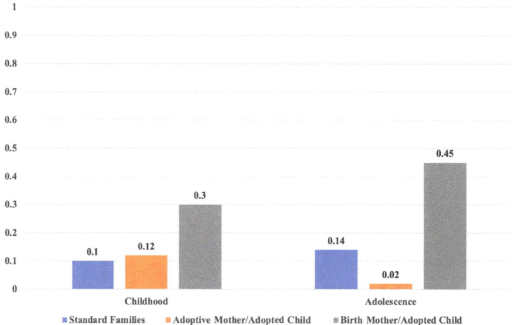

Figure 6.1 Parent–child IQ correlations in three groups for two age periods from the Texas Adoption Project (TAP). The highest correlations were for the group of birth mother with adopted child, supporting genetic influence. Note that by adolescence, the correlation was near zero for adoptive mother with adoptive child, which does not support environment influence.
Adapted from Loehlin et al. (1997).

environment. That is why the next comparisons are important.

2. The correlation for the adoptive families was $r = 0.09$ when children were three years old and $r = 0.03$ when they were sixteen years old. Therefore, parent–children similarity was close to zero at all ages, even when they shared their family environment as did the standard families. In other words, if you eliminate genetic similarity, then you obtain intelligence similarity values almost identical to what would be expected from people taken at random from the general population. Apparently, daily and prolonged family contact fails to produce intellectual similarity.

3. Finally, the correlation for the adopted-away children and their biological parents was $r = 0.12$ when children were three years old and $r = 0.38$ when they were sixteen. These are essentially the same values as those obtained from the standard families. Because there was no family contact of any kind in this group, their intelligence similarity must be attributed to the fact that they are genetic relatives.

These results are powerful and essentially the same as for the TAP (Figure 6.1). Virtually all adoption studies in North America and Europe produce similar results. The pattern of correlations indicates that heritability for general intelligence is at least 50 percent for children. Nongenetic effects due to differences between families can be substantial early in childhood but decrease in adolescence and almost vanish in adulthood. This suggests a dissipation in the effects of the adopting home over a person's lifetime. The genetic inheritance, however, is retained throughout life. As we will see, twin data show the same thing:

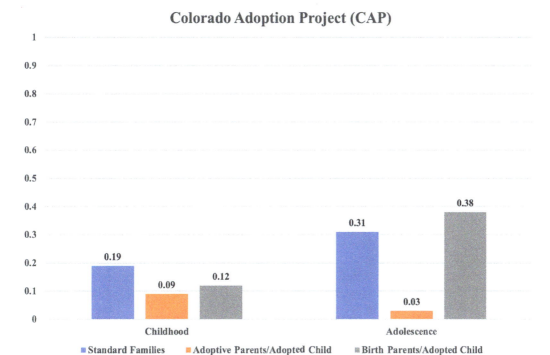

Figure 6.2 Parent–child IQ correlations in three groups for two age periods from the Colorado Adoption Project (CAP). The highest correlations were for the group of birth mother with adopted child, supporting genetic influence. Note that by adolescence, the correlation was near zero for adoptive mother with adopted child.
Adapted from Plomin et al. (1997, figure 1).

genetic contribution increases with age, unique environmental variance remains similar, and shared environment variance decreases to nearly zero.

Nevertheless, two qualifications should be considered with adoption data. First, these studies are about the effect of adoption on variation in measures of intelligence. They say nothing about the effect of adoption on the mean scores of adopted children, that is, about any benefit or general disadvantage of adoption. Mean values and correlations are different issues (Kendler et al., 2015; Sauce and Matzel, 2018). (We will say more about this in Section 7.3.2.) Second, can heritability coefficients from studies of adopted children be extrapolated to people in general? A case can be made that they cannot, on the grounds that adults who adopt children are less variable, within their own population, than parents in general. For example, the median age of women who adopt children is thirty.

Adults who adopt tend to be of higher SES and to have more education than typical parents. These tendencies work to reduce the environmental variability among adopting families to a lower value than population variability (McGue et al., 2007). If environmental variance is reduced, heritability coefficients will go up. The extent of the upward bias in heritability introduced by the homogeneity of adopting families is hard to estimate, but there is some bias to consider. Nonetheless, the weight of evidence from these adoption studies supported a substantial genetic influence on intelligence differences. Other adoption studies have more to offer about intelligence and we will consider them in Chapter 7.

6.3.3 *Twin Studies*

Research on twins is difficult due to twins' relative scarcity. MZ twins occur about once

Box 6.2 Quadruplets and Schizophrenia: A Case of Astronomical Chance

Nora, Iris, Myra, and Hester Genain were quadruplets who each developed schizophrenia. They were studied by a research team led by David Rosenthal at the National Institute of Mental Health (NIMH). The first and last names of these genetically identical quadruplets are pseudonyms (initials of first names are NIMH). These quads were born before the time of fertility drugs when identical quadruplets were a rarity and somewhat of an oddity. As children, their parents dressed the girls in identical costumes and took them on tour to sing and dance. In their late teens, each quad member developed schizophrenia – the only known such case. Each quad had different symptoms of schizophrenia and was diagnosed with a different subtype. They came to NIMH periodically for testing and evaluation, including PET scans and EEG mapping at their last visit (Buchsbaum et al., 1984). There is a classic book about the early research (Rosenthal, 1963) that captivated researchers at a time when many psychiatrists were dubious about any genetic component to schizophrenia. The quads are now all deceased. Although it is a fascinating case, they each had different experiences in life that may have contributed to her different subtype diagnosis. It was impossible for the researchers to disentangle genetic predisposition and which early life experiences could have been relevant to their illness. The odds of being identical quadruplets and having schizophrenia are astronomical. Had they also been adopted into different families at birth, the case would be even more fascinating. In fact, another case of astronomical chance is the equally compelling case described in Box 6.3.

in 250 births and DZ twins once in 100 births. Recall that MZ twins come from a single fertilized egg that splits into two during early development (hence the term monozygotic), or sometimes into three or four, producing triplets and quadruplets, respectively. These individuals are genetically identical to each other, except for random variations that may account for some phenotypic differences. Sometimes two or more different eggs are fertilized, producing DZ twins who share 50 percent of their genes.

MZ and DZ twins are sought out for genetic studies because they offer a natural experiment to test whether a trait or condition is related to genetics. If MZ twins are more concordant for a particular trait or condition (i.e., if one has it, the other has it too) than DZ twins, it is inferred that genetics plays a role. Concordance in MZ twins is often high, but rarely 100 percent for any trait or condition. Even in genetically identical individuals, genes are not the whole story. Box 6.2 tells an interesting case that illustrates this.

Although there are some general criticisms of twin research, many MZ and DZ twin studies from around the world validate the approach. A major study of meta-analyses published by Tinca Polderman and colleagues (2015) summarized findings from 2,748 reports (published between 1958 and 2012) that included almost 15 million twin pairs across the globe (thirty-nine countries from all continents) and considered almost 18,000 physical and cognitive traits. For all traits, the average similarity for MZ twins was $r = 0.64$, whereas the value for DZ was $r = 0.34$. Heritability was 0.49, and the value for shared environment was 0.17. The values for the cognitive traits were $r = 0.65$ ($N = 288,867$ MZ pairs) and $r = 0.37$ ($N = 304,720$ DZ pairs). The authors wrote, "This observed pattern of twin correlations is consistent with a simple and parsimonious

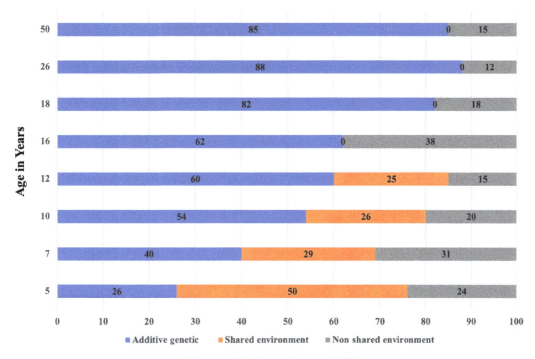

Percent Variance Accounted For

Figure 6.3 Percentage of the variance in intelligence test scores accounted for by genetic, shared environment, and nonshared environment influences (ACE model) shown by age. Genetic influence (blue) increases with age to about 85 percent at age fifty, whereas shared environment (orange) influences decrease to near zero by age sixteen, leaving about 15 percent attributable to unique, nonshared experiences (gray) by age fifty.
Based on the Dutch Twins Study, adapted from Posthuma et al. (2003).

underlying model of the absence of environmental effects shared by twin pairs and the presence of genetic effects that are entirely due to additive genetic variation" (p. 702). Specifically, for intelligence, the meta-analysis heritability estimates (based on the difference in correlations between MZ and DZ twins) indicate that about 50 percent of intelligence variance is due to genetic factors (combining ages at the time of testing), consistent with earlier single studies.

One of the early twin studies of intelligence assessed the ACE model described at the beginning of this chapter. The findings are shown in Figure 6.3 and are representative of other studies. Genetic influences on intelligence *increase* with age, while shared environment influences decrease to about zero with age.

Figure 6.3. illustrates three important findings: (1) intelligence is heritable starting at about age seven years old; (2) at age ten, genetic variance shows greater values than shared environmental variance; and (3) the increase in heritability with age ends about the time psychosocial maturity begins. These findings are difficult to explain if you assume that the environment is the primary critical factor in the development of intelligence. In fact, the near-zero effect of shared environment with age might be one of the most important findings in psychology. These age trends are robust, as evidenced by a major replication using 11,000 twin pairs from four countries (Haworth et al., 2010). The authors of this study, however, acknowledged nongenetic influence on intelligence when they wrote, "Why, despite life's 'slings and arrows of outrageous fortune,' do genetically driven differences increasingly account for differences in general cognitive ability? We suggest that the

Box 6.3 The Story of Identical Strangers

This unusual story about identical triplets came to wide public attention in a 2018 documentary produced by CNN. Each triplet grew up in an adoptive home without knowing he had siblings. One went off to college, where he encountered several students he did not know who spoke to him as if they knew him well, except they called him a different name. As it happened, they mistook him for another person who turned out to be his twin brother. This random meeting and mistaken identity led to the college freshman meeting his brother he never knew existed. Their story was featured in newspapers along with pictures of these two reunited brothers. Then another surprise – a third brother saw the pictures and recognized he must be related. In fact, the three were identical triplets who were adopted at birth into different families.

None of the boys or families were told of the existence of the other brothers. There was a happy reunion. But the story took another surprising and darker turn when it was discovered that these triplets were part of a secret experiment run by the adoption agency. Each brother was deliberately placed in an adoptive household chosen because they differed in SES. There was no informed consent for this decision (which would be required by today's ethical standards). Researchers running the experiment visited each household periodically to evaluate the boys on many factors, but none were told about their siblings. The story took a tragic turn when one of the brothers suffered major depression and committed suicide. What we see from the Genain quadruplets (Box 6.2) and these triplets is that identical genetic makeups are powerful, but not completely determinative. Nancy Segal (2021) published a book describing this controversial study.

answer lies with genotype–environment correlation: as children grow up, they increasingly select, modify and even create their own experiences in part based on their genetic propensities" (p. 1112).

Twins usually are raised in the same home by their biological parents. Conceptually, the almost perfect study of the heritability of IQ is the study of the intelligence of MZ and DZ twins raised in different households by adoptive parents, preferably without any interaction between the two twins. Pragmatically, such a study is hard to achieve. Twins are rare enough, and only a small fraction of all twins are adopted apart. And recall the story in Box 6.3.

The recent history of twins-reared-apart studies of intelligence usually starts in the mid-twentieth century with Sir Cyril Burt, who reported that intelligence test scores were strongly correlated in a sample of MZ

twin pairs separated at birth and raised in different families. He reported a correlation between these twin pairs of 0.77 in a series of relatively small samples. Burt's data became controversial after his death, with charges of fraud that were used by critics of genetic research to undermine the evidence for any genetic influence on IQ. However, modern research of twins reared apart overwhelmingly supports Burt's 0.77 correlation and his main conclusion. Averaged over other studies of twins reared apart from around the world, the IQ correlation is 0.75 (Plomin and Petrill, 1997). Key data came from Thomas Bouchard (1998, 2009) and his colleagues at the University of Minnesota, who conducted the classic Minnesota Study of Twins Raised Apart (MISTRA). Details of MISTRA are provided in Box 6.4.

A powerful meta-analysis of twins and adoptees confirmed average heritability

Box 6.4 Minnesota Study of Twins Raised Apart (MISTRA)

In 1979, researchers at the University of Minnesota, led by Thomas Bouchard, undertook a worldwide effort to find as many twins raised apart as possible. By the time recruitment was terminated in 2000, 139 twin pairs had been enrolled in the project (including at least one set of triplets).

In addition to the twins, Bouchard and his group interviewed and tested people related to the twins, including siblings and spouses. Eventually, more than 400 people participated. The testing took approximately fifty hours, over a weeklong visit to Minneapolis, and included personality tests, physical measurements, and collection of biographic data as well as extensive intelligence testing. In the intelligence part of the evaluation, participants took three previously developed batteries of cognitive tests, totaling forty-two tests of cognitive abilities (Johnson and Bouchard, 2005a, 2005b). The result has been the compilation of one of the most important and extensive databases on twins raised apart that exists.

Bouchard and his colleagues deserve great credit for their perseverance. Especially in the early years, government grant money for this important research was not forthcoming, largely because of political reaction against the idea that intelligence could be influenced by genetics. The early work was supported by foundations, and the University of Minnesota deserves credit for its willingness to support Bouchard and his group despite intense agitation against this project. In 2014, Bouchard received a Gold Medal for Lifetime Achievement in the Science of Psychology from the American Psychological Association. The citation read, "Bouchard has forever changed the way people understand individual differences in human behavior. His signature work, the Minnesota Study of Twins Reared Apart, demonstrated that genetic influence is pervasive, affecting virtually all measured traits. The study findings shaped the efforts of countless colleagues and graduate students, furthering and launching many careers and accomplishments."

estimates reported by Bouchard and other researchers (Tucker-Drob and Briley, 2014). It also confirmed the age effect on heritability. The values were obtained from fifteen independent longitudinal samples: 4,548 MZ twins raised together, 7,777 DZ twins raised together, 34 MZ raised apart, 78 DZ raised apart, 141 adoptive sibling pairs, and 143 nonadoptive sibling pairs, ranging in age from infancy to late adulthood. The MZ correlations were about 0.8, regardless of the twins' age. The DZ correlations were about 0.6 until adulthood, at which point they dropped to 0.4. This is consistent with previous ACE observations that the relative importance of genetic influences on intelligence *increases* in adulthood and that effects of environment decrease (Figure 6.3).

Overall, the similar findings from multiple adoption and twin studies in large samples gives us confidence that the best heritability estimates for intelligence are 50–60 percent. This demonstrates a substantial influence of genes, but also leaves at least 40 percent for other influences, as we discuss in Chapter 7. Before we go to those other influences, however, there is more to say about genetic data.

6.3.4 *Heritability of Information Processing*

In addition to general intelligence, there are studies of the genetic basis of two cognitive processes related to intelligence, as discussed in Chapter 4: speed of mental processing and working memory. Both show substantial heritability. The genetic correlations are nearly the same as for general intelligence. Here are some of the data from four studies.

Dutch twins, born in the early 1990s and evaluated at five and twelve years of age, were studied along with their siblings by Polderman et al. (2006). Participants took a comprehensive battery of tests to extract a psychometric *g* and tasks evaluating mental speed and the capacity of working memory. Heritability estimates for speed and working memory were in the 0.5–0.6 range, increasing slightly from age five to age twelve. Stability, in the sense of the extent to which scores at age five could predict subsequent scores, appeared to be largely mediated by genetic influences. The authors point out that this stability is impressive, because the brain undergoes substantial development from age five to age twelve.

In Australia, Luciano and colleagues (2001b) evaluated genetic contributions to information processing and intelligence test scores in sixteen-year-old twins and their siblings. Heritability estimates for choice reaction time tasks varied from 0.5 to 0.7, with heritability increasing as the number of choices increased (i.e., the tasks became more complex). A delayed reaction time task was used to evaluate working memory, showing a heritability estimate of 0.48.

In Japan, a study was conducted of young adult twins (aged fourteen to twenty-nine, mean age twenty). The twins completed an intelligence test and both verbal and visuo-spatial working memory tasks. The heritability coefficients for the working memory tasks ranged from 0.43 to 0.48, depending on the task (Ando et al., 2001).

In the United States, a twin study used records maintained by the US Department of Veterans Affairs (Kremen et al., 2007). Nearly 350 pairs of male twins, ranging in age from forty-one to fifty-eight, were given verbal working memory tasks. Combined, the estimate of heritability based on the difference in correlations between MZ and DZ twins was 0.58.

These studies, done with participants of different ages and conducted by different laboratories in different countries, have produced consistent results. Working memory, one of the key information-processing underpinnings of intelligence, is subject to substantial genetic influence. On the other hand, establishing the genetic basis for speed of mental processing is complicated. This cognitive process is a component of intelligence distinguishable from the storage and attentional control components of working memory (see Chapter 4). However, in above-average young adults (usually college students), the contribution of processing speed to conventional psychometric measures of intelligence is considerably smaller than the contribution of the working memory functions, as we have noted. By contrast, the decline in measures of psychometric intelligence from middle age onward is very largely associated with declines in speed of processing. Clearly, studies of college students may not tell all that we need to know about the importance of speed of processing in the population at large.

There are many ways to measure speed of mental processing. Most conventional intelligence tests use pencil-and-paper methods, where the dependent measure is how many easy tasks can be accomplished in a fixed time period. Researchers, especially those who have dealt with information-processing measures in contexts other than studies of individual difference, prefer the tighter control afforded by computer-controlled tasks, such as the choice reaction time and inspection time measures, as described in Chapter 4. Within each of these broad paradigms, and especially within the information-processing paradigm, there are many variations in test procedures.

Despite these difficulties, studies of the genetic basis of speed of mental processing have produced interesting results. A meta-analysis of several studies concluded that heritability was about 0.18 for easy timed tasks (which do not have high correlations with intelligence test scores) and 0.52 for hard tasks (Jensen, 2006; Beaujean, 2005). Looking at the details of a few of the studies is informative.

In the study of Dutch twins referred to earlier, the investigators found that, depending on the particular speed-of-processing task, heritability coefficients ranged from 0.4 to 0.5. Moving to adolescents, investigators in the CAP studied the genetics of speed of processing in sixteen-year-olds, using pencil-and-paper tasks. This study found a speed of processing heritability of 0.48 (Alarcon et al., 1999). The estimated heritability of inspection time was 0.80 in the Australian study of sixteen-year-old twins (Luciano et al., 2001a). One of the studies associated with the Dutch Twin Registry examined heritability in two cohorts, one aged twenty to thirty at the time of the examination and another aged forty to fifty-five. The heritability estimate for the inspection time measure was 0.46 and did not differ across cohorts (Posthuma et al., 2001).

A Swedish study of twins raised apart included participants from fifty to eighty years old. The study included pencil-and-paper measures of perceptual speed (Finkel and Pedersen, 2004; Reynolds et al., 2005). Because this was a repeated-measures study, it was possible to estimate the extent of heritability and environmental influences on both overall perceptual speed (as evaluated by pencil-and-paper testing) and changes in perceptual speed. The heritability coefficient was around 0.8 for overall speed (depending on which of two tests was used), but the major influence on change was the nonshared environmental difference. Overall, speed of processing has a genetic component, but, like all estimates of heritability, the size of the estimate varies across populations.

6.3.5 Heritability of Academic Skills

Conventional intelligence tests evaluate important cognitive abilities, but any comprehensive theory of intelligence should identify the genetic aspects of other mental abilities and skills that are important in the world but lie outside the conventional testing realm. Among the most important of these are skills in reading and elementary mathematics and science, because of their central role in success in school and many workplaces.

The Twins Early Development Study (TEDS) is an ongoing project in the United Kingdom focused on genetics and academic achievement, assessed separately from general intelligence (Rimfeld et al., 2019). Twins from more than 16,000 families born between 1994 and 1996 were originally recruited. Periodically, evaluations are made of children's progress in English-language studies, mathematics, and science. A comprehensive analysis reported that the estimated heritability for teacher ratings and tests of children's mathematics, reading, and science levels at three different ages ranged from 0.4 to 0.7 (Kovas et al., 2007). In all cases, the heritability estimate exceeded the estimates for variance associated with either shared or nonshared environments, and genetic sources of variance often exceeded total environmental sources.

The TEDS data indicated that substantially the same genetic influences were being expressed in all three topics – reading, math, and science. Not surprisingly, the genetic influences appear to express themselves more on their shared general component than on specific cognitive skills (Chapter 8 discusses a similar set of findings in high school students; Deary et al., 2007). The environmental influences, however, appeared to be unique to each of the three skills. Environmental effects appeared to control deviations from a genetically related stable path, either upward or downward, rather than individual differences in the average performance of a child over time. As has been the case with studies of genetic influence on intelligence test scores, the environmental contribution was largely due to unique nonshared environmental differences between twins or siblings, rather than to shared experiences within families.

Educators and parents have long debated the causes of poor school performance. To shed light on this question, the TEDS researchers asked whether the genetic models that applied to all children in the study also applied to those in the lowest fifth (or, alternatively, the lowest fifteenth) percentiles of performance on tests. This was the case, suggesting that poor academic performance is generally the result of an overall cognitive deficit, rather than a specific learning problem. It is true that some cases of impaired learning skills have been identified with certain genomes. However, specific genetic disabilities may account for only a small fraction of cases of poor reading or mathematics performance. Implications for these findings are discussed in Box 6.5.

Box 6.5 Implications of Findings from the Twins Early Development Study (TEDS)

TEDS found that, in a broad sense, genetic heritability was the single largest influence on test scores. There was no evidence that very poor performance, in the bottom 5 percent or 15 percent of the population, was due to some special condition (e.g., a genetic condition that impacted ability to acquire language skills). Putting this colloquially, according to the TEDS results, many of the children who would qualify for special education in reading and mathematics aren't really special; they are just at the lower end of the normal distribution of reading and mathematics skills.

Against this, we have to remember that there are specific genetic causes for a number of moderate mental disabilities, including reading disabilities. By lumping together all children below a certain ability level, the TEDS investigators may have developed a picture of disability that is accurate at a large scale but that overlooks some important, albeit uncommon, specific disabilities.

Environmental effects on mean test performances of certain groups are well established. For instance, intensive preschool educational programs can improve the academic readiness of children from very low SES groups. This indicates that better school environments can help on average, but whether these programs do anything to diminish the variation in performance among children is not supported by most data (see Chapter 13).

The fact that genetic influences on academic skills are stronger than nongenetic ones suggests that genetic analysis could serve as an early warning signal for identifying children who might have difficulty acquiring language and mathematical skills. The genetic analysis could be on the behavioral level by identifying a child whose relatives have had trouble with academics in the past, or it could be on a molecular level by identifying risk from a child's DNA sample. We are moving toward the possibility of the latter sort of testing, as extensively discussed by Robert Plomin (2018) in his thought-provoking book *Blueprint*.

The TEDS analyses of children at the low end of mathematics, reading, and science competency indicate that, in the vast majority of cases, these children represent the low end of normalcy rather than specific genetic or environmental problems. This may have no implications at all for education policy. The level at which schoolchildren are considered for special education, in any field, has to be determined by two things: (1) the level of competency the child requires to function in the society and (2) the money available to enroll students in special education programs. Why a particular child falls into a special class is of interest only if it serves as a guide to improvement, at least suggesting some strategy to the teacher. If most students requiring special education are from the low end of the normal distribution, there likely are few genetic constraints on various teaching methods that may be helpful. What those methods should be is a topic beyond the current discussion. (For detailed discussions about the implications of intelligence research on education, see Asbury and Fields [2021] and Wai and Bailey [2021].)

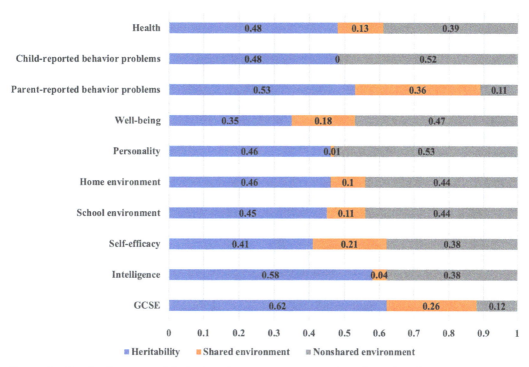

Figure 6.4 Results for genetic (blue), shared environment (orange), and nonshared environment (gray) components of variance for general academic achievement (GCSE) and nine related traits (y-axis). Total variance for each source sums to 100 percent (x-axis). N = 13,306 twins were assessed at age sixteen years.
Adapted from Krapohl et al. (2014).

Another study from TEDS used the ACE model and reported a noteworthy role for genetic influences on school achievement, intelligence, and other broad traits (not academic subjects) (Krapohl et al., 2014). They studied 13,306 twins at age sixteen who had completed eighty-three tests within nine domains (intelligence, self-efficacy, school environment, home environment, personality, well-being, parent-reported behavior problems, child-reported behavior problems, and health). The results are shown in Figure 6.4. The most genetic variance was found for general academic achievement and for general intelligence, not surprisingly. Genetic variance was also high for parent-reported and child-reported behavioral problems and for general health. Well-being and self-efficacy had lower genetic variance (but substantially greater than zero), and school and home environments and personality were slightly higher.

What does this pattern mean? The authors concluded, "To the extent that children's traits predict educational achievement, they do so largely for genetic reasons. ... Education is more than what happens to a child passively. Children are active participants in selecting, modifying, and creating their experiences that are correlated with their genetic propensities, known in genetics as genotype–environment correlation. ... It is important to recognize that children differ for genetic reasons in how easy and enjoyable they find learning" (p. 15276).

Implications of the TEDS findings for education are exhaustively discussed by Kathryn Asbury and Robert Plomin (2014) in their book *G Is for Genes: The Impact of Genetics on Education and Achievement*. This was their main take-home message: "By personalizing education, schools, through embracing the process of genotype–environment correlation, should draw out natural ability and

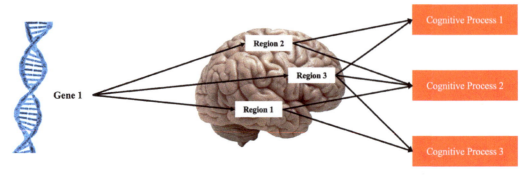

Figure 6.5 Illustration of the generalist genes hypothesis: the same gene affects many cognitive abilities. In this figure, one gene influences several areas of the brain, and each area affects several cognitive processes (Kovas and Plomin, 2006).

build individual education plans for every single child, based on pupils' specific abilities and interests. … Children come in all shapes and sizes, with all sorts of talents and personalities. It's time to use the lessons of behavioral genetics to create a school system that celebrates and encourages this wonderful diversity" (pp. 11, 187).

6.3.6 *Are Findings from Behavioral Genetics Convincing?*

Anyone familiar with the lack of replicability that plagues putative results from many social science studies must be struck by the consistency of behavior genetic findings (Plomin et al., 2016). This is no accident, because this field of research has benefited from evolving sophistication of research studies with respect to obtaining large samples and testing with multiple measures. Based on what we have already presented in this chapter, here are five broad considerations based on compelling weights of evidence from behavioral genetic data:

1. Additive genetic heritability accounts for 40–80 percent of the variance in virtually all cognitive traits.
2. About half the variance in mental tests is common to them. This strongly suggests that there are common genes that influence brain structures underlying virtually all cognitive performance; these are so-called generalist genes (Kovas and Plomin, 2006), as illustrated in Figure 6.5.

3. Environmental effects (including the prenatal environment) on mental abilities are strongest in early childhood and diminish thereafter.
4. Childhood environmental effects are primarily due to unique nonshared experiences that differ within families; shared experiences within families have less influence than most people imagine. Nonshared environmental influences are also the largest environmental influences in adults, although this refers to life experiences beyond childhood family environments.
5. Genetic influences are strong in establishing the trajectory of a person's cognitive development (and the later reduction of cognitive ability in old age), and variations from that trajectory are influenced by environmental factors.

Even though the data for these statements are compelling and present a consistent picture, there are two cautions to consider. First, heritability estimates are influenced (and in some cases dictated) by the contrast between MZ and DZ twins. As we noted, the rough and ready method of estimating the heritability of trait X is twice the difference between MZ and DZ correlations of the trait. This contrast assumes that MZ twins have identical genotypes. Also, it is assumed that DZ twins share half of their genomes identically by descent, at least on average. The assumption about MZ twins is mostly true; there are some differences mostly due to random chance effects. The assumption about DZ twins depends on the further assumption that the mothers and fathers show no correlation in the trait. This may be true in fruit flies and fish, but it is not in humans. The correlation between cognitive test scores for spouses generally is 0.2–0.3. This demonstrated tendency is called *assortative mating* (Reynolds et al., 2000; van Leeuwen et al., 2008; Colom et al., 2002). The effect of ignoring assortative mating on heritability estimates is to bias them downward, by raising the DZ correlation closer to the MZ correlation in the manner expected from shared environment influencing the trait.

The second caution is that analyses like ACE do not allow for gene–environment interactions. This is a deficiency of some behavioral genetic approaches. Why could this be a problem? For example, genetically influenced traits, such as alcoholism, involve a potential for a pathology that is released by environmental agents. Could the same kind of interaction be true for cognitive ability? The data are complex as revealed from international comparisons: "In U.S. studies, we found clear support for moderately sized gene × SES [social economic status] effects. In studies from Western Europe and Australia, where social policies ensure more uniform access to high-quality education and health care, gene × SES effects were zero or reversed" (Tucker-Drob and Bates, 2016, p. 138). A large study of twins in Florida ($N = 24,640$), however, found no evidence that SES moderated genetic influences on cognitive development (Figlio et al., 2017). The authors noted that future molecular genetic research could be informative: "As causal genetic variants for cognition become better established, one question will be whether they are more influential in some environments than others and whether there will be a more systematic genomic pattern of greater cumulative influence in higher-SES environments" (p. 13445).

The original observations that genetic heritability of cognitive ability was higher in advantaged groups, sometimes called the Scarr–Rowe interaction or hypothesis (Scarr-Salapatek, 1971; Turkheimer et al., 2003), have not proved easy to replicate. For now, the weight of evidence does not support SES–gene interactions for intelligence, so the impact on models like ACE may be less important than some critics believe. Results of future DNA-based studies likely will invigorate additional research.

Despite the lack of strong evidence, such interactions have been widely assumed to exist. It is worth discussing that several mechanisms have been postulated. One argument is that harsh social environments will restrict the expression of genetic potential and, therefore, genetic variation can express itself only in an environment that nurtures cognition (Sauce and Matzel, 2018; Bronfenbrenner and Ceci, 1994). There is an analogy to weight: in a famine-stricken town, everyone is emaciated; but when food is available, genetic tendencies toward obesity can express themselves. Substitute "learning" for "weight" and "bad schools" for "famine" and you have the idea.

A second argument is that society includes a number of positive feedback mechanisms that serve to increase variation, especially in the high range of a distribution. James Flynn has been a particularly strong advocate of this hypothesis (Flynn, 2007, 2012, 2018; Dickens and Flynn, 2001). One of his frequently used analogies is to basketball. Initially, tall and strong young people may be singled out for special playing and coaching opportunities. The better the player is, the more coaching the player gets, thus increasing the variance in skill between the initially mediocre and initially somewhat better players. Applied to cognitive skills, this suggests that, if a school system directs talented and untalented students into different tracks, with different qualities of instruction, one can anticipate a rich-get-richer, poor-get-poorer phenomenon, in both basketball and intelligence. Initial, genetically produced talent is nurtured by the environment.

Both these explanations of heritability–environment interactions differentiate between initial talent and opportunities to develop talent. There is a third argument that is equally reasonable but has an entirely different basis. The issue may not be that higher-SES environments permit the expression of greater genetic effects; it may be that lower-SES environments have greater environmental ranges in variables that are relevant to intelligence. In other words, the cognitive benefits of sending a child to a good middle-class preschool (or even of not sending the child to preschool at all but engaging the child at home) may result in the same benefits as sending a child to an exceptionally expensive preschool. The differences between keeping children in a chaotic atmosphere, with very little encouragement to explore or express themselves, compared to sending the same children to a modestly funded preschool three or four days a week may be profound. This sort of effect is *not* a gene–environment interaction. The effect is due to greater environmental variation in the lower-SES environment compared to the higher-SES environment. On this point, a study of families in extreme poverty reported that family chaos did not moderate genetic effects on cognitive abilities but did moderate shared environment influence (Hur and Bates, 2019).

These arguments about the importance of gene–environment interactions are reasonable, but do they invalidate twin and adoption analyses that do not account for interactions or that show no interactions? There may be little hope of identifying or untangling interactions without valid and reliable measures of the specific environmental variables that influence cognition. Without a theory of environmental action, it is hard to know where to begin. Given the dominance of the environmental perspective in psychology for decades, why do we still not have such a theory? Is quantifying the environment more difficult than quantifying a genome?

This is an enigma. Defending the research on twin studies and, by implication, most other behavioral genetic studies of intelligence, Bouchard (1997) pointed out that claims that additive heritability coefficients, like those in ACE, actually are due to hidden gene–environment interactive effects are arguments about what might be the case.

The arguments are sometimes accompanied by analogies to gene–environment interactions in plants, but seldom, if ever, by an example involving human intelligence. In this respect, Flynn's basketball example is interesting, but an actual example involving human intelligence with empirical data would be more convincing.

Bouchard made two arguments to counter the interactionist critique of twin and adoption studies. Both have to do with the strategy one follows in scientific research. First, he observed that many of the arguments appealed to what ought to have been measured in various studies. Any result can be explained by appeal to unmeasured variables, even if there are no good measures of those variables, as discussed in Chapter 2. Second, scientists generally accept a principle known as *Occam's razor*: given two equally accurate explanations of the same phenomenon, one should always prefer the simpler one. Behavioral genetic studies have shown that the assumption of substantial additive genetic variance can account for much of data about the covariation of intelligence across people of different degrees of genetic relatedness (Polderman et al., 2015; Hill et al., 2008).

Behavioral genetics tells us how much heritability there is for intelligence within a given population. We may have reached the limit of what behavioral genetics can contribute to intelligence research. The fact that genes are involved, now a certainty, is only a starting point for research. To advance our understanding, we have to find which genes are involved and how they influence brain structure, function, and development. It is a complex set of challenges on par with the most profound problems in science. Answers require melding psychology with biology and neuroscience to take intelligence research into the molecular fabric of DNA, the fundamental code of life.

6.4 DNA and Molecular Genetics

Mendel established the logical basis of genetic inheritance but knew nothing of its physical mechanism. How genes work is the fundamental topic of *molecular genetics*, a field that started at the beginning of the twentieth century with the discovery that chromosomes were the bearers of genetic material. Discovery after discovery followed (along with several Nobel Prizes), the key one being James Watson and Francis Crick's discovery of the structure of the genetic material, deoxyribose nucleic acid (DNA) (Watson and Crick, 1953; Watson, 1968). Watson's history and discussion of the twentieth-century discoveries is good background – *DNA: The Secret of Life* (Watson and Berry, 2003). Genetic inheritance turns out to be a nightmarishly complicated process with many mysteries still to be solved by researchers around the world. Fortunately, a simplified description of the process is sufficient for our purposes.

The material in a chromosome is a DNA molecule, which has the structure of a double helix, two strands of material twisted around each other. The strands are made up of four molecules called bases: adenine (A), thymine (T), guanine (G), and cytosine (C). The strands of the helix are bound together because the bases bind to each other, forming units called *base pairs*. The binding is unique such that an A always pairs with a T, a C always with a G. Logically, the DNA molecule can be thought of as a sequence of base pairs. The sequence determines the instructions for producing proteins and constitutes a person's *genome*. This is why the term *sequencing the genome* is sometimes used to refer to the process of identifying the base pairs in a genome. The term *gene* is often restricted to a special contiguous stretch of the genome encoding a protein.

Variations in base pair content at a specific site in the genome define *alleles*, and hence variations in the program for building the organism, including the brain. Variation is required for evolution because the process of natural selection identifies incremental variations that increase the probability of survival of the organism, including humans. Box 6.6 summarizes some basic information about genes based on current understanding.

6.4.1 *Genetic Basis of Intelligence Variation*

The molecular genetic analysis of variation in intelligence presents a different picture than the analysis of some pathological conditions or illnesses. The first thing to say, and to say loudly, is that there is no one gene, or even a small number of genes, responsible for variations in intelligence in the normal range. If there were, we would have found it by now with modern techniques. However, from behavioral genetic studies, we know that in the populations studied, at least half of the variation in intelligence is due to genetic variation. Apparently, we are dealing with a large number of genes, each having a small effect.

Multiple genes imply a polygenic inheritance model. Unless a sample is extremely large, the chances of detecting small effects are not good. The challenge is to collect DNA and measures of intelligence on samples larger than have ever been studied by individual research groups. Amazingly, with the advent of multinational consortia designed to collect and share data, new DNA research related to intelligence has reached powerful sample sizes of more than a million people. The rest of this chapter tells how the story unfolded and what we have learned.

6.4.2 *Hunting Genes for Intelligence*

The hunt for specific genes that influence intelligence started out earnestly and enthusiastically at the end of the twentieth century. One reason for the enthusiasm was the demonstration that genetically engineered mice could become excellent at problem solving their way through mazes after researchers had spliced a single gene (*NR2B*) into the DNA of ordinary mice embryos (Tang et al., 1999). This gene is related to the NMDA receptor at the synapse, generally involved in learning and memory. Could such genetic engineering work in humans to increase problem-solving ability? This specific experiment has not been done to our knowledge, but as we shall

Box 6.6 A Few Facts about the Genome

1. Base pairs are the building blocks of genes made from molecules of adenine (A), thymine (T), guanine (G), and cytosine (C). There are approximately 3 billion (3,000,000,000) base pairs in the human genome. The number of protein-encoding genes is estimated to be only about 20,000. Humans share about 99 percent of all base pairs with each other and about 96 percent with chimpanzees. All the differences within humans and between humans and chimps are in the 1–4 percent of the 3 billion base pairs. This leaves about 30 to 120 million potential genetic differences to sort through for finding those relevant to complex traits like intelligence. This endeavor presents enormous problems for data storage and analyses that fuel the field of bioinformatics.

2. Proteins are the building blocks of life, and each gene contributes to multiple proteins. Different proteins may be expressed in different parts of the body. Currently, an estimated one-third of human genes may have some expression in the brain.

3. A mutation occurs when there is a change in a base pair, creating a new allele (called a *polymorphism*). This can happen due to errors in the process of DNA replication, or mutations can be triggered by exposure to environmental hazards. These include radiation, ultraviolet light, and exposure to certain chemicals. Importantly, some mutations appear to occur by random chance events of unknown cause.

4. Some genes regulate the expression of other genes in the body.

Microenvironmental factors within the cell can also control genetic expression, but we do not have much understanding of how gene–gene or environment–gene regulation works. The research questions around these issues are among the most complex and challenging problems in biology.

5. Genetic influences can turn on and off during the life span. Some may not be expressed for some time after birth or, in some cases, may be suppressed throughout the individual's life.

6. Intelligence research focuses on differences in genomes. A difference that occurs at a single base pair in at least 1 percent of the population is called a common single-nucleotide polymorphism (SNP, or "snip"). There are about 10 million SNPs in the human genome. Blocks of alleles that are close together on the same chromosome, and tend to be inherited together, are called *haplotypes*. The SNPs themselves may be in a protein-encoding gene, in a region between genes, or part of the nonencoding portion of a gene. One of the major steps in research on genetic differences involves locating alleles or haplotypes that differ in frequency between populations of interest, such as individuals who do or do not have a particular form of mental disability, or high and low intelligence. SNPs can be assessed either from a portion of a full genome (as did the early studies with limited DNA assessment technology) or in the full genome by a newer method now widely available called genome-wide association study (GWAS). More about this is in Box 6.7.

see in Chapter 13, genetic engineering technology has advanced considerably, with both fantastic potential and intricate ethical issues.

After a frustrating decade of hunting intelligence-related genes in humans, researchers realized the hunt was going to be slower and more complex than originally anticipated. Early discoveries followed a discouraging pattern. Initially, putative genes emerged as reasonable candidates because of their involvement with neural efficiency, brain development, or some other key function that ought to be related to intelligence. But initial associations did not replicate in independent studies. The fact that this vexing pattern occurred so frequently led Plomin and colleagues (2006) to speculate that no single gene accounts for more than 0.5 percent of the variance in intelligence. He reiterated this theme in an address to the International Society for Intelligence Research, Amsterdam, in December 2007 and in an interview with the same society in Albuquerque in 2015.

At this stage of research, the complexity of the hunt has three basic aspects: (1) at least hundreds of genes are likely involved, and each one has only a tiny effect; (2) individual genes can affect more than one trait or condition (pleiotropy); and (3) the functional expression of genes (how they turn on and off across the life span) often depends on complex interactions with other genes, environmental factors like stress, and even random events that influence the developing brain.

In fact, some researchers believe the dynamic cascade of neurobiological events defined at these interactive levels from genes to brain development to traits is so complicated that it is not possible to establish any causal links at all, now or ever. Are they right? Well, there is progress.

Early attempts to find specific genes for intelligence were limited to needle-in-the-haystack methods. Essentially, the methods compared DNA samples for specific genes that were more frequent in a group defined by high IQ scores than in a low-IQ group. Candidate genes were identified for further study. One problem was that even in the best of these studies, no one candidate gene accounted for much variance. Not surprisingly, successful replications were minimal. After about fifteen years of trying these methods, it became apparent that the group sample sizes were far too small to overcome the statistical problem of multiple small

effects and random false positives. Estimates based on mathematical models indicated that sample sizes of at least a million people might be needed to find relevant genes, even if an entire genome could be tested. The cost would be astronomical. How could this research progress?

6.4.3 Cooperation, GWAS, and Polygenetic Scores to Predict Intelligence

Genetic researchers understood that the best solution to this problem was to form multinational consortia to pool data into large samples beyond the resources of any one research group. It sounds easier than it is to pool data from different research groups. Complex questions about standardized protocols, ownership of data, authorship of publications, logistical management of and access to huge data sets, individual institutional regulations, and other issues were all overcome. Moreover, to accelerate progress, these large data sets mostly are available to all researchers, not just limited to those in the consortia. By any standard, the emergence of these cooperative consortia is a major advance in science to celebrate.

One more critical development occurred that was a game-changer: genetic technology advanced dramatically so that researchers could genotype thousands or even millions of sites across the entire genome for each research participant with cost-effective methods. This astonishing technology has led to a new phase of GWAS and the analysis of errors in individual base pairs called SNPs (see Box 6.6, fact 6). Box 6.7 explains more about DNA methods, including GWAS.

GWAS have expanded rapidly from initial sample sizes of thousands of people to millions, and the resulting increase in statistical power has increased the probability of finding SNPs possibly related to intelligence measures (Davies et al., 2018). We will discuss some of the key studies later in this chapter. Note that GWAS findings are statistical associations, although GWAS findings can be used for causal inferences. Nonetheless, identifying statistical associations is important.

Box 6.7 Screening Methods for Identifying Genetic Effects

These are some research designs used to establish a correlation between variations in alleles (*polymorphisms*) and variations in a phenotypic trait like intelligence.

Differences between individual genomes (*polymorphisms*) occur when a sequence of base pairs differs at a specific position, or example, as in the sequence A C G T A A in one person compared to A C C T A A at the same location in another person. There is a difference between G and C in the third position. A variation such as this is called a *single-nucleotide polymorphism* (SNP, or "snip"). A SNP can occur inside or outside the boundaries of protein-encoding genes. The different base pairs that can be carried at a SNP are called *alleles*. The frequencies of the two alleles at a "common" SNP, by definition, must exceed some threshold (typically 1 percent). Many SNPs may have no functional consequence.

Special techniques are necessary to identify the order (or sequence) of all base pairs on each of the twenty-three chromosomes. The early sequencing techniques were slow and expensive, making sequencing an entire genome for an individual prohibitively costly and time consuming. Modern techniques have dramatically reduced the time and cost so that it is now possible to sequence the entire genome for many people in large research projects. Even so, at the current time, most GWASs do not employ whole-genome sequencing but instead genotype a fixed number of sites in the genome known to have common SNPs. Because SNPs in a given block of the genome tend to be highly correlated, it is possible to statistically impute an individual's genotypes at SNPs not assayed by the genotyping technology. In this way, a GWAS can study more or less all common SNPs in a given population.

Participants in a research study donate DNA samples (usually from saliva or blood) and are measured on the phenotypic trait like with an intelligence test. The DNA samples are then genotyped (or, more rarely, fully sequenced). The assayed and imputed SNPs are examined for correlations between the presence of alleles/SNPs and the values of the phenotypic trait. For instance, if one allele at a SNP is statistically associated with high IQ scores and another with low scores, then a genetic association with IQ is identified. As noted, until recently, only portions of the entire genome could be sampled for comparison due to cost.

Before GWAS, linkage analyses were done using samples of related individuals. For instance, registrants in the Dutch twin registry and their relatives have participated in linkage analyses. Linkage analysis is highly useful for tracing the genetic basis of discrete traits dominated by one or a few sites of large effect, such as eye color. It is less useful for continuous traits like intelligence. This is especially true if expression of the trait is affected by environmental as well as genetic factors. We know that this is the case for intelligence from behavioral genetic studies. Not surprisingly, linkage studies of intelligence were not informative.

Finding clusters of SNPs statistically associated with a complex trait like intelligence or a condition like autism can lead to identifying relevant SNPs and genes. This could be a key step toward finding treatments and cures for many adverse conditions.

Another advance is that all SNPs related to a trait or condition can be used to compute an individual score based on counting the relevant alleles found in a person. There are different ways of defining, summing, and weighting relevant SNPs into scores. The

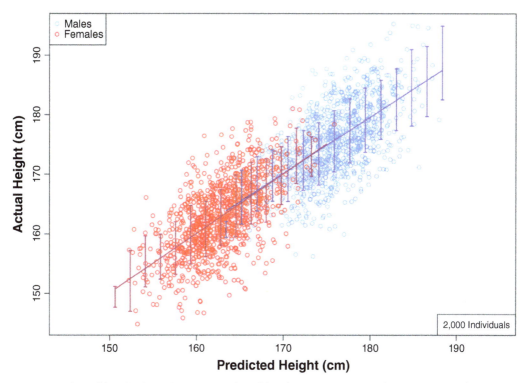

Figure 6.6 Actual height (*y*-axis) versus predicted height (*x*-axis) using polygenetic scores from 2,000 randomly selected individuals. The correlation of 0.65 means that the prediction accounts for about 40 percent of the variance (Lello et al., 2018).

best formulation for any dependent variable is an empirical question. Whatever way it is computed, the result is called a polygenic score (PGS). For example, many SNPs have been associated with height, each with a small effect. In a large sample, a polygenic score for height can be computed for each person. A result from an early study of 2,000 random people is shown in Figure 6.6. The correlation between the polygenic score and actual height is about $r = 0.65$, indicating that about 40 percent of variance can be predicted by genes (Lello et al., 2018; Yengo et al., 2022). Another study reports that as much as 82 percent of adult height variance can be predicted by PGS when parental height is included (You et al., 2021). Is the same true for intelligence variance? If so, this would be a major advance, like the ability to predict IQ scores from brain images.

There are now a number of GWAS reports related to intelligence from large sample consortia published in top scientific journals. Many papers include two samples. One is the discovery sample, where the main analyses are computed. The second independent sample is used to replicate the findings from the discovery sample. This important research design feature helps validate relevant SNPs and the method of computing a PGS as well as enhancing the robustness of any prediction results.

Many participants in these studies complete some form of intelligence testing as part of the project, but the vast majority do not. However, a variable substantially related to intelligence ($r = 0.55$) is available for nearly all GWAS participants. Educational attainment is the number of years of schooling a person completes. It is obtained on the general information questionnaire participants complete as part of these DNA projects. Educational attainment is not the ideal proxy for an intelligence test score, but it has been used in many studies. It is correlated to intelligence

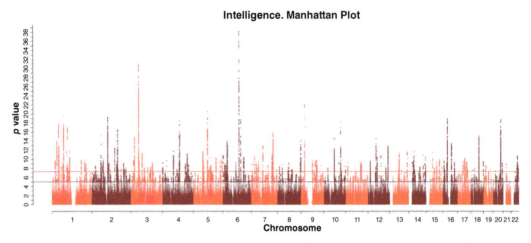

Figure 6.7 Manhattan plot showing candidate genes related to intelligence ($N = 248,482$). The x-axis is chromosomal position, and the y-axis is the statistical significance on a $-\log10$ scale. The horizontal lines mark statistical thresholds for genome-wide significance (Hill et al., 2018).

test scores either because more education leads to higher intelligence test scores or because brighter people spend more years in formal education, including completion of advanced degrees. Some combination of both possibilities is likely. When available, both intelligence test scores and educational attainment are reported in the same publication to identify overlapping results.

Plomin and von Stumm (2018) have written an accessible discussion of polygenic scores and how they relate to intelligence and educational attainment data. They estimated that a GWAS sample size of 1 million people likely would be necessary to build a polygenic score that predicts 10 percent of intelligence variance in an independent sample. How long would it take to reach this goal? Here's a hint: not that long.

GWAS reports are highly technical and complex, but we can distill the major findings for intelligence/education attainment reported so far. To give an appreciation for the accelerating pace and excitement of the hunt for genes and the use of PGSs, we will comment on studies published in rapid succession in 2017, 2018, and 2022.

A breakthrough report in 2017 was based on more than 78,000 individuals from different samples of European ancestry (Sniekers et al., 2017). The full sample included more than 12 million SNPs, and 336 of them met

stringent statistical criteria for possible association with intelligence measures. From these SNPs, twenty-two genes were identified; eleven had been implicated in previous studies. Some of the top genes were implicated in synapse formation (*SHANK3*), axon guidance and putamen volume (*DCC*), and neuron differentiation (*ZFHX3*). Using a variety of methods, a total of forty-seven genes were identified as possibility related to intelligence or educational attainment. Polygenic score analysis predicted up to about 4.8 percent of the variance in intelligence depending on the subsamples considered. This value demonstrated that genetic prediction of this complex trait was possible, at least to some extent.

Shortly thereafter, another GWAS combined samples from more than 248,000 individuals (Hill et al., 2018). Using stringent statistical techniques, they identified 538 genes possibly related to intelligence, as shown in Figure 6.7; since the plot looks somewhat like a city skyline, such plots are called Manhattan plots. Key genes in this group are implicated in a variety of brain functions, including neurogenesis and possibly myelination. Polygenic score predictions of intelligence variance were as high as 6.8 percent, depending on subsamples used. These researchers also noted that intelligence and educational attainment might have important different genetic architectures in

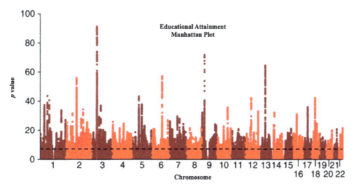

Figure 6.8 Manhattan plot showing candidate genes related to educational attainment. The x-axis is chromosomal position, and the y-axis is the significance on a $-\log 10$ scale. The dashed line marks the statistical threshold for genome-wide significance (Lee et al., 2018).

addition to some overlap, drawing attention to the need for including intelligence testing in the consortia protocols.

This report was followed quickly by two additional studies. One of more than 300,000 individuals identified 709 genes potentially related to general cognitive ability; 418 of them replicated previous studies (Davies et al., 2018). Their polygenic score predictions for general cognitive ability ranged up to 4.3 percent. The other had nearly 270,000 individuals (Savage et al., 2018) and identified 1,016 genes possibly related to intelligence. They reported polygenic scores predicting up to 5.2 percent of the variance in intelligence scores.

The landmark sample of more than 1.1 million individuals was reported in 2018 by Lee and colleagues, an astounding example of consortia success. These researchers identified 1,271 SNPs related to educational attainment that implicated as many as 1,838 genes, many involved in brain development and neuron communication (see Figure 6.8). Using subsamples, polygenic scores predicted 11–13 percent of the variance in educational attainment and 7–10 percent of the variance in cognitive test performance. These values are essentially what Plomin and von Stumm (2018) predicted just months before. Figure 6.9 illustrates key genes implicated in this study and their possible influences at the level of the synapse (drawn by Dr. Emily Willoughby, an

associate of Dr. Lee). The details in this figure are not important for the main point: there is extraordinary complexity at even just one of the many levels between genes and behavior. Figure 6.9 also evokes the beauty and even grandeur of the complexity.

Another study of 7,026 children and adolescents from the UK Biobank Project reported that PGSs predicted up to 11 percent of intelligence and 16 percent of educational attainment (Allegrini et al., 2019). Predictive power increased with age (as does heritability, as discussed earlier), and there were no sex differences. PGSs yielded the strongest predictions compared to multitrait genomic methods.

So, can polygenic scores predict individual differences in intelligence? At this time, the answer is a cautious yes, because the amount of variance predicted is relatively small. It is the same answer we had regarding the prediction of IQ from brain images. Polygenic scores and connectome predictions likely will improve as future studies refine the analyses, not necessarily by increasing dramatically sample sizes much further. For example, a study topped the 1.1 million sample with an analysis of 3 million individuals. It reported 12–16 percent of explained variance in educational attainment (Okbay et al., 2022). This was only a small increase from the values obtained in the 1.1-million-person study. As the authors concluded, "the sample size of the GWAS of educational attainment

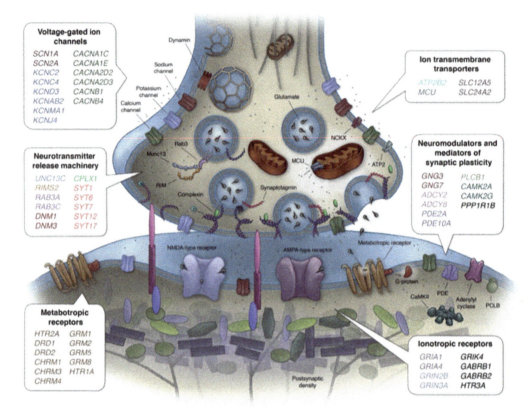

Figure 6.9 Genes implicated in Lee et al.'s (2018) study and the complexity of their influence at the level of the synapse.
Courtesy of Dr. Emily Willoughby.

reported in this paper is the largest published to date. For some purposes, such as attaining greater predictive power for the polygenic score, there are clearly diminishing returns" (p. 444). It should be no surprise by now that the story is complex and unsettled.

At some point, both genetics and brain imaging will be combined to see if an even stronger prediction is possible (Lee and Willoughby, 2021; Anderson and Holmes, 2021), although such studies will require large samples to detect small effects and, as always, independent replication (Uddén et al., 2019).

Even at this point, however, it must be noted that predicting 16 percent of the variance using DNA is an achievement that validates an important role for genetics. But, based on behavioral genetics research on twins and adoptees discussed earlier, we could expect that 50 percent of intelligence

variance should be predicted with PGSs. Whether there is "missing heritability" is an open question, since findings based on PGSs may be underestimated (Willoughby et al., 2021). Perhaps educational attainment has too many other factors that confound any *g* effects, and predictions may be better for some cognitive abilities than others (Genc et al., 2021). There are also other considerations about PGSs. One is that predictions might vary across populations of dissimilar genetic ancestry, and even geographic regions can influence results (Abdellaoui et al., 2022; Mitchell et al., 2022); both are empirical questions. Another one is whether PGSs can give meaningful insights into the molecular dynamics of how gene expression influences intelligence (Turkheimer, 2019). But there are other methods to use for mechanism questions that usually follow from predictive data. Robert Plomin and

Sophie von Stumm present an optimistic view about PGSs because they view *prediction* as useful scientifically even before causal mechanisms are discovered. They also argue the PGSs for intelligence can be used in some research designs in place of psychometric test scores (Plomin, 2018; von Stumm and Plomin, 2021). A similar view of the importance of PGSs has been expressed by Peter Visscher (2022).

Using PGSs might also enhance our understanding of gene–environment interplay. For example, an adoption study by Rosa Cheesman and colleagues (2020) reported results that indicate some progress in this regard. Their sample was 6,311 adoptees and 6,500 nonadopted individuals from the UK Biobank Project. They found that PGSs accounted for 10 percent of individual differences in educational attainment. They acknowledged that polygenic scores quantify genetic propensity plus family environmental factors. The key finding was that adoptees' genotypes were less related to their rearing environments than for the nonadopted sample. In fact, when polygenic scores were used for predicting years of education, the values were 0.074 for nonadopted individuals and 0.037 for adoptees. As the authors noted, "genetic influences on education are mediated via the home environment. ... The evidence presented in this article highlights the importance of the family environment to causal mechanisms influencing individual differences in educational attainment. This can be through possessing genes that shape the educational environment provided for offspring that also directly influence attainment in the child or through providing an educationally supportive environment for an adopted child" (pp. 1, 9). Other examples based on education attainment include Kong et al. (2018) and Bates et al. (2018). However, the same family–gene interaction may be less apparent for direct measures of intelligence (Willoughby et al., 2021; Okbay et al., 2022; Howe et al., 2022); this is another area where the weight of evidence is still accumulating.

The use of PGSs in intelligence research is in a nascent stage, but it underscores neuroscience approaches that potentially can expand scientific knowledge far beyond the limits of psychometrics and cognitive psychology. We must not underestimate the complexities, but, based on the history of scientific advances in the face of daunting complexities in other fields, we are optimistic that future research may exceed the limited expectations of skeptics and resolve current difficulties with genetic research on intelligence, including the missing variance. We see steady progress chipping away at the complexities bit by bit.

6.4.4 Genes to Neurons to Brains to Intelligence

Figure 6.9 vividly demonstrates only one of the complexity nightmares of neuroscientists trying to explicate how gene expression ultimately influences why some people are smarter than others. But this figure also captures the beauty and grandeur of the deep brain. How events at this level influence intelligence, as they must surely do, is the fascination of neuroscientists, not unlike the impact images from the James Webb Space Telescope have on cosmologists. In science, complexity is a temporary challenge, not a permanent roadblock.

We are making great progress in identifying genes associated with variations in normal intelligence. It is a hard problem, but GWAS have expanded and quickened the pace of progress well beyond the limitations of twin and adoption studies. We can now navigate many levels of explanation, from the identification of SNPs to genes to gene expression to brain development to brain structure and function to variation in cognitive processes to differences in psychometric test scores. It may turn out that the strategy of looking for gene–intelligence correlations directly can be improved. An alternate approach, for example, would be to identify the brain mechanisms that are associated with intelligence, and then find the genes that apparently influence the development of these mechanisms. So much more is possible.

As discussed in Chapter 5, there is emerging evidence associating efficiency of neurons to intelligence test scores (Genc et al., 2018; Goriounova et al., 2018). Mapping techniques for synapses and neurotransmitter activity are evolving. These studies are early examples of how neuroscience is addressing fundamental questions about intelligence. Much more can be expected. It is possible, for instance, to associate GWAS data to types of brain cells and tissue-specific transcriptome (RNA molecules) data from postmortem human brains (Ardlie et al., 2015). This methodology has the potential to associate specific gene expression to proteins that work in cell function and brain development. These approaches are taking intelligence research even deeper into the brain (Goriounova and Mansvelder, 2019) and suggest possible causal links between neurobiology and intelligence. Linking the molecular mechanisms at each level of explanation to each other will be the great challenge for the next decade or more. To meet this challenge, new research programs focused on the molecular genetics and biology of intelligence will build on and rival molecular research programs for learning and memory.

At least within the scientific realm, we are light-years past the old controversies about whether intelligence can be studied scientifically or whether genes have any influence at all. The findings discussed in this chapter are compelling, and we hope you feel the excitement in the field.

6.5 Concluding Thoughts

Variation in human intelligence is influenced by genetics. As we have detailed, substantial heritability coefficients are found in studies from around the world. The exact value for heritability coefficients may vary somewhat from study to study, but that is to be expected. It is never 0, and it is never 1 (100 percent) (Plomin et al., 2016). The heredity–environment debate too often is posed as a contest. It is not. We need to find causal mechanisms for both. A relatively unacknowledged problem that hampers the field is that, while we have an excellent theory of

genetic variation, we do not have a comparable theory of environmental variation (Scarr, 1997). It is easy to point to good and bad environments for the development of cognitive ability; it is much harder to provide some sort of quantifiable metric to assess how good or how bad a particular environment is. We need serious efforts to fix this situation so that research on both environmental and genetic effects on intelligence can progress, especially with respect to identifying interactions of specific variables.

Meanwhile, the current frontier has shifted from behavioral genetic research showing that intelligence is inherited to a significant degree to molecular genetic research aimed at finding the mechanisms of inheritance. There is a long history of hunting for the genetic causes of low intelligence and serious cognitive deficiencies. Public opinion generally accepts that it is reasonable to look for genetic causes for extreme mental deficiency (IQ less than 70), just as it is sensible to look for a genetic cause for medical conditions like heart disease. Such research justifiably receives strong public, political, and financial support.

There is considerably more ambiguity about efforts to identify a genetic basis for intelligence in the normal range. Perhaps there are two primary reasons for this:

1. Intelligence is associated with elitism. Any suggestion that cognitive competence is mostly inherited, whether through SES or genetic influence, goes against the popular ideal of democratic equal opportunity. This is especially so for genetics, because it is widely but erroneously believed that genes are deterministic and might permanently limit someone's potential for life success. This view is often expressed as the belief that you can be anything you want to be, if you work hard enough. The view that genetics denies the importance of individual opportunity and effort is frustrating to those who understand the importance of the partial genetic inheritance of intelligence, because no competent

behavioral geneticist or intelligence researcher argues that mental competence is completely inherited. Behavior geneticists generally agree with educators that socially important cognitive skills can, within broad genetic limits, be influenced by a combination of education, experience, and effort. An argument for an inherited component of intelligence is *not* necessarily an argument against the value of education and effort (Asbury and Plomin, 2014). This point often is not appreciated by those who attack genetic models of intelligence.

2. Discussion of the evidence for a genetic basis for intelligence variance differences among individuals is often conflated with discussions of average mean differences between populations. Claims of genetic influences on intelligence are seen by some as tantamount to a sort of Darwinian justification of economic and social inequalities. In fact, these are separate issues. The vast majority of studies in behavioral genetics have, at best, only indirect implications for issues surrounding any average population differences in intelligence scores. This is so simply because they are studies of the variation of genes and intelligence within rather than across populations. The causes of within-group variation are not necessarily the same as the causes of between-group variation.

Some of the confusion comes from Jensen's (1998) default hypothesis *to test* (not conclude) whether the causes of intelligence variation within groups were the same as the causes of between-group average differences. Behavioral genetics researchers generally are skeptical that this can ever be tested because of complex differences among groups that cannot be controlled in any comparisons of average test score data. We discuss the issues about population difference research in Chapter 9, and, in this context, Chapter 12 is relevant. The point here is that even the suggestion of a genetic component to group differences may contribute to a lack of enthusiasm for genetic research on intelligence in general. This is true for the public and for academics from fields unfamiliar with the findings presented in this chapter and throughout this book.

The search for the genes that influence intelligence might not be a hopeless project. Solid results from quantitative behavior genetics suggest that target genes exist, even though there likely are hundreds to consider. The technology for associating genes with traits is powerful and getting even better. Identifying causal pathways of gene expression for complex traits is becoming possible (Mountjoy et al., 2021). So is an understanding of molecular gene influences on brain development (Bhaduri et al., 2021), and so is mapping neurotransmitter systems to brain organization (Hansen et al., 2021). Ultimately, science may find relevant genes and understand how they work. This would be a huge advance for molecular genetic research seeking to understand the neurobiological basis of intelligence, with important implications for possibly enhancing intelligence (Chapter 13).

There are signs of increasing public awareness of genetic research on individual differences. In the afterword to the paperback edition of *Blueprint*, Plomin (2018) writes, "The public reaction has been positive beyond my wildest dreams. ... People are excited and enthusiastic about *Blueprint*. A typical comment is that the book was an eye-opener. People say they were not really opposed to genetic influences on individual differences, even for psychological traits. They just hadn't known much about genetics, and *Blueprint* helped them see its relevance for their lives."

6.6 Summary

This chapter introduced what we know about the genetics of individual differences in intelligence and explained how we know it. By the year 2000, twin and adoption data from behavioral genetic studies cemented the fact that genes play a major role, and this has been replicated by even stronger studies in the twenty-first century. Genetic research has witnessed rapid progress in the last two decades as DNA analyses have created a nascent field of the molecular genetics of intelligence to address how genes might influence individual differences in intelligence. There are exciting and thought-provoking ways to use DNA to predict intelligence and related variables. These findings have helped move intelligence research from classic psychometrics to advanced neuroscience approaches. Whether the vast complexities of gene–gene and gene–environment interactions will overwhelm scientists to stop progress remains to be seen.

6.7 Questions for Discussion

6.1   Why do you think a strong genetic influence on individual differences in intelligence continues to be controversial?

6.2   If a trait is highly heritable, does that prove a strong genetic component?

6.3   How would having a good measure of environmental similarity influence twin and adoption studies?

6.4   If it becomes possible to predict academic performance from a DNA test at least as good as current tests, would you prefer a DNA test for college entrance instead of standardized tests?

6.5   Given the complexity of understanding how genes influence intelligence, do you think the sequence of cascading, interacting steps can ever be understood well enough to modify it to enhance intelligence?

References

Abdellaoui, A., Dolan, C. V., Verweij, K. J. H., & Nivard, M. G. 2022. Gene–environment correlations across geographic regions affect genome-wide association studies. *Nature Genetics*, 54, 1345–1354.

Alarcon, M., Plomin, R., Fulker, D. W., Corley, R., & Defries, J. C. 1999. Molarity not modularity: Multivariate genetic analysis of specific cognitive abilities in parents and their 16-year-old children in the Colorado Adoption Project. *Cognitive Development*, 14, 175–193.

Allegrini, A. G., Selzam, S., Rimfeld, K., et al. 2019. Genomic prediction of cognitive traits in childhood and adolescence. *Molecular Psychiatry*, 24, 819–827.

Anderson, K. M., & Holmes, A. J. 2021. Predicting individual differences in cognitive ability from brain imaging and genetics. In Barbey, A. K., Karama, S., & Haier, R. J. (eds.), *The Cambridge handbook of intelligence and cognitive neuroscience*. New York: Cambridge University Press.

Ando, J., Ono, Y., & Wright, M. J. 2001. Genetic structure of spatial and verbal working memory. *Behavior Genetics*, 31, 615–624.

Ardlie, K. G., Deluca, D. S., Segre, A. V., et al. 2015. The Genotype-Tissue Expression (GTEx) pilot analysis: Multitissue gene regulation in humans. *Science*, 348, 648–660.

Asbury, K., & Fields, D. 2021. Implications of biological research on intelligence for education and public policy. In Barbey, A., Karama, S., & Haier, R. J. (eds.), *The Cambridge handbook of intelligence and cognitive neuroscience*. New York: Cambridge University Press.

Asbury, K., & Plomin, R. 2014. *G is for genes: The impact of genetics on education and achievement*. Chichester, UK: Wiley-Blackwell.

Bates, T. C., Maher, B. S., Medland, S. E., et al. 2018. The nature of nurture: Using a virtual-parent design to test parenting effects on children's educational attainment in genotyped families. *Twin Research and Human Genetics*, 21, 73–83.

Beaujean, A. A. 2005. Heritability of cognitive abilities as measured by mental chronometric tasks: A meta-analysis. *Intelligence*, 33, 187–201.

Bhaduri, A., Sandoval-Espinosa, C., Otero-Garcia, M., et al. 2021. An atlas of cortical arealization identifies dynamic molecular signatures. *Nature*, 598, 200–204.

Bouchard, T. J., Jr. 1997. Twin studies of behavior: New and old findings. In Schmitt, A., Atzwanger, K., Grammer, K., & Schäfer, K. (eds.), *New aspects of human ethology*. New York: Plenum Press.

Bouchard, T. J., Jr. 1998. Genetic and environmental influences on adult intelligence and special mental abilities. *Human Biology*, 70, 257–279.

Bouchard, T. J., Jr. 2009. Genetic influence on human intelligence (Spearman's *g*): How much? *Annals of Human Biology*, 36, 527–544.

Bronfenbrenner, U., & Ceci, S. J. 1994. Nature–nurture reconceptualized in developmental perspective – a bioecological model. *Psychological Review*, 101, 568–586.

Buchsbaum, M. S., Mirsky, A. F., Delisi, L. E., et al. 1984. The Genain quadruplets: Electrophysiological, positron emission, and x-ray tomographic studies. *Psychiatry Research*, 13, 95–108.

Cheesman, R., Hunjan, A., Coleman, J. R. I., et al. 2020. Comparison of adopted and nonadopted individuals reveals gene–environment interplay for education in the UK Biobank. *Psychological Science*, 31, 582–591.

Colom, R., Aluja-Fabregat, A., & García-López, O. 2002. Assortative mating in intelligence, psychoticism, extraversion, and neuroticism. *Psicothema*, 14, 154–158.

Davies, G., Lam, M., Harris, S. E., et al. 2018. Study of 300,486 individuals identifies 148 independent genetic loci influencing general cognitive function. *Nature Communications*, 9, 2098.

Deary, I. J., Strand, S., Smith, P., & Fernandes, C. 2007. Intelligence and educational achievement. *Intelligence*, 35, 13–21.

Defries, J. C., Plomin, R., Vandenberg, S. G., & Kuse, A. R. 1981. Parent–offspring resemblance for cognitive-abilities in the Colorado Adoption Project: Biological, adoptive, and control parents and one-year-old children. *Intelligence*, 5, 245–277.

Dickens, W. T., & Flynn, J. R. 2001. Heritability estimates versus large environmental effects: The IQ paradox resolved. *Psychological Review*, 108, 346–369.

Figlio, D. N., Freese, J., Karbownik, K., & Roth, J. 2017. Socioeconomic status and genetic influences on cognitive development. *Proceedings of the National Academy of Sciences of the United States of America*, 114, 13441–13446.

Finkel, D., & Pedersen, N. L. 2004. Processing speed and longitudinal trajectories of change for cognitive abilities: The Swedish Adoption/Twin Study of Aging. *Aging Neuropsychology and Cognition*, 11, 325–345.

Flynn, J. 2007. The latest thinking on intelligence. *Psychologist*, 20, 356–357.

Flynn, J. R. 2012. *Are we getting smarter? Rising IQ in the twenty-first century*. Cambridge: Cambridge University Press.

Flynn, J. R. 2018. Reflections about intelligence over 40 years. *Intelligence*, 70, 73–83.

Genc, E., Fraenz, C., Schluter, C., et al. 2018. Diffusion markers of dendritic density and arborization in gray matter predict differences in intelligence. *Nature Communications*, 9, 1905.

Genc, E., Schluter, C., Fraenz, C., et al. 2021. Polygenic scores for cognitive abilities and their association with different aspects of general intelligence: A deep phenotyping approach. *Molecular Neurobiology*, 58, 4145–4156.

Goriounova, N. A., Heyer, D. B., Wilbers, R., et al. 2018. Large and fast human pyramidal neurons associate with intelligence. *Elife*, 7, e41714.

Goriounova, N. A., & Mansvelder, H. D. 2019. Genes, cells and brain areas of intelligence. *Frontiers in Human Neuroscience*, 13, 44.

Hansen, J. Y., Shafiei, G., Markello, R. D., et al. 2021. Mapping neurotransmitter systems to the structural and functional organization of the human neocortex. *bioRxiv*, 2021.10.28.466336.

Haworth, C. M. A., Wright, M. J., Luciano, M., et al. 2010. The heritability of general cognitive ability increases linearly from childhood to young adulthood. *Molecular Psychiatry*, 15, 1112–1120.

Hill, W. D., Arslan, R. C., Xia, C., et al. 2018. Genomic analysis of family data reveals additional genetic effects on intelligence and personality. *Molecular Psychiatry*, 23, 2347–2362.

Hill, W. G., Goddard, M. E., & Visscher, P. M. 2008. Data and theory point to mainly additive genetic variance for complex traits. *PLoS Genetics*, 4, e1000008.

Horn, J. M. 1983. The Texas Adoption Project: Adopted children and their intellectual resemblance to biological and adoptive parents. *Child Development*, 54, 268–275.

Howe, L. J., Nivard, M. G., Morris, T. T., et al. 2022. Within-sibship genome-wide association analyses decrease bias in estimates of direct genetic effects. *Nature Genetics*, 54, 581–592.

Hur, Y.-M., & Bates, T. 2019. Genetic and environmental influences on cognitive abilities in extreme poverty. *Twin Research and Human Genetics*, 22, 297–301.

Jensen, A. R. 1998. *The g factor: The science of mental ability*. Westport, CT: Praeger.

Jensen, A. R. 2006. *Clocking the mind: Mental chronometry and individual differences*. New York: Elsevier.

Johnson, W., & Bouchard, T. J., Jr. 2005a. Constructive replication of the visual–perceptual–image rotation model in Thurstone's (1941) battery of 60 tests of mental ability. *Intelligence*, 33, 417–430.

Johnson, W., & Bouchard, T. J., Jr. 2005b. The structure of human intelligence: It is verbal, perceptual, and image rotation (VPR), not fluid and crystallized. *Intelligence*, 33, 393–416.

Kendler, K. S., Turkheimer, E., Ohlsson, H., Sundquist, J., & Sundquist, K. 2015. Family environment and the malleability of cognitive ability: A Swedish national home-reared and adopted-away cosibling control study. *Proceedings of the National Academy of Sciences of the United States of America*, 112, 4612–4617.

Kong, A., Thorleifsson, G., Frigge, M. L., et al. 2018. The nature of nurture: Effects of parental genotypes. *Science*, 359, 424–428.

Kovas, Y., Haworth, C. M. A., Dale, P. S., & Plomin, R. 2007. The genetic and environmental origins of learning abilities and disabilities in the early school years. *Monographs of the Society for Research in Child Development*, 72, 1–144.

Kovas, Y., & Plomin, R. 2006. Generalist genes: Implications for the cognitive sciences. *Trends in Cognitive Sciences*, 10, 198–203.

Krapohl, E., Rimfeld, K., Shakeshaft, N. G., et al. 2014. The high heritability of educational achievement reflects many genetically influenced traits, not just intelligence. *Proceedings of the National Academy of Sciences of the United States of America*, 111, 15273–15278.

Kremen, W. S., Jacobsen, K. C., Xian, H., et al. 2007. Genetics of verbal working memory processes: A twin study of middle-aged men. *Neuropsychology*, 21, 569–580.

Lawrence, E. M. 1931. An investigation into the relation between intelligence and inheritance. *British Journal of Psychology*, 16(Suppl.), 80.

Lee, J. J., Wedow, R., Okbay, A., et al. 2018. Gene discovery and polygenic prediction from a genome-wide association study of educational attainment in 1.1 million individuals. *Nature Genetics*, 50, 1112.

Lee, J. J., & Willoughby, E. A. 2021. Predicting cognitive-ability differences from genetic and brain-imaging data. In Barbey, A. K., Karama, S., & Haier, R. J. (eds.), *The Cambridge handbook of intelligence and cognitive neuroscience*. New York: Cambridge University Press.

Lello, L., Avery, S. G., Tellier, L., et al. 2018. Accurate genomic prediction of human height. *Genetics*, 210, 477–497.

Loehlin, J. C., Horn, J. M., & Ernst, J. L. 2007. Genetic and environmental influences on adult life outcomes: Evidence from the Texas Adoption Project. *Behavior Genetics*, 37, 463–476.

Loehlin, J. C., Horn, J. M. & Willerman, L. 1989. Modeling IQ change: Evidence from the Texas Adoption Project. *Child Development*, 60, 993–1004.

Loehlin, J. C., Horn, J. M., & Willerman, L. 1997. Heredity, environment and IQ in the Texas Adoption Project. In Sternberg, R. A., & Grigorenko, E. (eds.), *Intelligence, heredity, and environment*. Cambridge: Cambridge University Press.

Luciano, M., Smith, G. A., Wright, M. J., et al. 2001a. On the heritability of inspection time and its covariance with IQ: A twin study. *Intelligence*, 29, 443–457.

Luciano, M., Wright, M. J., Smith, G. A., et al. 2001b. Genetic covariance among measures of information processing speed, working memory, and IQ. *Behavior Genetics*, 31, 581–592.

McGue, M., Keyes, M., Sharma, A., et al. 2007. The environments of adopted and non-adopted youth:

Evidence on range restriction from the Sibling Interaction and Behavior Study (SIBS). *Behavior Genetics*, 37, 449–62.

Mitchell, B. L., Hansell, N. K., Mcaloney, K., et al. 2022. Polygenic influences associated with adolescent cognitive skills. *Intelligence*, 94, 101680.

Mitchell, K. J. 2018. *Innate: How the wiring of our brains shapes who we are*. Princeton, NJ: Princeton University Press.

Mountjoy, E., Schmidt, E. M., Carmona, M., et al. 2021. An open approach to systematically prioritize causal variants and genes at all published human GWAS trait-associated loci. *Nature Genetics*, 53, 1527–1533.

Okbay, A., Wu, Y., Wang, N., et al. 2022. Polygenic prediction of educational attainment within and between families from genome-wide association analyses in 3 million individuals. *Nature Genetics*, 54, 437–449.

Petrill, S. A., & Deater-Deckard, K. 2004. The heritability of general cognitive ability: A within-family adoption design. *Intelligence*, 32, 403–409.

Petrill, S. A., Lipton, P. A., Hewitt, J. K., et al. 2004. Genetic and environmental contributions to general cognitive ability through the first 16 years of life. *Developmental Psychology*, 40, 805–812.

Pinker, S. 2002. *The blank slate: The modern denial of human nature*. New York: Viking.

Plomin, R. 1999. Genetics and general cognitive ability. *Nature*, 402, C25–C29.

Plomin, R. 2018. *Blueprint: How DNA makes us who we are*. Cambridge, MA: MIT Press.

Plomin, R., Defries, J. C., Knopik, V. S., & Neiderhiser, J. M. 2016. Top 10 replicated findings from behavioral genetics. *Perspectives on Psychological Science*, 11, 3–23.

Plomin, R., Fulker, D. W., Corley, R., & Defries, J. C. 1997. Nature, nurture, and cognitive development from 1 to 16 years: A parent–offspring adoption study. *Psychological Science*, 8, 442–447.

Plomin, R., Kennedy, J. K. J., & Craig, I. W. 2006. The quest for quantitative trait loci associated with intelligence. *Intelligence*, 34, 513–526.

Plomin, R., & Petrill, S. A. 1997. Genetics and intelligence: What's new? *Intelligence*, 24, 53–77.

Plomin, R., & von Stumm, S. 2018. The new genetics of intelligence. *Nature Reviews Genetics*, 19, 148–159.

Polderman, T. J. C., Benyamin, B., De Leeuw, C. A., et al. 2015. Meta-analysis of the heritability of human traits based on fifty years of twin studies. *Nature Genetics*, 47, 702–709.

Polderman, T. J. C., Gosso, M. F., Posthuma, D., et al. 2006. A longitudinal twin study on IQ, executive functioning, and attention problems during childhood and early adolescence. *Acta Neurologica Belgica*, 106, 191–207.

Posthuma, D., de Geus, E. J. C., & Boomsma, D. I. 2001. Perceptual speed and IQ are associated through common genetic factors. *Behavior Genetics*, 31, 593–602.

Posthuma, D., de Geus, E. J. C., & Boomsma, D. 2003. Genetic contributions to anatomical, behavioral, and neurophysiological indices of cognition. In Plomin, R., Defries, J., Craig, I. W., & Mcguffin, P. (eds.), *Behavioral genetics in the postgenomic era*. Washington, DC: American Psychological Association.

Reynolds, C. A., Baker, L. A., & Pedersen, N. L. 2000. Multivariate models of mixed assortment: Phenotypic assortment and social homogamy for education and fluid ability. *Behavior Genetics*, 30, 455–476.

Reynolds, C. A., Finkel, D., McArdle, J. J., et al. 2005. Quantitative genetic analysis of latent growth curve models of cognitive abilities in adulthood. *Developmental Psychology*, 41, 3–16.

Rimfeld, K., Malanchini, M., Spargo, T., et al. 2019. Twins Early Development Study: A genetically sensitive investigation into behavioral and cognitive development from infancy to emerging adulthood. *Twin Research and Human Genetics*, 22, 508–513.

Rosenthal, D. 1963. *The Genain quadruplets: A case study and theoretical analysis of heredity and environment in schizophrenia*, New York: Basic Books.

Sauce, B., & Matzel, L. D. 2018. The paradox of intelligence: Heritability and malleability coexist in hidden gene–environment interplay. *Psychological Bulletin*, 144, 26–47.

Savage, J. E., Jansen, P. R., Stringer, S., et al. 2018. Genome-wide association meta-analysis in 269,867 individuals identifies new genetic and functional links to intelligence. *Nature Genetics*, 50, 912–919.

Scarr, S. 1997. Behavior-genetic and socialization theories of intelligence: Truce and reconciliation. In Sternberg, R. J., & Grigorenko, E. (eds.), *Intelligence, heredity, and environment*. Cambridge: Cambridge University Press.

Scarr-Salapatek, S. 1971. Race, social class, and IQ. *Science*, 174, 1285–1295.

Segal, N. L. 2021. *Deliberately divided: Inside the controversial study of twins and triplets adopted apart*. Lanham, MD: Rowman and Littlefield.

Sniekers, S., Stringer, S., Watanabe, K., et al. 2017. Genome-wide association meta-analysis of 78,308 individuals identifies new loci and genes influencing human intelligence. *Nature Genetics*, 49, 1107–1112.

Tang, Y. P., Shimizu, E., Dube, G. R., et al. 1999. Genetic enhancement of learning and memory in mice. *Nature*, 401, 63–69.

Tucker-Drob, E. M., & Bates, T. C. 2016. Large cross-national differences in gene × socioeconomic status interaction on intelligence. *Psychological Science*, 27, 138–149.

Tucker-Drob, E. M., & Briley, D. A. 2014. Continuity of genetic and environmental influences on cognition across the life span: A meta-analysis of longitudinal twin and adoption studies. *Psychological Bulletin*, 140, 949–979.

Turkheimer, E. 2019. Genetics and human agency: The philosophy of behavior genetics – introduction to the special issue. *Behavior Genetics*, 49, 123–127.

Turkheimer, E., Haley, A., Waldron, M., D'onofrio, B., & Gottesman, II. 2003. Socioeconomic status modifies heritability of IQ in young children. *Psychological Science*, 14, 623–628.

Uddén, J., Hultén, A., Bendtz, K., et al. 2019. Toward robust functional neuroimaging genetics of cognition. *Journal of Neuroscience*, 39, 8778–8787.

van Leeuwen, M., van den Berg, S. M., & Boomsma, D. I. 2008. A twin-family study of general IQ. *Learning and Individual Differences*, 18, 76–88.

Visscher, P. M. 2022. Genetics of cognitive performance, education and learning: From research to policy? *NPJ Science of Learning*, 7, 8.

von Stumm, S., & Plomin, R. 2021. Using DNA to predict intelligence. *Intelligence*, 86, 101530.

Wai, J., & Bailey, D. H. 2021. How intelligence research can inform education and public policy. In Barbey, A., Karama, S., & Haier, R. J. (eds.), *The Cambridge handbook of intelligence and cognitive neuroscience*. New York: Cambridge University Press.

Watson, J. D. 1968. *The double helix: A personal account of the discovery of the structure of DNA*, London: Weidenfeld and Nicolson.

Watson, J. D., & Berry, A. J. 2003. *DNA: The secret of life*. New York: Alfred A. Knopf.

Watson, J. D., & Crick, F. H. 1953. Molecular structure of nucleic acids: A structure for deoxyribose nucleic acid. *Nature*, 171, 737–738.

Willoughby, E. A., McGue, M., Iacono, W. G., & Lee, J. J. 2021. Genetic and environmental contributions to IQ in adoptive and biological families with 30-year-old offspring. *Intelligence*, 88, 1–9.

Yengo, L., Vedantam, S., Marouli, E., et al. 2022. A saturated map of common genetic variants associated with human height from 5.4 million individuals of diverse ancestries. *bioRxiv*, 2022.01.07.475305.

You, C., Zhou, Z., Wen, J., et al. 2021. Polygenic scores and parental predictors: An adult height study based on the United Kingdom Biobank and the Framingham Heart Study. *Frontiers in Genetics*, 12, 669441.

CHAPTER 7

Experience and Intelligence

7.1 Introduction

Every experience we have leaves an imprint in our brains. Physical and social experiences can change the brain. We refer to *experience* because *environment* (1) is a catch-all term and (2) suggests that humans are passive entities. We know that this is not the case, as carefully discussed almost a century ago by Louis Leon Thurstone (1923) during the behaviorism academic tidal wave that simplified all human behavior as comprising nothing more than responses to stimuli without any role for motivation, intention, or any other nonobservable construct (Watson, 1919). With today's historical perspective, behaviorism was much less generalized and influential than usually discussed in textbooks about the history of psychology (Braat et al., 2020). Thurstone's active view is illustrated in the bottom of Figure 7.1. Critiquing the passive approach of

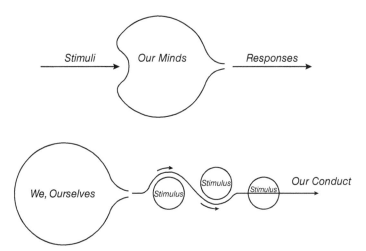

Figure 7.1 (top) The classic passive model of stimulus-organism-response. (bottom) The active model of organism-stimulus-response proposed by L. L. Thurstone in 1923 and revived by R. Plomin in 1994.

behaviorism, he wrote in "The Stimulus–Response Fallacy in Psychology,"

I suggest that we dethrone the stimulus. He is only nominally the ruler of the domain of psychology. The real ruler of the domain which psychology studies is the individual and his motives, desires, wants, ambitions, cravings, aspirations. The stimulus is merely the more or less accidental fact in the environment which becomes a stimulus only when it serves as a tool for somebody's purposes. When it does not serve as a tool for getting us what we want, it is no longer a stimulus. It is not a cause. It is simply a means by which we achieve our own ends, not those of the stimulus. The psychological act which is the central subject-matter of psychology becomes then the course of events, primarily mental, which intervene between the motive and the successful neutralization or satisfaction of that motive. The stimulus appears somewhere between the provocative and overt terminals of the psychological act. Mental life consists primarily in the approximate formulation of the motives leading toward overt expression. To the extent that mental life is of a relatively high order these approximate formulations of the motives become more and more tentative, deliberate, inhibited,

delayed, and subject to choose before precipitating into their final overt form. (pp. 354–369)

Robert Plomin (1994) revived Thurstone's psychologically active perspective seven decades later in his essay "Genetics and Experience": "The active model could be called an organism-stimulus-response (O-S-R) model [Figure 7.1, bottom]. Here the emphasis is heavily on the 'O,' especially on the organism's active role in constructing environments. The O-S-R perspective is best captured by the word 'experience.' The O-S-R model emphasizes the active selection, modification, and construction of environments. The same event could be appraised as a threat by one person and as a challenge by another" (pp. 29–31).

In this chapter, we distinguish between physical and social experiences. Physical experiences include anything that makes itself felt by direct physical action – variables like air pollutants or nutrition. Social experiences involve opportunities for self-development of cognitive skills, social actions that enhance or threaten security or education. It is a pragmatic distinction, because we know that both are blended in the brain with information coming from the genome. These interactions result in behaviors that scientists measure for understanding human behavior (Colom, 2016). Methods for doing so are discussed in Box 7.1.

Physical and social experiences may modify brain structure and function, but the brain level of analysis is not always informative. A teacher in elementary school does not care whether a vocabulary-learning exercise will transfer brain activity related to word recognition from the frontal cortex

Box 7.1 Methodological Considerations

Genetic and nongenetic factors interact. The research goal is to identify the extent to which these factors contribute to psychological outcomes, but pursuing this goal is not a simple task. Three concepts have to be kept in mind when interpreting data: (1) *reaction range*, (2) *distal–proximal causation*, and (3) *collinearity*.

Reaction Range

We introduced this concept in Chapter 2 (see Figure 2.3). Genes set a probable potential for the expression of a trait, but in many situations, the extent to which the trait is actually expressed (if at all) might be influenced by nongenetic factors. At one extreme, we have eye color, for which there is virtually no reaction range. At the other extreme are disorders like alcoholism and Alzheimer's dementia, where a person inherits a risk that can be triggered and promoted by nongenetic factors. Complex traits like intelligence likely have reaction ranges, although some cognitive factors, such as g, may have more limited ranges than others, such as verbal or numeral cognitive abilities.

We can think of nongenetic factors as influencing where persons will operate within their genetically specified (but unknown) reaction ranges. Because people will operate in different environments, psychological variations will be influenced in part both by differences between people in reaction range and differences in environmental factors. This is the reason that heritability estimates can change in different settings, as underscored in Chapter 6; it reflects the relative importance of differences in reaction ranges and differences in environmental conditions that operate within the potential afforded by genomes.

Nongenetic environmental effects are often illustrated by studies in which two relatively extreme environments are contrasted with each other. Such a study shows what might happen under certain extreme conditions. While this information can be important, it does not tell us what is likely to happen under the vast majority of normal conditions of environmental variation. This raises a measurement issue. To define the extent of environmental variation, we must have a metric specifying how close two environments are to each other. Unfortunately, and surprisingly, no such metric exists, largely because we do not have sound theories of environmental variation.

Distal (Genetic) and Proximal (Psychological) Causation

As any parent of two or more children knows, people influence their own environments (or at least, they try to). Environments vary in the extent to which they encourage the acquisition of skills and knowledge. For instance, parents in the higher socioeconomic status (SES) range may be more likely to encourage their children to engage in exploratory and problem-solving activities than are parents in the lower SES ranges (Nisbett, 2009). Suppose that variations in the initial tendency to explore an environment have a partially genetic basis. How are we to interpret a study that demonstrates that children show greater cognitive development if they are exposed to enriched environments that encourage exploration?

Box 7.1 *(continued)*

Would this be a distal (genetic) or proximal (psychological) effect?

The extent to which children are raised in an environment that encourages exploration varies with SES. Generally, low-SES children are raised in more restrictive environments than are the children of upper- and middle-SES families. Upper- and middle-SES children also, on average, have higher intelligence scores than lower-SES children. But it may be that the children in upper- and middle-SES families are more likely to be genetically predisposed to explore than are lower-SES children. It might be that the upper- and middle-SES parents are genetically predisposed to interact with children in a way that encourages exploration. And there are many other possibilities. For example, it could be that higher-SES families are subject to fewer social stresses and therefore are better able to create an encouraging environment.

Here is an example using a prospective longitudinal approach. Adrian Raine and colleagues (2002) tested the prediction that high-stimulation-seeking three-year-olds would score higher on intelligence in early adolescence (eleven years old). Based on 1,795 children, results showed that three-year-old high stimulation seekers (1) scored 12 points higher on intelligence tests and (2) had greater scholastic performance. The main implication was that stimulation seekers are more prone to create enriched environments and enhance their own cognitive development. The finding applied to boys and girls, as well as to different groups (Native Americans and Creoles). Furthermore, they were not mediated by parental education and occupation. As acknowledged by the authors, up to 63 percent of the variance in stimulation seeking is accounted for by genetic factors. Moreover, (1) stimulation seeking might be viewed as a component of the intelligence construct and (2) individual differences at the brain level may be the link connecting stimulation seeking and cognitive ability. Outside of the laboratory, proximal and distal variables are often thoroughly confounded. In interpreting studies of the causes of intelligence, this caution should be kept in mind.

Collinearity

Several possible causes of a phenomenon might be correlated. Collinearity presents serious difficulties for anyone interested in determining nongenetic influences on intelligence. To illustrate, there is a positive correlation between family income and children's intelligence test scores. Nisbett (2009) has suggested a simple economic solution to the problem: providing poor people with monetary subsidies to improve their children's life situation. However, he has also pointed out that this would not help immediately. Family income is correlated with a number of other potential factors of children's development, including nutrition and child-rearing practices. Nisbett does not stress the point, but SES, which includes education and income, is also correlated with genetic factors. People tend to marry people of their own general social class, especially with respect to education (Blackwell and Lichter, 2000; Murray, 2012). Providing poor people with a subsidized income for child-rearing might improve nutrition and other aspects of the environment immediately, might improve child-rearing practices over generations (a point Nisbett stresses), and would probably have little effect on genetic makeup.

Across nations, poor nutrition is associated with low intelligence scores. Poor nutrition is most likely to occur in countries that also have poor school systems and mostly to affect children whose parents have low intelligence scores (Rindermann, 2018). So, what is causing what?

Box 7.1 *(continued)*

Look at the issue symbolically. Let I be intelligence and $C_1 \ldots C_k$ be a set of potential causes of intelligence. We observe a correlation between I and C_1. But C_1 is correlated with many of the other causes, $C_2 \ldots C_k$, so the observed $I \times C_1$ correlation might be due to any of the other causes (Rohrer, 2018).

For ethical and practical reasons, it is seldom possible to avoid the collinearity problem by conducting controlled experiments, where the levels of various causes are manipulated by an experimenter. When controlled experiments can be conducted, there is often reason to question whether the laboratory results can be extrapolated to actual day-to-day situations.

The same criticism can be levied against simple predictive (regression) models, where one variable is "held constant" by statistical means. In these studies, investigators assess the influence of causal variable C_i on the residual variance in intelligence (I) after removing variation in I due to all other causal variables of interest. This technique is a useful way of determining how much of the variation in the intelligence scores can be associated with each of the predictors. The statistical model implies that each causal variable can be manipulated independently, but, in practice, this may not be possible. Studies relying on structural equation modeling may help disentangle collinearities, but the resulting models are quite complex and intricate, with a number of limitations.

In summary, given these methodological issues, it is obvious that the goal of understanding the interplay between genetic and nongenetic factors is hard work – but it is not impossible.

---

to the posterior temporal cortex; the teacher wants to know if the exercise will improve the way the students use words. It is important to keep discussions at the appropriate level of analysis. This is what we designated before as Brunswickian symmetry (Box 4.4).

7.2 Experience and the Physical World

Several features of the physical world are associated with intelligence levels. These include prenatal and infant experiences, injuries, exposure to atmospheric pollutants, health practices, nutrition, and substance use and abuse. In some cases, the direction of causality is clear-cut – any physical agent that damages the structure of the brain or interferes with brain processing may show negative impact on people's intelligence. In other cases, there is a statistically reliable association between intelligence and some agent, but causality is difficult to establish.

There might be a chicken-and-egg problem: which came first, intelligence levels or exposure to the agent? Linda Gottfredson (2007), who has discussed the social implications of intelligence differences, has pointed out that the task of following health and safety regimens is itself cognitively challenging. To take a specific case, intelligence scores are predictors of involvement in motor vehicle accidents, and such accidents are one of the most common causes of head injury (O'Toole, 1990; Smith and Kirkham, 1982).

Collinearity makes it hard to isolate causality (Box 7.1). Exposure to dangerous conditions and intelligence scores are correlated with many other variables, such as SES, parental intelligence, and inadequate access to good schools, that could themselves affect intelligence. Picking out just one of these factors as the cause of low intelligence cannot be justified. To conclude that a particular incident or environmental condition has affected intelligence, it is necessary to show

(1) that intelligence at baseline did not contribute to the incident or condition and (2) that other possibly relevant conditions have been measured or controlled experimentally or statistically. In the ideal case, premorbid measures of intelligence, taken before the incident or exposure, must be compared to measures taken after the outcome of interest.

7.2.1 *Prenatal and Infant Experiences*

A great deal of attention has been devoted to prenatal and infant experiences because this is when the foundations of cognition are established. In their first year, infants acquire considerable information about language and social interactions (Gopnik et al., 1999). To what extent do individual differences in cognitive function in infancy predict later indices of intelligence? Infancy is also a period of physiological vulnerability. To what extent are these early indicators sensitive to nongenetic factors?

Infant development can be assessed by the Bayley scales, originally designed by Nancy Bayley (2005) and her colleagues at the University of California, Berkeley, and updated periodically. These scales document the age of normal occurrence of a variety of behaviors, such as crawling, toddling, vocalization, and reactivity. This approach resembles the first intelligence scale, developed by Alfred Binet in France (Chapter 1).

Middle- and upper-class parents are especially concerned that their children be on schedule. This concern may be overdone. The correlation between scores on developmental scales that are based on activity and vocalization over the first thirty-six months of life and scores on adult intelligence tests is nearly zero (Bayley, 1968). However, this does not mean that the development of information-processing and verbal measures can be disregarded. By the age of three, tests that involve vocabulary correlate in the 0.5–0.7 range with IQ scores for young adults. Whether a large vocabulary is a cause or consequence of intelligence is the question.

Infants must recognize stability and change in the environment. The infant's ability to do this can be evaluated by measuring habituation, the ability to discriminate between novel and repeated stimuli (Brody, 1992; Fagan et al., 2007). Joseph Fagan and his colleagues (2007) measured habituation by showing infants an interesting picture, often a picture of another infant. After the infant has viewed one picture, it is removed, and then presented again, along with a new picture. The measure of habituation is the extent to which the infant looks at the new picture (i.e., habituated to the old picture). This simple test for six- to twelve-month-old infants has a correlation of 0.59 with measures of adult intelligence at twenty-one years of age (when corrected for reliability of measures).

Infant cognition can be also assessed testing reasoning processes. Erno Teglas and colleagues (2011) demonstrated that twelve-month-old infants have reasoning ability, contrary to the developmental model proposed by Swiss psychologist Jean Piaget. As they wrote, "the ability to flexibly combine multiple sources of information and knowledge to predict how a complex situation will unfold is at the core of human intelligence" (p. 1054).

Their clever experiments were based on probabilistic inference: greater looking times were regarded as increased violation of the infant's expectations regarding their previous knowledge, or increased novelty regarding their interpretation of habituation stimuli.

Infants viewed movies of four objects belonging to two categories (shape and color) bounced randomly inside a container with an opening on its lower side (Figure 7.2). Twelve versions of movies were produced, manipulating (1) number of objects of each type in the scene (three instances of one type and one of the other types), (2) their physical arrangement (objects of one type were always closer to the exit before occlusion than object of the other type), and (3) the duration of occlusion (zero, one, or two seconds).

Predictions about which object type will exit first must be related to (1) the number and the physical arrangement of the object types and (2) occlusion duration. The findings showed that twelve-month-old infants

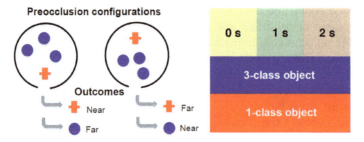

Figure 7.2 Infant perception experiment. Infants saw three objects of one type and one object of another type bouncing randomly inside a container. After some time, an occluder masked the objects, and one of four outcomes occurred: an object exited the container through the bottom opening that was either of the common object kind or the unique object, with a position before occlusion that was either far from or near to the exit (Teglas et al., 2011).

are able to represent spatiotemporal and logical features of dynamic scenes "with multiple objects in motion and integrate these cues with optimal context-sensitive weights to form rational expectations" (p. 1057).

Another commonly used index of prenatal health is variation in birth weight. These variations are correlated with intelligence for both premature and normal-term infants. For example, Matte and colleagues (2001) estimated that there is an increase of between 0.3 to 0.5 IQ points for every 100 g of birth weight over 2,500 g. This finding, however, is ambiguous, because maternal intelligence is positively correlated to the birth weight of the child. When maternal intelligence is controlled statistically, the relationship between intelligence and birth weight decreases but does not disappear (Deary et al., 2005). The finding is not confined to cases of abnormally low birth weight. Differences in intelligence between heavier and lighter babies in the normal range (2,000 g and above) have been reported in studies in a variety of North American and European countries, as well as in populations of European and African ancestry in the United States (Dombrowski et al., 2007).

The relationship does not solely reflect any tendency for bright mothers to have children that are both heavier and have a higher genetic potential for intelligence, as separate effects. This was shown in an elegant study of two large samples with twins, one in New Zealand and one in the United Kingdom (Newcombe et al., 2007). Because of the sample sizes, it was possible to compare birth weights and IQ scores in identical twins, thereby controlling for genetic influences. The heavier twins had higher intelligence scores, demonstrating a role for the prenatal environment independent of genotype.

We can also consider extreme cases by looking at "preemies," newborns with a gestational age of less than thirty-two weeks (eight months), many of whom have birth weights well below 2,000 g. These children, on average, have IQ scores almost 1 standard deviation below control groups (Esbjorn et al., 2006). In general, nonverbal functioning and abstract reasoning are more affected than verbal functioning. Considering the links between information processing and g, it is not surprising to find that premature infants do poorly in childhood (at eight to twelve years old) on tests of working memory and cognitive control (Bayless and Stevenson, 2007). Because birth weight statistically predicts future intelligence, it is of interest to know what variables may lead to low birth weight. By far the largest risk factor is premature birth, although a myriad of other causes can affect maternal and fetal health. Women over forty years old are at increased risk for premature birth (Fuchs et al., 2018).

All these factors are associated with lower intelligence, but is there any way to improve intelligence, within the normal range, during this period of development? So far, there is no objective evidence that various diets, exercise regimens, or extra cognitive stimulation have positive long-term effects on intelligence (see Chapter 13). It is easier to do harm than to do good.

7.2.2 *Direct Insults to the Brain*

Damage to the brain may result in decreased intelligence and, if the injury is sufficiently focal, may disrupt certain intellectual functions and not others, as we discussed in Chapter 5 (see Figures 5.4 and 5.5). Damage to the forebrain-parietal system will disrupt the working memory–cognitive control complex, thus damaging the functions underlying general reasoning ability. Damage in the medial temporal region, and especially to the hippocampus, will disrupt the ability to form new memories (anterograde amnesia). The disability can be severe enough that the affected person has to be in custodial care for life. Paradoxically, some have claimed that this does not affect intelligence, on the ground that intelligence test scores many not be lowered. Because any reasonable definition of intelligence has to include the ability to learn from experience, the fact that profound anterograde amnesia does not influence scores on standardized intelligence tests simply shows that these scores fail to tell the whole story.

Less dramatic, but important, losses of cognitive functions can occur when the brain is subjected to apparently minor physical damage. Closed head injury (concussion) is of special interest because of its prevalence (mostly from accidents and sports); there are more than a million cases annually just in the United States. In the European Union, the overall incidence rate is 262 per 100,000 for traumatic brain injury (Peeters et al., 2015).

Closed head injuries are often followed by a period of disorientation. In most cases, this subsides. When conventional intelligence tests are given, a year or more later, major effects are often not found. When effects are found, they are usually on abstract, nonverbal reasoning tests. Tests that emphasize verbal ability appear to be much less sensitive to the aftermath of severe concussions. So, if we were to restrict ourselves to the intelligence test data alone, we might conclude that a knock on the head is not all that serious. However, when patients are tested using laboratory tasks or neuropsychology tests that evaluate working memory functions, effects are often found. These tasks are considerably more sensitive than the memory evaluations included in most intelligence tests. Injured people often do not report problems in everyday life, even if they have trouble with the laboratory tasks. It would be wrong to conclude on the basis of intelligence tests alone that the residual effects of concussion are not serious enough to be of concern.

Investigators at the United Kingdom's Applied Psychology Unit (Cambridge) even went a step further (Sunderland et al., 1983). They questioned the partners of people who had suffered concussions. The partners reported affirmative examples of deficits, and the severity of the deficit was correlated to the difficulty the affected person had with the laboratory tasks. This study showed the importance of obtaining information about a person's everyday performance, as well as observing performance on laboratory tasks or in a conventional testing session.

A second study showed how concussion can act as a distal influence that increases the risk of incurring a condition that acts as a proximal influence on intelligence. The people studied were elderly pairs of twins (over sixty years), where one twin suffered from Parkinson's disease, which can result in a loss of intelligence in the latter stages, and the other did not (discordant twins). The risk of incurring Parkinson's disease tripled for those who had experienced head injuries, even though the head injuries had occurred, on average, thirty-seven years before the study was conducted (Goldman et al., 2006).

These studies illustrate that long-term damage to cognition and intelligence can result from what are apparently minor injuries to the brain.

7.2.3 *Atmospheric Lead*

Industrial pollutants in the air, water, and soil are a longtime problem. Mercury nitrate, a neurotoxin, was used in hat making until the twentieth century. A century earlier, the phrase "mad as a hatter" was in common use, for hat makers were thought to have delusions and poor motor control. Today mercury's dangers are well known, and industrial exposure is tightly regulated, although regulations among countries vary. There is still considerable concern about the presence of trace amounts of mercury in the food chain, especially in certain fish.

Lead, a much more common pollutant, presents a larger problem worldwide. Lead was one of the first metals mined. Analysis of human remains shows that prior to the beginning of metallurgy, about 4000 BCE, the concentration of lead in the human body was 0.0016 micrograms per deciliter (μg/dL). In 1975–1980, the concentration in US children was estimated to be fifteen micrograms per deciliter, more than 1,000 times the natural level. The concentration has fallen markedly since then, due largely to the banning of tetraethyl, lead-added gasoline (Hubbs-Tait et al., 2005).

Today's concerns are about the cumulative effects of exposures to much lower concentrations of atmospheric lead and, in particular, the effects on children. These concerns were heightened by results reported by Herbert Needleman and colleagues (1979). They conducted a large-scale epidemiological survey (more than 2,000 children) of lead concentrations in children's bodies by collecting schoolchildren's baby teeth after they fell out and analyzing them for lead content. They compared the Wechsler (WISC-R) test performance of first- and second-grade children with high and low concentrations of lead in their baby teeth, allowing for control variables like parental SES. Higher concentrations were associated with lower IQ scores of about 6–7 IQ points.

Collinearity posed a difficult problem. Was the poor test performance of children with higher lead concentrations caused by the concentration of lead, or did they perform poorly because they tended to come from poorer homes and therefore would be expected on statistical grounds to have lower scores on intelligence tests? The best way to investigate this issue is by replication of a result in situations where the collinearity problems plaguing the original finding do not occur.

Several such replications have been conducted. One example is a longitudinal study conducted in Kosovo in the 1990s (Wasserman et al., 2000). Two separate populations were studied: (1) children living in a relatively large town near a lead smelter and (2) children in a similar town, twenty-five miles away, where the atmospheric lead exposure level was much lower. Pregnant mothers were enrolled in the study, prenatal lead concentrations were estimated, and children were followed until they were seven years old in order to assess the cumulative effects of lead in the body. Both prenatal levels of lead and postnatal increases of lead were associated with decreased intelligence test scores after all other variables were considered. Lead concentrations were statistically associated with about 4 percent of the variance in children's intelligence test scores. There was no evidence of threshold effects, which provides an argument against maintaining a permissible concentration level of ten micrograms per deciliter, the typical recommendation. Similar results have been obtained in other studies in the United States and in South America (Chiodo et al., 2007; Counter et al., 2005).

In 2005, a consortium of researchers involved in these studies published an analysis of the international findings (Lanphear et al., 2005). They concluded that there is a nonlinear relationship between intelligence test scores and the level of lead in the blood. The decreases are cumulative, so the expected loss for a child with a thirty micrograms per deciliter concentration of lead in

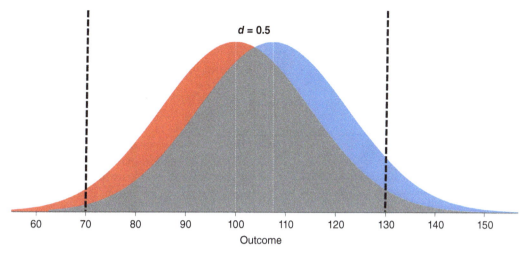

Figure 7.3 Population effect of a drop of half a standard deviation on the IQ scale (mean = 100, SD = 15). Vertical lines help visualize the effect of the average mean differences at the extremes of the distribution (>2.0 and <−2.0 standard deviations); *d* is the statistical symbol for effect size.

the blood was almost half a standard deviation (6.9 IQ points), consistent with Needleman's original estimate.

How serious is a loss of 7 IQ points? There is not much difference in the intellectual abilities of two children, one with an IQ of 102 and the other with an IQ of 95. On a population basis, however, this would be a serious issue because of changes in the frequency of exceptionally high and low levels of intelligence with fewer people in the former category and more people in the latter one, as illustrated in Figure 7.3.

Consider a population of 1,000 children. If the mean IQ in the population is 100 and the standard deviation is 15, we would expect to find approximately fifty children with IQs below 70, which is often considered an indication for enrollment in special education programs, and an equal number of children with IQs above 130, often the cutoff for entry into a gifted education curriculum. If the mean of the population dropped to 95, we would expect to find about one hundred children with IQs below 70 and only twenty-five with IQs above 130. In other words, a drop of five IQ points in the population average would be associated with a doubling of the number of children in the special education program, while the

number of children eligible for the gifted program would drop by half.

These research findings have had an effect on public policy, at least in the industrial and postindustrial countries. In 1960, the divide between safe and unsafe lead exposure levels in blood was sixty micrograms per deciliter, twice the level used as a point of concern in the 2005 comprehensive review. Actions to reduce lead concentrations are now recommended by the US Centers for Disease Control and Prevention (CDC), as of 2012, when children's blood levels exceed five micrograms per deciliter. The 2012 CDC advisory committee report maintained that blood concentrations higher than forty-five micrograms per deciliter require medical treatment (see www.cdc .gov/nceh/lead/prevention/blood-lead-levels.htm).

Bruce Lanphear published a perspective article in 2017 with the title "Low-Level Toxicity of Chemicals: No Acceptable Levels?" The key message was that societies need to achieve zero exposure to chemicals such as lead, radon, airborne particles, asbestos, and benzene because "the amount of toxic chemical linked with the development of a disease is proportionately greater at the lowest dose or levels of exposure" (p. 1). Figure 7.4 depicts the function relating

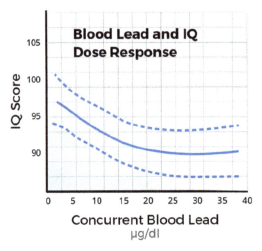

Figure 7.4 Decelerating dose–response curve relating blood lead and IQ score (Lanphear, 2017).

blood lead levels and IQ scores and the sobering main conclusion: "An increase in blood lead from <1 μ/dL to 30 μ/dL was associated with a 9.2 IQ deficit, but the largest fraction of this deficit (6.2 IQ points) occurred below 10 μ/dL. ... Despite the dramatic decline in blood lead levels, lead exposure accounts for a loss of 32 million IQ points in a 6-year birth cohort of US children" (p. 3).

Unfortunately, significantly high levels of atmospheric lead are found in some countries. For instance, in the industrial port of Callao, near Lima, Peru, lead storage areas are located near one of the poorest residential areas of the city. A survey found that approximately half of the early elementary school children had blood lead levels of twenty micrograms per deciliter or higher (Guerrero, 2009). Lead in the atmosphere clearly represents one of the major threats to the development of intelligence in some areas of the world today. Lead levels in water have not been as ubiquitous a problem, but the lead poisoning of the municipal water system in, for instance, Flint, Michigan, due to governmental negligence carried renewed attention to water sources of lead. Whether lead finds its way into the bloodstream from breathing the air or drinking the water, higher blood levels are associated with lower intelligence test scores.

Governmental action can and must fix this (Bellinger et al., 2017).

7.2.4 *Nutrition*

It makes sense that nutrition influences intelligence, especially during periods of neural growth. However, it is difficult to prove this. We distinguish between temporary effects during a period of malnutrition and permanent effects exhibited following recovery from malnutrition. It is also necessary to distinguish between the effects of brief periods of malnutrition, the timing of these periods during development, and the effects of chronic malnutrition. The type of malnutrition is also an issue. A diet may have adequate caloric intake and still be deficient in protein, iron, or further substances important for neural development. There is also the problem of distinguishing between cognitive and attentional effects. Malnourished people are physically weak and have trouble concentrating. This may lead to underperformance both in a testing session and, more importantly, in any situation involving cognitive demands over a long period of time. These are both conceptual and practical problems.

It would be unethical to subject people to malnutrition in a controlled experimental study. Correlational and epidemiologic studies are informative, with limitations such as that reliable records of nutritional intake may not be available, and collinearity is usually an issue. Also, malnutrition to a level that would influence neural development, and hence intelligence, is rare in developed countries. When it does occur, it is likely to be accompanied by poor parenting practices and a lack of social support for families. In low-income, less developed countries, malnutrition tends to occur in the poorest areas with the least access to schools and medical facilities. In general, more intelligent, better-educated mothers provide better nutritional environments for their children (Wachs and McCabe, 2001). Genetic and social effects on the child's intelligence can be incorrectly attributed to nutrition, unless care is taken to measure these factors as covariates in research designs.

Here are some examples. Most of these have involved infants or young children, as they are perceived to be most vulnerable to the effects of malnutrition. There are fewer studies of the relation between nutrition and adult cognition.

Brief periods of malnutrition do not appear to leave permanent effects, even when the malnutrition occurs in utero or in infancy. The evidence for this comes from a large, quasi-experimental study that occurred during a period of starvation in the northern region of the Netherlands as a by-product of a Nazi blockade during World War II. At one point. the estimated caloric content of rations was down to 640 calories per adult per day. Meanwhile, the southern Dutch cities, under Allied control, received adequate food. The starvation period lasted from September until March, when US troops liberated the northern Dutch cities and supplied food. The war in Europe ended shortly thereafter.

In the Netherlands postwar, men register for military service at age nineteen. All registrants complete a progressive matrix test. Zena Stein and her colleagues (1972) from Columbia University compared the scores of men who were in utero or neonates while living in the northern Dutch cities during the months of the starvation period to the scores of men who had been born at the same time but had lived in the cities under Allied control. Other records of mental ability, such as the frequency of mental disability, were also compared across the two populations. There was no difference in any indicator of mental competence. This study of more than 125,000 men provides striking evidence of the resilience of human cognition. Apparently, any effects of severe short-term malnutrition are not permanent. The key results are shown in Figure 7.5.

Chronic, prolonged malnutrition shows different results. Studies in low-income nations have provided evidence that prolonged malnutrition is associated with low cognitive test scores, particularly on nonverbal tests. A closer look reveals an interesting pattern. Malnourished children are described as less attentive, more impulsive, and easily distracted. Researchers primarily interested in nutrition describe this as a confounding variable, saying that it is unclear whether the effects of malnutrition are on intelligence or whether they are on attention (Sigman and Whaley, 1998).

If we consider these observations in the light of studies of the role of basic information processing in intelligence described in Chapter 4, a different conclusion is possible.

The behaviors that characterize malnourished children are indications of lack of cognitive control, which is a vital part of the working memory system (Burgoyne et al., 2022). As we saw in Chapter 4, individual differences in working memory are strongly related to intelligence. Prolonged malnutrition also influences habituation, which, as we have seen, is an indicator of infant intelligence.

Viewed in this light, malnutrition might directly influence intelligence scores because the act of taking the test requires exercise of the working memory system. The same deficiencies in information processing influence young children's school performance. This produces a deficiency in the knowledge-based aspects of intelligence. A physical agent like malnutrition results in an inability to benefit from environmental supports that may contribute to better cognition.

This conclusion is reinforced by studies of the beneficial effects of intervention. Two sources of intervention have been tried: (1) improvements in general nutrition and (2) improvements in intake of specific nutrients, primarily iron and protein. Iron is important because iron-deficiency anemia leads to general listlessness and inability to focus attention. Proteins are considered necessary for development of the neural system.

The results of a study conducted in rural Guatemala are particularly informative. Protein supplementation was provided for some children, while a nonprotein supplement was offered to others. Protein supplementation improved various measures of cognitive performance. In addition, there was an important interaction: the greatest gains occurred when protein supplementation

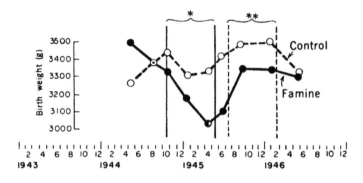

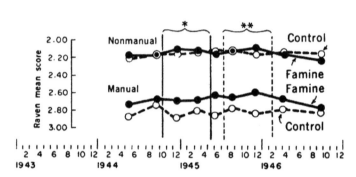

Figure 7.5 Famine, birth weight, and intelligence. (top) Mean birth weight in maternity hospitals selected from famine and control cities by cohort birth. (bottom) Mean scores on Raven's Progressive Matrices of Netherlands men examined at age nineteen, for manual and nonmanual classes according to father's occupation, by cohort of birth in famine cities and control cities. Solid vertical lines bracket the period of famine, and dashed vertical lines show the period of births of children conceived during famine (Stein et al., 1972).

was combined with school attendance (Pollitt et al., 1993).

This is consistent with the argument that appropriate nutrition will benefit working memory and related cognitive processes. These effects make it possible to take advantage of an environment that supports the development of cognitive abilities and skills. A policy implication is that, for optimal effect, a supplementation program to help children recover from malnutrition should be combined with an educational program.

In a comprehensive study, Tamara Daley and colleagues (2003) analyzed data obtained in two cohorts of rural Kenyan seven-year-old children over a fourteen-year period (1984 cohort and 1998 cohort). A total of 118 children were tested in 1984, and another 537 children were tested in 1998. Both

cohorts completed the same cognitive test battery, including the Raven's Colored Progressive Matrices, a verbal meaning test similar to the Peabody Picture Vocabulary Test, and a digit span (short-term memory) test. For both groups, food intake and information on the children's health (including blood and stool samples) were also measured. Finally, parental and household measures were obtained: parents' age and marital status, composition of the household, number of years of parents' schooling, parents' literacy, and so on. The reported IQ gains in the fourteen-year period on the three cognitive measures were 25 points for the Raven's, 7 for verbal, and 3 for digit span.

Considering all the measures for both cohorts, three features were highlighted for explaining the observed cognitive gains: (1)

parents' literacy, (2) family structure, and (c) children's nutrition and health. The authors wrote, "A shift in values toward an emphasis on schooling may have occurred during the period of our study as a result of a combination of increased parental education and an emphasis on school-related skills and knowledge (but) changes in children's nutrition and health may have also contributed to increased cognitive scores" (p. 219). The astute reader has recognized that parents' literacy may be confounded with parental intelligence.

7.2.5 Alcohol

Alcohol is the most common and most abused recreational drug in the world. While definitions of abuse have shifted over time and place, psychiatric criteria usually distinguish two forms of alcoholism: (1) alcohol dependence, in which the affected person can scarcely go without a drink, and (2) alcohol abuse, which refers to people who consume alcohol frequently and heavily but can go for periods of time without drinking.

Alcoholism is diagnosed using four criteria:

1. *Craving*: A strong compulsion to drink. This goes well beyond thinking it would be nice to have wine with your meal.
2. *Loss of control*: The inability to cease drinking once drinking has begun.
3. *Withdrawal symptoms*: Nausea, sweating, shakiness, and anxiety when alcohol use is stopped.
4. *Tolerance*: The need to drink ever greater amounts of alcohol to experience symptoms of intoxication, including relaxation and euphoria.

Brain imaging studies of alcohol abusers compared to age-matched controls have suggested that prolonged alcoholism damages the frontal lobes (Moselhy et al., 2001). This damage is accompanied by decrements in abstract reasoning and in visuospatial reasoning. Both are sorts of behaviors that one associates with decreased working memory, loss of the ability to plan actions,

and diminished cognitive control (Schottenbauer et al., 2007). Extremely heavy, prolonged drinking also can produce damage to the limbic system with an associated loss of the ability to store new memories (*Korsakoff's syndrome*). Korsakoff's syndrome patients are unable to function in society and require hospitalization.

The effects of social drinking are less clear. Surveys of social drinkers indicate a small negative association between moderate social drinking and intelligence scores. Not surprisingly, the effect is strongest among those who consume alcohol more than four times a week and who regularly consume more than forty to fifty milliliters of alcohol per occasion (Parker et al., 1991). This is roughly equivalent to three or four drinks of hard liquor, glasses of wine, or bottles of beer.

There are marked individual differences in tolerance to alcohol. The critical amount varies with weight and with sex, women generally being more sensitive. In terms of psychometric models, alcohol appears to affect the general factor of intelligence (g). However, collinearity makes it hard to define causality. The extent of social drinking varies greatly among different demographic groups. In some circles, social drinking is an accepted and almost expected practice. In others, any use of alcohol is frowned on. It is difficult to disentangle the effects of social drinking from inherited intellectual potential, health practices, and other lifestyle variables. There is the chicken-and-egg problem: does heavy social drinking reduce intelligence, or is it the case that the intelligent person is less likely to drink heavily on a regular basis?

This question is hard to answer because there are data indicating that childhood intelligence scores (measures of cognitive ability taken before people begin to drink) predict drinking patterns. People with higher intelligence scores are less likely to drink to the point of having a hangover and are more likely to consume wine, which is generally drunk more slowly and with meals, than whiskey or beer (Batty et al., 2006; Mortensen et al., 2005). This distinction is important, because the toxic effects of alcohol are related to a buildup of alcohol metabolites in the bloodstream. This occurs when alcohol is taken in at a faster rate than it can be processed by the liver. When alcohol is taken with food, the rate of absorption from the stomach to the bloodstream is reduced, thus lessening its impact.

Although the issue is not clear, the evidence favors the hypothesis that repeated heavy drinking (to the point of feeling high but not necessarily to the point of losing consciousness or motor control) leads to some cognitive deficit. The effects of alcohol on intelligence are important on a population basis for similar reasons to those of air pollutants and changes to the frequency of individuals at both ends of the IQ distribution.

Who should pay attention to these findings? Currently, these are the top ten countries with the highest levels of alcohol consumption: Belarus, Lithuania, Grenada, Czech Republic, France, Russia, Ireland, Luxembourg, Slovakia, and Germany (according to https:// worldpopulationreview.com/countries/alcoh ol-consumption-by-country/).

7.3 Experience and the Social World

In his encyclopedic book *Human Accomplishment: The Pursuit of Excellence in the Arts and Sciences (800 BC to 1950)*, Charles Murray (2003, p. 209) referred to the evolution of certain ways of thinking as *meta-inventions*: "the introduction of a new cognitive tool for dealing with the world around us (artistic realism, linear perspective, artistic abstraction, polyphony, drama, the novel, meditation, logic, ethics, Arabic numerals, the mathematical proof, the calibration of uncertainty, the secular observation of nature, and the scientific method)."

Literacy may be the most important of these meta-inventions, because it fosters abstract thinking and provides continuity with the past. Mathematical reasoning and scientific approaches to problem solving also improve thinking. In our society, the intelligent person has a good grasp of these tools of thought. Individuals and societies who use these ways of thinking have some advantage over those that do not. Parents in modern nations spend time and effort trying to ensure that their children become familiar with these ways of thinking, and the effort begins at home. For example, there is some evidence that viewing quality children's television programs, such as *Sesame Street*, improves cognitive skills in preschoolers (Anderson, 1998).

Variables in the social environment may influence the development of intelligence, but we still do not know specifically how this occurs. There are likely hundreds of factors, each with a tiny influence, as we have seen when describing the genome in Chapter 6. Similarly, in the case of socioenvironmental factors, this complicates identifying the precise paths from social experiences to intelligence differences. At some point, the mechanisms of influence must be biological, because the impact inevitably takes place through the brain.

Furthermore, social variables are easy to name but hard to define. Good parenting is something we all applaud, but what is a good parent? SES is real, but how do scientists measure it? Income is often used as a proxy. The president of the European Parliament has a salary of approximately €306,000 per year, whereas Leo Messi, the celebrated soccer player, has a salary of €565,000 per week. This is not an error: per year and per week, respectively, are correct. Who has higher socioeconomic status, the president or the soccer player?

Measurement is not the only problem. As noted, we do not have a theory of the social environment that approaches the clarity of genetic models. Nor do we have any broad theoretical approach to environmental issues that can play a role similar to the role that Darwin's theory of evolution has in biology. Lack of a comprehensive theory of just how environments influence cognition has resulted in numerous ad hoc studies without much systematic knowledge accumulated. The lack of an adequate theory has also made it hard to understand the relationships that have been observed in studies. Social variables are highly collinear (1) with each other, (2) with genetic measurements, and (3) with measurements of the physical world. Nutrition is linked to SES, especially in low-income nations. Parents who produce favorable home environments for their children are likely to seek out the best school environments. In middle- and high-income countries (and in urban districts worldwide), residence is closely tied to SES and sometimes to ancestry. This constrains the composition of children's play groups. Without a theory, it is difficult to design models that can guide the selection of variables to guide scientific inquiry. Nonetheless, there is progress.

7.3.1 Socioeconomic Status (SES)

SES refers to the nebulous concept of social classes. There is no generally agreed-on definition of a class. The most common practice is to define anywhere from three to six different classes, based on combinations of (1) income, (2) education, and (3) occupational prestige. It is not unusual to find studies in which a single variable, such as income or education, is used as a proxy for SES. Given the variety of definitions, it is a bit surprising to find that SES is correlated 0.30–0.40 with performance on such diverse variables as the Wechsler Adult Intelligence Scale (WAIS), the SAT, and the Raven's Progressive Matrices (Ceci and Williams, 1997; Raven, 1989).

As a further complication, intelligence test scores and SES are both correlated with various parental advantages, including parental (and hence one's own) genetic constitution. Parental advantage, carried forward across generations, could, in theory, be a cause for one's own socioeconomic success. And just to make things even more difficult, cognitive tests like the SAT are used as screening devices in education, raising the possibility that measures of intelligence act as gatekeepers for access to resources that contribute to social success. Charles Murray (2020, p. 237) stated the issue clearly: "In the United States, the poster child for the indictment of tests is

the relationship between parental SES and performance on college admissions test such as the SAT: the higher the parental education and income, the higher the scores of the children. How much of this relationship is causal? How much of it is a reflection of the uncomfortable possibility that smart parents attain high SES and also produce smart children?" An analysis by Marks and O'Connell (2021, p. 24) also addresses this important issue, concluding that "SES effects are likely to be, to a considerable extent, proxies for the effects of parental ability."

In most studies, SES is confounded with intelligence since SES measures like income and education are correlated to intelligence. Research assessing both SES and intelligence find that SES measures do not add much predictive power beyond what intelligence predicts (see Chapter 8). The weight of evidence indicates that intelligence is a stronger predictor of SES than the other way around, even though each has some influence on the other (Sackett et al., 2009; Teasdale and Owen, 1986; Zwick and Green, 2007; Herrnstein and Murray, 1994; Colom and Flores-Mendoza, 2007; Lemos et al., 2011). Box 7.2 describes a classic study that illuminates the problem.

It is apparent that SES is too global a measure of either parental influence or one's own success in life. To understand the contribution of social experiences to intelligence, we have to take a finer look at phenomena that underlie the correlation. Three classes of studies have been used to evaluate the effects of the social world: (1) adoption studies, (2) studies involving social interventions, and (3) multivariate analyses of specific nongenetic features.

7.3.2 Adoption

As we saw in Chapter 6, adoption studies support a genetic contribution to intelligence differences: correlations between intelligence measures in adoptees and their biological parents are higher than the correlations between adoptees and their adoptive parents. However, correlations between two

Box 7.2. A Unique Social Experiment

Anna Firkowska and colleagues (1978) reported findings from a unique social experiment that are relevant for disentangling collinearity issues related to personal intelligence and social factors. This was the key question: what is the effect of a mandated social environment change to SES on the mental performance of schoolchildren?

In the study, factors intrinsic to family social structure and position (personal attributes of family members, e.g., parental occupation and education and family size) were distinguished from factors extrinsic to family social structure and position (the environment external to family social life, e.g., schools, housing, health and welfare services, recreation, criminality, and employment rates).

The distinction was possible because of a unique situation in Warsaw, Poland. Under communist control, the city was a carefully planned creation of the post–World War II period. Seventy percent of Warsaw was destroyed. New buildings were constructed, and housing was distributed by the government on egalitarian principles. Residential districts did not differ on key aspects of social class. Living standards were aggressively equalized. People of all levels of education and all types of occupations lived in similar apartments, shopped in the same stores, and shared the same cultural centers. Schools and health care centers had equivalent equipment, and they were equally accessible to everyone. In short, most extrinsic factors were neutralized.

What about the intrinsic factors? They were not so similar. Thus, for instance, educational levels and occupation types showed considerable variation. Nevertheless, the key prediction was that the thirty-year period of equalization would reduce any cognitive differences that were influenced (or caused) by social class.

All children born in 1963 were recruited for the study ($N = 14,238$). Their mental ability was tested in 1974 (age range between 10.3 and 11.6 years) using the Raven's Progressive Matrices. Figure 7.6 shows the correlation between the Raven's score and SES estimated by parental education and occupation. The pattern is striking: the higher the SES value was, the greater the children scored on the Raven's Progressive Matrices. The correlation was almost perfect ($r = 0.98$).

The family group of variables showed higher correlations to intelligence scores than did the district or school group variables: "the situation in Warsaw meets the conditions for a conclusive answer to a crucial question. Since social policy has distributed families of varying occupation and education fairly evenly across the ecological milieu of the city, we are in a position to assign a value to the association of mental performance with occupation and education. Despite this social policy of equalization, the association persists in a form characteristic of more traditional societies (nevertheless) we are not able to apportion these effects among social and genetic factors" (p. 1362).

The authors of this fascinating study close their article underscoring that schooling may reinforce individual differences in mental ability already present in the family settings (possibly due in part to genetic influences). This finding about intrinsic factors should be considered by education programs aimed at increasing cognitive ability to minimize average group differences if we want to increase the likelihood of success (see Box 7.6 and Chapter 13).

variables and the mean levels of these variables are different statistics, and they tell different stories. The mean intelligence test scores of adoptees might well support the relevance of nongenetic factors (Sauce and Matzel, 2018).

For example, a classic adoption study of Marie Skodak and Harold Skeels (1949)

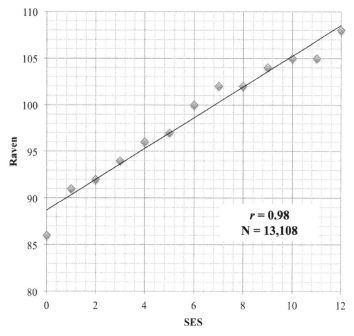

Figure 7.6 Correlation between intelligence scores (Raven's Progressive Matrices) and SES estimated by intrinsic family variables (not by extrinsic variables like school or teacher quality). Adapted from Firkowska et al. (1978, table 6).

produced results that were typical of later studies. The biological parents were all working class, and the mothers, as a group, had an estimated mean IQ of 86. The adopting parents were all described as well educated, presumably with higher IQs than the biological parents. The mean IQs of the adoptees were 117 at age two and 108 at age thirteen. This increase in means could be evidence for an effect of the childhood home environment. However, the correlations between the biological mothers' levels of education and the adopted-away child's IQ scores were 0.04 at age two and 0.31 at age thirteen. This finding supports a role of genetics and reflects similar findings in later studies showing that measures of genetic contribution to intelligence increase as children grow older (as discussed in Chapter 6).

In the 1970s, Sandra Scarr conducted another classic study, the Minnesota Trans-Racial Adoption (MTRA) study of US children of African and European ancestry. Children of either ancestry had been adopted into upper-middle-class homes where all

adoptive parents were of European ancestry. Both adoptees and biological children of the adoptive families were tested, first when the adoptees were preschoolers, and then ten years later, when the adoptees were adolescents. Figure 7.7 shows the results (Scarr and Weinberg, 1976; Weinberg et al., 1992).

Biological children outscored adopted children in the transracial group (adopted African children) but not in the intraracial group (adopted European group). All test scores declined from childhood to adolescence, which could be caused by a variety of factors, including the fact that different tests were used over time. Importantly, the adoptees of African ancestry had an average score around 100 in adolescence (the population mean), which is above the mean score of 85 typically found in US populations of African ancestry on this type of test.

The research design of comparing the adoptees' scores to maternal scores, or to population expectations, has limitations. Suppose that the birth mother's IQ is 90. This does not mean that a child would be

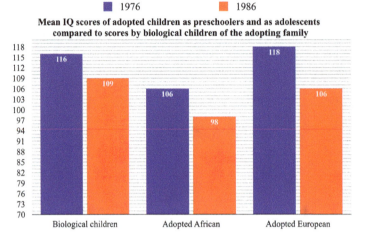

Figure 7.7 Mean IQ scores of adopted children as preschoolers
(1976, blue) and ten years later as adolescents (1986, red) compared
to scores of biological children of the adopting family (Weinberg
et al., 1992).

expected to have an IQ of 90, for two reasons. First, no allowance has been made for
the father's intelligence, which is often not
known in research studies. In addition, statistical regression to the mean implies that
the child's IQ will be closer to the appropriate population mean than to the average
parent IQ. But what is the appropriate
mean? Mothers who give up their children
for adoption are not a randomly selected
group of all women, or even of all women
in the ancestry or educational group.
A better way is needed to estimate the
expected IQ of adopted children.

One way of doing this is to compare the
cognitive performance of adoptees to the cognitive performance of children who have not
been adopted, but who might have been. The
test scores of adopted children can be compared either to those of unadopted siblings or
to unrelated unadopted children in the same
pool of potential adoptees. When this is done,
the results show a positive adoption effect of
slightly less than 1 standard deviation unit,
both for test results and for measures of school
performance (van Ijzendoorn et al., 2005).

This is probably an overestimate of the
adoption effect, however, because the comparison includes the effects of any tendency
leading to adoption of apparently more favorable children. Nevertheless, given that most

children are adopted as infants, and that standard developmental inventories do not predict later intelligence very well, this effect is
likely to be small. Except for pathological
cases, adoption agencies would have difficulty
identifying infants who were going to be
smart or not fifteen years later.

Understandably, such studies often stress
the positive effects of adoption, but negative
effects also should be considered. If intelligence can be enhanced by favorable adoption into higher-SES families, then the
opposite could also be true. Such an analysis
requires an unusual situation, for high-SES
parents are less likely to give up their children than are low-SES parents, and adoption agencies are more likely to place
adoptees with high- rather than low-SES
families. However, the unusual sometimes
happens (Box 6.3).

French researchers were able to locate a
relatively small number of young children
where the birth parents were of higher SES
than were the adoptive parents (Capron and
Duyme, 1989, 1996). This group and controls had small sample sizes (eight to ten per
group), so the results are presented with
caution. Figure 7.8 shows the IQ scores on
the Wechsler Intelligence Scale for
Children–Revised (WISC-R) at an average
age of fourteen years.

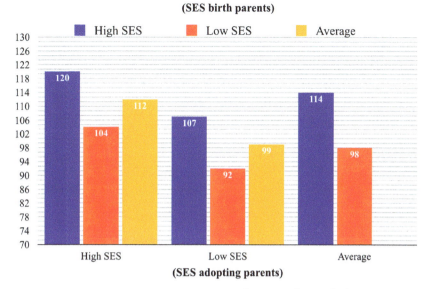

Figure 7.8 Mean WISC-R scores (*y*-axis) as a function of SES of adopting parents (*x*-axis) and SES of birth parents (bars).
Based on data reported by Capron and Duyme (1989).

Taken at face value, the data in Figure 7.8 indicate a high birth–SES comparison to low birth–SES effect of about 16 IQ points in the high-SES adoptive parents (120 vs. 104). When high birth–SES children are adopted into low-SES homes, the IQ difference is 13 points (120 vs. 107). In other words, the data are consistent with both genetic and nongenetic effects on intelligence and make the important point that the two causes are not mutually exclusive. It would be unwise, however, to over-interpret these encouraging findings. The study is a small one, and it is not clear that the difference between the adoptive high- and low-SES groups was equivalent to the difference between the high- and low-SES birth parents.

The results of adoption studies are used to argue for both genetic and nongenetic influences on intelligence differences. Those who want to emphasize genetic causes cite the fact that indices of adoptees' intelligence test scores are better predicted by biological parents' scores than by adoptive parents' scores. Those who want to emphasize environmental causes cite average gains in intelligence observed in some groups of adoptees. Both arguments are reasonable and have merit.

However, debate about interpretation of these findings can be biased. In some of the studies that support the genetic position, the authors do not report changes in mean scores. In some studies that support the environmental position, the authors do not report parent–adoptee correlations. This practice more resembles the behavior of a lawyer presenting the best evidence for a client rather than the behavior of a scientist reporting data for evaluating theories. This is not good practice in science.

In fact, there is not necessarily conflict between the two interpretations. Erik Turkheimer (1991) offered an analysis that brings both these results into the same framework. His analysis was based on the concept of reaction range. He developed a mathematical model that separates the genetic and nongenetic effects and used it to reanalyze a study contrasting adopted and nonadopted siblings (Schiff and Lewontin, 1986). The original authors had concluded that there was a major effect of social class, on the grounds that

(1) adoptees had higher intelligence scores than their nonadopted siblings and that (2) the adopting parents had higher occupational status than the biological parents. According to Turkheimer's analysis (which treated the within-group and between-group effects within a single framework), there was no reliable effect of a direct measure of adoptive SES (father's occupational status), but there was a large (and unexplained) effect in favor of the adopted children.

Various explanations of this finding have been proposed, but ambiguities remain. Progress may result from more attention to nongenetic proximal variables, as described in Box 7.3.

7.3.3 The Home: Parenting

Parenting practices are evaluated by rating homes on such things as (1) the general orderliness of the home, (2) the amount of reading or other explicitly educational

---

Box 7.3 Malleability of Intelligence after Adoption

Kenneth Kendler and colleagues (2015) compared home-reared and adopted-away co-siblings ($N = 436$) to estimate the contribution of the rearing environment to intelligence. The key finding revealed that the adopted siblings showed an average advantage of 4 IQ points over the nonadopted siblings. After addressing several considerations, the authors concluded that educational levels of adoptive parents may explain this increase, because adoptive parents had higher educational levels, on average, than biological parents.

As we have underscored, mean values of variables x and y and the correlation between x and y tell different stories. The malleability issue relates to mean values. Values in x can change, even dramatically, while the correlation between x and y remains exactly the same value.

These were the results from this powerful version of the adoption design:

1. The mean IQ of the individuals reared at home and those adopted away was 92 and 97, respectively – a difference of 5 IQ points on average in favor of the adoptive homes (before correcting for the differential in parental educational levels).
2. When the mean adoptive parental educational status was 2.5 points higher than biological parental educational status, the adopted-away sibling showed 7.6 IQ points higher than his home-reared sibling. The differential of 2.5 points is roughly equivalent to a difference between no high school and some postsecondary education.
3. When the mean biological parental education status was 2 points higher than adoptive parental educational status, the adopted-away sibling showed 3.8 IQ points lower than his home-reared sibling.

In the authors' own words, "relations with adoptive-parent education behave exactly as one would expect if rearing environment has a causal effect on ability: offspring placed in the best educated homes had the highest scores, whereas those placed in homes less educated than the family of origin actually performed worse than their nonadopted siblings. ... Despite being demonstrably related to genetic endowment, cognitive ability is environmentally malleable, and the malleability shows plausible dose–response relations with the magnitude of the environmental differences" (p. 4616).

material available, (3) the number and quality of interactions between parents and children, and (4) the extent to which children are encouraged to work out the answers to questions and puzzles, as opposed to being told how to do so.

It appears that the best environment for intellectual development is one in which the child is encouraged to work out problems with guidance and support from parents, as opposed to an authoritarian setting in which the parent tells the child what to do or a laissez-faire setting in which children are pretty well left on their own (Kotchick and Forehand, 2002). Parenting styles, and especially substandard parenting practices, are statistically associated with indices of socioeconomic status and family solidarity, such as income and whether the child is in a one-parent, father-absent, or conventional mother-father home. However, these are general findings, and nothing is simple. Kotchick and Forehand make the interesting point that if a family lives in a potentially threatening environment, an authoritarian, controlling style of raising young children may be adaptive, even though it does not foster intellectual development, because of the need to protect the child.

The enumerated variables of parenting practices are important because the home environment can influence the intellectual development of young children. In one such study, Victoria Molfese and her colleagues (1997) followed 121 children from age three years old to age five years. The children completed intelligence tests annually. Family SES was assessed by combining indices of education, occupation, and income. The home environment was assessed by observation and rating, using the Home Observation for Measurement of the Environment (HOME) scale. It requires observation of the child's home, including the provision and use of reading material for children and reports of the manner in which adults interact with the child (Bradley, 1993). The results showed slight trends of decreasing influence of home environment and increasing influence of SES as the children aged. The trends are consistent with behavior genetic studies that show generally increasing influences of genetic factors and lowered influence of home environment on intelligence as people age (Chapter 6), albeit over a much greater time span than was the case in this study.

Similar findings were reported from a study in the Philippines (Church and Katigbak, 1991) and a study from Kosovo (Wasserman et al., 2000). The American Midwest, the rural Philippines, and the Balkans are very different places. When relationships between any variable and intelligence are consistently related over such diverse settings, one has to pay attention. The environment of the home influences intelligence differences during the early childhood years. Better home environments apparently give children some advantage as they enter school. Unfortunately, these studies typically do not control for the confounding effects of parental intelligence, so definitive conclusions cannot be made from them.

7.3.4 The Home: Resources

As discussed in Chapter 6, behavior genetic studies find that nonshared environmental experiences are a relevant source of nongenetic effects on intelligence in children (see Figure 6.3; Jensen, 1997; Plomin et al., 2016). Whereas family factors like SES can be counted as part of shared environment, each child within a family has unique (nonshared) experiences (e.g., friends in school and hobbies). This is especially so in larger families. Having brothers and sisters is not like being an only child. The extreme might be a medieval royal family, in which the first-born child, as potential heir to the throne, received special treatment.

Children raised in large families tend, on average, to have lower intelligence test scores than children raised in small families. This trend is shown in Figure 7.9, which presents data from the National Longitudinal Study of Youth in 1979 (NLSY79) (see Box 7.4). Similar findings have been obtained in other studies. Correlations between family size and children's intelligence test scores usually fall between −0.15 and −0.20 (the greater the family size is, the lower are the scores)

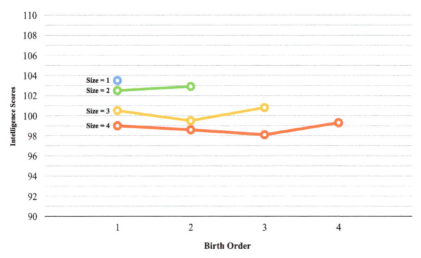

Figure 7.9 Mean intelligence estimates (y-axis) of young children born to mothers who were in the NLSY79 survey. The intelligence estimate is based on the Peabody test, which is suitable for children in the preschool and elementary school age range. Families of one, two, three, and four children are shown. Adapted from Rodgers et al. (2000, table 4).

(Herrnstein and Murray, 1994; Lynn, 1998; Lynn and Harvey, 2008; Lynn and Van Court, 2004), but if there are no data on parental intelligence, these correlations are difficult to interpret.

Finally, in addition to family size, birth order has been studied as a possible correlate of intelligence. A number of early studies suggested that first-born children, on average, had higher intelligence test scores than later-born children, but the findings typically were weak. A consensus is emerging that any birth order effect on intelligence is quite small (and effects on personality cannot be demonstrated at all) (Damian and Roberts, 2015b). For example, in a large representative US sample, after controlling for family size and other variables like age and sex, correlations for birth order and intelligence ranged only between 0.00 (for spatial ability) and 0.08 (for verbal ability) (Damian and Roberts, 2015a). Given these results, birth order does not appear to be an important factor for explaining family size correlations with intelligence.

7.3.5 *Early Interventions in At-Risk Populations*

There have been numerous attempts to improve the environments of children thought to be at risk for showing compromised intellectual development. For instance, since the 1960s, the US government has funded Head Start programs for preschoolers from low-SES families. Head Start was motivated by a concern that low-SES children, and particularly US individuals of African ancestry, entered primary school lacking a number of cognitive and social skills that were important in children's adjustment to the school experience. These programs typically involve preschool training for a half day, five days a week, for the year prior to entering school. Many Head Start programs also include some form of parental education, because inadequate parental support may be a factor leading to the children's poor performance in schools.

Even before Head Start, there were numerous similar programs. Do such compensatory programs increase cognitive abilities? In 1969, Arthur Jensen published an article in the *Harvard Educational Review* that reviewed the data on such programs (Head Start was not included because it had just begun; we discuss this more in Chapter 13). He began the article with this famous sentence: "Compensatory education has been tried and apparently it has failed." Later in the article, he wrote, "The evidence

Box 7.4 The National Longitudinal Studies

The National Longitudinal Studies are surveys carried out by the US Department of Labor to obtain an accurate picture of the demographics, social structure, and economics of the United States. Such information is needed by policy makers. The surveys also provide valuable sources of data for economists, sociologists, educational researchers, and psychologists. In general, the surveys are conducted on a random sample of US citizens relevant to the purpose of the survey. On occasion, certain groups of interest are intentionally oversampled to provide more accurate information about them.

The National Longitudinal Study of Youth 1979 (NLSY79) was a survey of more than 10,000 young men and women aged fourteen to twenty-two when the survey was initiated in 1979. This sample has been followed periodically with additional surveys. Information on health, economics, and social status is obtained. At the time of the first survey, a fortuitous event occurred. The Department of Defense needed to update its normative sample of the Armed Services Vocational Aptitude Battery (ASVAB) and the associated Armed Forces Qualification Test (AFQT). Accordingly, a large percentage of the NLSY79 participants took the ASVAB. The result was an important prospective study of intelligence since the ASVAB scores of the fourteen- to twenty-two-year-old participants can be related to their subsequent progress through life. Herrnstein and Murray made extensive use of the NLSY79 database in their 1994 book *The Bell Curve*.

The NLSY has spawned a second survey, the NLSY Children and Young Adults study, in which data are gathered on children of the original NLSY79 participants. This provides social scientists with valuable data on changes in the population over two generations.

A completely new sample was obtained in 1997 to repeat the NLSY 1979 survey. This new NLSY97 included more than 9,000 young men and women born between 1980 and 1984; they are interviewed on an annual basis. The purpose is to track social issues concerned with the transition from youth to adulthood. Comparisons to the NLSY79 data, when possible, provide a way of comparing cohorts as they transit from youth to adult life.

so far suggests the tentative conclusion that the payoff of preschool and compensatory programs in terms of IQ gains is small" (p. 108). And further down on the same page: "The techniques of raising intelligence per se in the sense of *g*, probably lie more in the province of the biological sciences than in psychology and education." Jensen maintained this perspective in a comprehensive book on intelligence published almost thirty years later – *The g Factor: The Science of Mental Ability* (Jensen, 1998). Herrnstein and Murray (1994) echoed Jensen's conclusion about raising intelligence in schools. They wrote, "The school is not a promising place to try to raise intelligence or to reduce intellectual differences, given the constraints on school budgets and the state of educational science" (p. 414).

A different view was expressed by Edward Zigler (Zigler and Styfco, 1997). As a government official, he became known as the "Father of Head Start." Zigler observed that a variety of behaviors are improved by Head Start programs. These include (1) good study habits, (2) cooperative work skills, and (3) improved parental support. Improving these characteristics was, according to Zigler, at least as important as improving the cognitive skills measured by intelligence test scores.

The debate over Head Start is more nuanced than it appears when people simply recite mantras that Head Start programs do or do not work. Skeptics like Jensen and Herrnstein and Murray generally focus on IQ scores and similar estimates of intelligence (especially *g*) because this was the original primary goal of the program. Supporters have a broader view of intelligence, to include things like knowing how to manage time and how to learn arbitrary material. The supporters also stress improvement in the child's general social situation, as Zigler's comment about parenting skills indicates.

There is another qualification. Jensen and Herrnstein and Murray did not say that preschool programs will not work; they said that those programs that were economically realistic did not work. That is an important distinction. "What early childhood programs improve intelligence?" is a question for educational psychologists. "What could work at a cost of *x* dollars per child?" is an important question for educational policy makers. The two viewpoints are not the same.

However, there is some evidence of progress (Barnett and Hustedt, 2005). There is a great deal of variation among preschool intervention programs. At the low end, we have reasonably well-run but minimal Head Start programs, which children attend two to five times a week for one or two years. At the high end, some intense programs have included staff–client ratios as high as one well-trained teacher for every three students, parental counseling, and interventions for five years or more. Two meta-analyses that address such differences are noteworthy (Gorey, 2001; Barnett, 1998).

In general, these meta-analyses show that the less intense programs can have short-term results but that these results (in terms of IQ scores and school accomplishment) fade away. By contrast, the more intense programs have results for both intelligence test scores and school achievement that can last for years. Fade-out is discussed in Box 7.5. One meta-analysis reports an improvement of 9 IQ points for the intensive programs five years after the programs ended. This would be somewhere toward the end of elementary school for most participants. Only a few studies have followed students as far as high school, although two

very intense intervention studies traced participants into adulthood. These studies report substantial positive effects on social behaviors, such as encounters with the law, but smaller effects on IQ scores more than ten years after preschool participation.

Possibly the most intensive program, the ABCDerian project, was aimed at a deeply impoverished group of children in North Carolina. This study reported positive results, compared to a control group, at age twenty-one (Campbell et al., 2001). Because this is generally conceded to be one of the most effective (and expensive) of the preschool programs, it is worth looking at the study in more detail.

The participants were children from low-SES families in North Carolina. All the children were of African ancestry. Slightly more than 100 children were assigned to the special program or to a control group. The intervention began at an average age of 4.4 months and continued until the children entered kindergarten. Depending on the period, there was one instructor for every three to six children. In addition to instruction and supervision, a nutritional program was developed and offered to both the experimental and control groups. The preschool lasted virtually the entire day, five days a week, until age five years. Participants were followed until they were twenty-one years old.

When the program ended at age five to six, the children took the Wechsler Preschool Intelligence (WIPSI) test. The experimental group had a mean IQ of 100 and the control group a mean IQ of 94. Figure 7.10 shows this and the results of subsequent testing. There are two striking features in this figure: both groups showed a steady decline in test scores over the years after leaving the project; at the same time, the experimental group maintained a roughly 5 IQ point advantage over the control group throughout. This increase is encouraging, but most intervention programs are not as intensive and show little if any increase (Box 7.5).

Although IQ scores may not increase appreciably in the long run, the picture for

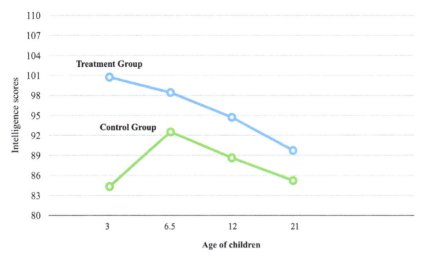

Figure 7.10 Mean IQ scores obtained by the participants in the ABCDerian preschool project (*N* = ~50) and in a randomly chosen control group (*N* = ~50). The intervention ended when the children entered school at ages five to six, and they were followed until age twenty-one.
Data are from Campbell et al. (2001, table 1).

some aspects of academic achievement is more encouraging. One meta-analysis indicated that 80 percent of the children who participated in intensive programs had a higher level of school performance than did the median participant in the control group five years or more after the intervention ended. Participants also experienced fewer social and behavioral problems. Entry to college was also higher in participants of these programs compared to control groups (Barnett, 1998).

The decline shown in Figure 7.10 is not surprising. IQ tests for children under the age of five have limited reliability and are only moderately predictive of scores at a later age. This could be because the tests for very young children are tests of cognitive functions different from those evaluated by later tests, or it could be because the cognitive systems important for test performance (such as working memory) are only slightly developed in young children. Maturing of the cognitive system likely affects test performance differences.

So, what is the overall picture? If you think that the role of early childhood intervention and special education is to improve intelligence, as measured by test scores, the

critics are right that at best, marginal improvements can be made (Detterman and Thompson, 1997; Herrnstein and Murray, 1994; Jensen, 1998). However, the programs apparently improve a number of academically relevant behaviors, ranging from study skills to improvement in interacting with students and teachers. Such behavior leads to improved learning in formal educational settings. These are not negligible benefits and deserve attention.

Nevertheless, scaling up a project like the ABCDerian project to cover a substantial portion of low-SES children would require a huge financial investment and pose a formidable problem of appropriate staffing. One can argue over the relative values of the costs and benefits of such programs and how they should be assessed. These are educational policy issues, not issues in the scientific study of intelligence. Herrnstein and Murray's (1994, p. 415) conclusion to their chapter devoted to raising intelligence is worth consideration: "the debate over Head Start should move away from frivolous claims about how many dollars it will save in the long run, none of which stands up to examination, and focus instead on the degree to which it is actually serving the laudable and

Box 7.5 The Fade-out Effect: A Meta-analysis

John Protzko (2015) calculated a comprehensive meta-analysis based on 7,585 individuals from nearly forty randomized controlled trials. He applied growth curve modeling, and the findings revealed that (1) interventions are successful for raising intelligence, but (2) the observed positive effect fades away gradually once the intervention is terminated. After discussing various alternative explanations for this fade-out effect, Protzko favored the unidirectional-reactive model: gains fade away because children return to an impoverished environment – "the intelligence of children will react to the demands of the environments they are placed in, for good and for ill" (p. 209).

The meta-analysis considered any type of intervention: training studies, educational, nutritional, and so on. In the calculated growth curves, the intercept corresponds to the time when the intervention ends, and the time variable is years from the end of intervention (slope). The average growth curve fits to the trends of the individual participants. As depicted in Figure 7.11, interventions increased intelligence about 5.5 IQ points. Interestingly, (1) interventions starting at early ages did not show greater impact and (2) intervention time was unrelated to the observed fade-out effect.

The main conclusion from this meta-analysis is consistent with what Herrnstein and Murray (1994, p. 389) concluded in *The Bell Curve*: "in principle, intelligence can be raised environmentally to unknown limits (however) taken together, the story of attempts to raise intelligence is one of high hopes, flamboyant claims, and disappointing results."

The meta-analysis indicates that intervention results are clear: improvement in intelligence test scores does not last. Why this is so is not clear. Protzko rejects appealing reciprocal models because children with higher intelligence scores failed to use their cognitive advantage for seeking or creating environments with greater cognitive challenges. These models conflict with the fade-out effect. He very much prefers a unidirectional relationship "environment → intelligence" (unidirectional-reactive model): "intelligence reacts to the demands of the environment, growing when the demands increase and shrinking when the demands decrease" (p. 209).

Perhaps the crucial message from this meta-analysis is that it may be unrealistic to expect long-lasting effects from limited interventions. Enhancing intelligence is a complex challenge that is likely beyond any relatively simple intervention, as noted by David Moreau and colleagues (2019) and discussed in Chapter 13 (see also Haier, 2017).

---

more fundamental function of rescuing small children from unsuitable, joyless, and dangerous environments."

7.3.6 *A Value Judgment*

Family environments may contribute to children's development of intelligence, but, within the range of normal family environments in more middle- and high-income nations, these effects appear to be modest and ephemeral. They virtually disappear in

adulthood. The situation is quite different when we look at more extreme environments, such as the low-SES homes of the children at risk for lowered cognitive development like those who participated in the ABCDerian project.

High-cost intervention projects have improved some intellectual performance, but unfortunately, the effects are far less than most reformers would like to see. Many less expensive projects, such as the typical Head Start program, show little if

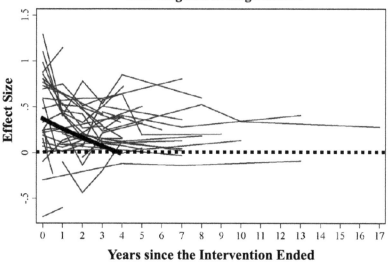

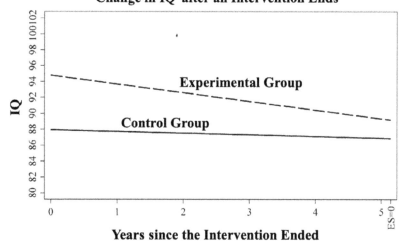

Figure 7.11 Fade-out effect assessed by meta-analysis. (top) Effect sizes over time in different studies after an intervention has ended. (bottom) Average IQ changes decline to zero in all interventions.
From Protzko (2015).

any lasting effect on intelligence test scores, or on other cognitive factors. However, there are indications that these programs prepare at-risk children for the social experience of school, and that this may facilitate later school development.

The last point is important. In industrial and postindustrial societies, schooling is a major factor in a person's contribution to the general society. People who finish high school earn more than people who do not, and generally, the more years of education a person has, the more likely it is that they will have stable employment and higher salaries throughout their lifetime. Other social problems, such as poor health and criminal convictions, also are negatively correlated with the possession of a diploma. Any program that increases the likelihood that children in at-risk situations will complete schooling is a socially valuable program. It has also been estimated that even the more expensive programs are cost-effective, when the total cost to society over a long period of time is considered (Barnett, 1998), but these projections are based on a number of assumptions that may or may not hold.

Many child advocates have pointed out that a child born to a poor family with limited resources is often in a bleak environment. It is a good thing to provide programs offering such children better nutrition, a safe place to interact with adults and other children, and an interesting, challenging environment. If the resulting program also improves their cognitive abilities, their intelligence in the important, conceptual sense, then that is a very nice benefit of having done a good thing. If the program just improves test scores, but does not improve cognitive abilities in the more general sense, then that is just a curiosity.

It is reasonable for society to provide the best possible social environment for all children, especially if disadvantaged. Costs and benefits are relevant when comparing programs; there is no sense in spending more than you have to. However, society has a duty to provide such programs, just as much as it has a duty to provide military defense or security for the aged.

7.4 Experience, Formal Education, and Intelligence

Most countries make widespread use of formal school systems and a broad education. This is a marked contrast to the situation less than 100 years ago, when most children learned what they needed to know by some form of apprenticeship, augmented by a small amount of formal training.

Upward of 50 percent of citizens in the middle- and high-income nations receive some form of postsecondary education. While there are some discussions of the importance of a liberal education, most postsecondary instruction is oriented toward training specialists in fields ranging from welding to the law or medicine. The result is an extension of the time between the end of childhood and the point at which a fully trained adult enters society. For perspective, Alexander the Great was twenty-two when he led the Macedonian army into Asia. Napoleon was appointed brigadier general at twenty-four.

7.4.1 *Does Education Contribute to Intelligence?*

Educated people are generally capable of solving problems that uneducated people cannot. In the broad conceptual sense, part of a person's intelligence includes solving everyday practical problems. Education is necessary to deal with things from reading and understanding newspapers to balancing checkbooks and comprehending the terms of mortgages.

What about the effects of education on intelligence in the narrow sense of improving scores on standardized intelligence tests? This question has a long history of conflicting and inconsistent research findings (Ceci and Williams, 1997; Bronfenbrenner et al., 1996; Ceci, 1990). Among intelligence researchers, there is an emerging consensus that schools and teachers do not have as much effect on the development of intelligence as might be expected. This was first reported in a comprehensive US national study (Coleman, 1966) and reinforced by recent studies (see Box 7.6).

If teacher and school quality do not have much impact on intelligence test scores, does education in general have an effect on any aspect of intelligence? We discussed in Chapter 3 the psychometric models in which intellectual abilities are hierarchically organized from specific abilities to broad factors to the g-factor. From this perspective, it is reasonable to ask if education has any impact on each level of the intelligence hierarchy, and if so, to what degree. Stuart Ritchie and colleagues (2015) designed a longitudinal study to address the question "Is education associated with improvements in general cognitive ability (g) or in specific abilities and skills?"

Box 7.6 Pity the Poor Teacher

In the international seminar that took place in Madrid in 2016 ("Advances in Intelligence Research: What Should Be Expected in the XXI Century?"), Douglas K. Detterman (2016) presented evidence leading to a surprising conclusion: only 10 percent of school achievement variance can be attributed to school factors (including teachers); the remaining 90 percent is associated with students' personal features, including intelligence. This was his main message: "as long as educational research fails to focus on students' characteristics, we will never understand education or be able to improve it." There may no better example of ignoring this warning than the $575 million effort funded from the Gates Foundation ($212 million) and school districts ($363 million) to improve student achievement by increasing teacher effectiveness. It failed for both individuals and for increasing group averages (Stecher et al., 2018).

Detterman proposed this thought experiment:

1. Begin with two groups of fifty randomly selected teachers and 1,000 randomly selected students. Teacher quality and students' intelligence are known.
2. In the first experimental condition (teacher-quality condition), twenty students are randomly assigned to each of the fifty teachers.
3. In the second experimental condition (student-ability condition), the students are rank ordered by intelligence score. Students are divided into groups of twenty, and the result is twenty groups of students from the most to the least able. Teachers are then randomly assigned to these groups of students.
4. Obtain a measure of student achievement for each group in each condition before and after teaching some academic subject.
5. Teach each group using the same content, and measure learning outcomes of each student.
6. Answer the question, will students with better teachers do better than teachers with smarter students? For the teacher-quality condition, the answer can be obtained by correlating teacher quality with learning outcomes. For the student-ability condition, learning outcomes can be correlated with student ability.
7. Detterman predicts that the first correlation will be 0.30 in the best scenario, whereas the second correlation will be around 0.95. He then went on to review the relevant empirical evidence about his predictions.

This is a perfect example of the distinction between the correlation between two variables (x and y) and the mean values of these variables. Detterman focuses on the first for supporting the conclusion that personal features are responsible for most of the variance in educational outcome. This is correct, and the weight of the evidence is overwhelming. However, students can learn many relevant things in school (y) without changing their intelligence levels (x). As a consequence, y will move upward, while x remains at exactly the same values. Brighter students will still show greater learning outcomes, but not-so-bright students will now have more knowledge and skills than what was observed before the educational intervention.

This fact was appreciated by Arthur Jensen (1981, p. 241): "The student who wants to play basketball doesn't think about trying to increase his height, but works at developing the specific skills that will improve his actual performance in basketball, and by so doing he will become a better player. The very few who become champions will have done much the same, and, in addition, will have exceptional physical advantages (including being very tall) for which they (or their parents) should take no personal

Box 7.6 *(continued)*

credit." Change height to *g* and the point is clear.

This line of thought has been highlighted by David Moreau and colleagues (2019, p. 4) in their article "Overstating the Role of Environmental Factors in Success: A Cautionary Note":

"recognizing and understanding individual differences, rather than denying or undermining their importance, leads to politics of equity – providing individuals with the means to thrive – rather than equality – treating everyone uniformly regardless of their specific needs."

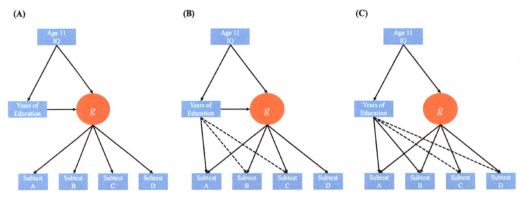

Figure 7.12 Predicting effects of education on intelligence. (left) Model A: education (years) directly impacts *g*, and this influences the specific ability measures (subtests). (middle) Model B: in addition to the impact on *g*, there is also a direct impact of education on the subtests. (right) Model C: education does not impact *g*, but there is an impact of education on the subtests. From Ritchie et al. (2015).

These researchers analyzed a sample of 1,091 individuals from the Scottish study whose intelligence levels were measured at eleven years of age (with the Moray House Test) and tested again at age seventy with a battery of intelligence tests comprising ten subtests tapping various cognitive abilities. The models tested are depicted in Figure 7.12. Each model predicts test performance at age seventy from test scores at age eleven.

It is reasonable to predict that individuals with higher childhood intelligence will remain within the educational setting for a longer period of time. If this is the case, then educational duration will not show any impact on intelligence. Alternatively, education might enhance intelligence through the acquisition of the cognitive tools required for the efficient resolution of the more

complex items included on the standardized intelligence test batteries. The interesting feature of Ritchie et al.'s investigation is that the tested models allowed controlling for intelligence levels measured at eleven years of age, before the introduction of variations in educational duration.

Not unexpectedly, the findings showed the following:

1. Intelligence at age eleven correlated with years of education ($r = 0.42$). The higher the intelligence at eleven years of age was, the longer the time was within the educational system.

2. Factor analyses supported the presence of a single general factor (*g*) derived from the intelligence subtests at seventy years of age.

The main finding was that model C, with no path from education to g, showed better statistical fit to the data than models A and B. Therefore, education showed domain-specific effects on intelligence. This might help explain the lack of transfer effects endemic in the educational settings (Protzko, 2016) plus further issues about increasing intelligence we will address in Chapter 13.

Is model C the key to the discussion regarding the formative or causal nature of g? This is a controversial topic (Kan et al., 2019). As we have seen in Chapter 4, devoted to the cognitive models of intelligence, there is a case that the general factor of intelligence is a formative concept: g is an effect, not a cause, of several diverse independent processes. However, if education improves intellectual specific abilities but fails to move g, it is almost inevitable to conclude that g is a cause, not an effect, of what happens in the lower levels of the intelligence hierarchy. This conclusion is reinforced by findings observed in studies that have related focal brain lesions to the hierarchical structure of human intelligence (Protzko and Colom, 2021).

7.4.2 How Much Does Education Improve Intelligence?

Education may fail to influence g (as in model C in Figure 7.12), but it still can influence other intellectual abilities. A comprehensive meta-analysis reported by Stuart Ritchie and Elliot Tucker-Drob (2018) helps answer a key question: do increases in educational duration after early childhood have beneficial causal impact on later intellectual abilities? The number of participants considered in the meta-analysis was 615,812, and they were recruited from forty-two independent data sets.

The research analyses considered three categories of studies for testing the causal effect of education on intelligence. In human research, not all variables can be experimentally manipulated, so each category was based on a quasi-experimental design:

Category 1. *Controlled prior intelligence (CPI)*: Used a longitudinal design where intelligence scores were assessed before and after different number of years of education.

Category 2. *Policy change (PC)*: Could differences in test scores be due to changes in education policy?

Category 3. *School age cutoff (SAC)*: Some schools apply a date-of-birth cutoff for school entry, so it is possible to compare children of different chronological ages attending the same school year.

Each study in each category had its own strengths and weaknesses, but overall, this study found that greater educational duration was a causal factor for higher intelligence scores. The authors estimated the impact of one year of education using data from each category. For CPI, the effect was equivalent to 1.2 IQ points; for PC, it was 2.1 IQ points; and for SAC, the estimated effect was 5.2 IQ points. The education effect was an average increase of 3.4 IQ points. This is a relatively small increase for an individual, but it is an important increase for the population, shifting the normal distribution to the right.

The authors estimated that education has a causal effect on intelligence, as assessed by standardized tests (between 1 and 5 IQ points), but they noted that these estimates should be viewed with caution. Five key questions could not be answered even within this comprehensive study. For instance, no study provided evidence about the preservation into adulthood of the largest gains observed in the SAC category. Furthermore, there is no evidence with respect to whether the effects are additive for additional years of education beyond a single year; whether the biggest gains were in children who started with the highest scores (i.e., are there individual differences in the benefit of additional years of education?); or whether increases in IQ scores include g, because most studies included in the analyses did not use a battery of tests, so

g could not be extracted as a latent variable (however, other studies indicate that g is not influenced by education; Ritchie et al., 2015). There are also the questions why nonspecific aspects of education result in some lasting effect (CPI and PC), whereas targeted intervention programs do not (Protzko, 2015), and what are the psychological mechanisms of the educational effect on intelligence.

A large longitudinal study from Denmark (Hegelund et al., 2020) also found that education attainment (years) was related to IQ scores in young adulthood ($N = 11,108$) and in midlife ($N = 2,486$). The results showed average IQ score differences between groups defined by years of education (controlling for IQ at age twelve). For example, in young adulthood, the average difference for the group with seven years of education and the group with seventeen years was 22 IQ points. The same comparison at midlife was 13 IQ points. The increases were greater for those with IQs less than 90 at age twelve. However, further analyses showed that most of the variance in the IQ scores at both young adulthood and midlife was explained by IQ at age twelve (46 percent). Adding educational attainment increased this to 53 percent. The small influence of education is consistent with other studies, but there are still questions about whether this is a g effect. The authors were cautious in their conclusion: "education might constitute a promising method for raising intelligence, especially among the least advantaged individuals (but) replication of our study findings is needed to be able to inform policy" (pp. 6–7).

7.5 Summary

As we saw in Chapter 6, genes do not account for all the variance in intelligence, so there must be nongenetic effects. Lead intake, poor nutrition, and other environmental hazards lower IQ scores. However, there is little progress identifying specific aspects of the environment that foster intellectual development. Doing harm is much easier than doing good.

While some preschool programs have shown long-term results for some important outcomes, increasing intelligence is not one of them. Some intensive programs may help some aspects of cognitive development for the most disadvantaged students, but large-scale projects have not yet been done. Education has some impact on intelligence into adulthood, but not as much as we might think. Unfortunately, whether education increases the g-factor is not established.

Intelligence is associated with a tendency to become intellectually engaged with challenging tasks. Although correlation does not mean causality, it is arguable that people who engage in intellectually challenging activities improve some of their cognitive abilities, but not necessarily g. This may be why education in general is associated with small increases in IQ scores.

The science of human intelligence suggests that when we focus on nongenetic factors related to observed individual differences in cognitive ability, it can be a cost-efficient plan first to address those factors we already know do harm (e.g., lead levels), then to focus on complex ways to do good (provide better nutrition and health care) or on unproven expensive interventions claimed to promote intelligence development, such as scholastic environments challenging pupils' cognitive resources. These are complementary actions that require evidence-based policies.

7.6 Questions for Discussion

7.1 Why does the difference between distal and proximate causes matter?

7.2 How do adoption data support both genetic and nongenetic factors linked to intelligence?

7.3 How are SES and intelligence confounded?

7.4 What is the relationship between education and intelligence?

References

Anderson, D. R. 1998. Educational television is not an oxymoron. *Annals of the American Academy of Political and Social Sciences*, 557, 24–38.

Barnett, W. S. 1998. Long-term cognitive and academic effects of early childhood education of children in poverty. *Preventive Medicine*, 27, 204–207.

Barnett, W. S., & Hustedt, J. T. 2005. Head Start's lasting benefits. *Infants and Young Children*, 18, 16–24.

Batty, G. D., Deary, I. J., & Macintyre, S. 2006. Childhood IQ and life course socioeconomic position in relation to alcohol induced hangovers in adulthood: The Aberdeen children of the 1950s study. *Journal of Epidemiology and Community Health*, 60, 872–874.

Bayless, S., & Stevenson, J. 2007. Executive functions in school-age children born very prematurely. *Early Human Development*, 83, 247–254.

Bayley, N. 1968. Behavioral correlates of mental growth: Birth to 36 years. *American Psychologist*, 23, 1–17.

Bayley, N. 2005. *Bayley Scales of Infant and Toddler Development.* 3rd ed. San Antonio, TX: Harcourt Assessment.

Bellinger, D. C., Chen, A., & Lanphear, B. P. 2017. Establishing and achieving national goals for preventing lead toxicity and exposure in children. *JAMA Pediatrics*, 171, 616–618.

Blackwell, D. L., & Lichter, D. T. 2000. Mate selection among married and cohabiting couples. *Journal of Family Issues*, 21, 275–302.

Braat, M., Engelen, J., Van Gemert, T., & Verhaegh, S. 2020. The rise and fall of behaviorism: The narrative and the numbers. *History of Psychology*, 23, 252–280.

Bradley, R. H. 1993. Children's home environments, health, behavior, and intervention efforts – a review using the Home Inventory as a marker measure. *Genetic Social and General Psychology Monographs*, 119, 439–490.

Brody, N. 1992. *Intelligence.* San Diego, CA: Academic Press.

Bronfenbrenner, U., Mcclelland, P., Wethington, E., et al. 1996. *The state of Americans: This generation and the next.* New York: Free Press.

Burgoyne, A. P., Mashburn, C. A., Tsukahara, J. S., & Engle, R. W. 2022. Attention control and process overlap theory: Searching for cognitive processes underpinning the positive manifold. *Intelligence*, 91, 1–11.

Campbell, F. A., Pungello, E., Miller-Johnson, S., Burchinal, M., & Ramey, C. T. 2001. The development of cognitive and academic abilities: Growth curves from an early childhood educational experiment. *Developmental Psychology*, 37, 231–242.

Capron, C., & Duyme, M. 1989. Assessment of effects of socio-economic status on IQ in a full cross-fostering study. *Nature*, 340, 552–554.

Capron, C., & Duyme, M. 1996. Effect of socioeconomic status of biological and adoptive parents on WISC-R subtest scores of their French adopted children. *Intelligence*, 22, 259–275.

Ceci, S. J. 1990. *On intelligence … more or less: A bioecological treatment on intellectual development.* Englewood Cliffs, NJ: Prentice Hall.

Ceci, S. J., & Williams, W. M. 1997. Schooling, intelligence, and income. *American Psychologist*, 52, 1051–1058.

Chiodo, L. M., Covington, C., Sokol, R. J., et al. 2007. Blood lead levels and specific attention effects in young children. *Neurotoxicology and Teratology*, 29, 538–546.

Church, A. T., & Katigbak, M. S. 1991. Home-environment, nutritional-status, and maternal intelligence as determinants of intellectual-development in rural Philippine preschool-children. *Intelligence*, 15, 49–78.

Coleman, J. S. 1966. *Equality of educational opportunity.* Summary report. Washington, DC: US Department of Health, Education, and Welfare, Office of Education.

Colom, R. 2016. Advances in intelligence research: What should be expected in the XXI century (questions and answers). *Spanish Journal of Psychology*, 19, 1–8.

Colom, R., & Flores-Mendoza, C. E. 2007. Intelligence predicts scholastic achievement irrespective of SES factors: Evidence from Brazil. *Intelligence*, 35, 243–251.

Counter, S. A., Buchanan, L. H., & Ortega, F. 2005. Neurocognitive impairment in lead-exposed children of Andean lead-glazing workers. *Journal of Occupational and Environmental Medicine*, 47, 306–312.

Daley, T. C., Whaley, S. E., Sigman, M. D., Espinosa, M. P., & Neumann, C. 2003. IQ on the rise: The Flynn effect in rural Kenyan children. *Psychological Science*, 14, 215–219.

Damian, R. I., & Roberts, B. W. 2015a. The associations of birth order with personality and intelligence in a representative sample of US high school students. *Journal of Research in Personality*, 58, 96–105.

Damian, R. I., & Roberts, B. W. 2015b. Settling the debate on birth order and personality. *Proceedings of the National Academy of Sciences of the United States of America*, 112, 14119–14120.

Deary, I. J., Der, G., & Shenkin, S. D. 2005. Does mother's IQ explain the association between birth weight and cognitive ability in childhood? *Intelligence*, 33, 445–454.

Detterman, D. K. 2016. Education and intelligence: Pity the poor teacher because student characteristics are more significant than teachers or schools. *Spanish Journal of Psychology*, 19, 1–11.

Detterman, D. K., & Thompson, L. A. 1997. What is so special about special education? *American Psychologist*, 52, 1082–1090.

Dombrowski, S. C., Noonan, K., & Martin, R. P. 2007. Low birth weight and cognitive outcomes: Evidence for a gradient relationship in an urban, poor, African American birth cohort. *School Psychology Quarterly*, 22, 26–43.

Esbjorn, B. H., Hansen, B. M., & Mortensen, E. L. 2006. Intellectual development in a Danish cohort of prematurely born preschool children: Specific or general difficulties? *Journal of Developmental and Behavioral Pediatrics*, 27, 477–484.

Fagan, J. F., Holland, C. R., & Wheeler, K. 2007. The prediction, from infancy, of adult IQ and achievement. *Intelligence*, 35, 225–231.

Firkowska, A., Ostrowska, A., Sokolowska, M., et al. 1978. Cognitive development and social policy. *Science*, 200, 1357–1362.

Fuchs, F., Monet, B., Ducruet, T., Chaillet, N., & Audibert, F. 2018. Effect of maternal age on the risk of preterm birth: A large cohort study. *PLoS ONE*, 13, e0191002.

Goldman, S. M., Tanner, C. M., Oakes, D., et al. 2006. Head injury and Parkinson's disease risk in twins. *Annals of Neurology*, 60, 65–72.

Gopnik, A., Meltzoff, A. N., & Kuhl, P. K. 1999. *The scientist in the crib: Mind, brains, and how children learn.* New York: Morrow.

Gorey, K. M. 2001. Early childhood education: A meta-analytic affirmation of the short- and long-term benefits of educational opportunity. *School Psychology Quarterly*, 16, 9–30.

Gottfredson, L. S. 2007. Innovation, fatal accidents, and the evolution of general intelligence. In Roberts, M. J. (ed.), *Integrating the mind: Domain general vs domain specific processes in higher cognition.* New York: Psychology Press.

Guerrero, L. M. K. 2009. *The impact of lead contamination in children of Callao, Peru.* Lima, Peru: Fundacion Cayetano Heredia.

Haier, R. J. 2017. *The neuroscience of intelligence.* Cambridge: Cambridge University Press.

Hegelund, E. R., Gronkjaer, M., Osler, M., et al. 2020. The influence of educational attainment on intelligence. *Intelligence*, 78, 1–7.

Herrnstein, R. J., & Murray, C. 1994. *The bell curve: Intelligence and class structure in American life.* New York: Free Press.

Hubbs-Tait, L., Nation, J. R., Krebs, N. F., & Bellinger, D. C. 2005. Neurotoxicants, micronutrients, and social environments: Individual and combined effects on children's development. *Psychological Science in the Public Interest*, 6, 57–121.

Jensen, A. R. 1969. How much can we boost IQ and scholastic achievement? *Harvard Educational Review*, 39, 1–123.

Jensen, A. R. 1981. *Straight talk about mental tests.* New York: Free Press.

Jensen, A. R. 1997. The puzzle of non-genetic variance. In Sternberg, R. J., & Grigorenko, E. (eds.), *Intelligence, heredity, and environment.* Cambridge: Cambridge University Press.

Jensen, A. R. 1998. *The g factor: The science of mental ability.* Westport, CT: Praeger.

Kan, K. J., van der Maas, H. L. J., & Levine, S. Z. 2019. Extending psychometric network analysis: Empirical evidence against g in favor of mutualism? *Intelligence*, 73, 52–62.

Kendler, K. S., Turkheimer, E., Ohlsson, H., Sundquist, J., & Sundquist, K. 2015. Family environment and the malleability of cognitive ability: A Swedish national home-reared and adopted-away cosibling control study. *Proceedings of the National Academy of Sciences of the United States of America*, 112, 4612–4617.

Kotchick, B. A., & Forehand, R. 2002. Putting parenting in perspective: A discussion of the contextual factors that shape parenting practices. *Journal of Child and Family Studies*, 11, 255–269.

Lanphear, B. P. 2017. Low-level toxicity of chemicals: No acceptable levels? *PLoS Biology*, 15, e2003066.

Lanphear, B. P., Hornung, R., Khoury, J., et al. 2005. Low-level environmental lead exposure and children's intellectual function: An international pooled analysis. *Environmental Health Perspectives*, 113, 894–899.

Lemos, G. C., Almeida, L. S., & Colom, R. 2011. Intelligence of adolescents is related to their parents' educational level but not to family income. *Personality and Individual Differences*, 50, 1062–1067.

Lynn, R. 1998. Dysgenics. *American Psychologist*, 53, 1232.

Lynn, R., & Harvey, J. 2008. The decline of the world's IQ. *Intelligence*, 36, 112–120.

Lynn, R., & Van Court, M. 2004. New evidence of dysgenic fertility for intelligence in the United States. *Intelligence*, 32, 193–201.

Marks, G. N., & O'Connell, M. 2021. Inadequacies in the SES-achievement model: Evidence from PISA and other studies. *Review of Education*, 9.

Matte, T. D., Bresnahan, M., Begg, M. D., & Susser, E. 2001. Influence of variation in birth weight within normal range and within sibships on IQ at age 7 years: Cohort study. *British Medical Journal*, 323, 310–314.

Molfese, V. J., Dilalla, L. F., & Bunce, D. 1997. Prediction of the intelligence test scores of 3- to 8-year-old children by home environment, socioeconomic status, and biomedical risks. *Merrill Palmer Quarterly Journal of Developmental Psychology*, 43, 219–234.

Moreau, D., Macnamara, B. N., & Hambrick, D. Z. 2019. Overstating the role of environmental factors in success: A cautionary note. *Current Directions in Psychological Science*, 28, 28–33.

Mortensen, L. H., Sorensen, T. I. A., & Gronbaek, M. 2005. Intelligence in relation to later beverage preference and alcohol intake. *Addiction*, 100, 1445–1452.

Moselhy, H. F., Georgiou, G., & Kahn, A. 2001. Frontal lobe changes in alcoholism: A review of the literature. *Alcohol and Alcoholism*, 36, 357–368.

Murray, C. 2003. *Human accomplishment: The pursuit of excellence in the arts and sciences, 800 BC to 1950*, New York: HarperCollins.

Murray, C. 2012. Coming apart: The state of white America, 1960–2010. *New York Review of Books*, 59, 21–23.

Murray, C. 2020. *Human diversity: The biology of gender, race, and class*. New York: Twelve.

Needleman, H. L., Gunnoe, C., Leviton, A., et al. 1979. Deficits in psychologic and classroom performance of children with elevated dentine lead levels. *New England Journal of Medicine*, 300, 689–695.

Newcombe, R., Milne, B. J., Caspi, A., Poulton, R., & Moffitt, T. E. 2007. Birthweight predicts IQ: Fact or artefact? *Twin Research and Human Genetics*, 10, 581–586.

Nisbett, R. E. 2009. *Intelligence and how to get it: Why schools and cultures count*. New York: W. W. Norton.

O'Toole, B. I. 1990. Intelligence and behaviour and motor vehicle accident mortality. *Accident Analysis and Prevention*, 22, 211–221.

Parker, E. S., Parker, D. A., & Harford, T. C. 1991. Specifying the relationship between alcohol use and cognitive loss: The effects of frequency of consumption and psychological distress. *Journal of Studies on Alcohol*, 52, 366–373.

Peeters, W., van den Brande, R., Polinder, S., et al. 2015. Epidemiology of traumatic brain injury in Europe. *Acta Neurochirurgica*, 157, 1683–1696.

Plomin, R. 1994. *Genetics and experience: The interplay between nature and nurture*. Thousand Oaks, CA: SAGE.

Plomin, R., Defries, J. C., Knopik, V. S., & Neiderhiser, J. M. 2016. Top 10 replicated findings from behavioral genetics. *Perspectives on Psychological Science*, 11, 3–23.

Pollitt, E., Gorman, K., Engle, P. L., Martorell, R., & Rivera, J. 1993. Early supplementary feeding and cognition – effects over 2 decades. *Monographs of the Society for Research in Child Development*, 58, R5–R98.

Protzko, J. 2015. The environment in raising early intelligence: A meta-analysis of the fadeout effect. *Intelligence*, 53, 202–210.

Protzko, J. 2016. Does the raising IQ–raising *g* distinction explain the fadeout effect? *Intelligence*, 56, 65–71.

Protzko, J., & Colom, R. 2021. A new beginning of intelligence research: Designing the playground. *Intelligence*, 87, 101559.

Raine, A., Reynolds, C., Venables, P. H., & Mednick, S. A. 2002. Stimulation seeking and intelligence: A prospective longitudinal study. *Journal of Personality and Social Psychology*, 82, 663–674.

Raven, J. 1989. The Raven Progressive Matrices: A review of national norming studies and ethnic and socioeconomic variation within the United States. *Journal of Educational Measurement*, 26, 1–16.

Rindermann, H. 2018. *Cognitive capitalism: Human capital and the wellbeing of nations*. Cambridge: Cambridge University Press.

Ritchie, S. J., Bates, T. C., & Deary, I. J. 2015. Is education associated with improvements in general cognitive ability, or in specific skills? *Developmental Psychology*, 51, 573–582.

Ritchie, S. J., & Tucker-Drob, E. M. 2018. How much does education improve intelligence? A meta-analysis. *Psychological Science*, 29, 1358–1369.

Rodgers, J. L., Cleveland, H. H., van den Ourd, E., & Rowe, D. C. 2000. Resolving the debate over birth order, family size, and intelligence. *American Psychologist*, 55, 599–612.

Rohrer, J. M. 2018. Thinking clearly about correlations and causation: Graphical causal models for observational data. *Advances in Methods and Practices in Psychological Science*, 1, 27–42.

Sackett, P. R., Kuncel, N. R., Arneson, J. J., Cooper, S. R., & Waters, S. D. 2009. Does socioeconomic status explain the relationship between admissions tests and post-secondary academic performance? *Psychological Bulletin*, 135, 1–22.

Sauce, B., & Matzel, L. D. 2018. The paradox of intelligence: Heritability and malleability coexist in hidden gene–environment interplay. *Psychological Bulletin*, 144, 26–47.

Scarr, S., & Weinberg, R. A. 1976. IQ test performance of black children adopted by white families. *American Psychologist*, 31, 726–739.

Schiff, M., & Lewontin, R. C. 1986. *Education and class: The irrelevance of IQ genetic studies*. Oxford: Oxford University Press.

Schottenbauer, M. A., Momenan, R., Kerick, M., & Hommer, D. W. 2007. Relationships among aging, IQ, and intracranial volume in alcoholics and control subjects. *Neuropsychology*, 21, 337–345.

Sigman, M., & Whaley, S. E. 1998. The role of nutrition in the development of intelligence. In Neisser, U. (ed.), *The rising curve: Long-term gains in IQ and related measures*. Washington, DC: American Psychological Association.

Skodak, M., & Skeels, H. M. 1949. A final follow-up study of 100 adopted children. *Journal of Genetic Psychology*, 75, 85–125.

Smith, D. I., & Kirkham, R. W. 1982. Relationship between intelligence and driving record. *Accident Analysis and Prevention*, 14, 439–442.

Stecher, B. M., Holtzman, D. J., Garet, M. S., et al. 2018. *Improving teaching effectiveness – final report: The intensive partnerships for effective teaching through 2015–2016*. Santa Monica, CA: RAND Corporation. www.rand.org/pubs/research_reports/RR2242.html.

Stein, Z., Susser, M., Saenger, G., & Marolla, F. 1972. Nutrition and mental performance. *Science*, 178, 708–713.

Sunderland, A., Harris, J. E., & Baddeley, A. D. 1983. Do laboratory tests predict everyday memory? A neuropsychological study. *Journal of Verbal Learning and Verbal Behavior*, 22, 341–357.

Teasdale, T. W., & Owen, D. R. 1986. The influence of paternal social class on intelligence and educational level in male adoptees and non-adoptees. *British Journal of Educational Psychology*, 56, 3–12.

Teglas, E., Vul, E., Girotto, V., et al. 2011. Pure reasoning in 12-month-old infants as probabilistic inference. *Science*, 332, 1054–1059.

Thurstone, L. L. 1923. The stimulus–response fallacy in psychology. *Psychological Review*, 30, 354–369.

Turkheimer, E. 1991. Individual and group-differences in adoption studies of IQ. *Psychological Bulletin*, 110, 392–405.

van Ijzendoorn, M. H., Juffer, F., & Poelhuis, C. W. K. 2005. Adoption and cognitive development: A meta-analytic comparison of adopted and nonadopted children's IQ and school performance. *Psychological Bulletin*, 131, 301–316.

Wachs, T. D., & McCabe, G. 2001. Relation of maternal intelligence and schooling to offspring nutritional intake. *International Journal of Behavioral Development*, 25, 444–449.

Wasserman, G. A., Liu, X., Popovac, D., et al. 2000. The Yugoslavia Prospective Lead Study: Contributions of prenatal and postnatal lead exposure to early intelligence. *Neurotoxicology and Teratology*, 22, 811–818.

Watson, J. B. 1919. *Psychology from the standpoint of a behaviorist*. Philadelphia: Lippincott.

Weinberg, R. A., Scarr, S., & Waldman, I. D. 1992. The Minnesota Transracial Adoption Study: A follow-up of IQ test-performance at adolescence. *Intelligence*, 16, 117–135.

Zigler, E., & Styfco, S. J. 1997. A "Head Start" in what pursuit? IQ versus social competence as the objective of early intervention. In Devlin, B., Fienberg, S. E., Resnick, D. P., & Roeder, K. (eds.), *Intelligence, genes, and success: Scientists respond to* The Bell Curve. New York: Springer.

Zwick, R., & Green, J. G. 2007. New perspectives on the correlation of SAT scores, high school grades, and socioeconomic factors. *Journal of Educational Measurement*, 44, 23–45.

CHAPTER 8

Intelligence and Everyday Life

8.1 Introduction

Many laypeople, as well as social scientists, subscribe to the belief that the abilities required for success in the real world differ substantially from what is needed to achieve success in the classroom. Yet, this belief is not empirically or theoretically supported. A century of scientific research has shown that general cognitive ability, or g, predicts a broad spectrum of important life outcomes, behaviors, and performances. (Kuncel et al., 2004, p. 148)

In no realm of life is g all that matters, but neither does it seem irrelevant in any. In the vast toolkit of human abilities, none has been found as broadly useful – as general – as g. (Gottfredson, 2002, p. 332)

As of the end of the twentieth century, the United States is run by rules that are congenial to people with high IQ and that make life more difficult for everyone else. (Herrnstein and Murray, 1994, p. 541)

These are statements about the importance of intelligence for everyday life. They seem consistent with our own experiences, but they also could imply a cognitive elitism that many find uncomfortable. Nonetheless, compelling data support these statements and imply that everyday life can be seen as one long continuous intelligence test, and that the test is getting harder as modern life becomes more complex (Gottfredson, 1997; Gordon, 1997; Hunt, 1995).

The evidence can be contentious, especially if one believes that intelligence is immutably genetic and that high intelligence is a requirement for any success one might have irrespective of how hard one works. As discussed in previous chapters, neither of these beliefs is consistent with the empirical data.

What role does intelligence actually play in everyday life, and is that role alarming? This chapter examines research findings about the relationships between intelligence and performance in three broad domains chosen from a large possible set: academics, the workplace, and personal adjustment (further examples can be found in Jensen, 1998; Warne, 2020). The findings are among the most robust in all of psychological science, which may be largely due to the sophisticated psychometric methods used to overcome problems inherent to such a complex research problem.

8.2 Investigating Intelligence and Success in Life

We cannot study the totality of academia, the workplace, or personal life. We have to select slices of them where practical measures can be obtained. These slices are, however, not random samples of the broad domains, and the measures used must be understood statistically. This kind of research is not easy for at least three key reasons. First, there is the conceptual criterion problem of specifying what we mean by success in each domain. Second, we have to select measures of success that are amenable to quantitative analyses, as discussed in Chapter 2. Observable measures, such as grade point average (GPA) or money earned, are only partially satisfactory measures because they often have undesirable statistical and measurement properties that hinder analysis and interpretation. Third, there is the problem of research design limitations. The optimal research designs for investigating intelligence–success relationships often are longitudinal rather than cross-sectional. For practical reasons of time and cost, longitudinal studies are rare (Belsky et al., 2020).

The rest of this section describes these three key problems in more detail to help readers critically evaluate individual research studies. We also show how these problems can be overcome by combining studies to establish a weight of evidence that transcends any one study, even if it has problematic aspects. Remember, even less-than-ideal studies can be informative.

8.2.1 *The Conceptual Criterion Problem*

We estimate the extent to which intelligence, measured by standardized tests, is related to some criterion variable or index of success by calculating *predictive validity*, which is defined as the correlation between a measure of intelligence and the criterion measure. The higher the correlation is, the better, although the value can be moderate, or even small, and still have important practical relevance (Funder and Ozer, 2019).

Is there an acceptable definition of success that can be used in scientific research? In the academic arena, a student is considered successful based on a degree of learning. The most common measure of academic success is a person's GPA across classes (within and across subject areas). Another measure is graduation rate. However, neither measure is easily comparable across classes, subjects, or institutions. These criterion measures of learning may be the best we can do, but they introduce unwanted sources of variance that produce observed correlations with intelligence that likely underestimate the actual relationship.

It is even harder to define success in the workplace. Within an industry or occupation, income partially captures the idea of success, but incomes across occupations are hard to compare. Incomes are also often influenced by variables unrelated to intelligence, such as seniority of employment. Similarly, supervisor ratings are not reliable unless the raters are trained and the criteria for rating have been agreed on. Objective measures of employee output are often hard to come by and generally capture only a part of a person's job. For example, a company might track the number of checkouts per hour that each of their clerks handles but does not measure things like customers' reactions to a clerk's manner.

Defining success in life is still harder. We can measure extreme personal/social adjustment, which can vary from achieving a civic prize to going to jail, but most people do neither. Success in life is a multifaceted thing. Informative studies have been conducted of the relation between intelligence and particular aspects of life success, such as maintaining good health or going on welfare, but trying to relate intelligence, or virtually any other trait, to such a multifaceted concept as "life success" always requires cautious interpretation. Please keep this warning in mind.

Nevertheless, even with imperfect criterion measures, predictive validity is useful in many circumstances where selecting applicants is required. This is illustrated in general in Figure 8.1 from a review of the predictive value of general intelligence (Sackett et al., 2020). The upper left panel shows that selecting applicants using a test with no criterion validity, essentially random selection, results in average mean performance of the selected group. The upper right panel shows that selection with a perfect criterion results in high mean performance. Modest criterion validities (0.5 in lower left panel; 0.2 in lower right panel) also result in better than average performance when compared to the upper left panel (baseline). This illustrates that even imperfect criterion measures can improve the selected group performance mean.

8.2.2 Three Statistical Problems

Establishing the relationships between intelligence and other measures is sensitive to three statistical concepts: reliability, range restriction, and generalization.

RELIABILITY

A correlation between any two variables is affected by the reliability of each variable. Consider that any measurement contains two elements, a "true value" and a residual term. While the residual term is frequently referred to as "error," it is not necessarily error in the sense of a mistake. It refers to the sum of all influences on the measured variable that are statistically independent of the true value. To take an example, consider the way in which weight is measured during the typical annual physical. Examinees are told to take off their shoes and stand on a scale. Measured weight is then shown on the scale. The measured weight has the following components:

$$\begin{aligned} \text{Measured Weight} = \ & \text{Actual Body Weight} \\ & + (\text{Weight of Clothes} \\ & + \text{Scale Bias}), \end{aligned}$$

where Scale Bias refers to any tendency of the scale to weigh high or low. The terms in parentheses, here Weight of Clothes and Scale Bias, are residual effects, uncorrelated with the examinee's actual weight. If an examinee were to be weighed on a different scale, wearing different clothes, measured weight might change, even though actual body weight remained the same. Measured weight is said to be *reliable* to the extent that the same measure is obtained across comparable conditions. This reasoning applies to intelligence testing and to criterion measures.

An intelligence test score includes the examinee's "real" intelligence and a residual term that is unique to the examination of that person at that time. Exactly the same thing can be said of any criterion measure.

In both the academic and workplace settings, the difference between the observed correlation and the true correlation can be

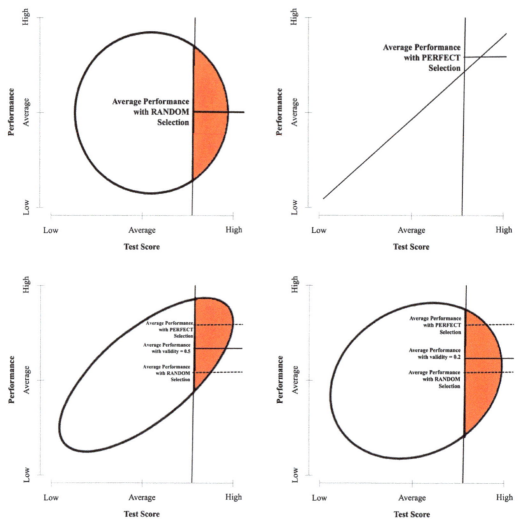

Figure 8.1 Interpreting criterion-related validity when the test score has four different relationships to the criterion variable. Vertical lines indicate a 10 percent cutoff score for all applicants. Horizontal lines show the mean performance of applicants who pass the cutoff. The upper left example shows a result of average test performance when the test has no relationship to the criterion of performance; the upper right shows high performance when there is a perfect relationship between selection test and performance criterion; and the lower left and lower right show different performance averages using different test score/performance criteria (Sackett et al., 2020).

substantial. Professionally developed intelligence tests generally have reliabilities of 0.85 or above. This is because of careful item selection based on a sophisticated set of psychometric tools and methods. Large-scale academic achievement tests, such as those used in the United States to assess educational progress on a statewide basis, or PISA assessments, have similar reliability coefficients. Within-class, teacher-assigned grades

are less reliable; in the best scenarios, the reliabilities are in the 0.6–0.8 range. Supervisor ratings typically are about 0.6 or lower (Sackett et al., 2021).

The lower the reliability of the measures is, the more the true correlation between them is underestimated by the observed correlation. There are statistical methods for correcting an observed correlation using the actual reliabilities of the measures. The general point is that the evaluation of the strength of an observed correlation should include knowing whether it is corrected for the reliabilities of the measures. A low observed correlation that is not corrected might obscure a stronger true correlation and lead to an unwarranted conclusion of a weak (or no) relationship between the measures.

Note that a correction for reliability is appropriate when one's interest is in theoretical constructs underlying measures – for instance, whether intelligence as a concept is related to academic ability as a concept. The correction is not appropriate when one's interest is in whether one set of scores predicts the value of another set of scores – for instance, if you wanted to know whether the SAT predicts first-year college GPA. There are many kinds of reliability, and a full treatment is beyond our scope here other than to call attention to this general problem.

RESTRICTION OF RANGE

Correlations are also affected by the range of values for each variable. In most studies, the variability of intelligence in the group actually studied will be smaller than the range in the population to which we wish to generalize. The problem of estimating intelligence-grade relations in elementary schools illustrates the situation. Elementary schools generally draw students from the neighborhood immediately around them. In most industrially developed countries, neighborhoods tend to have distinct socioeconomic and sometimes demographic characteristics. Therefore, the student demographics in a single elementary school will be more homogeneous than in the district or state. We say that the scores are subject to *range restriction*. In the case of the elementary schools, we would expect there to be less variability in a measure of intelligence within a school than across a state.

Range restriction in a sample typically results in a smaller correlation coefficient than would be present in the population with unrestricted range. *Selection restriction* is an important special case of range restriction. Selection restriction occurs whenever an applicant population takes a predictor

test, here some sort of intelligence test, as part of an application for a job or educational opportunity. All applicants above a given *cut score* are then accepted, and their performance on the job or in the school is recorded. For example, suppose a university uses an entrance examination and admits the top 10 percent of the applicants. To validate the entrance examination, university officials would want to know if it was a good predictor of the grades that an applicant would obtain. However, grades are available only for the admitted students. Since, obviously, the top 10 percent of the applicants will have less variation in examination scores than the entire group of applicants, the correlation between examination scores and grades in the applicant population can be estimated by computing the examination–grades correlation in the admitted group and then correcting for range restriction.

Uncorrected selection restriction of range underestimates a true correlation. This is illustrated in general in Figure 8.2, which shows that the observed correlation (labeled "validity" in the figure) is progressively smaller than the true correlation (also labeled "validity") when based on a sample drawn only from part of the distribution (upper right, lower right, and left panels). The problem is most severe when the sample is most restricted (lower right panel).

This is the case, for example, when an admissions officer at a highly selective university proclaims that there is little or no correlation between SAT scores of admitted students and their subsequent GPAs. Sometimes this is used as an argument for dropping the SAT, but it is not a valid argument due to selection restriction of range. The lower right panel in Figure 8.2 represents such a case, finding a weak correlation when the sample is drawn only from accepted students at the upper end of the distribution.

STATISTICAL POWER

Statistical power refers to how large a sample should be to maximize the likelihood of finding a significant observed correlation at

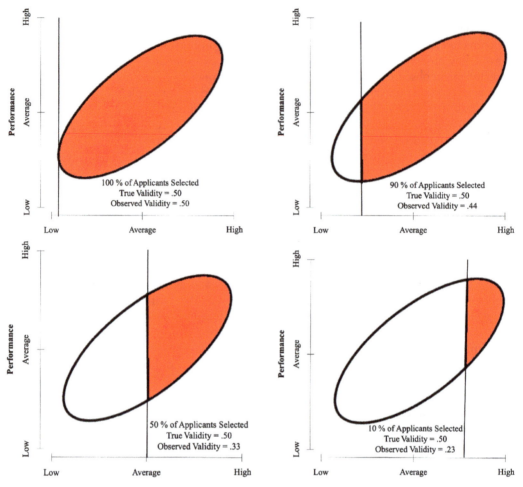

Figure 8.2 Illustrations of the impact of restricted range on correlations. The upper left panel shows no restricted range, and the lower right shows the most restricted range. Correlations between actual performance and test scores for people in the red areas are highest in the unrestricted case (observed validity = 0.50) and lower in the most restricted case, shown in the lower right (0.23), when holding constant the actual correlation between test scores and the criterion of performance (true validity = 0.50). The upper right and lower left are intermediate examples of restricted ranges and their effects on observed correlations (Sackett et al., 2020).

a particular level of significance if that correlation is in fact significant in the population being studied. Suppose that the actual population correlation is 0.50. If we set the significance level at $p < 0.05$, 16 out of 100 samples will *not* reach a value reliably greater than zero, even though the population correlation is a substantial 0.50. In other words, there is a good chance of mistakenly concluding from your sample that no correlation exists, when, in fact, there is a strong correlation. This is an example of an underpowered study. You have not

rejected the null hypothesis, but it would be incorrect to accept that there is no relationship.

Power increases with sample size. There are statistical tables that show the power of detecting significant findings at different sample sizes. In the example just given, if the sample is increased from 25 to 100, the power is greater than 0.995, so there is only a 0.005 chance of missing a significant observed correlation. More power is better than less, and that is why very small effects can be detected only in very large samples (see Chapter 6 regarding GWAS results). Researchers should compute power before they design their study to ensure that the sample is large enough to detect a significant relationship, if one actually exists. Often, however, this is not done, and an observed correlation is less than the critical significance level (e.g., 0.05 or 0.01). It would be incorrect to conclude that the two variables were unrelated. In other words, we cannot *reject* the null hypothesis. But *"not rejecting" is not the same as accepting*. This is just as critical a mantra as "correlation is not causation."

It is also possible to find spurious significant correlations in small-sample studies because the chances of random influences are greater. This is a primary reason that many studies in psychology fail to replicate. The larger the sample size is, the less random chance effects can influence the result. According to a meta-analysis, "intelligence research does show signs of low power and publication bias, but these problems seem less severe than in many other scientific fields" (Nuijten et al., 2020, p. 1).

The power problem becomes critical when it is combined with the problem of criterion reliability. Grades within a class and employer rating systems will often have a reliability of around 0.60, and intelligence tests will have a reliability of about 0.85. Suppose that the true correlation between intelligence and academic ability is 0.50 in the population. The expected population correlation between test scores and grades is 0.26 after correction for attenuation. Setting the significance level at 0.05, the statistical power of a study with twenty-five participants is approximately 0.36 (power estimates are based on Cohen, 1988, table 3.3.2). This means that about two out of three studies of this size would *not* provide strong enough evidence to reject the null hypothesis. even though it is false. If the sample size were to be increased to 100, power would increase to about 0.75. In this case, failure to reach statistical significance would be reasonable evidence against the hypothesis that a "true score" was 0.50 or larger.

What these examples show is that power is produced by an interaction between the reliability of the measures and the size of the study. This interaction has to be considered when evaluating the lack of a significant observed correlation (null result). Do researchers actually fail to do this? The answer is, quite often, yes. Box 8.1 presents the case of a widely cited older study in which no consideration was given to these issues.

Finally, we note that the traditional use of $p < 0.05$ and $p < 0.01$ as critical values for statistical significance has come under considerable criticism in the context of the failure to replicate many "classic" findings in psychology (Cumming, 2014; Funder and Ozer, 2019). The replication crisis fundamentally is caused by emphasizing statistically "significant" findings at a relatively low level of probability (like $p < 0.05$) obtained in small sample sizes that lack sufficient power to find a relationship, if one actually exists, and that capitalizes on chance effects that create spurious correlations. These problems make many findings potentially unreliable.

8.2.3 The Research Design Problem

In addition to the criterion and statistical problems, the third area concerns research design. The ideal research design is a *prospective* study, in which the investigator obtains data on the intelligence of people at some point in their lives, ideally before they enter an academic program or the workforce, and then assesses how well they succeed. This is by far the easiest kind of study to interpret but the hardest to conduct. It is possible only if the investigator has some way of testing a large number of people, and then following them for a reasonably long period. A few studies have done this. However, such studies are expensive and difficult, so there are not many of them, even though they provide some of the most compelling findings.

Prospective studies can sometimes be conducted by examination of government records. Studies of this sort have been

Box 8.1 A Day at the Races: A Failure to Consider Power and Reliability

In 1986, two Cornell University psychologists, Stephen Ceci and J. K. Liker, published an eye-catching article titled "A Day at the Races." They reported a four-year study of the expertise of a group of thirty habitual bettors on harness racing. Ceci and Liker did not study the accuracy with which these bettors predicted winners because, as they said, and as many horse racing fans know, the winners are often determined by unpredictable events. Instead, they studied the accuracy with which the bettors were able to predict the favorite and top three favorites at post time (the start of the race), given the extensive information about each horse that was contained in the daily racing form, which is available to bettors prior to a race. Although secondary references often misinterpret this, what Ceci and Liker actually studied was their participants' ability to predict how other bettors would place their bets, not which horse would win.

Ceci and Liker found that the participants' decision was a mathematically complex function of the information contained on the racing form and that the accuracy of the predictions had a correlation of −0.03, essentially zero, with the participants' IQ scores on the short form of the WAIS (this figure is a correction to the original value, provided in Liker and Ceci, 1987). Ceci and Liker drew the following strong conclusion:

> (a) IQ is unrelated to performance at the racetrack but, more important (b) IQ is unrelated to real-world forms of cognitive complexity. (p. 255)

These are strong words indeed. The null finding was claimed to be reliable, and the task, something that is related to but not the same as picking the winners in a race, was unhesitatingly generalized to the universe of complex tasks. Nowhere in the article was there any mention of reliability or power.

Douglas Detterman and his colleague Kathleen Spry (1988) wrote a detailed critique of the Ceci and Liker study. Among other things, they observed that Ceci and Liker's criterion, the ability to predict the odds at post time, had a reliability of at most 0.41. What does this mean? Suppose that the correlation between the underlying abilities, intelligence and skill at setting the odds, is a perfect 1.0. The reliability of the short form of the WAIS is known to be 0.85. Therefore, the expected value of the correlation in the sample would be $0.85 \times 0.41 = 0.35$. If $N = 30$, the power of the Ceci and Liker study would be approximately 0.5, which means that a study like theirs should fail to reach the conventional 0.05 level of statistical significance five out of ten times, even if the underlying correlation were a perfect 1.0.

Of course, nobody thinks that the correlation between intelligence and racetrack betting is the ideal 1.0. Based on meta-analysis, a widely quoted estimate of the correlation between intelligence and performance on a cognitively complex task is 0.5 (Schmidt and Hunter, 1998). To be generous, increase this to 0.6. Then, solely on the basis of reliability, the expected sample correlation would be 0.21. Using this estimation, the power of the Ceci and Liker study was a mere 0.20; studies like theirs should fail to reach the 0.05 level of significance four out of five times.

The Ceci and Liker study presents us with good news and bad news. The good news is that when a published study contains major flaws, other scientists point out the errors. The bad news is that almost no one notices the correction. The Ceci and Liker study has been cited 148 times as evidence that intelligence, as measured by the IQ tests, is not related to real-world cognition. The Detterman and Spry study has been cited thirteen times (per an ISI Web of Knowledge citation search on May 3, 2022).

carried out in those countries in which eighteen- to twenty-year-old men have to register and be tested for potential military enlistment. (As far as we know, Israel is the only country that requires registration for both men and women.) Valuable information can be gained if some of the registrants can be reinterviewed (or tested) later in life, to determine how well they have fared. In some countries, this can be done without actually contacting the individuals because the government keeps extensive records of the health, education, and income of its citizens. Legal and ethical issues concerning access to such data have to be resolved, but the important point is that such studies can be done.

The alternative to a prospective study of the relation between intelligence and success is a *retrospective* study. In a retrospective study, a group of people are identified who have varying degrees of success. The investigator then estimates their intelligence, either by direct testing or by examination of relevant records. Studies of eminence or genius often fall into this category. The investigator identifies a group of individuals who meet some criterion for accomplishment and then tries to identify any common characteristics of the group. One of the most ambitious and well known of such studies was Simonton's (2006) determination of the correlation between a measure of intellectual capacity, reconstructed from historical records, and historians' ratings of the performance of the forty-two US presidents, from George Washington through George Bush. The correlation was 0.56.

8.2.4 *Are These Problems Fatal?*

In many cases, investigators must accept less-than-ideal measures, such as using educational attainment or a brief vocabulary test as a proxy measurement of intelligence, or using place of residence as the sole measure of a participant's socioeconomic status. Such compromises are not fatal errors; they are things that must be considered when evaluating results. This is also the case for studies that used statistical corrections for reliability of measures, restricted ranges, power, and other issues of data analysis that plagued many early studies (Richardson and Norgate, 2015).

Perhaps the most effective way to address all these problems is to use a statistical technique called *meta-analysis* to draw conclusions from multiple studies (Sackett, 2021; Hunter and Schmidt, 1990). Special statistical methods are used to identify trends that may not be clear from focusing on the details of any one study. Finding results based on combining studies (and weighting the findings more from the largest and best studies) can be compelling, but there are also problems with this approach that should not be ignored (Lakens et al., 2017).

Meta-analyses come in different forms and make different assumptions, so they are not immune from careful scrutiny. The individual studies reviewed in a meta-analysis will, inevitably, vary in the quality of the measurements taken and the procedures used. These considerations are judgments that have to be made by considering the details of each study. They do not lend themselves to statistical treatment. Some meta-analyses have attempted to deal with this problem by classifying studies by their perceived quality and then analyzing high- and low-quality studies separately, to see if this makes any difference. A finding that appeared only in low-quality studies would certainly be treated with suspicion. Overall, however, meta-analyses often provide a solid basis for establishing a weight of evidence.

For example, Table 8.1 is based on meta-analyses and lists a number of aspects of life outcomes that are correlated to intelligence test scores (Strenze, 2015). These data are convincing about the important role of intelligence in everyday life. At the same time, they demonstrate that intelligence is far from a perfect predictor; the highest correlation is 0.58. Nonetheless, no other variable in psychological research has as many strong or replicated correlations across a wider range of variables/social outcomes.

In the next sections, we will look in more detail at some findings that have established

Table 8.1 Relationship between intelligence and social outcomes (results of meta-analyses)

Measure of success	Effect size (r)
Academic performance in primary education	0.58
Educational attainment	0.56
Job performance (supervisory ratings)	0.53
Occupational attainment	0.43
Job performance (work sample)	0.38
Skill acquisition in work training	0.38
Degree attainment speed in graduate school	0.35
Group leadership success (group productivity)	0.33
Promotions at work	0.28
Interview success (interviewer rating of applicant)	0.27
Reading performance among problem children	0.26
Becoming a leader in group	0.25
Academic performance in secondary education	0.24
Academic performance in tertiary education	0.23
Income	0.20
Having anorexia nervosa	0.20
Research productivity in graduate school	0.19
Participation in group activities	0.18
Group leadership success (group member rating)	0.17
Creativity	0.17
Popularity among group members	0.10
Happiness	0.05
Procrastination	0.03
Changing jobs	0.01
Physical attractiveness	−0.04
Recidivism (criminal behavior)	−0.07
Number of children	−0.11
Traffic accident involvement	−0.12
Conformity to persuasion	−0.12
Communication anxiety	−0.13
Having schizophrenia	−0.26

Note. From Strenze (2015).

relationships between intelligence test scores and key variables. Knowing these details will help readers evaluate new research studies.

8.3 Intelligence and Academic Achievement

Binet's motivation for constructing the original intelligence test was to identify children who were at risk for failing in the standard academic system, as discussed in Chapter 1. For example, failure to graduate high school is associated with lower scores on intelligence tests in many studies (Wechsler and Matarazzo, 1972). An

example of the relationship based on a large-sample study is shown in Figure 8.3. Graduation rates, however, are only a slice of school achievement, and there are many other factors associated with dropouts (Heckman et al., 2000; Heckman et al., 1998). Subsequent psychometric research has generalized Binet's goal to predicting degrees of success at all levels of education. Many of these studies overcome the typical problems we have enumerated in the previous sections. What do they find? Psychometric research has generalized Binet's goal to predicting degrees of success at all levels of education (Box 8.2).

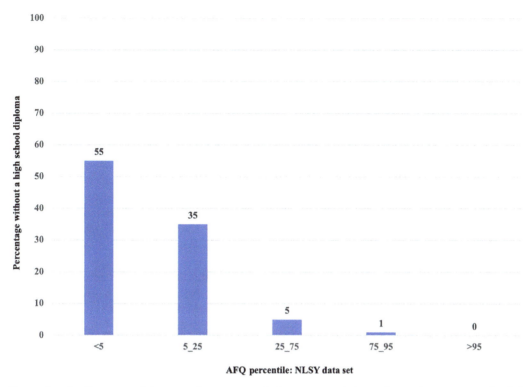

Figure 8.3 Intelligence and dropouts. Percentage of white young adults in the NLSY79 survey who did not complete high school, plotted as a function of their percentile scores on the Armed Forces Qualifying Test.
Data are from Herrnstein and Murray (1994).

8.3.1 *Intelligence in Grades K–12*

Many studies show that intelligence test scores are related to academic achievement. One of the best examples is a comprehensive, prospective study from the United Kingdom (Deary et al., 2007). This study sampled an entire population with a battery of intelligence tests and standard achievement assessments in different school subject areas – all in a prospective design. More than 70,000 students completed the Cognitive Abilities Test (CAT, a standardized test battery) in English schools during a school year. The test takers included almost all the eleven-year-olds in England, so range restriction is not relevant. At age sixteen, all the same students took nationwide examinations in a variety of subjects. The national examinations are subject to much more careful psychometric evaluation than is typically the case for locally generated (and

certainly for teacher-generated) examinations, so reliability of the criterion variable was not a major concern.

The researchers extracted a general intelligence (*g*) factor from the CAT scores at age eleven and a general academic achievement factor from the scholastic examination scores at age sixteen (hence the prospective design). The correlation between the two was 0.81 (Figure 8.4). There was substantial variation between associations with the *g*-factor and educational accomplishment within individual subject areas. Correlations ranged from a high of 0.77 for mathematics to 0.43 for art and design. In general, the topics usually considered the academic core courses – the humanities, mathematics, and the sciences – had correlations in the 0.50–0.75 range, while "practical" topics, such as art and design, music, and textiles, had correlations in the 0.4–0.5 range.

Box 8.2 Intelligence and School Grades

Bettina Roth and colleagues (2015) published a large-scale meta-analysis considering 240 independent samples totaling 105,185 participants (average age = 13.9 years, SD = 4) which were obtained from 162 primary studies from thirty-three countries (Australia, Austria, Brazil, Canada, Central Philippines, China, Croatia, Czech Republic, Dubai, Estonia, Finland, France, Germany, Great Britain, Guatemala, India, Iran, Iraq, Italy, Kenya, Lebanon, Luxembourg, Netherlands, Poland, Portugal, Russia, Slovenia, South Africa, Spain, Sweden, Switzerland, United States, and Yemen). The years covered by the studies ranged from 1922 to 2014.

The average correlation between intelligence and school grades for the full sample was r = 0.54. By academic subject, the correlations were as follows: (1) math and science, r = 0.49; (2) language, r = 0.44; (3) social sciences, r = 0.43; (4) fine art and music, r = 0.31; and (5) sports, r = 0.09. By grade level, the correlations were as follows: (1) elementary school, r = 0.45; (2) middle school, r = 0.54; and (3) high school, r = 0.58. This finding is surprising given the reasonable prediction that range restriction will be present at higher levels. Nevertheless, the authors note, "As the content becomes more demanding throughout grade levels it should be increasingly difficult to compensate for intelligence deficits through practice alone" (p. 126).

Regarding the type of intelligence test, the values were as follows: (1) verbal, r = 0.53; (2) nonverbal, r = 0.44; and (3) mixed, r = 0.60. "A broad measure of intelligence or g is the best predictor of school grades" (p. 126). The values for boys and girls were exactly the same (r = 0.58).

Still another interesting finding to mention was how the predictive validity changed across the century. Before 1983, the value was 0.68, whereas after 1983, it decreased to 0.47. In the authors' own words, "a possible explanation might involve grade inflation, which describes the observation that throughout the last decades progressively better grades are awarded for work that would have received lower grades in the past" (p. 126).

The value after 1983 is consistent with findings from the meta-analysis by Sakhavat Mammadov (2021) based on 267 independent samples (N = 413,074) recruited for research purposes in the last thirty years. The correlation value between intelligence and school grades was r = 0.42. This researcher also computed the meta-analytic correlation between the Big Five dimensions of personality and school grades. In this case, the values were r = 0.27 (conscientiousness), r = 0.16 (openness to experience), r = 0.09 (agreeableness), r = 0.02 (neuroticism), and r = 0.01 (extraversion). The author concluded, "Cognitive ability was the most important predictor with a relative importance of 64%. Conscientiousness emerged as a strong and robust predictor of performance, even when controlling for cognitive ability, and accounted for 28% of the explained variance in academic performance" (p. 1).

Having this information is important because school grades open or close doors for accessing further scholastic and occupational opportunities, which likely will greatly impact a person's life.

This classic study provides some of the strongest evidence for a robust relationship between general intelligence (g) and general academic achievement. The results also show that other factors must be involved. Do other cognitive factors account for any part of this remaining variance?

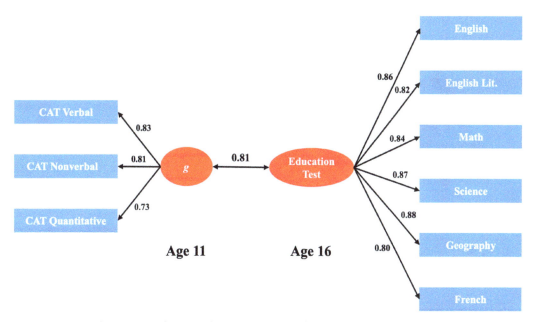

Figure 8.4 Correlation (0.81) between latent intelligence factor (*g*; assessed at age eleven) and latent academic grade factor (education test factor; assessed at age sixteen). CAT is the Cognitive Abilities Test battery (Deary et al., 2007).

This question was addressed in a meta-analysis based on studies that tested the Cattell–Horn–Carroll (CHC) theory and academic achievement (Zaboski et al., 2018). We discussed this theory in Chapters 3 and 4. Measures included a *g*-factor, fluid reasoning (Gf), comprehension knowledge (Gc), long-term storage and retrieval (Glr), visual processing (Gv), auditory processing (Ga), short-term memory (Gsm), and processing speed (Gs). Table 8.2 shows the main results for the full sample for two academic areas. The *g*-factor always predicted more variance than the other factors – up to 0.76 for reading comprehension. For the older group (aged fourteen to nineteen years), the most variance predicted by *g* was for basic reading (*r* = 0.84). The authors concluded, "Across all academic domains and for all age groups, *g* had by far the strongest relations, with large effect sizes. The variance explained by *g* was typically more than that accounted for by all the broad abilities combined" (p. 52). Beyond *g*, this study did not find interesting relationships between academic performance and other cognitive factors.

In the K–12 system, cognitive test scores often are used to identify students whose low scores indicate that they may need to be assigned to special education classes; this is consistent with Binet's original motivation. Tests are also used to assign students to accelerated programs for the gifted. In both cases, other factors are also considered by most Western school systems, although decisions by test scores alone are frequent in other parts of the world. The majority of students fall somewhere between these extremes, and for most of them, the test scores have no practical consequence because no educational decisions are made on the basis of middle-range scores. In college and university entrance decisions, however, test scores matter across the entire range of scores, as discussed in the next section.

8.3.2 *Intelligence and Selection in Higher Education*

Since World War II, US colleges and universities have incorporated two major testing programs, the SAT and the American College Testing Program (ACT), into the admissions process. These tests are revised and validated periodically by

Table 8.2 Meta-analysis results for broad cognitive abilities included in the CHC model by academic achievement area

Scholastic achievement area	Effect size (r)
Basic reading	
g	0.74
Gf	0.07
Gc	0.45
Glr	0.23
Gv	0.01
Ga	0.34
Gsm	0.28
Gs	0.21
Basic math	
g	0.72
Gf	0.39
Gc	0.30
Glr	0.09
Gv	0.02
Ga	0.03
Gsm	0.09
Gs	0.23

Note. From Zaboski et al. (2018).

Table 8.3 Correlations between tests used for college/university selection, the ASVAB general factor or AFQT, and Raven's Advanced Progressive Matrices

Test	SAT	ACT
ACT	0.87[a]	
ASVAB	0.86; 0.90[a]	0.77; 0.92[a]
RAPM	0.72[b]	0.71[c]

[a] Data are from Coyle and Pillow (2008) (NLSY97 data). [b] Data are from Frey and Detterman (2004). [c] Data are from Koenig et al. (2008).

correlating test scores with first-year grade point average (GPA1), cumulative grade point average (GPAC), or probability of graduation within a specified period of time after matriculation (usually four to six years). The use of these tests is not without controversy. In May 2020, the University of California (UC) system has decided to stop using the SAT and the ACT despite a recommendation by a faculty committee to keep them in the admissions process. The faculty review found that they were useful predictors of academic success and were not biased against any social group. To the contrary,

the tests help identify disadvantaged students who might otherwise not meet admissions criteria. Here we concentrate on technical issues like the ones considered by the UC faculty review that should inform policy issues about the best use of these tests, although policies often are considered from nonscientific perspectives.

As introduced in Chapter 3, the first portion of the current SAT, referred to officially as the SAT-I, contains sections stressing verbal comprehension and logical reasoning. By tradition (and officially, in earlier versions), these two sections have been referred to as the SAT-V and SAT-M. Both sections represent attempts to evaluate comprehension and reasoning without tying questions to specific high school curricula.

The ACT program takes a different approach (see Chapter 3). It develops tests that are specifically tied to curricular material, such as history and mathematics. The idea is to predict what a student will learn in college by determining how much he or she has learned in high school. The second part of the SAT, the SAT-II, does the same thing.

Both tests are good estimates of the g-factor; the tests are essentially interchangeable for making admissions decisions. Table 8.3 presents estimates of the correlations between the SAT, the ACT (summary score), and the general factor derived from the Armed Services Vocational Aptitude Battery (ASVAB), which is closely approximated by the Armed Forces Qualifying Test (AFQT). The correlations with Raven's Advanced Progressive Matrices are also included, to show the relation between the educational tests and a highly *g*-loaded test

The correlations are high. The correlation between the SAT and the ACT approaches the reliabilities of the two tests. This suggests that the true correlation between the two tests is nearly perfect. A study using NLSY97 data found that both academic aptitude tests had loadings of about 0.9 on a general factor derived from the ASVAB (Coyle and Pillow, 2008). The finding is important because the ASVAB general factor is a measure of crystallized intelligence (Gc), rather than of fluid intelligence, or Gf (Roberts et al., 2000). Therefore, these data support using these standardized selection tests as proxy measures for general intelligence.

These tests were designed to go beyond Binet's original intent for children and predict academic achievement in higher education. They have become an industrial force that touches most applicants to college and university. How well do they do?

A classic meta-analysis study by Paul Sackett and his colleagues (2009) at the University of Minnesota yielded important information. They analyzed data provided by the College Board for forty-one colleges and universities where the SAT was used in 1995–1997. More than 155,000 test takers were involved. Three SAT–GPA1 correlations were calculated:

1. *The uncorrected correlation between SAT and GPA1 in admitted students, calculated within institutions and then averaged across institutions.* It was 0.35.
2. *The correlation between SAT and GPA1 corrected for restriction of range within the applicant population for each institution, and then averaged.* This is the predictive correlation that would be of interest to admission officers in each institution. It was 0.47.
3. *The correlation between SAT and GPA1 corrected for restriction of range of SAT*

scores across all institutions. This can be thought of as the predictive correlation to be used to determine the benefit of using the test across all participating institutions. It was 0.53.

These strong psychometric findings are based on GPA1. Box 8.3 addresses other criteria that could be used and problems associated with them. The key question is whether correlations like the ones with GPA1 justify the use of these tests for admissions decisions.

If standardized tests used for admissions have limited predictive value for important aspects of future academic success other than GPA1, why are they used? Probably, the short answer is that they are proxies for general intelligence, and high scores increase the probability of completing a program successfully. But success depends on many other factors too. That is why most admissions decisions, at least in Western countries, do not rely on test scores alone. The long answer has to do with the statistics of personnel selection, a well-developed field with sophisticated techniques designed to maximize the efficiency of selecting individuals likely to succeed at a school or job. Box 8.4 explains the basic concepts and what is gained by using admissions tests. The main point is that prediction can be useful even if it is not perfect. Remember that perfection is not a reasonable goal.

8.4 Intelligence in the Workplace

Earl Hunt's (1995) book title asked a provocative question: *Will We Be Smart Enough? A Cognitive Analysis of the Coming Workforce.* In it he presciently notes, "Three technological changes – computers, communications, and transportation – have combined to produce a workplace where there is an increasingly sharp demarcation between a few good jobs and a large number of mediocre ones. What each of these technologies does is to multiply the effectiveness of a smart person" (p. 284). Even earlier, Robert Reich (1991), an economist who

Box 8.3 What Does the SAT Do?

Freshman GPAs indicate a student's initial reaction to college. What about predicting later performance or graduation? Beyond the first year, there is great variation in courses college students take, and there are also substantial differences in grading practices across disciplines. This muddies the situation.

For example, there is a negative correlation between the SATs of students within an academic program and the mean grade point assigned by that program. This is because mathematics and science programs, which assign relatively low grades, tend to draw the students with the highest SATs, while humanities and education programs, which assign high grades, draw students with lower SATs. The effect is quite large. A study involving more than 200,000 students from thirty-eight public universities during the 1990s found that the difference in SAT scores between the discipline with the highest entering scores, engineering, and the one with the lowest scores, education, was, conservatively, 0.92 standard deviation units. The negative correlation between the rigorousness of grading within a discipline and the SATs of the entering students will reduce the correlation between overall GPAs and entering SATs, calculated over the institution as a whole (Kroc et al., 1997).

What about graduation rates? The probability of graduation is influenced by many variables, including intelligence. Herrnstein and Murray's analysis of the NLSY79 database showed that, as of the 1980s, approximately 70 percent of the survey participants in the top decile of AFQT scores obtained bachelor's degrees. This fell to 30 percent in the eighth decile and to 10 percent in the fifth decile. A detailed report from the College Board (Kroc et al., 1997) found that graduation rates are nonlinearly (logistic) related to an index composed of SAT, high school GPA, and several demographic variables, including gender and race. People with relatively low scores on the index generally were unlikely to graduate, people with high scores were highly likely to graduate, and the probability of graduation changed markedly between "low average" and "high average" scores.

As was the case for the K–12 system, the findings on the correlation between test scores and college/university success are strikingly consistent. The SAT, the most widely used test in the United States, has a predictive validity of about 0.5. This is probably an underestimate of the correlation between the SAT and an abstract measure of academic ability. This is because students with high SAT scores are more likely to enroll in "tough-grading" courses than students with low SATs, and the SAT–GPA1 correlations will be depressed below what they would have been if all students took the same courses.

In 2021, the UC system dropped the SAT and similar standardized tests from the admissions process. It had been under pressure to do so for years based on concerns that the scores may not be fair to many students. However, a faculty task force that studied the concerns reported, "Test scores are predictive for all demographic groups and disciplines, even after controlling for HSGPA [high school grade point average]. In fact, test scores are better predictors of success for students who are Underrepresented Minority students (URMs), who are first-generation, or whose families are low-income: that is, test scores explain more of the variance in UGPA [undergraduate grade point average] and completion rates for students in these groups." They also noted "the average differences in test scores among groups and expected to find that test score differences explain differences in admission rates. That is not what we found. Instead, the [task force] found that UC admissions practices compensated well

Box 8.3 *(continued)*

for the observed differences in average test scores among demographic groups. This likely reflects UC's use of comprehensive review, as well as UC's practice of referencing each student's performance to the context of their school" (https://senate .universityofcalifornia.edu/_files/underre view/sttf-report.pdf).

These empirical findings appear to undermine the administrative decision that ended the use of these tests. For more about this issue, see Wai et al. (2019).

Box 8.4 Cognitive Tests and Selection Decisions

The college/university admissions process is an example of a personnel selection decision. How useful are entrance examinations, such as the SAT (a high-*g* test that can be used as a proxy measure of intelligence; Table 8.3), in making such decisions? This raises the question of how high a correlation has to be to be useful in practice, whether or not it is "statistically significant." This depends on how the correlation is to be used.

One way of evaluating the size of a correlation is to square it, and then to report it as the proportion of variance accounted for in either variable by predictions using the other variable. In the admissions case, this would be the proportion of variance in grades that could be associated with variance in an admissions test. This is approximately $0.5^2 = 0.25$ multiplied by 100, so one could say that 25 percent of the variance in grades is accounted for by variance in the test score. If the same calculation is applied to the uncorrected correlation between grades and test scores (0.35) in the population of admitted students, only 11 percent of the variance of grades is associated with variance in test scores. However, that is misleading, because the uncorrected correlation is not the appropriate value for this purpose.

But does even the 25 percent figure justify using the test? If the selector uses a screening test, it is possible to predict grades or workplace performance and accept people in order of predicted performance. Unless prediction is perfect (predictive validity = 1), people with the same predicted performance usually turn out to have different actual performances. Students with identical SATs do not all have identical grades. In statistical terms, there is variance around the predicted performance level, and the greater the variance is, the less accurate is the prediction.

However, variance around the predicted performance can never be greater than the variance in the applicant population. So, variance in the *applicant* population can be used to scale the extent to which the prediction is *not* accurate. The ratio I = (Variance around Predicted Value of Aptitude) / (Variance of Aptitude in Applicant Population) represents an "inaccuracy" index, relative to the inaccuracy that would be achieved without using a selection test. It follows that the complement of I, $1 - I$, is an index of accuracy. It can be interpreted as the relative reduction in inaccuracy achieved by using a predictor test. The I index is related to predictive validity by the equation

$$r_p^2 = 1 - I,$$

Box 8.4 *(continued)*

where $p = 1$ or 2, depending on whether you are interested in within-institution or across-institution predictivity. Multiplied by 100, r_p is the percent increase in efficiency achieved by using a screening test. If, as is the case, $r_p = 0.5$, the increase in efficiency is 25 percent.

At this point we can see an argument brewing between the admissions committee and the rejected applicants. Suppose an applicant is rejected and then learns that among accepted applicants, the correlation between SAT and grades is 0.35. How dare the committee reject an applicant on the basis of a test that is only 11 percent better than chance?

The committee's first reply can be that the correlation is not really 0.35; it is 0.5. The applicant's rejoinder is that a 25 percent improvement over chance still is not good enough. But this is not the admissions committee's real argument.

Traditionally, the admissions committee is less interested in the accuracy of individual predictions and more interested in selecting the best possible entering class (although some institutions may pay more attention to the individual case than others). Suppose that the institution has room for only 10 percent of its applicants (a rejection rate of 90 percent). Insofar as is possible, the committee wants to select the top 10 percent of the applicants in terms of academic ability. However, the committee knows only about the top 10 percent of the test scorers, and not the top 10 percent based on actual ability. If $r_p = 1$, the two "top 10 percents" will be the same people; to the

extent that r_p is less than 1, there will be some disagreement.

The success of the selection process will be determined by both the accuracy of the test, r_p, and the rejection rate. If the rejection rate is zero, everyone who wants to enter gets to enter. The accuracy of the test does not matter, because no decision is going to be made using the test score. At the other extreme, suppose there is room for just one person. The person accepted will be the one with the highest test score, and the probability of that person being the person with the highest ability in the applicant pool will depend on the accuracy of the test.

Between these two extremes, the expected quality of accepted applicants is determined by an interaction between r_p and the rejection rate. If the rejection rate is low, the value of the predictive correlation matters very little. If the rejection rate is high, it matters a lot. For example, if the rejection rate is 90 percent, as it is for some elite universities, the use of an entrance examination with a predictive correlation of 0.47 can improve the mean level of aptitude in the entering class from the fiftieth percentile of ability in the applicant population (no test used) to about the seventy-seventh percentile.

Exactly the same reasoning applies to industrial hiring. If the rejection rate is high, a screening examination with predictive validity in the 0.4–0.5 range can substantially improve the selection process, as seen by the employer, the one that will pay the employees' salaries.

served as Secretary of Labor in the Clinton administration, wrote that work has shifted from emphasizing the manipulation of objects to manipulating abstract ideas, varying from programming a robot to analyzing a financial system.

Skill in manipulating complex abstract concepts is a basic definition of intelligence, especially the *g*-factor. It certainly appears that intelligence has become progressively more valuable, at least in terms of employment and better incomes. In fact, the large

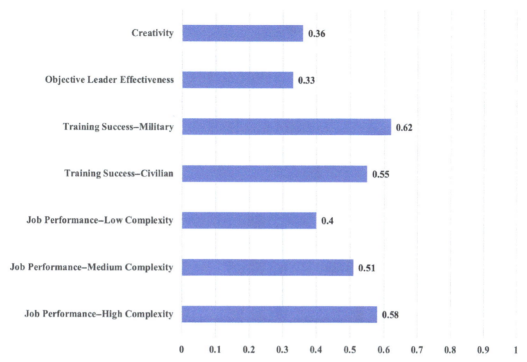

Figure 8.5 Correlations between cognitive ability and measures of work performance. Coefficients are corrected for measurement error and restriction of range, where possible (Kuncel et al., 2004; Ones et al., 2005).

income/wealth gap increases we see today likely result in part from individual differences in intelligence and valuable cognitive abilities (Sandel, 2020; Herrnstein and Murray, 1994).

As we have noted for academic achievement, success in the workplace does not depend solely on intelligence, but how important is it? Figure 8.5 gives an overview; clearly it is important. In the next sections, we will discuss compelling evidence from studies in the military and civilian workplaces.

8.4.1 Studies of Military Enlisted Performance

In the 1980s, the US Department of Defense conducted extensive studies of the prediction and assessment of the job performance of enlisted personnel (Wigdor et al., 1991; Campbell and Knapp, 2001). The predictive measurements taken included cognitive and personality tests and biographical statements of interest. Occupational assessments included job performance ratings and records of promotions, commendations, and disciplinary

actions. Both pencil-and-paper and hands-on performance tests were given. Examinees had to demonstrate their general skills and knowledge as soldiers, sailors, marines, or airmen and their proficiency in their specific occupations. The occupations chosen varied from strictly military positions, such as infantrymen and artillerymen, to jobs with exact counterparts in the civilian world, such as automobile mechanics, clerks, and cooks.

Five dimensions of job performance were identified. Two, *general military proficiency* and *technical proficiency* in one's specialty, were "can do" measures. They evaluated how well a person could do their job when they knew that they were being evaluated. Three factors were "will do" measures. *Discipline* referred to whether the individuals followed regulations and could be relied on to be ready to do their job. *Leadership* referred to the ability to encourage others and to take initiative. *Fitness* referred to personal bearing, appearance, and physical fitness. Generally, these dimensions apply to both the military and civilian workplaces.

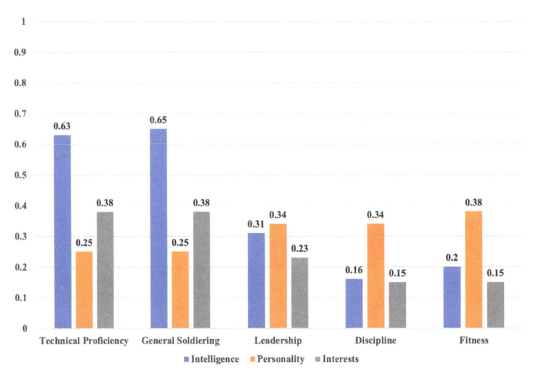

Figure 8.6 Correlations (*y*-axis) between predictors (*x*-axis) and criterion measures in the US Army study of enlisted performance.
Data are from McHenry et al. (1990, table 4).

Figure 8.6 shows the relation between the five factors and measures of personality, biographical interests, and cognitive performance (including scores derived from the ASVAB). The cognitive measures were the best predictors, by far, of the two "can do" factors. Interest and personality measures were the best predictors of the "will do" aspects of job performance.

Steven Hunt, an industrial and organizational psychologist, has pointed out that the first two steps in developing an assessment program in industry are to define the job that you expect employees to do and to determine how you are going to decide whether their performance measures up to these expectations (Hunt and Society for Human Resource Management, 2007). It is not reasonable to expect anyone to excel in all aspects of performance. To the extent that the required job skills are themselves not correlated, it is impossible for one predictor to predict them all. The results shown in Figure 8.6 illustrate Steven Hunt's point.

"Can do" without "will do" is only part of predicting success.

Malcolm J. Ree and colleagues used large samples of air force personal to investigate the importance of the *g*-factor. They reported that in a sample of more than 78,000 enlistees in eighty-two jobs, *g* was the best predictor of training success and job performance. This was not surprising, but they also showed that no other variables added to the prediction appreciably, including variables related to specific jobs (Ree and Earles, 1991; see also Ree et al., 1994). Ree and Carretta (2022) published an update after thirty years of research on general and specific cognitive abilities. The title summarizes the main conclusion: "Still Not Much More than *g*."

The military provides a highly structured workplace, and the workforce is younger than the civilian workforce. What are the relationships between intelligence and performance in the broader civilian workplace?

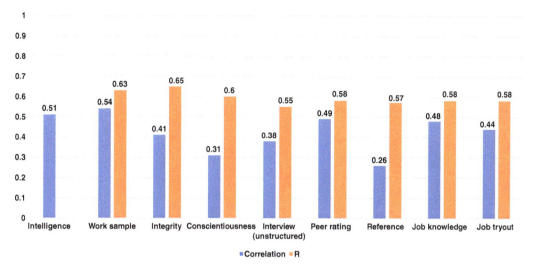

Figure 8.7 The correlations between measures of job performance and measures of general intelligence, and a variety of other assessment measures (*x*-axis). The values shown (*y*-axis) are for the correlation between job performance and the assessment (correlation, blue) and the correlation between job performance and an optimum weighting of the assessment and the assessment of general intelligence (multiple *R*, orange).
Data are from Schmidt and Hunter (1998, table 1).

8.4.2 *Studies of the Civilian Workplace*

Two American industrial-organizational psychologists, John Hunter and Frank Schmidt, have conducted a number of widely cited meta-analyses of predicting workplace success. Figure 8.7, taken from one of their best-known studies (Schmidt and Hunter, 1998), shows that in the blue-collar, clerical, and administrative occupations, the predictive validity of general intelligence alone, averaged over all studies, is 0.51 (corrected for range restriction and unreliability in the job performance criterion; see also Schmidt and Hunter, 2004; Schmidt et al., 2016). Predictive ability can be raised to a maximum validity by adding a test of integrity to general intelligence (multiple *R* = 0.65); conscientiousness is a close second (multiple *R* = 0.60). This illustrates the combined importance of "can do" and "will do" traits.

Schmidt et al.'s (2016) last review of the evidence revealed that the two combinations with the highest multivariate validity for predicting job performance were general intelligence + integrity (*R* = 0.78) and general intelligence + structured interview (*R* = 0.76). The effect size for general intelligence alone was *r* = 0.65 for overall job performance, whereas the value was *r* = 0.67 for occupational training performance. He concluded with a warning: "By using selection methods with low validity, an organization can lose millions in reduced production, reducing revenue and profits. Many employers throughout the world are currently using suboptimal selection methods. ... In a competitive world, these organizations are unnecessarily creating a competitive disadvantage for themselves" (p. 49).

Table 8.4 shows another example of relationships between job skills within a specific industry and general reasoning ability (assessed as a general factor score extracted from a test battery).

The correspondence between the military and civilian data shows that the findings are robust over different situations and different methods of evaluation. The military data were gathered by direct observation of young adults; the civilian figures were based on a meta-analysis of dozens of small studies, covering all age ranges, but none as comprehensive or rigorous as the military studies.

Table 8.4 Selected job skills and their factor
loadings for general intelligence

Items	Factor Loading
Deal with unexpected situations	0.75
Able to learn and recall job-related information	0.71
Able to reason and make judgments	0.69
Able to identify problem situations quickly	0.69
React swiftly when unexpected problems occur	0.67
Able to apply common sense to solve problems	0.66
Able to learn new procedures quickly	0.66
Alert and quick to understand things	0.55
Able to compare information from two or more sources to reach a conclusion	0.49

Note. Developed from 140 petrochemical jobs (Arvey, 1986, adapted from their table 1).

The results are clear. General intelligence tests have a predictive validity of about 0.50 in the workplace, just as they do in academia. No other method of assessment does appreciably better, although many are widely used. Here are five examples:

1. *Work sample.* This is the only type of test with predictive ability about equal to a test of general intelligence. However, it can only be used in some situations. For instance, when musicians audition for places in major symphony orchestras, they play their instruments (sometimes behind a curtain, so that the judges do not know who the candidate is). Work samples have two highly desirable qualities:

they are statistically valid and they are easily justified when assessment methods are challenged. Their drawbacks are that they can be rather expensive and that they can be used only if the candidates for a job have already been trained to do the job.

Combining a work sample and a general intelligence measure increases predictive validity to about 0.63. The increase is not surprising, for by combining a general intelligence measure with a work sample, the employer is simultaneously informed about the prospective employee's general reasoning powers and specific job knowledge.

In personnel selection situations a test this accurate, combined with a high rejection rate, can greatly increase the quality of the employed workforce. Recall that if no screening test is used, the average person hired should have an ability level equal to the fiftieth percentile (median) of the applicant population, regardless of the rejection rate.

If a predictive validity of 0.63 is combined with a rejection rate of 50 percent (half the applicants are hired), the average ability level of a person hired will be at the sixty-ninth percentile of the applicant population.

2. *Unstructured interview.* This is another widely used alternative where the recruiter and the candidate "just chat," so that the recruiter can get a feel for the candidate. The unstructured interview is not very good on its own ($r = 0.38$, corrected) and adds very little to the information gained using a test of general intelligence.

3. *Structured interview* (not shown in Figure 8.7 but included in Schmidt and Hunter's 1998 analyses). This is an interview in which the recruiter has decided beforehand what topics are to be discussed and what information must be provided. The technique requires a careful analysis of the requirements of the position to be filled, before searching for candidates. Structured interviews have good predictive validity,

both on their own and when combined with a test of general intelligence ($r = 0.51$, combined $R = 0.63$).

4. *Job knowledge.* This is usually assessed by performance on a written test, where the questions are chosen to reflect what a job holder should know. This is a face-valid measure; we can reasonably expect bus drivers to know the rules of the road and firefighters to know how to use various pieces of equipment. Job knowledge is not as good a predictor as is general intelligence, but in some circumstances, it adds predictive validity beyond a general intelligence score ($r = 0.48$, combined $R = 0.58$). In terms of the Gf–Gc model of intelligence, a job knowledge test assesses what the applicant knows about the particular situation in which they will be working. Some tests of practical intelligence designed by Robert Sternberg are similar to job knowledge tests, and they show some incremental validity (Sternberg, 2003), but there is controversy about this (Gottfredson, 2003b). For instance, one practical intelligence test designed for Alaskan hunters asked what different pieces of evidence mean as indicators of coming weather (Grigorenko et al., 2004). Such questions measure crystallized intelligence (Gc) within a specialized context (for expansions on this point, see Gottfredson, 2003a; Hunt, 2008).

5. *Situational judgment test.* An examinee is asked what they would do in a realistic, difficult situation. An applicant for a middle management position, for example, could be asked how they would inform their supervisor that the supervisor's pet project was not working. This kind of test draws on both a cognitive skill narrowly defined and the examinee's social skills. Situational judgment tests add an additional 0.06 to the predictive validity achieved by a cognitive test alone – not a large amount, but perhaps enough to be worthwhile in some large-scale assessment programs (Schmidt et al., 2016). It is worth noting that a situational judgment test asks the examinee what they would do in a hypothetical situation. It does not immediately follow that that is what the examinee would do if placed in an actual situation with emotional and/or job retention risk.

The data presented so far are based mostly on studies of blue-collar and white-collar jobs, up to the lower managerial level. In this population of occupations, the correlation between general intelligence test scores and job performance generally increases with increasing job complexity (Gottfredson, 1997, 2002). Given this fact, it would be reasonable to expect the correlation to be still higher for high-level managerial, executive, and professional positions where complex, abstract reasoning is even more important. However, there are reasons not to assume a straightforward extrapolation of the results to the managerial/professional class.

Many studies of high-level occupations report the observed correlation between test scores and measures of job performance but cannot correct for selection restriction because there are no data on the applicant population. This is a serious problem, because selection effects are likely large. High-level positions are quite competitive and are virtually always filled by people in the upper end of the intelligence range. It is also difficult to find a measure of how well a professional or executive is doing, beyond gross judgments of satisfactory or unsatisfactory performance. As *Fortune* magazine repeatedly shows in its annual survey of executive salaries, the correlations between executive compensation and objective measures of company performance are close to zero. Physicians, attorneys, and other professionals are evaluated periodically, but the ratings are often limited to certification of competence without any further differentiation.

There is also the problem of the multidimensionality of the criterion. People who occupy high-level positions are typically asked to do a number of different tasks. These range from high-level planning to public relations, face-to-face leadership, and negotiations. The relative importance of different tasks varies greatly across occupations and even within an occupation. It is not surprising that there is resistance to the idea that any unidimensional measure could predict performance at high levels.

One way to evaluate highly intelligent people using the conventional psychometric paradigm is to use harder tests like the Raven's Advanced Progressive Matrices (RAPM) and the Miller Analogy Test. These high g-loaded tests predict job performance with a predictive ability on the order of about 0.40 (Kuncel et al., 2004). This is less than the level of prediction obtained for other skilled work but still shows that intelligence is an important factor for predicting success at this level of employment.

8.4.3 *General Intelligence and Specific Jobs*

In her classic paper "Why *g* Matters: The Complexity of Everyday Life," Gottfredson (1997) used intelligence test scores from job applicants to construct a "life's chances" chart for various occupations. She estimated average IQ (or its equivalent) of applicants for different classes of occupations, as shown in Table 8.5.

The table also includes estimates for comparable occupations based on Thorndike and Hagen's (1959) data taken forty years earlier from USAF aviation cadets and for the 2003 revision of the Wonderlic test (Wonderlic-Corporation, 2007). Wonderlic scores have been converted to estimated IQ scores (see Dodrill, 1981; Dodrill and Warner, 1988).

There is a striking similarity between Gottfredson's estimates, the Wonderlic estimates, and the estimates that Thorndike and Hagen had made forty years earlier. The data were gathered using different sampling methods, in different workplaces separated by over half a century, and the tests used were quite different. (Thorndike and Hagen's general reasoning factor was extracted from a battery of subtests that took several hours to complete; Gottfredson's estimates were based on data from the Wonderlic Personnel Test, WPT; the Wonderlic estimates are based on the revised version [WPT-R, with data from 2003]. Nevertheless, the estimates are basically the same.)

Both educational level and intelligence are substantial predictors of accomplishment in the workplace. The two are highly correlated. This is hardly surprising, since both measures

Table 8.5 Examples of intelligence levels associated with different occupations and methods of training

Training and qualification method	Typical position	Gottfredson's estimated IQ score	Estimates based on Thorndike and Hagen data	Wonderlic with comparison occupation
Explicit, hands-on training	Assembler	80–90	<95	88–104, electro-mechanical assembler
Mastery learning, hands-on training	Police officer	95–105	95–100	100–114, police officer and sheriff
Formal college instruction	Accountant	110–120	105–110	102–120, accountant
Graduate instruction, gathers own information	Attorney	120+	110–115	110–124, executive

Note. The two columns on the right show similar estimates from Thorndike and Hagen (1959) and Wonderlic scores.

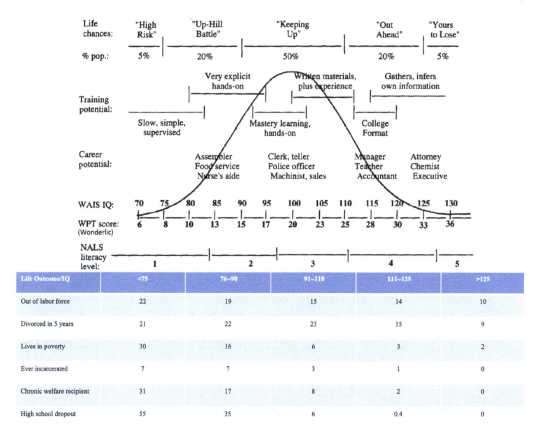

Figure 8.8 Some practical implications of IQ scores across the normal distribution for learning and vocations.
Adapted from Gottfredson (1997).

Life Outcome/IQ	<75	76–90	91–110	111–125	>125
Out of labor force	22	19	15	14	10
Divorced in 5 years	21	22	23	15	9
Lives in poverty	30	16	6	3	2
Ever incarcerated	7	7	3	1	0
Chronic welfare recipient	31	17	8	2	0
High school dropout	55	35	6	0.4	0

are based on the development and display of cognitive abilities and skills.

This is especially true across fields, for the vast majority of the more lucrative occupations have as an entry requirement at least a college education. The best-paid professional occupations require substantial graduate-level training. Even entry routes to skilled trades, such as auto mechanics, involve academic certification through community college or other professional training programs. Because educational attainment is strongly related to intelligence, any variable that is correlated with educational attainment will also correlate with intelligence. To the extent that intelligence is required to get through a rigorous training program, intelligence and education are entwined

What do the estimates in Table 8.5 prove? Not much – they merely illustrate that some

occupations, on average, have applicants with higher general intelligence scores than other occupations. To be successful in an occupation, however, is another issue.

Another way of determining the importance of intelligence in the workplace is to analyze the relative value of cognitive skills for different jobs. The US Department of Labor (DOL) describes more than 12,000 jobs and rates the extent to which they require certain skills. The skills range from general reasoning ability to finger dexterity. The rating system was originally incorporated into a descriptive volume called the *Dictionary of Occupational Titles* (DOT). The DOT has been superseded by an online, interactive system called O*NET, a considerable expansion over the DOT designed primarily to help job seekers. As a side benefit, it contains a massive amount of data

Table 8.6 Economic outcomes at different levels of National Adult Literacy Survey literacy (age 16–65 years)

Prose level	Scores	Out of labor force	Lives in poverty	Uses food stamps	Employed full-time	Median weekly wages	Employed in professional/ managerial job
1	<225	52	43	17	30	240	5
2	226–275	35	23	13	43	281	12
3	276–325	25	12	6	54	339	23
4	326–375	17	8	3	64	465	46
5	376–500	11	4	1	72	650	70

Note. Values are percentages. From Kirsch et al. (1993).

available to researchers interested in issues involving workforce skills.

Gottfredson (1986) used the original DOT system to construct a space of jobs. She identified five main classes of occupations – those dealing with physical relations, social and economic relations, maintaining bureaucratic order, and performing. Within each class, there were clusters of jobs that required similar skills. Individual jobs within a cluster could be associated with a pattern of mental abilities, as defined by the DOL's General Aptitude Test Battery and its four dimensions: general reasoning, verbal reasoning, numerical skills, and spatial skills.

The analysis indicated that general intelligence was by far the biggest driver of variations in cognitive skills across occupations. Verbal, numerical, and visuospatial skills were important in some occupations, but they accounted for much less of the variation in the descriptions of occupational requirements than did differences in requirements for general reasoning.

Figure 8.8 illustrates "life chances" and the relationships among IQ, training methods, and jobs estimated by Gottfredson and by the Wonderlic Personnel Test (WPT) as well as the National Literacy Survey (NLS). The NLS has five categories that estimate a rank order with IQ scores. Table 8.6 shows the relationship between the NLS categories (category 5 corresponds to higher IQ) and

several occupational and social variables (Kirsch et al., 1993).

This work remains the most comprehensive demonstration of how g is related to jobs. It is consistent with the military data and subsequent research in personnel and organizational psychology. Higher intelligence generally is rewarded in the workplace with higher incomes because more jobs require complex thinking. The data also show that success requires other factors.

The evidence discussed so far in this chapter has been known for decades within the scientific research community. The referenced meta-analyses and empirical reviews led most researchers to conclude that general intelligence is the best single predictor of occupational performance. However, Sackett and colleagues (2021) revisited the meta-analytic estimates of validity in personnel selection. Their key conclusion was that there has been systematic overcorrection for restriction of range. When computations are redone using updated criteria and new studies, in their view, intelligence is no longer the best predictor in personnel selection: "most of the same selection procedures that ranked high in prior summaries remain high in rank, but mean validities reduced from .10 to .20 points" (p. 2040). The usual selection procedures are still useful, but the relationships are smaller. For example, cognitive ability tests showed a validity estimate of 0.31 instead of 0.51, structured interviews showed a validity estimate of 0.42 instead of 0.51, and integrity tests showed a value of 0.31 instead of 0.41. Their findings suggest a reframing of past research: "while Schmidt and Hunter positioned cognitive ability as the focal predictor, with others evaluated in terms of their incremental validity over cognitive ability, one might propose structured interview as the focal predictor" (p. 2062). This is an interesting suggestion, but it must be noted that job interviews are job-specific measures. Also, it is reasonable to expect that researchers will react soon to the authors' invitation: "we eagerly await the availability of the data that will permit further refinement of our estimates" (p. 2066). This kind of periodic reevaluation of the weight of evidence is an important aspect of how science works.

In the next section, we consider additional aspects of personal and social adjustment.

8.5 Health, Personal/Social Adjustment, and "Emotional Intelligence"

All the problems we noted at the beginning of this chapter for defining criteria of success are even more difficult for the general category of personal and social adjustment. Like academic and workplace success, all the variables used as indicators of adjustment are multidetermined. Intelligence is only one of them, and some variance in intelligence measures may be influenced by these other variables.

Table 8.7 shows that levels of intelligence are associated with several indicators of personal/social adjustment. Whether such data provide insights into the causes of adjustment problems remains an open question, since other potentially confounding variables, like years of education, are not shown. Nonetheless, we think any discussion of personal and social adjustment benefits from

Table 8.7 Young adults with particular life outcomes by IQ level

Life outcome	<75, very low	76–90, low	91–110, medium	111–125, high	>125, very high	Ratio of low to high
Married by age 30	72	81	81	72	67	8:7
Out of labor force 1+ mo/yr (men)	22	19	15	14	10	4:3
Unemployed 1+ mo/yr (men)	12	10	7	7	2	3:2
Divorced in 5 years	21	22	23	15	9	3:2
Children below IQ 75 (mothers)	39	17	6	7	–	2:1
Had child when not married (women)	32	17	8	4	2	4:1
Lives in poverty	30	16	6	3	2	5:1
Went on welfare after first child (women)	55	21	12	4	1	5:1
Ever incarcerated (men)	7	7	3	1	0	7:1
Chronic welfare recipient (mothers)	31	17	8	2	0	8:1
High school dropout	55	35	6	0.4	0	88:1

Note. Values are percentages. From Herrnstein and Murray (1994).

considering individual differences in general intelligence, especially when addressing ways to minimize negative adjustment outcomes.

There is no more definitive health outcome than death, and as we saw in Chapter 1 and Figure 1.2, IQ scores predict mortality. That finding was from the Scottish Mental Survey, which tested virtually every eleven-year-old student in Scotland in 1921 and 1936. A follow-up more than seventy years later identified who was still alive and when and how the others had died. Figure 8.9 shows that each cause of death was higher in the lowest IQ group.

These were surprising differences, since Scotland has a national health care system that included access for all these individuals. Several other explanations for such associations have been suggested, including a genetic commonality between overall health and IQ (Arden et al., 2015). For now, any causal relationships are a matter of speculation (Hill et al., 2019), albeit the empirical fact of their association is robust enough.

In addition to intelligence, personality traits are among the most studied and psychometrically sophisticated category of variables that affect successful navigation of school, work, and personal/social adjustment. There is a long history of personality research that has morphed to include the study of emotional intelligence (EI; Salovey and Mayer, 1990). Generally, EI is defined as "a type of social intelligence that involves the ability to monitor one's own and others' emotions, to discriminate among them, and to use the information to guide one's thinking and actions" (Salovey and Mayer, 1990, p. 190; see also Mayer and Salovey, 1993). There are a variety of self-report and trait-like measures to assess EI.

There is a vast research literature on EI that makes many claims about its importance for success in many endeavors. However, this literature is based on different ways to assess EI. Results of meta-analyses present a complex picture, but generally, EI measures show weak correlations to intelligence measures, especially for g (O'Connor et al., 2019).

The most validated and widely used framework of core personality traits, derived from extensive factor analyses, is the so-called Big Five: *openness to experience, agreeableness, conscientiousness, extraversion,* and *neuroticism.* Only openness consistently shows a modest correlation to the g-factor (Stankov, 2018).

Since personality/EI and intelligence measures appear to be tapping weakly related domains, it is important to investigate whether they can be combined to increase predictions of success. Many studies show correlations between various measures of personality/EI and numerous criterion measures of success or adjustment in many contexts. Generally, based on meta-analyses, these correlations are smaller than the ones for intelligence test scores (Strickhouser et al., 2017; Trapmann et al., 2007).

In this chapter, our interest is narrowly focused on personality/EI studies that also include a measure of cognitive intelligence, especially g. Without measures from both the personality/EI and intelligence domains, we cannot determine which variables account for more predictive variance. We already noted some workplace studies with both cognitive and personality variables that showed little if any additional predictive value to the personality measures.

Meta-analyses of studies with variables from both domains would be particularly valuable for establishing a weight of evidence. There are not many such studies. In one comprehensive study, composite measures of EI added a small amount of predicted variance in job performance compared to measures of cognitive ability; personality measures added virtually none (O'Boyle et al., 2011).

Manfred Amelang and Ricarda Steinmayr (2006) published a study titled "Is There a Validity Increment for Tests of Emotional Intelligence in Explaining the Variance of Performance Criteria?" Their key finding was that EI "could not explain any variance in the criteria (educational level and social status) beyond psychometric intelligence and conscientiousness" (p. 459; Figure 8.10).

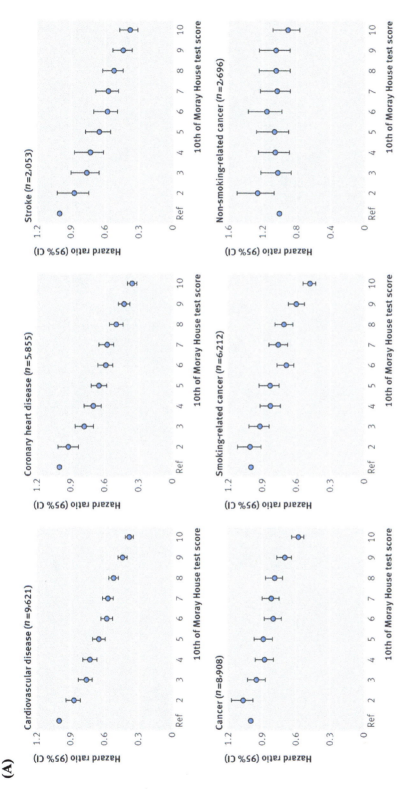

Figure 8.9 Association between intelligence at age eleven and major causes of death (age- and sex-adjusted hazard ratios and 95 percent confidence intervals) to age seventy-nine in the Scottish Mental Survey 1947 (Calvin et al., 2017).

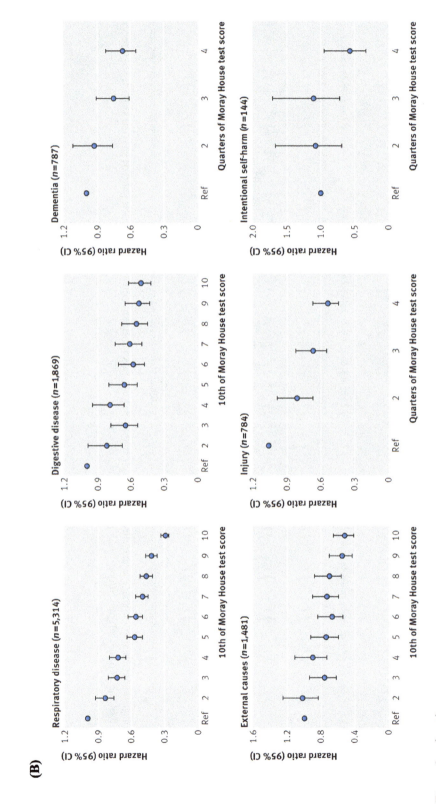

Figure 8.9 (cont.)

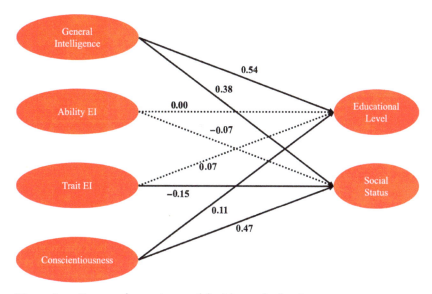

Figure 8.10 Structural equation model with standardized parameter estimates that predict education level and social status from measures of general intelligence, emotional intelligence (EI), trait EI, and conscientiousness. Neither EI measure is predictive. General intelligence predicts both variables with large effect sizes (0.54 and 0.38), and conscientiousness predicts social status (0.47), but the prediction of education level is weak (0.11). Dotted lines indicate statistically nonsignificant results.

From Amelang and Steinmayr (2006).

That is not to say that personality/EI variables are unimportant, but it does indicate caution to avoid overinterpreting their empirical importance relative to intelligence. For the focus of this chapter on intelligence in everyday life, the data on personality/EI are of interest as additional predictive variables to consider. They might be important for interpersonal relationships across contexts in their own right, irrespective of their relationship to intelligence.

A meta-analysis by MacCann and colleagues (2020) included 42,529 individuals who completed measures of intelligence, personality, and EI. Their criterion measure was academic performance. Findings indicated that intelligence explained between 58 and 69 percent of the common variance (25–47 percent) in the criterion. Conscientiousness explained around 20 percent of this common variance, whereas EI explained between 4 and 15 percent. These results show some influence for EI, but not much.

Another large-scale meta-analysis, published by Dimitri van der Linden and

colleagues (2017), found a strong correlation between the general factor of personality (GPF) and EI ($r = 0.86$) (Figure 8.11). What might this result mean? In their own words, "high-GFP [general factor of personality] individuals score higher on the MSCEIT [a widely administered measure of emotional intelligence] suggesting greater social knowledge and an ability to regulate behavior in order to achieve social goals" (p. 45). This sounds like crystallized intelligence (Gc).

The findings by Margaret Beier and Phillip Ackerman (2003) addressing the determinants of health knowledge are consistent with this presumption (Figure 8.12). They reported that the correlation between Gc and health knowledge was $r = 0.90$, whereas SES (as estimated through education and income) failed to predict health knowledge. Their main conclusion was this: "The domain of health knowledge is similar to other domains including job knowledge and academic knowledge (therefore) health education interventions and health care instructions must be accessible to those of

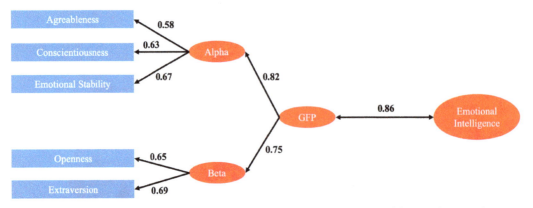

Figure 8.11 Hierarchical model showing the correlation between the general factor of personality (GFP), derived from the Big Five traits (blue), and emotional intelligence. Alpha summarizes personality traits related to the tendency to act in socially desirable ways; beta summarizes personality traits related to the tendency to seek novel and reinforcing experiences (van der Linden et al., 2017).

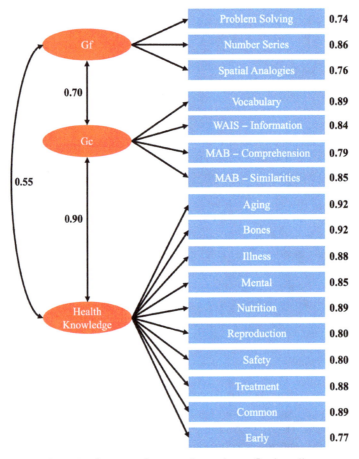

Figure 8.12 Confirmatory factor analysis relating fluid intelligence (Gf), crystallized intelligence (Gc), and health knowledge. This research shows that the latent health knowledge factor is highly correlated (r = 0.90) to crystallized intelligence, although the standardized tests and everyday life information items (blue) from which they are derived appear quite different (Beier and Ackerman, 2003).

all ability levels to be effective" (p. 447). This perspective is also discussed extensively by Linda Gottfredson (2004) in her article "Intelligence: Is It the Epidemiologists' Elusive 'Fundamental Cause' of Social Class Inequalities in Health?" In the same vein, the potential role of general intelligence for explaining lower or greater general vulnerability to psychopathology (*p*-factor) is addressed by Avshalom Caspi and Terrie Moffitt (2018) in their thought-provoking review article "All for One and One for All: Mental Disorders in One Dimension." They write, "Individuals with higher levels of *p* fare less well on tests requiring attention, concentration, and mental control, as well as visual-perceptual processing speed and visual-motor coordination. These deficits are not simply a consequence of lifelong disorders: they are already present in early life, before the onset of most disorders" (p. 835).

The content of many life problems may be less important than the cognitive complexity required for solving them and reaching the pursued goals. And success in this regard is likely mediated by the integrative psychological factor we call general intelligence.

8.6 Conclusion

Anyone who asserts that intelligence, assessed by standard cognitive tests, is irrelevant to performance in either academia or the workplace, or for general adjustment, is simply wrong. So is someone who says that intelligence is less important than personality and emotional characteristics. The data reviewed in this chapter show that general intelligence is the single best predictor of academic and occupational success. Nevertheless, cognitive tests are not perfect predictors. Predictive validity correlations are in the 0.4–0.6 range, far from perfect. Personality and EI apparently do not account for as much of the rest as commonly thought. Hard work, luck, and many other factors, difficult to quantify, surely are important contributors to any measure of success and everyday life performance.

The same message can be extracted from studies of extremes at the ends of the normal distribution of IQ. On average, people with high IQ get better jobs and make more money. As we will discuss in Box 10.1, the Terman and Hopkins studies undermined the stereotype of a gifted person as a neurotic introvert with health problems. The contrast between the Hopkins study of 1 in 200 and 1 in 10,000 students shows that cognitive tests have predictive power even at the highest levels.

At the opposite end of the distribution, people in the low range generally have difficulty with school, especially if they are placed in an academic track, and usually take occupations that do not make high cognitive demands. They earn less than people in the average to high intelligence range and are more likely to require some form of governmental assistance.

It cannot be stressed too strongly that these are trends. Every study of the extremes comes up with exceptions. There are people who do quite well although they had modest cognitive test scores (McGue et al., 2022), and there are stunning examples of people with high scores who never live up to their promise (Kell et al., 2022). The basic point is that considerations about how best to increase any person's probability of success should include what we know about intelligence.

Given these empirical findings, why is there a common belief that intelligence does not count for very much in life? Perhaps people without statistical training have difficulty grasping the concept of something that increases the probability of an event but does not establish its certainty. Thus, if we can think of examples of people with high test scores doing stupid things, or people with low test scores doing smart things, that is taken as proof that the predictors do not work.

People are heavily influenced by their personal experiences. Charles Murray (2012), in his book *Coming Apart*, points out that we live in a society that is sharply stratified by intelligence. College-educated people, by and large, interact with other college-educated people, and people with high school educations interact with each other. This is segregation by IQ – formally measured or not, it doesn't matter. Within the restricted range of intelligence that people observe directly in their circle of friends, other variables may account for more variation in performance than does intelligence. It is only when we step back and look at the big picture across the full range that the importance of intelligence becomes clear. It is important to keep this perspective in mind.

People in postindustrial societies often hold the belief that if one works hard, one can be a success (Sandel, 2020). This reasonable sentiment has morphed into "you can be anything you want to be" if you work hard enough. Hard work can lead to success, but an overemphasis on hard work is a trap for frustration for individuals without the cognitive resources necessary for success in specific endeavors they may (unrealistically) choose (Moreau et al., 2019). The point was stressed by Asbury and Plomin (2014, p. 86): "motivation has been shown to predict school course choice over and above grades. In other words, children pursue subjects they enjoy rather than ones they are good at."

These elements often collide with the laudable desire for equal opportunity for every person. The data on the role of intelligence in everyday life strongly imply that provision of equal opportunity for all individuals will not produce an equal distribution of social and economic rewards for everyone.

People who are deeply committed to social equality find such a conclusion offensive. It is difficult for them to argue about the empirical facts of differential distribution of intelligence test scores and its importance. Their counternarrative often is based on the denial of any validity of "intelligence" or its assessment and any relevance to social outcomes. The evidence presented in this book, however, undermines that counternarrative.

This conclusion has been clear for some time. Frank Schmidt (2002), a pioneer in industrial/organizational (I/O) psychology, reviewed the evidence about intelligence in the workplace and concluded, "The purely empirical research evidence in I/O psychology showing a strong link between GCA [general cognitive ability] and job performance is so massive that there is no basis for questioning the validity of GCA as a predictor of job performance. . . . These findings do not reflect the kind of world most of us were hoping for and hence are not welcome to many people. As a result, we see many attempts – desperate attempts – to somehow circumvent these research findings and reach more palatable conclusions. . . . These attempts are in many ways understandable; years ago, I was guilty of this myself. However, in light of the evidence that we now have, these attempts are unlikely to succeed. There comes a time when you just have to come out of denial and objectively accept the evidence" (pp. 207–208).

Solving or ameliorating the myriad of social problems requires a sound basis of factual information. What we know about intelligence can be a springboard, not a barrier. Fixing problems requires facing reality and the willingness to discuss empirical data that may lead in uncomfortable directions. It does no one any good to ignore or delegitimize what we know about the *probabilistic* role intelligence plays in everyday life for all individuals.

8.7 Summary

Everyday life presents many complex challenges. Intelligence helps navigate them. More intelligence, however, is no guarantee of better navigation, but the available data indicate that more intelligence is more useful in everyday life than less intelligence. This empirical fact raises provocative questions: Can an individual's intelligence be increased? Do we have a moral obligation to do so if we know how? Is increasing g the ultimate answer to Hunt's question about keeping a country's workforce competitive in the modern global economy (Hunt, 1995)? In some ways, the answers are what this entire book is about, and they are the focus of Chapter 13. But first, the next chapters tackle important and delicate topics that are socially relevant: intelligence and sex, age, and country/ancestry differences.

8.8 Questions for Discussion

8.1 What do you think about the statement that life is one long intelligence test?

8.2 Can you explain why intelligence tests predict so many aspects of everyday life?

8.3 Have you ever considered that you were not smart enough (or too smart) for a specific job or career path?

8.4 What do you think about the number of people attaining each level in the National Literacy Survey?

8.5 Why do you think intelligence is related to health outcomes, including mortality?

References

Amelang, M., & Steinmayr, R. 2006. Is there a validity increment for tests of emotional intelligence in explaining the variance of performance criteria? *Intelligence*, 34, 459–468.

Arden, R., Luciano, M., Deary, I. J., et al. 2015. The association between intelligence and lifespan is mostly genetic. *International Journal of Epidemiology*, 45, 178–185.

Arvey, R. D. 1986. General ability in employment – a discussion. *Journal of Vocational Behavior*, 29, 415–420.

Asbury, K., & Plomin, R. 2014. *G is for genes: The impact of genetics on education and achievement.* Chichester, UK: Wiley-Blackwell.

Beier, M. E., & Ackerman, P. L. 2003. Determinants of health knowledge: An investigation of age, gender, abilities, personality, and interests. *Journal of Personality and Social Psychology*, 84, 439–448.

Belsky, J., Caspi, A., Moffitt, T. E., & Poulton, R. 2020. *The origins of you: How childhood shapes later life.* Cambridge, MA: Harvard University Press.

Calvin, C. M., Batty, G. D., Der, G., et al. 2017. Childhood intelligence in relation to major causes of death in 68 year follow-up: Prospective population study. *British Medical Journal*, 357, 1–14.

Campbell, J. P., & Knapp, D. J. 2001. *Exploring the limits of personnel selection and classification.* Mahwah, NJ: Erlbaum.

Caspi, A., & Moffitt, T. E. 2018. All for one and one for all: Mental disorders in one dimension. *American Journal of Psychiatry*, 175, 831–844.

Ceci, S. J., & Liker, J. K. 1986. A day at the races: A study of IQ, expertise, and cognitive-complexity.

Journal of Experimental Psychology: General, 115, 255–266.

Cohen, J. 1988. *Statistical power analysis for the behavioral sciences.* Hillsdale, NJ: Erlbaum.

Coyle, T. R., & Pillow, D. R. 2008. SAT and ACT predict college GPA after removing g. *Intelligence*, 36, 719–729.

Cumming, G. 2014. The new statistics: Why and how. *Psychological Science*, 25, 7–29.

Deary, I. J., Strand, S., Smith, P., & Fernandes, C. 2007. Intelligence and educational achievement. *Intelligence*, 35, 13–21.

Detterman, D. K., & Spry, K. M. 1988. Is it smart to play the horses – comment. *Journal of Experimental Psychology: General*, 117, 91–95.

Dodrill, C. B. 1981. An economical method for the evaluation of general intelligence in adults. *Journal of Consulting and Clinical Psychology*, 49, 668–673.

Dodrill, C. B., & Warner, M. H. 1988. Further studies of the Wonderlic Personnel Test as a brief measure of intelligence. *Journal of Consulting and Clinical Psychology*, 56, 145–147.

Frey, M. C., & Detterman, D. K. 2004. Scholastic assessment or g? The relationship between the scholastic assessment test and general cognitive ability. *Psychological Science*, 15, 641.

Funder, D. C., & Ozer, D. J. 2019. Evaluating effect size in psychological research: Sense and nonsense. *Advances in Methods and Practices in Psychological Science*, 2, 156–168.

Gordon, R. A. 1997. Everyday life as an intelligence test: Effects of intelligence and intelligence context. *Intelligence*, 24, 203–320.

Gottfredson, L. S. 1986. Occupational aptitude patterns map: Development and implications for a theory of job aptitude requirements. *Journal of Vocational Behavior*, 29, 254–291.

Gottfredson, L. S. 1997. Why g matters: The complexity of everyday life. *Intelligence*, 24, 79–132.

Gottfredson, L. S. 2002. g: Highly general and highly practical. In Sternberg, R. J., & Grigorenko, E. (eds.), *The general factor of intelligence: How general is it?* Mahwah, NJ: Erlbaum.

Gottfredson, L. S. 2003a. Dissecting practical intelligence theory: Its claims and evidence. *Intelligence*, 31, 343–397.

Gottfredson, L. S. 2003b. On Sternberg's "Reply to Gottfredson." *Intelligence*, 31, 415–424.

Gottfredson, L. S. 2004. Intelligence: Is it the epidemiologists' elusive "fundamental cause" of social class inequalities in health? *Journal of Personality and Social Psychology*, 86, 174–199.

Grigorenko, E. L., Meier, E., Lipka, J., et al. 2004. Academic and practical intelligence: A case study of the Yup'ik in Alaska. *Learning and Individual Differences*, 14, 183–207.

Heckman, J., Hohmann, N., Smith, J., & Khoo, M. 2000. Substitution and dropout bias in social experiments: A study of an influential social experiment. *Quarterly Journal of Economics*, 115, 651–694.

Heckman, J., Smith, J., & Taber, C. 1998. Accounting for dropouts in evaluations of social programs. *Review of Economics and Statistics*, 80, 1–14.

Herrnstein, R. J., & Murray, C. 1994. *The bell curve: Intelligence and class structure in American life*. New York: Free Press.

Hill, W. D., Harris, S. E., & Deary, I. J. 2019. What genome-wide association studies reveal about the association between intelligence and mental health. *Current Opinion in Psychology*, 27, 25–30.

Hunt, E. 1995. *Will we be smart enough? A cognitive analysis of the coming workforce*. New York: Russell Sage Foundation.

Hunt, E. 2008. Applying the theory of successful intelligence to education: The good, the bad, and the ogre: Commentary on Sternberg et al. (2008). *Perspectives on Psychological Science*, 3, 509–515.

Hunt, S. T., & Society for Human Resource Management. 2007. *Hiring success: The art and science of staffing assessment and employee selection*. San Francisco, CA: John Wiley.

Hunter, J. E., & Schmidt, F. L. 1990. *Methods of meta-analysis: Correcting error and bias in research findings*. Newbury Park, CA: SAGE.

Jensen, A. R. 1998. *The g factor: The science of mental ability*. Westport, CT: Praeger.

Kell, H. J., McCabe, K. O., Lubinski, D., & Benbow, C. P. 2022. Wrecked by success? Not to worry. *Perspectives on Psychological Science*, 17, 1291–1321.

Kirsch, I. S., Jungeblut, A., Jenkins, L., & Kolstad, A. 1993. *Adult literacy in America: A first look at the results of the National Adult Literacy Survey*. Washington, DC: US Department of Education.

Koenig, K. A., Frey, M. C., & Detterman, D. K. 2008. ACT and general cognitive ability. *Intelligence*, 36, 153–160.

Kroc, R., Howard, R., Hull, P., & Woodard, D. 1997. Graduation rates: Do students' academic program choices make a difference? https://files.eric.ed.gov/fulltext/ED417677.pdf.

Kuncel, N. R., Hezlett, S. A., & Ones, D. S. 2004. Academic performance, career potential, creativity, and job performance: Can one construct predict them all? *Journal of Personality and Social Psychology*, 86, 148–161.

Lakens, D., Page-Gould, E., Van Assen, M. A. L. M., et al. 2017. Examining the reproducibility of meta-analyses in psychology: A preliminary report. https://doi.org/10.31222/osf.io/xfbjf.

Liker, J. K., & Ceci, S. J. 1987. IQ and reasoning complexity: The role of experience. *Journal of Experimental Psychology: General*, 116, 304–306.

MacCann, C., Jiang, Y., Brown, L. E. R., et al. 2020. Emotional intelligence predicts academic performance: A meta-analysis. *Psychological Bulletin*, 146, 150–186.

Mammadov, S. 2021. Big Five personality traits and academic performance: A meta-analysis. *Journal of Personality*, 90, 1–34.

Mayer, J. D., & Salovey, P. 1993. The intelligence of emotional intelligence. *Intelligence*, 17, 433–442.

McGue, M., Anderson, E. L., Willoughby, E., et al. 2022. Not by g alone: The benefits of a college education among individuals with low levels of general cognitive ability. *Intelligence*, 92, 101642.

McHenry, J. J., Hough, L. M., Toquam, J. L., Hanson, M. A., & Ashworth, S. 1990. Project A validity results: The relationship between predictor and criterion domains. *Personnel Psychology*, 43, 335–354.

Moreau, D., Macnamara, B. N., & Hambrick, D. Z. 2019. Overstating the role of environmental factors in success: A cautionary note. *Current Directions in Psychological Science*, 28, 28–33.

Murray, C. 2012. Coming apart: The state of white America, 1960–2010. *New York Review of Books*, 59, 21–23.

Nuijten, M. B., Van Assen, M. A. L. M., Augusteijn, H. E. M., Crompvoets, E. A. V., & Wicherts, J. M. 2020. Effect sizes, power, and biases in intelligence research: A meta-meta-analysis. *Journal of Intelligence*, 8, 1–24.

O'Boyle, E. H., Humphrey, R. H., Pollack, J. M., Hawver, T. H., & Story, P. A. 2011. The relation between emotional intelligence and job performance: A meta-analysis. *Journal of Organizational Behavior*, 32, 788–818.

O'Connor, P. J., Hill, A., Kaya, M., & Martin, B. 2019. The measurement of emotional intelligence: A critical review of the literature and recommendations for researchers and practitioners. *Frontiers in Psychology*, 10, 1–19.

Ones, D. S., Viswesvaran, C., & Dilchert, S. 2005. Personality at work: Raising awareness and correcting misconceptions. *Human Performance*, 18, 389–404.

Ree, M. J., & Carretta, T. R. 2022. Thirty years of research on general and specific abilities: Still not much more than g. *Intelligence*, 91, 101617.

Ree, M. J., & Earles, J. A. 1991. The stability of g across different methods of estimation. *Intelligence*, 15, 271–278.

Ree, M. J., Earles, J. A., & Teachout, M. S. 1994. Predicting job performance: Not much more than g. *Journal of Applied Psychology*, 79, 518–524.

Reich, R. B. 1991. *The work of nations: Preparing ourselves for 21st-century capitalism.* New York: Alfred A. Knopf.

Richardson, K., & Norgate, S. H. 2015. Does IQ really predict job performance? *Applied Developmental Science*, 19, 153–169.

Roberts, R. D., Goff, G. N., Anjoul, F., et al. 2000. The Armed Services Vocational Aptitude Battery (ASVAB): Little more than acculturated learning (Gc)!? *Learning and Individual Differences*, 12, 81–103.

Roth, B., Becker, N., Romeyke, S., et al. 2015. Intelligence and school grades: A meta-analysis. *Intelligence*, 53, 118–137.

Sackett, P. R. 2021. Reflections on a career studying individual differences in the workplace. *Annual Review of Organizational Psychology and Organizational Behavior*, 8, 1–18.

Sackett, P. R., Kuncel, N. R., Arneson, J. J., Cooper, S. R., & Waters, S. D. 2009. Does socioeconomic status explain the relationship between admissions tests and post-secondary academic performance? *Psychological Bulletin*, 135, 1–22.

Sackett, P. R., Shewach, O. R., & Dahlke, J. 2020. The predictive value of general intelligence. In Sternberg, R. J. (ed.), *Human intelligence: An introduction.* Cambridge: Cambridge University Press.

Sackett, P. R., Zhang, C., Berry, C. M., & Lievens, F. 2021. Revisiting meta-analytic estimates of validity in personnel selection: Addressing systematic overcorrection for restriction of range. *Journal of Applied Psychology*, 107, 2040–2068.

Salovey, P., & Mayer, J. D. 1990. Emotional intelligence. *Imagination, Cognition and Personality*, 9, 185–211.

Sandel, M. J. 2020. *The tyranny of merit: What's become of the common good?* New York: Farrar, Straus, and Giroux.

Schmidt, F. L. 2002. The role of general cognitive ability and job performance: Why there cannot be a debate. *Human Performance*, 15, 187–210.

Schmidt, F. L., & Hunter, J. E. 1998. The validity and utility of selection methods in personnel psychology: Practical and theoretical implications of 85 years of research findings. *Psychological Bulletin*, 124, 262–274.

Schmidt, F. L., & Hunter, J. 2004. General mental ability in the world of work: Occupational attainment and job performance. *Journal of Personality and Social Psychology*, 86, 162–173.

Schmidt, F., Oh, I.-S., & Shaffer, J. A. 2016. The validity and utility of selection methods in personnel psychology: Practical and theoretical implications of 100 years of research findings (Working paper). http://home.ubalt.edu/tmitch/645/session%204/Schmidt%20&%20Oh%20MKUP%20validity%20and%20util%20100%20yrs%20of%20research%20Wk%20PPR%202016.pdf

Simonton, D. K. 2006. Presidential IQ, openness, intellectual brilliance, and leadership: Estimates and correlations for 42 US chief executives. *Political Psychology*, 27, 511–526.

Stankov, L. 2018. Low correlations between intelligence and Big Five personality traits: Need to broaden the domain of personality. *Journal of Intelligence*, 6, 1–12.

Sternberg, R. J. 2003. Our research program validating the triarchic theory of successful intelligence: Reply to Gottfredson. *Intelligence*, 31, 399–413.

Strenze, T. 2015. Intelligence and success. In Goldstein, S., Princiotta, D., & Naglieri, J. A. (eds.), *Handbook of intelligence: Evolutionary theory, historical perspective, and current concepts.* New York: Springer.

Strickhouser, J. E., Zell, E., & Krizan, Z. 2017. Does personality predict health and well-being? A metasynthesis. *Health Psychology*, 36, 797–810.

Thorndike, R. L., & Hagen, E. P. 1959. *Ten thousand careers.* New York: John Wiley.

Trapmann, S., Hell, B., Hirn, J. O. W., & Schuler, H. 2007. Meta-analysis of the relationship between the Big Five and academic success at university. *Journal of Psychology*, 215, 132–151.

van der Linden, D., Pekaar, K. A., Bakker, A. B., et al. 2017. Overlap between the general factor of personality and emotional intelligence: A meta-analysis. *Psychological Bulletin*, 143, 36–52.

Wai, J., Brown, M., & Chabris, C. F. 2019. No one likes the SAT. It's still the fairest thing about admissions. *Washington Post.*

Warne, R. T. 2020. *In the know: Debunking 35 myths about human intelligence.* New York: Cambridge University Press.

Wechsler, D., & Matarazzo, J. D. 1972. *Wechsler's measurement and appraisal of adult intelligence.* Baltimore: Williams and Wilkins.

Wigdor, A. K., Green, B. F., & National Research Council Committee on the Performance of Military Personnel. 1991. *Performance assessment for the workplace.* Washington, DC: National Academy Press.

Wonderlic-Corporation. 2007. *Wonderlic Personnel Test normative report.* Libertyville, IL: Wonderlic Inc.

Zaboski, B. A., Kranzler, J. H., & Gage, N. A. 2018. Meta-analysis of the relationship between academic achievement and broad abilities of the Cattell–Horn–Carroll theory. *Journal of School Psychology*, 71, 42–56.

CHAPTER 9

Introduction to the Scientific Study of Population Differences

9.1 Introduction

The first eight chapters of this book focused on individual differences in intelligence. In Chapter 1, we introduced issues about group comparisons (Box 1.2) and some of the sensitive issues surrounding them. Before we discuss findings about sex (Chapter 10), age (Chapter 11), and differences around the world (Chapter 12), this chapter describes and discusses a number of essential issues required for properly interpreting data at the group or population level.

Groups and populations are not as easy to define as you may think. There are men and women (based on chromosomes or self-defined gender identity), young and old people (where is the cutoff?), and people of different nationalities and ancestries (based on arbitrary borders or on DNA). People are identified with, and self-identify with, all these groups. The identification typically is flexible, fluid, and inexact.

Historically and culturally, thinking about group and population differences has changed many times. For example, historical records suggest that many past societies generally assumed that men have greater cognitive abilities than women and acted accordingly. Currently, many societies generally presume equality between women and men and are attempting to act accordingly. However, equal does not mean identical. Should we regard relatively low numbers of women in corporate law practice as evidence of prejudice? Should the percentage of women in corporate law practice be the same as the percentage of women working as helicopter pilots? These are examples of important and practical questions that potentially can benefit from the weight of the empirical evidence on sex differences. As Alice Dreger (2015, p. 11) wrote, "science and social justice require each other to be healthy, and both are critically important to human freedom. ... Justice and morality require the empirical pursuit. ... Evidence really is an ethical issue, the most important ethical issue in a modern democracy."

We also make many distinctions in society based on age, like the minimum age to vote, to drink alcohol, or to draw a pension. Sometimes these distinctions are based on the perception that intelligence and cognition change over the life span. This is generally correct on average, although the extent and timing of changes vary considerably among individuals (Elliott et al., 2019; Tucker-Drob, 2019; Tucker-Drob et al., 2022). Nonetheless, there are mandatory retirement age requirements for air traffic controllers and commercial aviators, largely because of society concerns about declines in their cognitive abilities and in their physical health.

Many distinctions among groups are arguable and controversial. None are as troublesome as whether groups based on ancestry (designated as ethnicity or nationality) differ on intelligence (Ceci and Williams, 2018; Murray, 2020; Cofnas, 2020). For some people, even the suggestion of an average group difference is evidence of prejudice of any researchers who report such differences. For others, group differences support their own preexisting prejudices.

The fact is that empirical data from different tests of cognitive ability and educational achievement tests show average group differences. A scientific understanding of general intelligence and related cognitive abilities must include investigation of the extent to which and why these differences occur. The answers potentially can inform social policy makers, but this is not the goal of scientific inquiry. The next three chapters address the scientific evidence regarding the relationship of intelligence to sex, age, and ancestry-related populations. But, first, we look at some important issues that will help interpret what group data mean and do not mean.

9.2 The Issues Involved

There are no simple answers to any of the issues surrounding population differences. For appropriate discussion, it is necessary to (1) keep in mind that hypotheses are not the same as actual findings and (2) keep discussions about data interpretation separate from discussions about any potential policy implications. For example, research can inform lawmakers about the empirical data that as people age, on average, there are decreases in mental processing speed (indexed by reaction times) and the ability to control attention. Whether there should be upper age limits on automobile driving is a separate policy decision.

Because the study of population differences typically is contentious, it will help to set ground rules for discussing some problems of analysis and interpretation. The rules fall into two categories: general principles and issues of interpretation. Box 9.1 sets forth general principles, and the rest of this section focuses on issues that might affect interpretation of data, such as motivation of participants, recruitment and attrition of samples, and establishing causation.

9.2.1 *Motivation*

One reason there may be performance differences on any test is motivation. One view suggests that people with a mind-set that expects low achievement will not work hard to become high achievers (Dweck, 2006; Yeager et al., 2019). This may look like common sense, but independent research finds that mind-set has little if any effect on

Box 9.1 General Principles for the Study of Population Differences

Earl Hunt and Jerry Carlson (2007) made a number of suggestions about conducting research on intelligence and population differences:

1. The measures of intelligence must have construct validity. That is, they must measure what they purport to measure.
2. Intelligence measurements must be valid within the populations compared.
3. The fact that a score on an intelligence measure can be changed by training is not evidence against an inherent population difference unless the altered score is as valid a measure as the original score.
4. Generalization to populations at large depends crucially on the relation with the studied sample to the population of interest. Convenient samples may not be informative. A generalization from observations of a population difference in college students, for instance, to a conclusion about population differences for people of all ages is not likely valid.
5. Summations of the evidence must be done carefully, with special attention to research results that do not conform to the reviewer's conclusions. Complete objectivity might be impossible, but discussants in a scientific debate should strive for this ideal.
6. Alternative hypotheses and models should be explicitly considered.
7. The alternatives should duly represent the original authors' ideas. Straw models should not be set up in order to be knocked down. This has been a particular problem in the study of population differences. For instance, people sometimes attack the position that these differences are either entirely genetic or entirely nongenetic, whereas the real question is how much the genetic and nongenetic factors may contribute to a difference.
8. Heritability coefficients are measures of the relative size of genetic and nongenetic contributions to phenotypic individual differences within a population and can change across populations.
9. When a policy recommendation is made, one's policy model, including attitudes about desirable consequences, should be stated.
10. Be willing to say "we don't know." In many cases, we do not know what causes a population difference. There are some cases, especially involving the evolution of intelligence, where it is unlikely that we shall ever know. Be willing to acknowledge such situations. Acknowledging ambiguity is not a sign of weakness; it is a sign of honesty.

Linda Gottfredson (2007) published a commentary to these principles based on the main thesis that they raise stricter standards than required for other lines of scientific inquiry. Some scientific journals may apply unwritten stricter guidelines for intelligence research to avoid controversy. She noted, "Hunt and Carlson's proposal would legitimize what many scholarly journals have been doing surreptitiously for decades" (p. 217). Gottfredson thinks that adherence to these principles could worsen controversy because unwelcome findings will be even "harder to dismiss on scientific grounds" (p. 217).

The journal *Nature Human Behavior* published an editorial in August 2022 listing new ethics guidance for addressing "potential harms for human population groups who may be harmed" by potentially sensitive research findings. This is an interesting document acknowledging that it is important to ensure "that ethically conducted research on individual differences and differences among human groups flourishes, and no research is discouraged simply because it may be socially or academically controversial. ...

Box 9.1 *(continued)*

Researchers should be free to pursue lines of inquiry and the communication of knowledge and ideas without fear of repression and censorship" (p. 1029). Nevertheless, the editorial suggested that editors may reject papers to "avoid preventable harms." Should scientific findings be censored if they might cause harm? William Estes (as cited in Haier, 2020) wrote, "Somehow a balance must be found between the need for free exchange of research results among scientists concerned with intelligence and the need to be sure that no segment of our society has reason to feel threatened by the research or its publication."

---

school achievement (Moreau et al., 2019; Sisk et al., 2018). Mind-set as it relates to intelligence is discussed in Chapter 13. Another view claims that some groups perform poorly on tests because they are threatened by common stereotypes, such as that girls cannot do math (Steele and Aronson, 1995, 1998). Research on stereotype threat, however, has not held up well (Flore et al., 2018; Zigerell, 2017; Finnigan and Corker, 2016). Compelling results from a well-designed comprehensive study of motivation on intelligence test scores finds negligible to minor influence based on six studies with a combined sample size of 4,208 (Bates and Gignac, 2022).

On the other hand, increased motivation to succeed induced by cultural considerations may partially underlie the overrepresentation of some ethnic groups at elite universities (Flynn, 1991; Sue and Okazaki, 1990; Nisbett et al., 2012).

9.2.2 Recruitment and Attrition Effects

Many psychological research studies recruit college students and other convenient samples. These are known as the "weirdest people in the world" –Western, educated, industrialized, rich, and democratic (WEIRD) (Henrich et al., 2010). Thus, most of the information about differences in cognitive abilities between men and women, for example, is based on these nonrepresentative samples. Such recruitment effects limit generalization of findings to more heterogeneous groups representative of the general population. These nonrepresentativeness effects on the relationship between two variables were discussed in Chapter 8 as part of range restriction.

Selective attrition is another consideration. David Arribas-Águila and colleagues (2019) tested a developmental theory of sex differences in intelligence that predicted that with increased age, boys will show slightly greater intelligence scores than girls because, on average, boys have larger brains than girls. However, because boys mature later than girls, their advantage will be not evident before fifteen to sixteen years of age (Lynn, 1999). Results are shown in Figure 9.1. The findings were consistent with what the developmental theory predicts. The average difference is almost null until fourteen years of age, but the difference at eighteen years of age is equivalent to approximately 5 IQ points, which is the difference that the developmental theory predicts based on the small average brain size advantage of boys.

However, as Arribas-Águila et al. noted, there is an alternative explanation based on selective attrition. If more boys quit high school when they are allowed to, it is likely that those more prone to leave show lower intelligence. This selective attrition of the less intelligent boys may account for the small average advantage observed in Figure 9.1 precisely at the time it is manifested. There is no need to resort to any brain feature or to sex differences in maturation stages. Furthermore, the evidence is

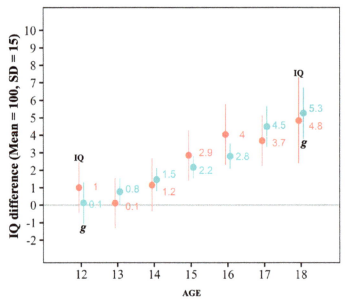

Figure 9.1 Average sex difference across age (*x*-axis) using both IQ scores (orange) and estimated *g* latent scores (blue). Each plotted difference is shown on an IQ scale (*y*-axis). Also shown are the confidence intervals around each data point (lines through circles). Results show a progressive increment of the average difference for both scores (Arribas-Águila et al., 2019). The difference values at eighteen years of age are consistent with the prediction of the developmental theory of sex differences, but also with attrition effects.

weak for delayed brain maturation of boys compared to girls (Wierenga et al., 2019).

9.2.3 *Establishing Causation*

University students learn to recite that "correlation does not mean causation." The rooster may crow before dawn, but we do not conclude that the rooster's crowing makes the sun rise. It is true that correlation does not mean causation, but correlations can help in understanding causal relationships (Rohrer, 2018).

In laboratory situations, scientists try to avoid confusing correlation with causation by conducting controlled experiments, where one possible causal factor is systematically changed as other factors are held constant, or by assigning participants randomly to experimental and control groups to average out random factors. However, this strategy does not work easily for the study of population

differences. We cannot assign people randomly to be men or women, young or old, Germans or Japanese. They come as they are. And they invariably come with many other differences and similarities besides the defined group membership. This is a challenge for understanding group differences.

There are collinearities (based on relationships among variables that are not independent), that are confounding factors, between group membership and a host of variables. Consider the relation between intelligence and national wealth. There is a substantial correlation between national wealth, measured by gross domestic product per person (GDP/c), and national average scores on standardized intelligence tests or PISA assessment scores, as we will describe in Chapter 12. There also are similar correlations between mean intelligence test scores and indices of physical health, scholastic achievement, and GDP/c. Can causal relationships be established? One could argue that intelligence causes wealth, or that school achievement causes wealth, or that wealth makes possible good schools, which in turn cause intelligence to rise – or all these might be the case to some extent.

Unfortunately, some studies simply ignore the problem of collinear relationships and confounding factors. They focus on a single putative causal variable; establish that it has a correlation with, say, intelligence; and spin a plausible tale about why this is so. Historically, this kind of reasoning led people to believe that malaria was due to dank, hot air because there was a correlation between humid conditions and outbreaks of the disease. We now know that the disease is borne by mosquitoes, which thrive in such conditions.

Collinearity issues are partly addressed in quasi-experimental designs (as we discussed in Chapter 7), in which we find two groups that appear to be equal on all the relevant variables except the one of interest, which may or may not be an experimental manipulation. This is the research design used when teaching methods are contrasted in comparable school districts, with one method being used in one district and another method in another.

Collinearity issues also can be addressed by means of a statistical technique known as *causal modeling*, which is designed to identify which of several explanations best fits the data. Causal modeling is a powerful technique, but it is not perfect. When all is said and done, causal models are based on correlations, and correlation is still not causation. Causal modeling does not establish what the truth is. The technique is useful to reject plausible models for better ones, not by rejecting the null hypothesis (Trafimow, 2017).

Many group difference studies use a classic experimental design: members of different populations are recruited, then within each group, participants are randomly assigned to experimental and control conditions, and the result of the experiment–control contrast is compared across groups. Studies using this design can show whether a particular manipulation has an effect on intelligence test scores. However, they cannot show whether the effect actually occurs outside the experimental situation or how large the effect is compared to other effects in the world outside the laboratory.

No method for investigating population differences in intelligence is perfect, but they all have something to contribute for establishing a weight of evidence. Keep in mind that "evidence" is not "proof." The strengths and weaknesses of the various designs must be considered when interpreting findings of population differences in intelligence. But remember that less-than-ideal research studies can be informative if they are properly framed and interpreted.

9.3 Statistics and Measurement

In evaluating studies of population differences, two statistical issues are critically important: (1) measurements of the size of a difference and (2) the appropriateness of using the same intelligence and cognitive assessments in different populations.

9.3.1 *The Size of Population Differences*

In the cases we consider in the next three chapters, differences of intelligence within a group are greater than differences between group average scores. This is true for most human traits, not just intelligence. Nonetheless, we need reliable statistical ways to measure the overlap between distributions of scores for any two populations. The difference between means in deviation units (d) provides a measure of how much two groups overlap: Cohen's d (Cohen, 1988). The larger the value of d is, the less is the overlap (i.e., the greater is the difference between the population means). This is illustrated in Figure 9.2, which shows the overlap between two groups that are ordered based on their mean differences (with identical standard deviations within each group), according to three values of Cohen's d (small, medium, and large).

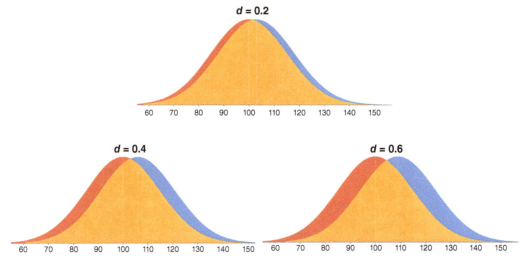

Figure 9.2 Illustration of overlap (orange) between the distributions of two populations (the blue group has a higher mean than the red group) for effect size (*d*) values of 0.2 (high overlap, small difference between populations), 0.4 (moderate overlap and difference), and 0.6 (small overlap, large difference between populations) (see more details at https://rpsychologist.com/d3/cohend/).

The *d* statistic quantifies the size of a difference between means, just as the correlation coefficient (*r*) is a measure of the strength of a relationship between two variables. Any value of *d* may or may not be statistically significant. However, just like a correlation, there must be an adequate sample size to ensure sufficient statistical power to detect the smallest difference of interest. All the concerns about power to detect meaningful findings that apply to *r* also apply to *d*. In large samples, a small value for *d* (or *r*) can be statistically significant but not especially relevant for any practical purpose. In small samples, values for *d* (or *r*) may be unreliable.

The key issue regarding the effect size refers to its proper interpretation. As noted, the usual conclusion derived from the comparison of two or more populations in human traits such as intelligence is that differences within populations (individual differences) are much greater than average differences between populations (group differences). However, effect sizes are not always appreciated when interpreting statistically significant results. A mean difference between two groups may be statistically significant, but the effect size may be too small and irrelevant to be of

practical importance. This is explained by David Funder and Daniel Ozer (2019) in their paper "Evaluating Effect Size in Psychological Research: Sense and Nonsense." Their article offers these recommendations for *r* values and their *d* equivalents (*r* and *d* values can be easily converted to each other; https://lbecker .uccs.edu):

- An effect size of *r* = 0.05 (*d* = 0.10) is small for interpreting a difference, but potentially important in the long run.
- An effect size of *r* = 0.10 (*d* = 0.20) is also small for interpreting differences, but potentially more important.
- An effect size of *r* = 0.20 (*d* = 0.40) is medium size, with some explanatory and practical use.
- An effect size of *r* = 0.30 (*d* = 0.60) is large and potentially powerful for explaining a difference finding.
- An effect size of *r* = 0.40 (*d* = 0.9) or greater is impressively large and important but rare in psychological research.

So, are small effect sizes worth taking seriously? Funder and Ozer wrote, "The relevance of the cumulation of small effects over time is particularly obvious for research on

---

Box 9.2 More about Sample Sizes and Effect Sizes

In Chapter 5, we described a meta-analysis that showed an increasing magnitude of positive correlations between brain size and intelligence with an increase in the quality of the studies included in the analysis (Gignac and Bates, 2017). The average correlation obtained from the excellent studies was higher than the average correlation obtained when all studies were combined (Pietschnig et al., 2015). Both the original and the revised meta-analyses were based on the same set of studies, but the strongest correlation was found in a smaller subset of excellent studies that used multiple tests to assess a latent intelligence variable (g) and excluded clinical samples. This example demonstrated the importance of study quality as a variable to consider.

It also demonstrated that smaller sample sizes sometimes are preferable if they are better quality. Gignac and Bates (2017, p. 27) note, "Researchers who administer more comprehensive cognitive ability test batteries require smaller sample sizes to achieve the same level of power [the probability of detecting an effect, if there is a true effect present to detect]. An investigator who planned to administer 9 cognitive ability tests (40 minutes testing time) would require a sample size of 49 to achieve a power of 0.80, based on expected correlation of 0.30. ... It is more efficient to administer a 40-minute comprehensive measure of intelligence across 49 participants in comparison to a relatively brief 20-minute measure across 146 participants."

Furthermore, Gignac and Bates caution against a tendency to give greater weight to results from studies with huge samples while ignoring small-scale research. As Paul Thompson and colleagues (2019, p. 25) observed in their review of a decade of research within the ENIGMA Consortium, "for effect sizes of d > 0.6, the reproducibility rate was higher than 90 percent even when including the datasets with sample sizes as low as 15, while it was impossible to obtain 70 percent reproducibility for small effects of $d < 0.2$ even with a relatively large minimum sample size threshold of 500."

Bigger samples without careful research designs are not always better.

---

individual differences, such as abilities and personality traits. ... Smaller effect sizes are not merely worth taking seriously. They are also more believable" (pp. 161, 166). Relatively small effects might show relevant consequences in the real world. Effect size and sample size are discussed more in Box 9.2.

9.3.2 *Using the Same Standardized Test for Measuring Different Populations*

If we want to measure population differences on a physical variable, such as height or weight, the principle is simple. We use a physical device to make a physical measurement. If we want to compare, say, the heights and weights of Asians to the heights and weights of Africans, there is no concern that tape measures and scales would behave differently across groups.

When it comes to intelligence and cognitive testing, different types of people may systematically respond to a test in different ways, possibly tapping different underlying psychological variables than those targeted. To illustrate this, imagine that we decide to develop a direction-taking test to measure people's ability to orient in space. The test might contain items like this:

Go to the flagpole to the north of your present position. Then go east to the church. Turn north at the church and proceed to the school.

We then see how long it takes people to get to the school and how many deviations they make from the prescribed route. Remember that we are not interested in direction taking per se; we are interested in performance on the direction-taking test as a measure of a more general ability to orient in space.

If we use a contrast between the speeds with which men and women arrive at the school, we have implicitly assumed that men and women will approach the test in the same way. But there is evidence this may not be the case. Men often approach problems in direction taking by constructing a mental map of the territory and the route through it. Women tend to memorize the directions and use them to follow a path (Hunt, 2002). Unintentionally, the test might evaluate a different underlying psychological trait in men and in women. Comparing scores would be valid if we were interested in how well people did in direction finding but invalid if we were interested in making an inference about where men and women rank on an underlying latent trait of orienting ability (Contreras et al., 2007).

To guard against such problems, we have to demonstrate that a test measures the same psychological trait in different groups. If we are dealing with a single test, and are concerned that different problems within the test tap different abilities in various groups, we look for evidence of differential item functioning (DIF). The idea behind DIF can be illustrated by another hypothetical test of baseball knowledge. The more you know about baseball, the better able you will be to answer questions about baseball. The question "How many players are there on a team?" can be answered by anyone with the slightest knowledge of the game, but the question "What is a squeeze play?" is more challenging. Many US citizens who have not thought about baseball since grammar school can answer the first question, but not the second; a sports reporter would be able to answer both. More generally, we can order questions in terms of their difficulty.

Now suppose we give the baseball test to two groups, US citizens and Cubans. Baseball is played in Cuba, so some Cubans know more about baseball than some US citizens. The question is, does the test measure the same thing, baseball knowledge, in both countries? If it does, then the order of difficulty of items should be the same in each country, even though Cubans, on average, might answer more questions correctly. Most probably, the order of item difficulty would be the same.

But go back to the direction-finding test, where the contrast is between men and women. Because the test measures orienting ability in men and verbal memory in women, there is no guarantee that the order of item difficulty would be the same. Suppose it is not. In that case, we say that the test shows DIF across groups. When DIF is found, conclusions about population differences on a psychological trait said to underlie a test cannot be drawn from observations of differences in test scores.

The critical point is that counting is not measuring. Although the simplest way to test for DIF is to calculate the rank order correlation (*rho*) between item difficulties within each group, scientists have developed much more sophisticated ways for testing DIF (Box 9.3).

9.3.3 Comparing Groups Using Test Batteries

DIF refers to items within a single test. However, some studies draw conclusions about populations based on summary scores extracted from a battery of tests. This would be the case, for instance, in a comparison of scores on the Wechsler or the Stanford–Binet, which are derived from subtests. Generally, the purpose is not to compare raw scores but to compare populations on some underlying latent factor (like *g*, verbal ability, or visuospatial ability) that is derived from the scores on the individual tests included in the battery. In those instances, the goal is to identify population differences on a latent trait from observations on the manifest test scores.

Box 9.3 Differential Item Functioning on the Raven's Advanced Progressive Matrices

Francisco J. Abad and colleagues (2004) compared performance on the items of the Raven's Advanced Progressive Matrices (RAPM) for 1,069 men and 901 women. The main prediction was that there will be items biased against women's performance because of the visuospatial format of the test.

They applied differential item functioning (DIF) methods for investigating sex differences in DIF for the thirty-six RAPM items. Several items were biased against women. The average difference between men and women on the complete test was equivalent to 4.1 IQ points, and the difference after deleting the biased items was equivalent to 3.3 IQ points.

The authors noted that even the unbiased items may be somewhat biased because they require visuospatial processing to some degree. If this is indeed the case, controlling for sex differences in visuospatial ability would eliminate the sex differences on the RAPM test. Colom and colleagues (2004) reported that this is indeed the case: men outperformed women in the RAPM and in the mental rotation test from the Primary Mental Abilities Battery. However, the men's average advantage on the RAPM vanished when differences in visuospatial ability were statistically removed.

Administering measurement instruments, getting scores, computing the average difference between groups, and concluding that group A is better on average than group B on trait X is not necessarily warranted. Before reaching this conclusion and reporting the findings, we must make sure that the same psychological construct (latent trait X) is tapped across groups by the measurement instruments. The technical term is that measurements must be *invariant* for the populations compared.

An example of such a comparison is the Spearman hypothesis (Jensen, 1998). It postulates that group differences on intelligence tests are mainly due to differences in g, a latent trait, rather than differences in specific factors or specific tests. By definition, latent traits cannot be observed. So how do we evaluate the hypothesis?

As we discussed in Chapter 3, general intelligence (g) is operationally defined as a factor underlying scores on several (observable) tests, such as the subtests of the Wechsler battery. Jensen (1998) designed a technique called the method of correlated vectors (MCV) as a way to test Spearman's hypothesis. Since this method has been used widely in the study of population differences, it is worth explaining.

If Spearman's hypothesis holds, then the more a test measures g, the larger the standardized group difference (d) should be on that test. The extent to which a test is a measure of g is determined by its factor loading obtained after computing a factor analysis from the correlation matrix defined by the tests in the battery. The extent to which groups differ on each test within a battery is computed from the d statistic for comparing groups on that test (the effect size of the difference; Figure 9.2). To apply the MCV, we compute the correlation between test loadings on g and the d statistics (Table 9.1). If the correlation is high enough, then the hypothesis is supported (usually this is a correlation between the rank orders of each variable instead of their raw scores; a rank order correlation is called *rho*).

Table 9.1 shows that Vocabulary has the highest g loading (0.95), whereas Symbol

Table 9.1 Subtests included in the Wechsler battery (WAIS-III), *g* loading of each test, and *d* statistic

WAIS-III subtest	g Loading	d Statistic
Vocabulary	0.95	1.5
Similarities	0.90	1.4
Information	0.85	1.3
Comprehension	0.80	1.2
Picture Completion	0.75	1.1
Block Design	0.70	1.0
Matrices	0.65	0.9
Picture Arrangement	0.60	0.8
Object Assembly	0.55	0.7
Arithmetic	0.50	0.6
Digit Span	0.45	0.5
Letter–Number Sequencing	0.40	0.4
Coding	0.35	0.3
Symbol Search	0.30	0.2

Note. Numbers are examples to show what happens in the ideal case in which the average difference between two groups on the tests can be attributed to the latent general factor of intelligence, *g*.

Search has the lowest *g* loading (0.30). The important point is that as the *g* loading increases, so does the *d* statistic; there is a positive correlation. This is what the Spearman hypothesis predicted, and we can represent the numerical pattern in Table 9.1 in a scatterplot (Figure 9.3).

Despite its simple concept and popularity, the MCV has serious problems. Depending on the circumstances, it can either overstate the size of group differences on *g* when they are actually small (Dolan et al., 2004; Dolan and Hamaker, 2001) or understate them when they are actually present (Ashton and Lee, 2005). This means that we have to rethink many of the

conclusions drawn by Jensen and other researchers based on MCV analyses. It does not mean that these conclusions are necessarily wrong; it just means that they have to be tested further with new analyses, as described in (Box 9.4).

9.4 What to Conclude?

As we will see in the next three chapters, there are average intelligence test score differences in groups defined by sex, age, and ancestry/country. We always are careful to use the word *tend* when referring to average group differences because the variance within a group is much larger than the average difference between groups. The most important conclusion is that any group difference does not apply to any individual member of the group. If, on average, there is a tendency of one group to score higher than another group on variable X, you cannot predict X for any individual in the group. This means that every person must be seen as an individual irrespective of any group membership. It also means that any average group differences should be considered carefully in discussions about complex social issues and policies.

Societies make legal and practical distinctions about groups all the time. The causes of population differences need investigation so that discussions are informed as much as possible by empirical evidence. We think that more information is better than less even if some data point in uncomfortable directions. We also think that research about population differences must be sensitive to possible misuses or misunderstandings that cause distress, encourage bad behavior, and become counterproductive.

9.5 Summary

We present key points to keep in mind when considering group difference data: (1) sweeping assertions about population differences must be treated with skepticism; (2) it

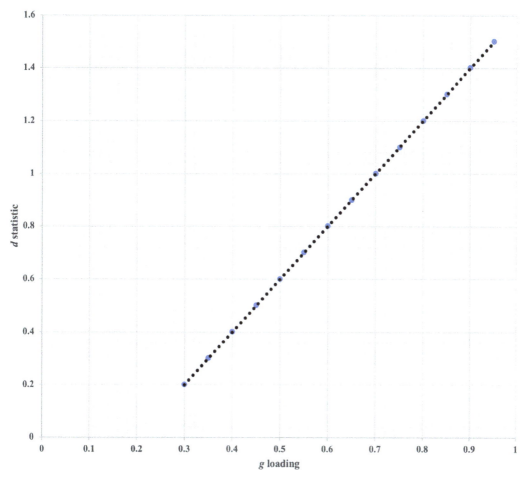

Figure 9.3 Scatterplot from values shown in Table 9.1. These values are selected to illustrate the ideal case of a perfect correlation.

Box 9.4 MCV and MGCFA for Comparing Populations

Multigroup confirmatory factor analysis (MGCFA) is statistically more complex than the MCV, but the results are much more stable and reliable. The first step in applying the MCV is to determine that the same relationships hold between tests within each group to be compared. This is evaluated by a statistical test called *congruence coefficient*, which assesses whether the two correlation (or covariance) matrices (among the tests) can be assumed to be identical in each group. If they are, the congruence condition is said to be satisfied. In a nonstatistical sense, what this means is that the observable subtest scores are related to each other in the same way in both groups.

Consider the hypothetical direction-taking test described in the main text. This test would be highly correlated to a test of verbal memory in women and to a test of rotational ability in men. High congruence would not be found. Jensen assumed that failure of congruence would be evidence that the tests within a battery were measuring different psychological constructs in each population and, therefore, that Spearman's hypothesis could not be tested.

Box 9.4 *(continued)*

Assuming that congruence is satisfied, the second step is to factor analyze the data; extract a first factor, which can be used as the operational definition of g; and correlate the factor loadings of each test with its associated d statistic. Colom and colleagues (2002) applied the MCV to the standardization sample of the WAIS-III for Spain. Table 9.2 shows the g loadings for the groups with primary and secondary school completed, along with the d values for each WAIS-III subtest. In this case, the correlation between g and d was -0.28 (not significant; $p = 0.33$), and therefore, g failed to explain the average difference across WAIS-III subtests distinguishing both educational groups. The key conclusion derived from this study was that more education goes with high intelligence scores (IQ) but a higher IQ is not necessarily a reflection of a higher g. However, Conor Dolan and colleagues (Dolan and Hamaker, 2001; Dolan et al., 2004) pointed out that to test Spearman's hypothesis, the data must satisfy a property called *measurement invariance*. The measurement invariance property is that the factor loadings for subtests should be identical within each group being compared. This ensures that two individuals, one from each group, who have identical scores on the latent traits (which is what we are interested in but cannot observe) also have identical scores on the subtests, which we can observe.

Dolan and his colleagues showed that the MCV can produce positive findings in situations where measurement invariance does not hold (Dolan and Hamaker, 2001; Dolan et al., 2004). Using a separate line of argument, Ashton and Lee (2005) have shown that the MCV can produce negative findings (the correlation between loadings and mean differences can approach zero) in situations in which there actually is a difference in g. That was the case in Colom et al.'s (2002) study.

To overcome these difficulties with the MCV, Dolan's group advocated using MGCFA to evaluate Spearman's hypothesis and similar conjectures. The technique is a variety of confirmatory factor analysis, the statistical method presented in Chapter 3.

MGCFA is better than the MCV in two ways. First, it involves explicit tests for measurement invariance. Second, importantly, it allows investigators to compare two hypotheses, rather than evaluating the weaker statement that a single hypothesis is better than chance description of the data. The MCV tests Spearman's hypothesis against chance, whereas the MGCFA compares competing hypotheses.

Figure 9.4 shows results derived from MGCFA using the same data set considered in the example described earlier regarding the WAIS-III (Table 9.2). Figure 9.4 depicts the factor loadings for the two groups studied (primary and secondary school completion). The loadings for each group are not nearly identical, so we suspect that the WAIS-III is not tapping the same constructs. For instance, the g loading of verbal comprehension is 0.76 for the primary study groups but 0.57 for the secondary study groups. However, these are qualitative observations. What we need is the quantitative comparison that MGCFA allows.

For MGCFA, the successive steps require checking if (1) the factor loadings can be considered equal across groups, (2) the means representing g performance of both groups can be considered equal, (3) the residual (unexplained) variances are equal, and, finally, (4) the latent variances are equal.

From these computations, we can estimate, for instance, the average difference in the general latent higher-order factor (g). In this case, the average difference was $d = 1.02$, meaning that the secondary

> **Box 9.4 *(continued)***
>
> school group obtained *g* scores 1 standard deviation above the *g* scores obtained by the primary school group. Therefore, while the MCV failed to find any average difference, MGCFA revealed that there was indeed a substantial group difference associated with educational level: the higher the educational level was, the greater were the *g* scores.
>
> The MGCFA technique has been known since the early 1980s. It requires relatively large samples from all groups and calls for expertise in the use of the required computer programs. Therefore, it is not used as often as it could be. Hopefully this will change as the flaws in the simpler MCV become more appreciated.

Table 9.2 WAIS-III subtests, *g* loading for two educational groups, and *d* statistics representing their average performance differences across subtests

	g Loading		
	Primary school	Secondary school	d Statistic
WAIS-III subtest			
Information	0.54	0.42	1.10
Similarities	0.64	0.49	0.96
Vocabulary	0.59	0.45	1.02
Comprehension	0.51	0.41	0.78
Picture Arrangement	0.74	0.54	0.75
Block Design	0.73	0.59	0.81
Matrix Reasoning	0.81	0.58	0.91
Picture Completion	0.69	0.39	0.81
Object Assembly	0.71	0.52	0.68
Arithmetic	0.67	0.52	0.82
Letter–Number Sequencing	0.75	0.61	0.83
Digit Span	0.64	0.51	0.74
Symbol Search	0.74	0.46	0.76
Coding	0.78	0.44	0.79
% Variance	47.21	25.14	

Note. From Colom et al. (2002).

is critical to distinguish scientific findings and political recommendations; (3) when population differences are observed, it may be unclear if they can be attributed to intelligence or to other (confounding) variables; (4) selective recruitment and attrition effects might be substantial; (5) high correlations between X (say, national wealth) and Y (say, national average scores on intelligence or educational tests) leave open the question of what causes what; (6) population overlap between groups is the norm; and (7) comparing scores across populations could be valid regarding their performance levels, but these scores would be not valid for making inferences about the groups' standing on the underlying trait of interest if individuals within one group apply different cognitive strategies to solving test problems than individuals in the other group. Valid group

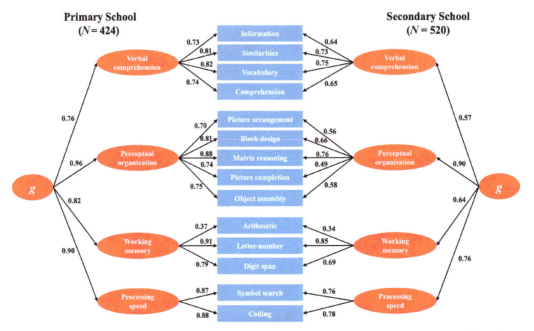

Figure 9.4 Results after computing MGCFA (see Box 9.4) using the same data set considered by Colom et al. (2002). Factor loadings are shown linking latent variables (ellipses) and tests (rectangles) for the two groups (primary and secondary school completed, left and right side, respectively). The loadings for each group are different enough to suspect that the WAIS-III battery may not tapping the same constructs in each group with the same precision.

Analyses are courtesy of F. J. Abad.

comparisons require sophisticated statistical methods that make simple explanations of any differences highly unlikely.

9.6 Questions for Discussion

9.1   Does motivation matter much when taking intelligence tests?

9.2   How do recruitment and attrition effects impact interpretation of data?

9.3   What is the implication of an average group difference for any individual in the group?

9.4   Do you think research on group differences should be limited in any way?

References

Abad, F. J., Colom, R., Rebollo, I., & Escorial, S. 2004. Sex differential item functioning in the Raven's Advanced Progressive Matrices: Evidence for bias. *Personality and Individual Differences*, 36, 1459–1470.

Arribas-Águila, D., Abad, F. J., & Colom, R. 2019. Testing the developmental theory of sex differences in intelligence using latent modeling: Evidence from the TEA Ability Battery (BAT-7). *Personality and Individual Differences*, 138, 212–218.

Ashton, M. C., & Lee, K. 2005. Problems with the method of correlated vectors. *Intelligence*, 33, 431–444.

Bates, T. C., & Gignac, G. E. 2022. Effort impacts IQ test scores in a minor way: A multi-study investigation with healthy adult volunteers. *Intelligence*, 92, 101652.

Ceci, S. J., & Williams, W. M. 2018. Who decides what is acceptable speech on campus? Why restricting free speech is not the answer. *Perspectives on Psychological Science*, 13, 299–323.

Cofnas, N. 2020. Research on group differences in intelligence: A defense of free inquiry. *Philosophical Psychology*, 33, 125–147.

Cohen, J. 1988. *Statistical power analysis for the behavioral sciences*. Hillsdale, NJ: Erlbaum.

Colom, R., Abad, F. J., Garcia, L. F., & Juan-Espinosa, M. 2002. Education, Wechsler's full scale IQ, and *g*. *Intelligence*, 30, 449–462.

Colom, R., Escorial, S., & Rebollo, I. 2004. Sex differences on the Progressive Matrices are influenced by sex differences on spatial ability. *Personality and Individual Differences*, 37, 1289–1293.

Contreras, M. J., Rubio, V. J., Peña, D., Colom, R., & Santacreu, J. 2007. Sex differences in dynamic spatial ability: The unsolved question of performance factors. *Memory and Cognition*, 35, 297–303.

Dolan, C. V., & Hamaker, E. L. 2001. Investigating black–white differences in psychometric IQ: Multigroup confirmatory factor analyses of the WISC-R and the K-ABC and a critique of the method of correlated vectors. In Columbus, F. (ed.), *Advances in psychology research*. Hauppauge, NY: Nova Science.

Dolan, C. V., Roorda, W., & Wicherts, J. M. 2004. Two failures of Spearman's hypothesis: The GATB in Holland and the JAT in South Africa. *Intelligence*, 32, 155–173.

Dreger, A. D. 2015. *Galileo's middle finger: Heretics, activists, and the search for justice in science*. New York: Penguin Press.

Dweck, C. S. 2006. *Mindset: The new psychology of success*. New York: Random House.

Editorial. 2022. Science must respect the dignity and rights of all humans. *Nature Human Behaviour*, 6, 1029–1031.

Elliott, M. L., Belsky, D. W., Knodt, A. R., et al. 2019. Brain-age in midlife is associated with accelerated biological aging and cognitive decline in a longitudinal birth cohort. *Molecular Psychiatry*, 26, 3829–3838.

Finnigan, K. M., & Corker, K. S. 2016. Do performance avoidance goals moderate the effect of different types of stereotype threat on women's math performance? *Journal of Research in Personality*, 63, 36–43.

Flore, P. C., Mulder, J., & Wicherts, J. M. 2018. The influence of gender stereotype threat on mathematics test scores of Dutch high school students: A registered report. *Comprehensive Results in Social Psychology*, 3, 140–174.

Flynn, J. R. 1991. *Asian Americans: Achievement beyond IQ*. Hillsdale, NJ: Erlbaum.

Funder, D. C., & Ozer, D. J. 2019. Evaluating effect size in psychological research: Sense and nonsense. *Advances in Methods and Practices in Psychological Science*, 2, 156–168.

Gignac, G. E., & Bates, T. C. 2017. Brain volume and intelligence: The moderating role of intelligence measurement quality. *Intelligence*, 64, 18–29.

Gottfredson, L. S. 2007. Applying double standards to "divisive" ideas: Commentary on Hunt and Carlson (2007). *Perspectives on Psychological Science*, 2, 216–220.

Haier, R. J. 2020. Academic freedom and social responsibility: Finding a balance. *Intelligence*, 82, 101482.

Henrich, J., Heine, S. J., & Norenzayan, A. 2010. The weirdest people in the world? *Behavioral and Brain Sciences*, 33, 61.

Hunt, E. B. 2002. *Précis of thoughts on thought*. Mahwah, NJ: Erlbaum.

Hunt, E., & Carlson, J. 2007. Considerations relating to the study of group differences in intelligence. *Perspectives on Psychological Science*, 2, 194–213.

Jensen, A. R. 1998. *The g factor: The science of mental ability*. Westport, CT: Praeger.

Lynn, R. 1999. Sex differences in intelligence and brain size: A developmental theory. *Intelligence*, 27, 1–12.

Moreau, D., Macnamara, B. N., & Hambrick, D. Z. 2019. Overstating the role of environmental factors in success: A cautionary note. *Current Directions in Psychological Science*, 28, 28–33.

Murray, C. 2020. *Human diversity: The biology of gender, race, and class*. New York: Twelve.

Nisbett, R. E., Aronson, J., Blair, C., et al. 2012. Intelligence: New findings and theoretical developments. *American Psychologist*, 67, 130–159.

Pietschnig, J., Penke, L., Wicherts, J. M., Zeiler, M., & Voracek, M. 2015. Meta-analysis of associations between human brain volume and intelligence differences: How strong are they and what do they mean? *Neuroscience and Biobehavioral Reviews*, 57, 411–432.

Rohrer, J. M. 2018. Thinking clearly about correlations and causation: Graphical causal models for observational data. *Advances in Methods and Practices in Psychological Science*, 1, 27–42.

Sisk, V. F., Burgoyne, A. P., Sun, J. Z., Butler, J. L., & Macnamara, B. N. 2018. To what extent and under which circumstances are growth mind-sets important to academic achievement? Two meta-analyses. *Psychological Science*, 29, 549–571.

Steele, C. M., & Aronson, J. 1995. Stereotype threat and the intellectual test-performance of African-Americans. *Journal of Personality and Social Psychology*, 69, 797–811.

Steele, C. M., & Aronson, J. 1998. Stereotype threat and the test performance of academically successful African Americans. In Jencks, C., & Meredith, P. (eds.), *The black–white test score gap*. Washington, DC: Brookings Institution Press.

Sue, S., & Okazaki, S. 1990. Asian-American educational achievements – a phenomenon in search of an explanation. *American Psychologist*, 45, 913–920.

Thompson, P., Jahanshad, N., Ching, C. R. K., et al. 2019. ENIGMA and global neuroscience: A decade of large-scale studies of the brain in health and disease across more than 40 countries. *Translational Psychiatry*, 10, 100.

Trafimow, D. 2017. The probability of simple versus complex causal models in causal analyses. *Behavior Research Methods*, 49, 739–746.

Tucker-Drob, E. M. 2019. Cognitive aging and dementia: A life-span perspective. *Annual Review of Developmental Psychology*, 1, 177–196.

Tucker-Drob, E. M., Fuente, J. D. L., Köhncke, Y., et al. 2022. A strong dependency between changes in fluid and crystallized abilities in human cognitive aging. *Science Advances*, 8, eabj2422.

Wierenga, L. M., Bos, M. G. N., van Rossenberg, F., & Crone, E. A. 2019. Sex effects on development of brain structure and executive functions: Greater variance than mean effects. *Journal of Cognitive Neuroscience*, 31, 730–753.

Yeager, D. S., Hanselman, P., Walton, G. M., et al. 2019. A national experiment reveals where a growth mindset improves achievement. *Nature*, 573, 364–369.

Zigerell, L. J. 2017. Potential publication bias in the stereotype threat literature: Comment on Nguyen and Ryan (2008). *Journal of Applied Psychology*, 102, 1159–1168.

CHAPTER 10

Sex Differences and Intelligence

10.1 Introduction: Why Sex Matters

Are women smarter than men? Or is it the other way around? Other than intelligence, are other mental abilities different between women and men, and if so, do these differences matter when it comes to education, vocations, or any other practical matters? Here is the short story: women and men do not differ much, if at all, on average g-factor scores, but there are differences on certain cognitive abilities that may be relevant to education, vocational choice, and success in various walks of life. Of course, the long story is more complex and includes compelling evidence about sex differences in the brain and the basic question about where the differences come from. And there is a practical question: should any of these findings inform social and educational policy in some way?

Before presenting the weight of evidence based on data that address sex differences, it is necessary to explain what we mean by men and women. This distinction was taken as obvious in previous books, but now there is an idea among some social thinkers that the biological sex categories of male and female are not valid for scientific research comparisons. Instead, the argument goes, these categories need to be replaced by a continuum of socially constructed genders of more than two varieties. In this book, we mean men and women as in the biological categories of male and female based on X and Y chromosomes that describe almost everyone on the planet. The small number of statistical exceptions of a mismatch between chromosomes and biological features are real and important, but to our knowledge, there are not sufficient empirical studies of intelligence or cognition to describe a weight of evidence for these exceptions.

Gender typically is defined as self-identified sex role orientation and can include categories in addition to male and female. Gender is a more complex and nuanced concept that has a relatively recent history in social theory. We are unaware of intelligence studies that compare gender categories in the research design. Even research studies that use the word "gender" instead of "sex" to describe participants typically are talking about males and females in the biological sense.

This distinction is important because, generally, if gender choice is the result of culture and social environment, as alleged by proponents of this theory, any gender differences in cognition, intelligence, and achievement, including disparities in vocations and careers, could be explained and addressed without reference to presumably less malleable biological differences between men and women. This may be a politically preferred point of view for some people, but without biological differences, evolution would not be relevant to understanding any brain-related differences within or between sexes. This seems unlikely since there is no reason for the brain to be immune to natural selection processes.

If biology is involved, as the evidence indicates, then genetics, brain imaging, and neuroscience research are worthwhile. This is certainly true for intelligence research (Haier, 2017; see Chapters 5 and 6) and for research on conditions like autism, schizophrenia, dyslexia, and attention deficit–hyperactivity disorder (both more frequent in males), as well as for Alzheimer's disease, major depression, anxiety, and eating disorders (all more common in females). Data suggest that it is also true for the cognitive abilities where science finds average sex differences.

Before describing key data, we note that in 2016, the National Institutes of Health adopted a policy called "Sex as a Biological Variable" (https://orwh.od.nih.gov/sex-gen der/nih-policy-sex-biological-variable), requiring all of its grantees to incorporate females into their research samples rather than to assume findings from male samples would generalize to females; this is also a consideration for animal studies (Clayton and Collins, 2014).

Here is a thought experiment to set the stage for discussion. Charles Murray compiled a statistical analysis of major contributions in various fields. He found that the ratio of eminent men to eminent women ranged from 50:1 (mathematics) to 10:1 (literature) (Murray, 2003). Do you think that the undisputed, widespread historical cultural and serious social limitations and discrimination in many forms against women are the mostly likely explanation for these disparities?

Now, consider what Diane Halpern (1986) had to say in the preface to the first edition of her classic book *Sex Differences in Cognitive Abilities*: "It seemed like a simple task when I started writing this book. ... At the time it seemed to me that any between-sex differences in thinking abilities were due to socialization practices, artifacts and mistakes in the research, and bias and prejudice. After reviewing a pile of journal articles that stood several feet high, and numerous books and chapters that dwarfed the stack of the journal articles, I changed my mind. The task I had undertaken certainly wasn't simple and the conclusions that I had expected to make had to be revised."

Note that was written back in 1986. Twenty-six years later, in the preface to the fourth edition of her book, Professor Halpern (2012) wrote, "The biological revolution has changed our understanding of the mind and behavior in general, but particularly in the way we think about cognitive sex differences. … Perhaps one of the greatest contributions from the biological revolution is that we can now see changes in the brain that result from experience. In a strange twist, modern biological techniques have advanced our understanding of the importance of environmental variables."

The next sections discuss what we now know about sex differences, intelligence, and cognitive abilities from the weight of evidence. As always, please keep an open mind.

10.2 Psychometric Studies of Intelligence Differences

The case for differences between men and women in general intelligence, or g, rests on three kinds of evidence: (1) the results from overall scores derived from battery-type tests, such as the IQ score derived from the Wechsler tests; (2) the results from factorial studies in which individual scores are computed for the g-factor derived from a variety of test batteries, including both avowed intelligence tests and batteries constructed for research purposes; and (3) the results from studies of individual tests, such as the Raven's Matrices, that are highly g loaded. We will look first at the results from the battery-type tests and then examine the results from the individual tests and some issues involving tests used for personnel screening.

10.2.1 *Evidence from Test Batteries*

Analyses of the adult standardization samples of the WAIS-III and WAIS-R generally show a small average difference in IQ in favor of men. The results are consistent across countries, running from 2–3 IQ points in the United States and Canada (Longman et al., 2007) (in deviation units, $d = 0.19$) to 4 points ($d = 0.27$) in China and Japan (Dai et al., 1991; Hattori and Lynn, 1997). These results are also close to the results obtained in earlier studies, showing consistency in time (Matarazzo et al., 1986; Snow and Weinstock, 1990).

These results are not exclusive to the Wechsler tests. Deary and his colleagues (2007a) reported an elegant example, involving data from the NLSY79 (see Box 7.4). They contrasted the scores of brothers and sisters of the seventeen to twenty-three-year-olds who had taken the test, thus controlling for family background. The male–female difference on the general score derived from the ASVAB part of the AFQT (a good measure of g), was −.02, a small, and not statistically reliable, finding in favor of females.

A different study by Deary and colleagues (2007b) was based on a large survey of more than 70,000 schoolchildren in the United Kingdom, who took the Cognitive Abilities Test battery at age eleven. Two factors were derived from this test: a general intelligence factor and a residual verbal factor (i.e., with g variance statistically removed from the verbal score). No male–female differences were found on the general factor, but there was a slight advantage for girls on the residual verbal factor. Brain and intellectual development continue after age eleven, however, so this result might change with increased age.

Overall, these studies indicate negligible sex differences. However, Richard Lynn (1999) has argued that there is actually a greater male advantage in intelligence than tests like the Wechsler reveal. His argument is based on two claims.

Lynn's first argument is that girls mature more rapidly than boys and that cognitive competence increases with physiological age, rather than with calendar age. The male–female difference might be small, and even negative (reflecting a female advantage) prior to puberty, but a male advantage would appear after adolescence and continue throughout adulthood.

Lynn is correct that male–female differences are smaller in childhood than in adulthood, although not very much smaller. It is not clear whether this should be regarded as an artifact or simply an observation. By analogy, male–female discrepancies in height are smaller (and can show a female advantage) in childhood and not in adulthood, but this is not an artifact of the way we measure height; girls do get closer to their adult height than boys do in their preteen and early teen years. To the extent that cognitive growth mirrors physical growth, one could argue that girls are, in fact, smarter than boys during the elementary and middle school years, a point that should be considered in situations where entry into higher levels of education is determined by performance in the elementary years. However, there is evidence that a brain age prediction model (including 270 brain measures) did not support delayed maturation in boys compared to girls (Wierenga et al., 2019).

Lynn's second argument is that at all ages, the tests are biased against men. He states,

> The adult male advantage of around 4 IQ points obtained by averaging the verbal comprehension, reasoning and spatial abilities is not generally found in the full-scale IQ of the Wechsler tests or in the overall IQ of similar tests because the spatial abilities are typically underrepresented in these tests. (p. 2)

To what extent is this argument plausible? Summary scores, such as the widely used WAIS Full Scale, Verbal, and Performance (FSIQ, VIQ, PIQ) scores, are calculated by a weighted combination of scores on subtests. If men have higher scores on some subtests, and women on other subtests, then, depending on the weights assigned to each subtest, you could produce a summary score that favored men over women, or vice versa, simply by manipulating the weights assigned to the subtests. And it is certainly true that if a test battery omits an important ability on which there are male–female differences, then the balance of men's and women's scores in an overall index will be different than it would have been had the omitted ability been evaluated.

Arthur Jensen (1998) argued that the way out of this dilemma is to compare men and women on g-factor scores. The argument is that the weighting of individual subtests would be by rational analysis of the data, rather than by using weights that were arbitrarily assigned to the subtests. This is the method of correlated vectors (MCV) we discussed in Chapter 9 and Box 9.4. When it is applied to standard test batteries such as the Wechsler, the results suggest that the minor average sex difference in IQ scores cannot be attributed to the g-factor.

The two approaches can lead to differences in the calculation of overall male–female differences. This is illustrated by Deary and colleagues' analysis of the NLSY79 data. Table 10.1 presents their data separately for male–female differences on the four subscales of the ASVAB used to compute the AFQT. As was mentioned earlier, the AFQT score showed essentially no difference between men and women. However, when Deary and his colleagues applied Jensen's MCV technique and computed male–female differences on a g index derived from factor analysis, there was a male advantage of 0.06 standard deviation units – not much, but still a reversal of the direction computed from the composite AFQT score.

Arribas-Águila and colleagues (2019) reported on 4,992 boys and 5,343 girls who completed the Batería de Aptitudes de TEA intelligence battery. They found no male–female differences in a g-factor latent estimate at age twelve, but at age eighteen, boys showed an advantage that translated to an estimated 5 IQ points, consistent with Lynn's argument. However, as noted when we described this study in Chapter 9, the findings can be explained by undetected attrition effects instead of by the developmental theory of sex differences proposed by Lynn. Exploring alternative explanations is required, especially if there is a potential

Table 10.1 Selected scores and male–female comparison for brother–sister pairs in the NLSY79 data

| Test | Mean | | Effect size in d |
	Male	Female	
Word knowledge	22.3	22.9	−0.07
Paragraph comprehension	9.2	10.0	−0.21
Arithmetic reasoning	15.9	14.7	+0.17
Mathematics knowledge	11.9	11.9	0.00
AFQT standardized score	−0.034	+.034	−0.02
AFQT g score	15.08	14.68	0.06

Note. Adapted from Deary et al. (2007a, table 1).

for using some findings to inform social policy. If there are any average male–female differences in general intelligence indices derived from the commonly used battery-type tests, those differences are negligible and have little, if any, practical impact. Although as we discuss later (Section 10.2.3), a difference in the variance around the mean can have an impact at the extreme ends of the cognitive distributions.

No statistical analysis can address a stronger form of Lynn's argument that appropriate subtests are not included at all. Women generally do better than men on verbal tests, and men may do markedly better than women on visuospatial tests, especially those tests evaluating the rotational aspect of visuospatial reasoning in the g-VPR model. But does this mean that those batteries that are now widely used, such as the WAIS, "underrepresent" visuospatial reasoning, as the quotation from Lynn implies? To answer this question, one would have to know what the proper representation of verbal, visuospatial, and other traits is.

If the test battery as a whole is to be validated by its ability to predict performance in an applied setting, then the appropriateness of adding or subtracting a particular subtest can be assessed by seeing if the addition improves accuracy of prediction. A powerful case can be made that adding visuospatial tests might do this in some situations, but might not in others (Humphreys and Lubinski, 1996). For that

matter, there are some situations in which an argument can be made for ignoring visuospatial ability tests in favor of a more extended evaluation of verbal testing.

If the test is intended to measure intelligence, in the abstract, questions about what to include in a test battery are unanswerable without a theory of what intelligence is, *defined independently of the tests*. This illustrates the intellectual poverty of de facto acceptance of the argument that "intelligence is what the intelligence test tests."

10.2.2 *Evidence from Individual Tests*

In theory, a way to study male–female differences while avoiding the problem of having to justify the composition of a test battery would be to look at men's and women's scores on a pure measure of g and compare the scores obtained in an accurate sample of a large population, such as the population of a country, where the possibility of differential recruitment of men and women into the population would not be at issue. In practice, it is difficult, if not impossible, to find such a study. No pure measure of g exists; the best we have are progressive matrix tests. The most widely used of these, the Raven's Progressive Matrices (RPM), do measure g, but in most populations, the scores contain a significant visuospatial reasoning component (Abad et al., 2004; Johnson and Bouchard, 2005).

Lynn (1999, p. 6) has put the problem succinctly: "Few people will be persuaded that general intelligence can be so narrowly defined as to consist solely of fluid ability measured by the Progressive Matrices. General intelligence is generally regarded as consisting of a broader range of cognitive abilities which would include the verbal and visuospatial second order factors."

This poses a problem when Raven's test scores are used to assess male–female differences in g. As will be documented later, there is substantial evidence that there are moderate to large male–female differences in some types of visuospatial reasoning. Any difference in Raven's scores between men and women will reflect a difference both in g and in visuospatial reasoning. Even if g is the predominant contributor to the RPM score, and there are no male–female differences in g, a moderate male–female difference in visuospatial reasoning would produce a small difference in RPM scores.

The second problem is that the conclusion is a statement about differences in g, or lack of them, between men and women in general. Such a statement can be justified only if it is based on studies where the participants can be thought of as reasonably close to a probability sample of some defined large population.

A review of studies conducted prior to 1980 concluded that there were no male–female differences in Raven's scores (Court, 1983). Lynn and Irwing (2004a) properly criticized this review for having included a large number of convenience samples that made no claim of being representative of any national population. They then conducted two meta-analyses of their own (Irwing and Lynn, 2005; Lynn and Irwing, 2004b) using both the RPM and the more difficult Raven's Advanced Progressive Matrices (RAPM). They concluded that in adults, men score higher than women by approximately 0.3 deviation units, equivalent to 4.5–5 points on the IQ scale.

However, their conclusions may be problematic. Lynn and Irwing (2004a) present their results as generalizations about male–female differences in RPM scores across countries. This means that the representation of male and female examinees in a study must be reasonably representative of males and females in the relevant country. At a minimum, the male:female ratio in the sample should be roughly equal to the likely male:female ratio in the population. Failure to meet this requirement is evidence that some unknown recruitment effect may be distorting the results. An unfortunate number of the studies on which Lynn and Irwing base their case do not meet this criterion, and therefore, their general conclusion is limited.

Here are a few more examples. A study said to represent Israel was actually a study of children in a kibbutz, surely a nonrepresentative sample of modern Israel. A study involving 200 people was offered as representative of India, a nation of more than 1 billion. A sample of Brazilians aged twenty to forty contained more than 1,900 men and 740 women – a huge distortion of the male:female ratio for that age group. In another study, a sample of "American college students" was actually taken from a single university that stresses its preeminence in engineering and agriculture (Lynn and Irwing, 2004a). The male:female ratio in the study was approximately 9:2. According to records obtained from the university's website, the enrollments of men and women were approximately equal during the time of the study. Such obvious deviations from representativeness make the application of meta-analytic techniques questionable. Additional criticisms have been made of the college student analysis on other grounds (Blinkhorn, 2005).

A somewhat different picture emerges from the rather scanty reports that have been made of standardization studies. In discussing the 1979 standardization of the Raven's Standard Progressive Matrices, which analyzed data from a city that had a demographic profile resembling the national profile rather than from a population sample, John Raven (2000) reported that there were no male–female differences in progressive matrix scores. No mention is made of male–female differences in his discussion of

British and American standardizations in the 1990s, but it is not clear whether a test for such differences was ever done. A collection of papers has been published that includes reports of several national standardizations of the Raven tests and a number of smaller studies, mostly of children and adolescents. No male–female differences were reported (Raven and Raven, 2008).

A major point in Lynn's argument is that the difference in RPM scores shifts toward a male advantage from childhood to adolescence. Statistically, this would amount to an age by sex interaction. In seven of the eight studies of children and adolescents in which a comparison between the age nine to ten and age fifteen to sixteen scores could be made, there was a shift toward better male performance with increasing age (Lynn and Irwing, 2004b, table 1).

Two conclusions can be drawn from these results. The first is that if there is any systematic average difference between men's and women's scores on the Raven's tests, it is a small one. Otherwise, it would show up much more clearly in studies that do approximate national samples. The second is that the difference, if it does exist, could be due to either the g or the visuospatial latent traits that underlie performance on progressive matrix items. Given these issues, the findings of little if any average g-

factor differences from large, well-done studies of test batteries take on even greater weight. The most likely conclusion is that there are not reliable average sex differences in the general factor of intelligence (g).

10.2.3 *The Importance of Variance*

Here is an important sex difference reported in numerous studies. Men's scores on measures of general intelligence are more variable than women's scores. What does this mean? A variance difference typically is reflected in respective standard deviations, a measure of the scatter or variability of scores around the mean of a distribution. This variance difference implies that there will be differences between men and women at both extremes of the intelligence distribution even if there is no mean difference (see Figure 10.1, left). Differences at the extreme will be exacerbated if there also is a mean difference as well, even a small one (Figure 10.1, right).

Figure 10.2 illustrates the sex difference for variance in general intelligence. It comes from a classic study of more than 80,000 children who completed IQ testing at age eleven as part of the 1932 Scottish Mental Survey (Deary et al., 2003). This is a nationally representative sample. The mean scores did not differ significantly for boys and girls, but the

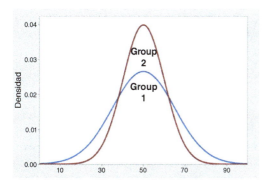
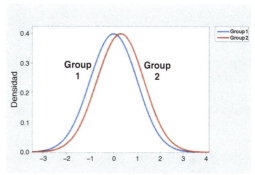

Figure 10.1 (left) Two distributions (group 1 and group 2) with the same mean, but each has a different standard deviation, a measure of variability around the mean. The distribution with greater variability (blue line) will have more individuals at both extreme tails of the distribution. (right) Two distributions with the same variabilities (standard deviations) but small mean differences. In this case, group 2 (greater mean; red line) has more individuals at the right tail and fewer at the left tail. Courtesy of Sergio Escorial.

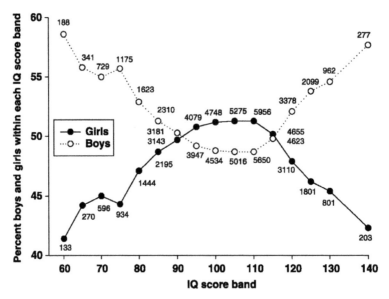

Figure 10.2 Boy–girl differences across IQ distribution, showing percentages (*y*-axis) of boys and girls found within each IQ score band (*x*-axis) of the Scottish population born in 1921 and tested in the Scottish Mental Survey in 1932 at age eleven (Deary et al., 2003). The *y*-axis represents the percentage of each sex in each 5-point band of IQ scores. The number beside each point represents the absolute number of boys and girls in each 5-point IQ score band. Note that the figure is drawn to highlight the differences at the extremes of IQ score bands.

variances did (14.1 for girls; 14.9 for boys), as assessed by the standard deviations for each group. This small difference nonetheless results in more boys (open circles) at both the low and high ends of the distributions.

In a different study, 320,000 eleven- to twelve-year-old children enrolled in UK schools took the Cognitive Assessment Test (CAT-III), a measure of general intelligence. The findings were similar to the Scottish sample – a small mean difference and more variability in the boys, as shown in Figure 10.3. The mean score across all sub-tests was 99.1 for boys and 99.9 for girls, an effect size (*d*) of 0.05 in favor of girls. The variance ratio was 1.13 (Strand et al., 2006).

Figure 10.3 uses the British study to illustrate what this difference in variances implies for the overall distribution of scores. In this study, there was a trivial difference in means between boys and girls but a reasonably large variance ratio. The figure shows the distribution of boys' and girls' scores in *stanines*. (A stanine score is a linear transformation and quantification of a standard score. Stanines are defined for intervals 1–9, with 5 representing the mean, equivalent to a standard score of zero and a standard deviation of 2.)

These examples from Scotland and the UK studies illustrate a general principle: slight differences in the means and variances of a distribution have little effect on the distribution of most of the scores but can have substantial effects on the upper and lower tails of the distribution. The combination of a slight difference in means and variances has very little effect on the distribution of intelligence test scores across men and women in the "generally normal" range, say, from IQ equivalents of 80–120, which is where 80 percent of all scores lie, but can produce substantial differences in the frequencies of men and women among the top and bottom 10 percent.

The different frequencies of men and women at the extreme ends of the distributions might have consequences for

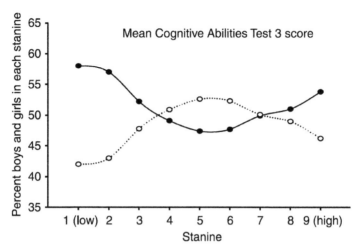

Figure 10.3 Boy–girl differences on average cognitive ability, showing percentages (*y*-axis) of boys and girls within each stanine score band (*x*-axis) for the overall mean CAT3 score. Boys' data are closed circles, and girls' data are open circles. Note the excess of boys at the extreme values (stanines 1–3 and 8–9) and the excess of girls in the middle stanines (4–6).
Data are from Strand et al. (2006).

educational programs. Boys markedly outnumber girls in special education programs, and in other indices of low but not abnormal intelligence, by a ratio of about 2:1. The ratio depends on the criterion used for admission to the program. In the United States, those states that offer fewer services, and hence have a stricter criterion for admission to the program, have higher male: female ratios (Coutinho and Oswald, 2005). Similar ratios have been reported for Europe (Skårbrevik, 2002).

At the other end of the academic spectrum, unless there is an administrative decision to require equal numbers of boys and girls, boys usually outnumber girls in programs for gifted students. This was true in the Study of Mathematically Precocious Youth (SMPY) begun at the Johns Hopkins University (Stanley et al., 1974; Stanley, 1997; Benbow et al., 1996). Part of this study involved three different cohorts that could be described as being in the top 1 in 100, 1 in 200, and 1 in 10,000 in the distribution of SAT scores. The corresponding male:female ratios were 1.5:1 (most recent), 2.1:1, and 11.2:1 (original). This

likely reflects increased educational opportunities for women. Further details about SMPY are in Box 10.1.

While differences in variance may be part of the explanation for the excess of men over women in the extremes of the intelligence distribution, this cannot be the whole story. On the positive end of the scale, a *d* scale value of 4 (IQ = 160) corresponds to the "1 in 10,000" cohort in the SMPY. Statistically, the expected male:female ratio at this point should be 3.5, which is not even close to the 11:1 observed ratio. The observed male:female ratio in special education classes (roughly equivalent to *d* =−2.5, IQ = 78 and below) is also higher than would be expected on the basis of differences in variance alone.

Wendy Johnson and her colleagues at the University of Edinburgh have suggested a reason for the overrepresentation of males at the low end of the distribution (Johnson et al., 2008, 2009). They assumed that the distribution of intelligence actually consists of two distributions: a distribution of the intelligence of normally developing individuals, which is centered slightly above the

Box 10.1 Builders of the Future

David Lubinski (2016) published an article in the *Review of Education* revisiting and discussing a century of findings on intellectual precocity defined by individuals being in the top 1 percent on measures of cognitive ability (in IQ units, from 137 to 200). He analyzed accomplishments of the intellectually precocious youth years after they were identified. Examples of these accomplishments are educational degrees, occupational income, academic tenure, patents, refereed publications, and prestigious awards, to name some of them.

His review was based mainly in two longitudinal studies that began a long journey with Lewis Terman and Julian Stanley. The Terman study began in 1921, while the Stanley study (SMPY) began fifty years later (1971). Findings from both studies have been published in six volumes across the years. Both are still active. Camila Benbow and David Lubinski began directing the SMPY in 1991.

Two key questions addressed in Lubinski's review were, (1) to what extent do individual differences within the top intelligence band matter for accomplishments of truly social significance? and (2) is there a threshold point at which differences in cognitive ability are no longer relevant and other factors become more important?

Figure 10.4 answers both questions with clarity. More ability matters. Nevertheless, the story is not exhausted by general cognitive ability. Specific abilities (verbal, mathematical, and spatial), along with personality, interests, preferences, drive, energy, and commitment, also contribute to predict social outcomes in the different realms of life.

In the final section of this informative article, Lubinski discusses the impact of our increasingly complex society. Top ability becomes more important in this world: "five great programmers can completely outperform one thousand mediocre programmers" (p. 32). The fact has huge implications across economic, political, and sociological areas.

The research group coordinated by Benbow and Lubinski has published several reports of great relevance for those interested in the significance of human capital for the development of nations in our postindustrial society. In 2013, Harrison J. Kell and colleagues asked, "Who rises to the top?" They considered 320 individuals (253 males and 67 females) showing verbal and mathematical cognitive abilities in the very top of the population distribution (top 1 in 10,000, those nominated as "scary smart") at approximately age thirteen, following their careers for three decades. The key findings revealed their accomplishments in real-life settings well beyond the educational environment: business, health care, law, or STEM (science, technology, engineering, mathematics) disciplines, to name some of them.

The same pattern emerged from another replication study from this group (Makel et al., 2016). These researchers analyzed an independent group of 259 individuals (214 males and 45 females) with the same cognitive extremely high performance at age thirteen (IQ > 160). In the follow-up (in their forties), 37 percent had earned doctorates, 7.5 percent were tenured professors, and 9 percent had registered patents (to provide some examples). Similar outcomes were observed in further life settings.

These scientists assumed that "being able to identify, attract, and develop human capital is increasingly critical for business, scientific, and technical organizations as they strive for a competitive edge" (Kell et al., 2013, p. 648). Nevertheless, it was also underscored that cognitive ability is not enough for achieving one proper understanding of exceptional human

Box 10.1 *(continued)*

potential. Motivation, engagement, and opportunity also play a role.

Similar findings have been observed by McCabe and colleagues (2020) and Bernstein et al. (2019). If replication is crucial for obtaining reliable findings, then the results reported by this research group across the years must be seen as solid knowledge. Those responsible for social policy should not be allowed to ignore the facts if they really care about those whom they represent (Kell et al., 2022). Their well-being is not a game: "Extraordinary economies are created by extraordinary minds. More than ever, the strength of countries and their competitiveness depends on exceptional human capital" (Makel et al., 2016, p. 1004).

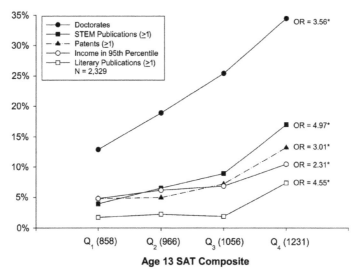

Figure 10.4 Later life achievements of precocious children. Individuals are divided by quartiles based on their SAT scores (*x*-axis) at age thirteen. These scores are shown in parentheses. Odds ratios (ORs) comparing the likelihood of each outcome (*y*-axis) in the top (Q4) and bottom (Q1) SAT quartiles are displayed at the end of each respective criterion line. The SAT assessments by age thirteen were conducted in the mid-1990s (Lubinski, 2009).

IQ = 100 point, and a distribution of individuals who have been subjected to either biological or environmental disturbances that disrupt normal development. This distribution, which is considerably smaller than the first, is centered on the IQ = 80 point. Assuming that both distributions have a standard deviation of 15 IQ points, about 75 percent of the individuals in the disrupted group would have IQs above 70, the usual criterion for the mentally disabled. Therefore, the disrupted-development population would consist largely of people whose intelligence was in the low normal, rather than the pathological, range. Johnson and colleagues further assumed that more males than females fall into the disrupted-development population. This is consistent with considerable other data showing that males are generally more at risk for biological disruption than females.

Johnson and her colleagues (2008, 2009) also pointed out that greater male variability would be expected if the (yet unidentified, but stay tuned Chapter 6) genes for general intelligence are located on the X chromosome, because the male genetic potential would then depend on a smaller, and hence more variable, sample of the alleles for intelligence than would be the case for women, with two X chromosomes. The assumption is not unreasonable, for we know that genes leading to severe cognitive pathologies are overrepresented on the X chromosome. A direct test of the hypothesis will have to wait until the genes for normal variations in the genetic potential are finally located. Johnson and colleagues' assumptions are sufficient to account for deviations from the normal distribution in low scores from two Scottish surveys of intelligence in eleven-year-olds, taken in 1932 and 1947. Similar excesses of low scores have been observed in other data sets.

10.3 Sex Differences in Cognitive Abilities

Although there is at most a minuscule average difference between men and women in general intelligence test scores, there are substantial average differences along some of the dimensions of intelligence. This is illustrated nicely in additional data from the UK study of more than 320,000 school-children (Strand et al., 2006). Figure 10.5 shows the boy and girl distributions for other intelligence factors. There are more girls at the high end for verbal reasoning and more boys at the low end. There are more boys at the high and low ends for quantitative reasoning, nonverbal reasoning, and the overall mean of the Cognitive Abilities Test (CAT3) score (not shown).

A similar pattern was found in studies of nationally representative adolescent populations. Table 10.2 shows the results from four national surveys of people who were tested

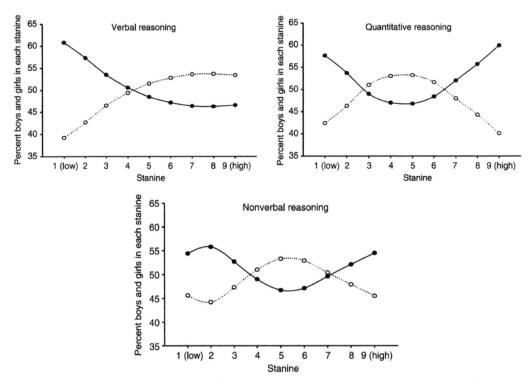

Figure 10.5 Boy–girl differences on specific cognitive factors, showing percentages (y-axis) of boys and girls within each stanine score band (x-axis) for the CAT3 battery scores (verbal, quantitative, and nonverbal reasoning). Boys' data are closed circles; girls' data are open circles.
From Strand et al. (2006).

Table 10.2 Male–female standard deviation unit (*d*) scores for effect size for different aspects of intelligence

Survey code	Test date	N	Reading	Math	Abstract reasoning	Spatial	Perceptual speed	Associative memory
Project Talent	1960	73,425	−0.15	0.12	0.04	0.13	n.a.	−0.32
NLS-72	1972	16,860	−0.05	0.24	−0.22	n.a.	−0.23	−0.26
HS&B	1980	25,069	0.002	0.22	n.a.	0.25	−0.21	−0.18
NLSY79	1980	11,914	−0.18	0.26	n.a.	n.a.	−0.43	n.a.
NELS:88	1992	24,599	−0.09	0.03	n.a.	n.a.	n.a.	n.a.

Note. The NLSY79 survey reported two scores for mathematics – arithmetical reasoning (use of numerical reasoning in problem solving) and mathematical knowledge. The same survey also reported two scores for the speed of conducting simple cognitive operations. One was a simple decoding operation; the other evaluated the examinee's speed of executing elementary numerical operations. n.a. = not applicable. Data are from Hedges and Nowell (1995, tables 1 and 2).

when they were in high school and who have since been followed through their early adult careers (Hedges and Nowell, 1995). While there is some discrepancy between the results, which is probably due to differences in content between different tests, two trends stand out. Women do better than men in tests of reading comprehension, speed of simple perceptual operations, and tests of associative memory, in which examinees have to recall arbitrary associations, such as associating a picture and a number. Men do better than women on tests of visuospatial reasoning and mathematics.

These conclusions refer to broadly defined abilities. Two psychometric research studies provide further detail and, in the case of the second study, establish a theoretical framework for thinking about these results.

First, the Differential Aptitude Battery (DAT) is a battery of tests developed by the Educational Testing Service (ETS) for research purposes. Adult male–female differences are consistent among samples from the United States, Spain, and the United Kingdom (Colom and Lynn, 2004; Strand et al., 2006). Women do somewhat better on tests of language skills (as opposed to reasoning about verbally presented material) and on tests of speed and accuracy of

simple operations. Men do markedly better on tests involving the manipulation of visual images. Men also do slightly better on tests of verbal and abstract reasoning.

Second, Johnson and Bouchard (2007a, 2007b) have placed similar results in the context of the *g*-VPR model. In previous work (reviewed in Chapter 4), Johnson and Bouchard had shown that the *g*-VPR model provides a good fit to the Minnesota Study of Twins Reared Apart data. They removed the variance in scores associated with the *g* dimension and analyzed the residual scores on the various tests. The residuals can be thought of as being the variation in test scores that *cannot* be ascribed to variations in general intelligence. The residual variation could be characterized by three orthogonal (independent) dimensions. The smallest of these, in terms of variance accounted for, was a "memory for content of passages" factor. This will not be discussed further. The other two factors are more interesting.

These two residual factors are *bipolar* factors, in the sense that tests tend to have either high or low loadings on each of them. Johnson and Bouchard refer to these two dimensions as *verbal-rotational* and *focused-diffuse*.

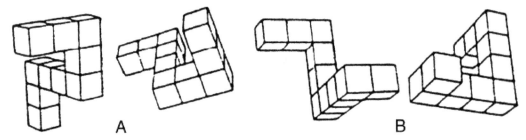

Figure 10.6 Illustration of a visuospatial reasoning task on which men outperform women on average. For each comparison (A and B), are the two objects the same (in different spatial orientations) or different? The answer is in footnote 1.

The bipolar factors need special interpretation, for they are not quite the same as the ability factors identified in the original g-VPR model. Vocabulary tests and rotational tests were good markers for the respective ends of the verbal-rotational factor. What the result says is that in the MISTRA sample of adults, *after general intelligence had been accounted for*, people who knew lots of words tended to be poor at manipulating mental images, and vice versa. The focused-diffuse dimension contrasted people who did well on tasks that require concentration on visual diagrams with people who did well on tasks that involve comprehension of verbal arguments and possession of information about the world.

Johnson and Bouchard measured male–female differences along each of the dimensions. They found only small, nonreliable differences on the g dimension. There were, however, large differences (d values larger than 0.5) on the subsidiary dimensions. Women tended strongly toward the verbal ends of both bipolar dimensions and tended to have superior memory for information presented during the testing session. Men tended to have higher scores on the "focused" (on visual objects) and "rotational" ends of the bipolar dimensions. Once again, though, it is important to remember that these differences refer to performance after removing variation due to general intelligence.

Johnson and Bouchard (2007b) have offered an interesting summary and

interpretation of their results. They argue that whenever a person solves a problem, they do so by combining general reasoning ability (i.e., g) with the particular mental tools the person has on hand. For instance, many ostensibly visuospatial problems can also be solved by verbal reasoning. The results just cited indicate that men and women differ somewhat in the quality of their mental tools; the verbal tools tend to be better for women, and the mental imaging and attention-focusing tools tend to be better for men. Rational application of general intelligence would thus lead men and women to adopt somewhat different strategies for problem solving. Therefore, even though men and women are essentially equal in general intelligence, they may differ in their performance on particular tests, because performance depends on both general intelligence (for which there is no sex difference) and the residual abilities, where sex differences may be substantial.

Both the national surveys and the psychometric research studies indicate that the big differences between men and women are on perceptual and visuospatial reasoning tasks – the P and R dimensions of the g-VPR model. Laboratory studies amplify these results. Men tend to be better than women at tasks involving the manipulation of mental images. The prototypical example is a mental rotation task, in which two figures must be compared by moving them about "in the mind's eye." Women take longer to do this, on average, and make more errors.

---

[1] Answers are the pair is the same for A and the pair is the same for B.

Target

Context A
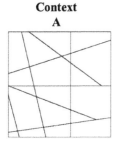

Context B
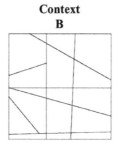

Context C
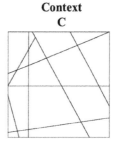

Examinees must choose which context contains the target

Figure 10.7 Detecting details in a complex picture. (top) A target figure. (bottom) Response options; only one contains the target, and two are distractors. This task requires analysis of a static visual figure. Women tend to outperform men on this sort of task. The answer is in footnote 2. From de-Wit et al. (2017).

Examples are shown in Figure 10.6. Men also do better than women on tasks involving judgment of real or imagined motion (Hunt et al., 1988; Law et al., 1993; Kimura, 1999, chapter 5).

By contrast, there are many visuospatial tasks that do not involve mental rotation. For instance, women tend to outperform men on tasks that require analysis of static visual figures. An example, where the task is to find a component within a picture, is shown in Figure 10.7.

A possibly related finding is that males do better than females in the ability to find locations and to maintain awareness of positions in the environment. This is true of both children and young adults and appears to be related to performance on spatial orientation tasks, although the relationship is not a strong one. This result holds for imagined routes through environments, actual orienting, and acquisition of knowledge of the environment through interaction with

computer-generated virtual environments (Choi and Silverman, 2003; Malinowski, 2001; Waller et al., 2001).

The skills evaluated by these tasks have applications in everyday life. There are very large individual differences in our ability to find our way about the world. And, as would be expected from Johnson and Bouchard's analysis, wayfinding can be accomplished by visualizing the surrounding environment or by using strategies that rely on verbal memory. There are also a number of practical situations in which a person must visualize motion or changes in perspective while viewing a static display. These vary from analyzing gear trains to assembling objects from diagrammatic instructions – a task that will be familiar and perhaps frustrating to anyone who has purchased to-be-assembled furniture kits.

Such observations do not mean that women can't find their way in the woods, read maps, be architects, fly helicopters, and

² Answer is C.

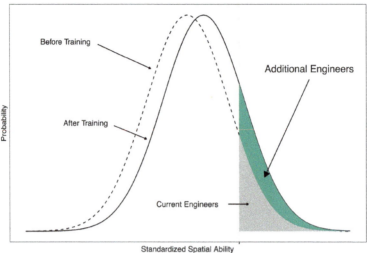

Figure 10.8 Consequences of implementing spatial training on the probability (*y*-axis) of individuals who would have the spatial skills (*x*-axis) associated with receiving a bachelor's degree in engineering. The dotted line represents the distribution of spatial skills in the population before training; the solid line represents the distribution after training. Shifting the distribution by *d* = 0.40 (a conservative estimate from the meta-analysis by Uttal et al. [2013]) would approximately double the number of people who have the level of spatial skills associated with receiving a bachelor's degree in engineering (green plus gray areas of the distribution).
Data are based on Wai et al. (2009) and Wai et al. (2010). Courtesy of Elizabeth Tipton and David Uttal.

so on as well as men. Virtually all tasks we encounter in daily life admit to multiple solutions. General intelligence is a far better predictor of performance than rotational ability, even though both are important. People can apply general intelligence to develop a problem-solving strategy that suits their particular cognitive strengths and compensates for their weaknesses.

Claims for better male performance in rotation and related visuospatial abilities refer to statistical trends. Assuming a half standard deviation difference on a rotation task (*d* = 0.5), we would still expect 30 percent of the women to outperform 50 percent of the men. The differences in the frequencies of men and women, in favor of men, would be more extreme at higher levels of performance given the variability ratios previously discussed (see Figure 10.1).

Keep in mind that visuospatial skills (like virtually all specific cognitive abilities) can be improved through training. However, training does not narrow the male–female visuospatial

gap, according to a comprehensive meta-analysis of 217 studies (Uttal et al., 2013). Nonetheless, these researchers note that widespread spatial skills training in schools could have impact on STEM careers overall, as illustrated in Figure 10.8, although this could accentuate the male–female STEM career gap.

10.3.1 *Sex Differences in Cognitive Traits and Education*

This section discusses cognitive differences between men and women that have a direct impact on education. The issue is important, because different educational avenues lead to markedly different careers in adulthood.

Psychologists and educators think about individual differences in different ways. Psychologists are interested in tasks that maximize individual differences in human behavior, try to characterize these differences, and, especially in recent years, try to relate behaviorally defined traits to biological systems that may or may not be influenced by genetics.

By contrast, educators think in terms of subject matter. They want to talk about cognitive traits associated with educational content. The biggest division of the curriculum is into topics that are broadly associated with language, the arts, and topics associated with mathematics. Accordingly, to an educator, the most interesting individual differences are differences in the ability to deal with language and mathematics. "General reasoning" is too amorphous a concept, and, to an educator, perceptual and visuospatial skills seem too microscopic.

These different perspectives have left the field of education without much incorporation of genetic and neuroscience findings about the general factor of intelligence (g). This is unfortunate because the g-factor alone reliably accounts for the largest amount of variance in school achievement. This has been demonstrated consistently in comprehensive studies from the Coleman Report (Coleman, 1966) to more contemporary examples (Deary et al., 2007b; Kaufman et al., 2012; Zaboski et al., 2018). Another consistent finding from these studies, hardly ever mentioned, is that teacher and school variables together account for only about 10 percent of the variance in school achievement (Detterman, 2016) (more details are in Box 7.6).

Coordinating these two views requires a sustained effort over time, as more education departments are just beginning to include courses on intelligence. In this section, we focus on the ways in which average sex differences impact education. A large study from Germany provides a good place to begin.

Martin Brunner and two colleagues (2008) at the Max Planck Institute conducted a psychometric analysis of the data from a German testing program involving more than 29,000 students, randomly selected from the seventeen-year-olds in the German school system. The study was exceptional both for the representativeness of the sample and for the care that the investigators paid to the technical issues concerning group differences that were described in Chapter 9. They concluded that the data were best fit by a hierarchical model, consisting of a general factor (g) and

a nested factor model, in which mathematics or verbal (reading) abilities contribute to overall mathematics or reading test scores. The traits of interest to educators, language and mathematical abilities, appear as specializations of a general reasoning factor, which is not a targeted educational variable.

Brunner and his colleagues then examined male–female differences along each dimension. The seventeen-year-old girls slightly outperformed the boys on the general reasoning and reading factors ($d = -0.09$ for both comparisons), but the boys markedly outperformed the girls on the mathematical factor ($d = 0.94$). This conforms to the observation that boys are better than girls on average at mathematics, although, as we will show, this finding requires qualification.

Brunner and his colleagues then considered what their findings might imply for mathematics. They argued that mathematical problems are attacked with a combination of general reasoning skills, in which men and women are essentially equivalent, and mathematics-specific skills, in which men, on average, exceed women. As a result, men should do better than women in those areas of the curriculum that emphasize mathematics, providing that the mathematics involved is sufficiently specialized to emphasize mathematical rather than general reasoning skills.

Applying Brunner and colleagues' (2008) reasoning to the typical educational progression, what we should see is a progressive sharpening of differences between men and women in educational accomplishment as we move from the general education curriculum through the undergraduate university years, and then on to specialized education and career achievements in the STEM fields. That is what happens, but there are some important qualifications, as described in the next sections.

10.3.2 Boys and Girls in the K–12 System

In the United States and many other developed countries around the globe, public education is available to everyone from kindergarten through the twelfth grade. While there is some variation, attendance is usually compulsory through age sixteen, and students are strongly

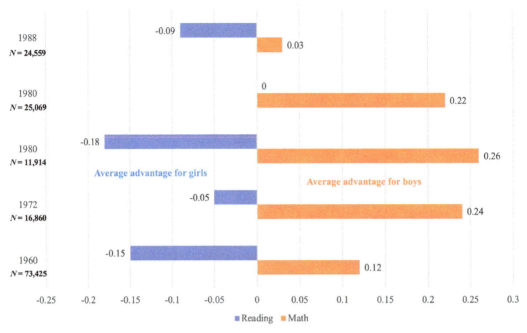

Figure 10.9 Average sex differences (d values) on tests of reading (blue) and mathematics (orange). All studies used large probability samples of the relevant US population. Positive numbers indicate a male advantage; negative numbers indicate a female advantage.
Data are from Hedges and Nowell (1995, table 2).

encouraged to complete the entire course, graduating at seventeen or eighteen.

Figure 10.9 presents a comparison of the scores achieved by US high school age boys and girls on the cognitive tests used in several Department of Labor surveys during the last half of the twentieth century (Hedges and Nowell, 1995). Throughout this period, girls consistently outscored boys on tests involving language use, while boys outscored girls by a somewhat larger margin on tests of mathematical skills. Similar results have been obtained by the National Assessment of Educational Progress. This test has been called "the nation's report card" for the evaluation of language, mathematics, and science skills. Somewhere between 70,000 and 100,000 students are evaluated each year. In all available reports, in the twelfth grade, girls outscore boys in reading, while boys outscore girls in mathematics ("Nation's Report Card," National Center for Education Statistics; https://nces.ed.gov/nationsreportcard/).

In the Program for International Student Assessment (PISA), representative schools of the Organisation for Economic Co-operation and Development (OECD) are selected, and fifteen-year-old children are evaluated on standardized tests involving reading, mathematics, science, and problem solving. The problem-solving section presents realistic problems beyond specific academic knowledge, such as finding an efficient route on a bus line. Figure 10.10 shows the scores for boys and girls on the PISA problem solving, reading, and mathematics tests, calculated for nine countries across continents. On average, boys score higher than girls on problem solving and mathematics, while girls score higher than boys on reading. Nonetheless, there is variation across countries. In the United Arab Emirates and Sweden, all differences are favorable to girls, no matter the content. For the remaining countries, the pattern follows the general rule: on average, boys show higher scores on problems solving and math, while girls show higher scores on reading. However, the specific values show considerable variation. Thus, for instance, the average sex difference in problem solving for Brazil and Japan is 22 and 19, respectively,

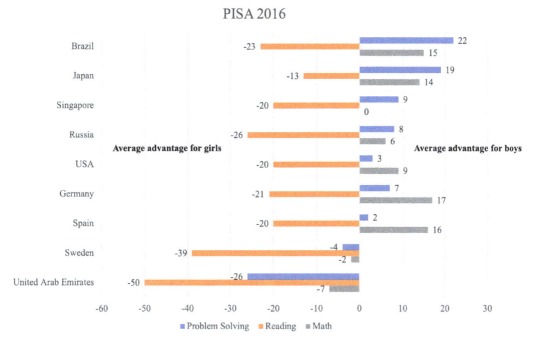

Figure 10.10 Sex difference raw scores (*x*-axis) on the PISA 2016 problem solving (blue), reading (orange), and mathematics (gray) subtests calculated for several countries (*y*-axis). Across countries, boys generally (but not always) score higher than girls in math and problem solving, while girls typically score higher than boys in reading.

while for Spain and the United States, it is 2 and 3, respectively.

The data for K–12 educational achievement show that, at the end of the standard (and often compulsory) school period, midteenage girls have greater language-related skills than boys of the same age, while the opposite is true of skills and knowledge in mathematics. It is important to remember that this is a statement about the extent to which the male and female score distributions differ with respect to each other. It is not a statement of differences in absolute skill, because there is no metric by which we can compare a difference in mathematical knowledge directly to a difference in language skill, even if PISA scores on language, math, and science reflect general intelligence (general learning ability) (Pokropek et al., 2021).

10.3.3 *College and University Undergraduate Education*

College and university students represent an important population, because this population of young adults contains most of the people who will be leaders of society, both in the dramatic sense of providing a few highly visible leaders and in the perhaps more important sense of providing the many business managers, entrepreneurs, technicians, and professionals who will constitute the economically and socially most productive segments of society (Gelade, 2008). In the last fifty years, there has been a tremendous expansion of social and economic opportunity for women within the college-educated segments of society. There is considerable reason to be interested in differences between the cognitive skills of men and women within this group.

It is important to remember, though, that results obtained by studies of people who are either in or about to enter undergraduate education do not necessarily generalize to the population at large. Since the early 1980s, more women have enrolled in college than men. As of 2018, women were about 56 percent of the undergraduate college population. This situation arises partly because at a given level of high school

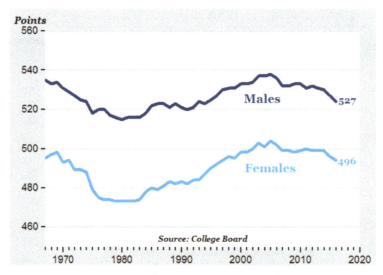

Figure 10.11 Mean SAT mathematics scores (*y*-axis) for men and women from 1967 to 2016 (*x*-axis).
From the College Board, https://reports.collegeboard.org/media/pdf/2016-total-group-sat-suite-assessments-annual-report.pdf.

academic achievement, a woman is more likely to take the first steps toward undergraduate education than a man (Hunt and Madhyastha, 2008).

One implication of this is that average sex differences will be algebraically larger in the undergraduate population than in the high school population. The effect will be further exacerbated by the variance effects described earlier. Men's scores will be increased, overall, because of a high male:female ratio in the population of people who have high test scores. The excess of men with low test scores will not affect the undergraduate means, because people with low test scores are unlikely to enroll in undergraduate programs.

These trends are seen in the test scores of entering students. In the high school population, as a whole, the women's mean reading score is above the men's mean by about 0.2 standard deviation units, and men exceed women in mathematics test scores by about 0.1 standard deviation units. For example, the 2008 SAT reading scores were 504 for men and 500 for women, while the mathematics scores were 533 for men and 504 for women. Using the nominal 100-point standard deviation for SAT sections, this translates into a trivial 0.04 *d* male advantage in

reading and a nontrivial 0.33 *d* advantage in the mathematics (now renamed "reasoning") portion of the test. This is the sort of magnification of sex differences that would be expected because of the differential recruitment issue (discussed in Chapter 9) of men and women from the high school to the undergraduate population.

The male–female discrepancy in SAT math scores is not a recent phenomenon. Figure 10.11 shows SAT-M scores of college-bound seniors from 1967 through 2016. Although the levels of scores for both men and women vary from year to year, the difference between means remains consistent.

In 2000, an approximately equal number of men and women earned bachelor's degrees in STEM fields. This was a change from thirty years earlier, when the ratio of men to women earning bachelor's degrees in these fields was approximately 2.5:1 (www.nsf.gov/statistics/2015/nsf15326/pdf/nsf15326.pdf). However, because the undergraduate male:female ratio is 3:4, the 1:1 ratio in the STEM fields implies that a man is about four-thirds more likely to major in a STEM field than is a woman. The extent to which this disparity is due to cognitive competence, personal interests, or social pressures is not clear. There is, however, research suggesting that it is a question of choice rather than of lack of the required ability (Wang et al., 2013), and we discuss this further later in this chapter.

Women are not entering the STEM fields and then dropping out because they find the work too difficult. Given constant SAT-M scores, women consistently outperform men in mathematics courses (Wainer and Steinberg, 1992; Lynn and Mau, 2001). The difference can be striking. Results from a large ($N \sim 49{,}000$) study of college students in a variety of US universities (Wainer and Steinberg, 1992) found that SAT-M scores increase as a function of the level of mathematics involved. This reflects an unsurprising fact: people with high SAT-M scores are more willing to enroll in mathematics courses than people with low scores. Within each type of course, SAT-M scores increase as the course grade increases. This shows that the SAT-M is a valid predictor of accomplishments in mathematics classes. Both of these trends hold for men and for women.

But there was a paradox. Within each course and grade level, women receive *higher* grades than men, even though they have lower SAT-M scores. In terms of educational outcomes, the difference can be substantial (Lynn and Mau, 2001). Women who receive B grades have SAT-M scores lower than the men who receive Cs and Ds in the same class. This is a striking example of a general tendency for SAT scores to underpredict women's educational achievements in the early undergraduate years.

Three explanations might account for the paradox. One is that the test is an objective measure of mathematics ability, while grades are a subjective measure based on the instructors' perceptions. Men might actually have more mathematics ability, but women are able to present a more favorable impression to instructors.

This explanation strikes us as a plausible argument for low correlations between test scores and grades in courses where there is a substantial subjective component to

grading, such as a course in English literature, but it is more difficult to see how the argument applies to lower-division college mathematics courses, where right answers are clearly defined.

A second argument is that women simply work harder to get grades: "the most probable explanation is that women's stronger work motivation compensates for their lower test score" (Lynn and Irwing, 2004b, p. 495). The motivational argument is not unreasonable. However, this implies a false dichotomy between intelligence and motivation as evidenced by behavior as a student. Talents such as good time management and establishing priorities among goals belong to the intelligence construct in the conceptual sense (Chapter 2) as much as the talents required for taking tests. This argument is also weakened by recent evidence that motivation has little appreciable influence on intelligence test scores (Bates and Gignac, 2022).

A third possibility is that tests of mathematical aptitude, such as the SAT-M, are influenced by a psychological trait on which there are sex differences, but this trait either does not contribute to mathematical performance outside the test, or does so, but has more influence on test performance than it does on in-class performance (Sackett, 2021). Visuospatial ability, the R dimension of the g-VPR model, has been suggested as a possibility, both for the SAT and for progressive matrix tests (Lynn and Irwing, 2004b; Casey et al., 1995). Alternatively, there might be some ability that is not evaluated by the test but that is important in the study of mathematics and is possessed by women more than by men. Time management is an example of such an explanation.

When all is said and done, we just do not know what the link is between male–female differences in test performance and in performance in mathematics. The problem becomes more acute as we look at discrepancies between men and women at a higher level of analysis, the pursuit of careers in the STEM fields. Let's see what the data indicate.

10.3.4 *Postgraduate Education and Career Development*

It is difficult to say anything succinct about male–female differences in postgraduate education and career development in general, because any statement has to be qualified by considering the field involved. Postgraduate education is itself so varied that general statements about how men and women progress through curricula as different as mathematics, education, medicine, and the law are equally suspect. What we can do, however, is look at some of the highly publicized differences in outcomes between men and women. Socially, what has been of particular concern is the scarcity of women in the STEM fields.

The disparity is long-standing. Charles Murray's (2003) statistical survey of 3,000 years of human accomplishment uncovered very few eminent woman scientists or mathematicians. This is hardly surprising, owing to restrictions on women's activities that were enforced by various human societies until modern times, and even now outside the industrially developed countries. Within these countries, lifting of the restrictions is quite recent. In 1925, Cecelia Payne-Gaposchkin (1900–1979) became the first woman to receive a doctorate (in astronomy, a STEM field) from Harvard University. A few years later, in 1934, Grace Hopper (1906–1992) became the first woman STEM graduate (in mathematics) at Yale University. (Hopper later developed COBOL, one of the early computer programming languages, and was the first person to use the term *bug* to describe an error in program execution.) In 2007, Drew Gilpin Faust became the first woman president of Harvard. Hanna Holburn Gray served as acting president of Yale for one year in 1977. Things have changed, but perhaps not at breakneck speed. Harvard was founded in 1636, Yale in 1701.

Figure 10.12 shows the percentages of women receiving doctorates in several relevant fields in 2016–2017. There has clearly been a great increase in the number of women

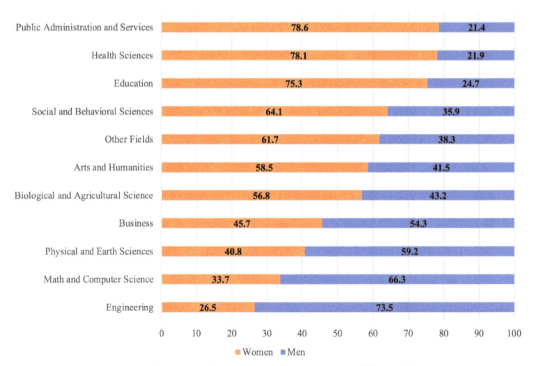

Figure 10.12 Percentages of women and men among the recipients of doctoral degrees, 2016–2017. From Council on Graduate Schools (2018).

receiving advanced degrees. However, the pattern for type of degree has changed very little over a thirty-year period marked by major advances in opportunities for women. Today, women greatly outnumber men who receive doctorates in education and are greatly outnumbered by them in mathematics, the mathematically oriented physical sciences, and engineering.

Why have the higher reaches of the STEM fields remained so weighted toward men? In a widely publicized speech, Faust's immediate predecessor at Harvard, the economist Larry Summers, proposed three reasons for the discrepancy. Two were social: the existence of conscious or unconscious prejudice on the part of hiring and promotion committees and a preference for a personal lifestyle that is not compatible with the workaholic standards Summers associated with very high productivity in the STEM fields. The third was biological. Summers observed that there is a striking disparity between the numbers of men and women who have very high scores on tests of mathematical ability and achievement. He then speculated that women's brains are organized in a way that makes acquiring the high level of analytic skills needed in STEM research more difficult, on average, for women than for men.

Summers's remarks created such a firestorm that he subsequently resigned from the Harvard presidency. The incident is described in Box 10.2. Going into all the possible social and biological reasons why women might be underrepresented in high-profile STEM positions is well beyond the scope of this book. Nevertheless, the controversy does provide a good entry point for a more general discussion in the next section of the possible origins of sex differences in some cognitive abilities.

Most researchers in the field were far more nuanced in their reaction to Summers's remarks than political figures and academic leaders. The American Psychological Association (APA) and the Association for Psychological Science (APS), the two professional organizations most involved, commissioned reports by people well known for their research in the field. The APS report took up an entire issue in one of its journals. Here are two sentences from one contribution: "There cannot be any single or simple answer to the many complex questions about sex differences in math and science" and "Early experience, biological constraints, educational policy and cultural context each have effects, and these effects add and interact in complex and sometimes unpredictable ways" (Halpern et al., 2007, p. 41).

The APA reports resulted in a book of contributed chapters titled *Why Aren't More Women in Science?* (Ceci and Williams, 2007). The editors concluded their final summary of the controversies by quoting from their chapter 8, "Brains, Bias and Biology: Follow the Data": "The challenge is to follow where the data lead, always cognizant of Orwellian fears and prejudiced misuse of knowledge balanced by the prospects of alleviating suffering from disorders and enhancing the quality of life for everyone. Along the way, controversy can only escalate as we constantly test new knowledge against old and comfortable ideas. This is the way science works and the way our culture evolves" (p. 234; see also Haier, 2007).

Still another edited volume surveyed both social/cultural and biological/genetic perspectives on women in science based on presentations at a symposium held at the American Enterprise Institute (AEI) (Sommers, 2009). The symposium was based on a report from the National Academy of Sciences, National Academy of Engineering, and Institute of Medicine (2007) that asserted with certainty that cognitive sex differences could not explain any of the male–female disparities in science and that no biological differences were relevant. This conclusion was the basis for far-reaching policy decisions about committing substantial National Science Foundation (NSF) funding to encourage more women in science. Did the data justify the policy? Like the other two collections of perspectives, the AEI volume acknowledged the history of bias and discrimination but focused on why there was

Box 10.2 The President of Harvard Said What?

On January 14, 2005, Lawrence Summers, at that time president of Harvard University, gave a speech addressing the fact that women are underrepresented in the STEM fields, and even more underrepresented at the top of those professions. Summers subsequently commented that he had been asked to be provocative. He was.

Summers proposed three causes for the discrepancy: (1) conscious or unconscious prejudice against women, (2) women's distaste for the intense professional commitments required to rise to the top of the STEM fields, and (3) women's difficulty in acquiring mathematical and scientific reasoning skills. Summers noted that a relatively small percentage of women achieve high scores on tests of mathematical reasoning and suggested that biological differences between men and women might contribute to the disparity in high-level mathematics skills.

Several prominent women scientists left the meeting in protest. The high-visibility magazine *Science* published a letter signed by seventy-three prominent academics protesting his statements (Muller et al., 2005). The Harvard faculty voted no-confidence in Summers, and in 2006, he resigned his post, returning to his position as a professor of economics. In January 2009, he was appointed chair of the Presidential Council of Economic Advisers, making him the chief White House adviser on economic matters within the Barack Obama administration.

The politics behind Summers's resignation are not relevant to the science of human intelligence. But the protest letter to *Science* is, because it indicates both the passions that are involved in discussions of group differences in intelligence and the beliefs held by highly influential people who likely are not familiar with the weight of empirical evidence or who seek to delegitimize it.

Here are two quotes from the letter to *Science*: "There is little evidence that those scoring in the very top of the range in standardized tests are likely to have more successful careers in science education" and "We are concerned by the suggestion that the status quo for women in science may be natural, inevitable, and unrelated to social factors" (Muller et al., 2005, p. 1043).

Although the list of seventy-three signers of the letter included prominent academic scientists and science administrators, it did not include any of the major figures who do research on individual differences in cognition. It is unlikely that many of them would have signed the letter, for the first statement is demonstrably false. By 2005, the results of the SMPY (i.e., people whose SAT-M scores were in the top 1 percent) were well known to professionals in the field. These results document the stunning success of people whose scores were "in the very top of the range in standardized tests" (Park et al., 2008).

What about the second statement? Summers never said anything about the status quo being natural, inevitable, or unrelated to social factors. On the contrary, he specifically listed two social factors that he thought contributed to the disparity: discrimination and conflicts between professional and family goals. The fact that Summers mentioned a possible biological explanation for differences in men's and women's intelligence was equated with a denial of social causes and a deterministic view of biology.

The academic leaders who signed the Muller letter either felt no need to consult with experts in the field, who would have been easily available to them, or decided to disregard far more nuanced expert opinions.

> **Box 10.2** *(continued)*
>
> The case of James Damore, a Google engineer, is another widely known instance of the risk taken when talking about empirical data of sex differences and the weight of evidence. He speculated that the paucity of women software engineers may have to do with a tendency to select careers that focus on people (like medicine or law) as opposed to things (like engineers). He was fired. We discuss the people/things effect in Section 10.4.1.

far more uncertainty in the data than the NSF acknowledged, especially regarding biological research. The introduction noted, "Nevertheless, the corrective to the history of damaging bias is not more bad science; it is good science, clear thinking, and open, fairminded discussion" (Sommers, 2009, p. 4).

The next sections present data about how cognitive sex differences might originate. Knowing about these perspectives is relevant for any discussions about whether the differences matter and for developing efficient social, educational, or vocational policy.

10.4 Possible Origins of Cognitive Sex Differences

It is unlikely that there is any one cause of the cognitive sex differences we have described. Social/cultural explanations are preferred by people who believe that these causes are more malleable than biological/genetic ones. The evidence for this belief, however, is rather weak. As discussed earlier, genes are expressed and influenced by other biological factors and by environmental factors including social and cultural contexts. The challenge is to investigate and understand how many factors interact. Some researchers have concluded that this may be impossible, especially when random events might drive key aspects of brain development (Mitchell, 2018). But there is room for more optimistic views given the history of scientific progress answering questions about complex processes.

An integrative framework for answering the question about the origins of sex differences in cognitive abilities was proposed by Halpern (2000) (Figure 10.13). Its focus is on biological, psychological, and social influences.

Consistent with the evidence discussed in previous chapters, nature versus nurture is an unwarranted oversimplification that departs from what the available scientific evidence shows so far. The biopsychosocial model supports a continuous loop in which the factors involved are closely linked. The next sections present the basic arguments for a few selected explanations of origins and illustrate new progress.

10.4.1 *Social/Cultural Influence and Interests*

What interests you? Summers thought that one of the reasons there are more men than women in the STEM fields is that work in these fields is simply more interesting to men than to women. If Summers was correct, men should have a higher participation rate than women in the STEM fields even within a population of talented men and women, where ability to enter the field is not an issue, but interest is.

The SMPY study identified such a group in the 1970s and 1980s when they were teenagers (see Section 10.2.3 and Box 10.1). They scored in the top 1 percent of their cohort in terms of the SAT-M test. They have been followed periodically with surveys about their careers, family lives, and interests (Benbow et al., 2000; Ferriman

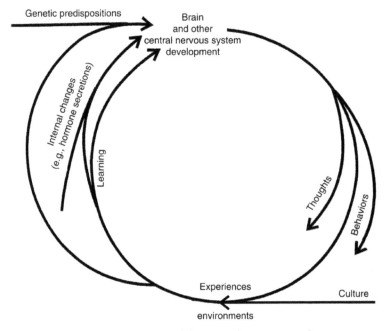

Figure 10.13 Biopsychosocial model. Genes, hormones, and experiences contribute to brain development and to how individuals select pieces from the environment following their predilections and past experiences. The latter also contribute to modifying their brains. Nature and nurture are related in a continuous feedback loop (Halpern, 2000, 2012).
Courtesy of Diane Halpern.

et al., 2009; Lubinski et al., 2006; Wai et al., 2005; Bernstein et al., 2019). Generally, participants who had primary interests in things (objects rather than people) and abstract ideas tended to follow careers in mathematics and the sciences. Participants who had interests in people and social issues followed careers in the humanitarian/social issues–oriented professions. Participants also differed in the extent to which they valued careers or families. Those participants who had strong family orientations were underrepresented in the STEM professions, which are notoriously demanding of time.

Men tended to fall more into the "things–ideas–career" pattern, while women tended toward the "people–family orientation" pattern (Benbow et al., 2000; Ferriman et al., 2009; Lubinski et al., 2006; Wai et al., 2005; Su and Rounds, 2015; Su et al., 2009). However, examples of each pattern occurred in both highly talented men and women. The codirectors of the SMPY, Camilla Benbow and David Lubinski, point

out that differences between men's and women's interests alone would create disparities in the extent to which men and women choose to work in the STEM fields (for a good discussion of these social issues, see Halpern et al., 2007, pp. 31–39).

Of course, the origin of interests is the issue. Are interests mostly influenced by the sociocultural milieu and/or by biological differences? The SMPY data do not address this question. However, there are data to test whether the degree of gender equality in nations is related to male–female disparities in STEM fields. It is reasonable to expect that more gender equality will be associated with smaller STEM disparities. This is because, generally, gender-equal countries give more educational and empowerment opportunities and do more to promote STEM fields to girls and women.

Stoet and Geary (2013) obtained PISA scores for reading comprehension, math, and science literacy from 472,242 students in sixty-seven nations or regions like Hong

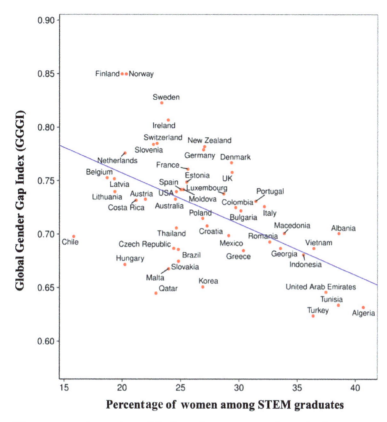

Figure 10.14 A paradoxical finding: the percentage (*x*-axis) of women getting STEM degrees was lower in more gender-equal countries according to the Global Gender Gap Index (*y*-axis) ($r = -0.47$). These data do not support the view that gender roles are determined solely by culture, since fewer women pursued and obtained STEM degrees in the most egalitarian counties.
Adapted from Stoet and Geary (2018).

Kong and Macao. Consistent with previous analyses of PISA data, across countries, boys generally scored better on math, and girls scored better on reading comprehension and science literacy. Each country was also scored on the Global Gender Gap Index (GGGI; World Economic Forum, 2015), which is based on fourteen key male–female difference indicators (e.g., earnings, life expectancy, seats in national legislatures). High GGGI scores denote more gender equality. Many other variables were also assessed (e.g., career interests, degrees obtained).

If social and cultural variables exert major influences on career choices, then in nations or regions with more egalitarian views about roles for men and women, the disparities in STEM degrees should be smaller. However, just the opposite was found. STEM degree disparities were greater in the most egalitarian countries (see Figure 10.14). Explanations for this paradoxical finding are not clear, but one possibility is that when given a fair choice, women prefer family over career and do not participate in STEM careers as much as their abilities might predict. This would be consistent with the people–things findings from the SMPY longitudinal studies.

There is also evidence that boys and girls have different learning experiences that might impact their acquisition of visuospatial skills, and concomitantly of mathematics.

A research literature review (Hyde, 2007) found evidence supporting that (1) in the United States, parents are more likely to provide analytical, causal-oriented explanations to boys than to girls, (2) middle school children generally accept the stereotyped view that females do not do well in mathematics and spatially oriented tasks, (3) male children begin to play and explore away from home at younger ages than girls do, and (4) mathematics and science teachers tend to direct their discussions more toward boys than toward girls.

Why might such differences arise? Is it because boys and girls draw forth different behaviors from the adults, or because the adults initiate these behaviors for either social or biological reasons? These possibilities are distal explanations; the proximal facts are that when people are provided differential learning experiences, they will learn different things. Psychologists cannot conduct an experiment to prove that this is a cause of differences in men's and women's performance on mathematical tasks – for people cannot be randomly assigned to lifestyles.

Another hypothesis involves different social expectations about math ability (or intelligence, or any other trait or ability) being greater in males than in females. If girls and women believe the stereotype of inferiority, won't this belief influence them in a way that transfers to poorer performance? This idea is called *stereotype threat* (Steele and Aronson, 1995, 1998), and we introduced it in Chapter 9. There is a research literature suggesting some validity to this hypothesis, but meta-analyses find that the effects generally are small or non-existent (Zigerell, 2017; Flore et al., 2018; Finnigan and Corker, 2016).

Social and cultural factors undoubtedly play some role in cognitive sex differences directly or indirectly. However, it is a difficult empirical problem to demonstrate how this might work. Picking out one of many bivariate correlations and offering it as *the* explanation for a phenomenon, without a serious consideration of alternative hypotheses, is not a compelling scientific argument.

Given the complexities, interactions, and difficulties of measurement that might influence learning, motivation, and interest, it might not be possible to identify specific social or cultural reasons for any sex differences in cognitive test scores. Is the task any easier for biological variables? Let's first consider evolution by natural selection, the driver of all biological processes directly or indirectly.

10.4.2 *Two Views of Evolution*

Evolutionary psychologists have offered a widely held distal biological explanation for differences in cognition between men and women: man is the hunter and woman is the gatherer. In every society, there are sex role differences that go beyond the childbearing difference forced by biology. Evolutionary psychologists suggest that these differences can be traced to a natural selection advantage held by groups that, in prehistoric times, assigned the hunting role to stronger males and the gatherer–childcare role to females.

The story is that men in prehistoric societies ranged widely as they hunted and that those men who were better hunters had a reproductive advantage, either because their success gave them more access to females or because their offspring were more likely to survive to reproductive age because good hunters could feed their children more reliably. This explanation also assumes that skills related to the R dimension of the *g*-VPR model, like the ability to judge the velocity of moving objects or to maintain orientation in space, aided in hunting. On the other hand, women are supposed to have primarily been gatherers of edible plants and small animals near the camp. A woman would have a reproductive advantage if she were a good gatherer and so be better able to feed her children. The final step is to assume that the ability to notice fine details, like an edible lizard in a bush or an edible berry amid shrubbery, would make for a better gatherer. Women's superior verbal abilities are explained by the assumption that women, being more dependent on others for protection of themselves and their offspring, had to be superior in social interchanges (Geary, 2005, 2007a, 2007b).

The story is plausible. Slight reproductive advantages associated with various abilities and skills, acting over thousands of generations, could produce a substantial sexual imbalance in skills. However, the story is a story, not a fact. There is no direct evidence for it, because we know little, and probably never will know much, about the behavioral characteristics of prehistoric *Homo sapiens*, let alone other hominid predecessors of our species. What we do know is inferred from indirect evidence, such as estimates about the rate of maturation of extinct hominids based on skeletal data, which then are used to infer the years required for protection of children, and then extrapolated to discussions of male–female roles in maintenance of children, foraging, and group protection. Analogies to the behaviors of existing human hunter-gatherer societies are frequently used, but they are obviously controversial.

There is another evolutionary story, based on social rather than biological influences. The human brain is designed for general learning, a trait that is very useful in a species that occupies multiple ecological niches. Certain behavioral practices that produce better social and economic organization can lead to a competitive advantage at the level of the group, rather than the individual. Groups that adopt such practices are more likely to survive (Campbell, 1975; Wilson and Wilson, 2007, 2008). These practices include differentiation of male and female roles. Since humans are learners, males and females will acquire different, role-appropriate cognitive abilities and skills. The tendency will be accentuated over time, but the accentuation will be due to social rather than biological evolution. The difference in cognitive skills that appears in late adolescence is due to different learning experiences that have evolved historically, not physiological differences in brain structure that have evolved genetically.

This hypothesis emphasizes social history. It can be given a biological twist. It

may be that boys learn more spatial skills because they explore more, and girls learn better verbal skills because they socialize more. This could be due to genetic influences on the tendency to explore each environment. Thus, sex differences in cognition (and other differences between people) could be due to differential learning experiences, but the differential learning experiences themselves could be under genetic influence (Bouchard, 1999, 2014).

There is no direct way to differentiate among these hypotheses. In fact, they are not mutually exclusive. It is tautologically true that all human variations in behavior will be within the range of variation permitted by the genome; the most dedicated geneticist does not deny that humans are superb learners. We have learned to live with this ambiguity with respect to physical behaviors. No one denies that males are genetically predisposed to be able to swing sticks more rapidly than females, and no one denies that today's training methods routinely produce women tennis players who hit the ball harder than the men's champions of yesteryear.

These evolutionary perspectives provide a rationale for investigating relationships between cognitive sex differences and the brain. The next section starts with a basic question.

10.4.3 Are There Male and Female Brains?

There are popular notions that one can (or cannot) characterize a brain as male or female. Is this a valid distinction according to the current scientific evidence? Here we consider some illustrative hormone and neuroimaging studies.

HORMONES
A vast research literature from animals and humans indicates that genetic potential interacts with hormonal balance during key periods of brain development to cause sex differences in behavior and cognition. There is an excellent summary and discussion of this research in chapter 5 of *Human*

Diversity (Murray, 2020), but here we briefly present only a few results. For example, manipulation of adrenal levels prenatally and postnatally can influence the display of typical male or typical female behavior in rats, including the extent of engagement in rough-and-tumble play and the patterns of behavior in maze exploration. Obviously, conducting an analogous experimental study on humans would not be ethical. What we can do is study certain medical conditions in which unusual hormonal concentrations occur. Remember that less-than-ideal studies can provide useful evidence.

Congenital adrenal hyperplasia (CAH) is a genetic condition in which the adrenal gland fails to generate a key enzyme, causing unusual sensitivity to male hormones. The condition can occur in both boys and girls. It is treated by restoring the normal hormone balance. Female CAH patients tend to have higher scores on spatial orientation tests than normal females. Males with CAH (a less-studied group) tend to have lower scores than normal males. There are indications that this result generalizes to other behavior patterns in women, for female CAH patients display more masculine behaviors and interests than do normal girls and women, including such things as preferences for "male-appropriate" or "female-appropriate" toys (Berenbaum and Resnick, 2007; Puts et al., 2008; Kimura, 1999, chapter 9).

These results refer to effects of hormones on the developing brain. There is also evidence that circulating hormonal levels in adults will influence human cognition. The results are somewhat inconsistent, as the studies are difficult to do and generally involve small numbers of participants, a condition that invites unstable findings. Nevertheless, certain results seem to be reasonably well established.

In women, high levels of circulating estrogens facilitate tasks involving verbal fluency and/or short-term memory. The evidence is mixed for performance on visuospatial reasoning, except for a consistent reduction in performance on mental rotation tasks. This has been established by two sources of data: studies of women tested at various times during their menstrual cycle and studies of postmenopausal women who either are or are not receiving estrogen replacement therapy (Hausmann et al., 2000; Halpern and Tan, 2001; Halpern, 2012; Levy and Kimura, 2009).

Complementary results have been found in studies of testosterone. Testosterone appears to have a nonmonotonic effect on visuospatial reasoning, enhancing it in women and men with low testosterone (a common condition in the elderly), but decreasing it in men with normal or high testosterone levels. The cognitive effects are complicated by the fact that circulating testosterone levels are associated with a myriad of other effects, including increases in impulsivity and aggressive behavior. To get some idea of the complexity of the effect, contrasting effects of testosterone on spatial rotation have been reported in England, the United States, and China (Yang et al., 2007). The authors suggest that this is because of different emphases on speed versus accuracy of response in different cultures. The more general point is that in visuospatial problem solving, a hormonal effect could be on either the brain mechanisms required for the task itself or the brain mechanisms involved in selecting a problem-solving strategy (Kimura, 1999).

NEUROIMAGING

Early neuroimaging studies in children and adults provided hints of male–female differences both in brain structure and in brain function that might be related to cognition (Mansour et al., 1996; Haier and Benbow, 1995; Haier et al., 2005; Jung et al., 2005; Luders et al., 2008; Schmithorst, 2009; Gur and Gur, 2007). These first studies were intriguing because they implied that not all brains work the same way, even when matched for cognitive ability (Haier and Benbow, 1995; Haier et al., 2005). This empirical observation sounds reasonable, but it is contrary to a basic assumption of many cognitive experimental researchers. That not all brains work the same way is a

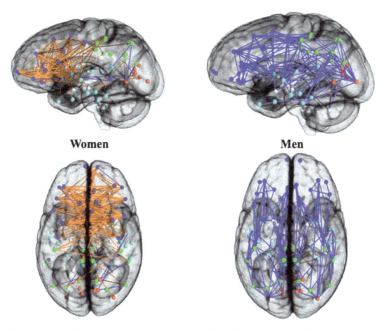

Figure 10.15 Brain connection–wise analysis for women and men. Brain
networks show increased connectivity in (left) females and (right) males.
Intrahemispheric connections are shown in blue, and interhemispheric
connections are shown in orange. The depicted edges are those that
survived permutation testing at $p = 0.05$. Node color representations are as
follows: light blue, frontal; cyan, temporal; green, parietal; red, occipital;
white, subcortical.
Adapted from Ingalhalikar et al. (2014b).

key concept for imaging studies of individual differences (Martínez et al., 2015).

These early studies, however, were limited by small samples and rudimentary
image analysis methods. There are now a
number of methodically more advanced
neuroimaging studies with larger, representative samples that compare men and
women. Brain differences are reported, but
their meaning remains unclear, especially
with respect to cognition differences
(Lotze et al., 2019).

For example, brain connectomes were
computed based on diffusion tensor imaging
(see Chapter 5) for 428 males and
521 females (Ingalhalikar et al., 2014b).
There were a number of differences, which
the researchers summarized as follows:
"Overall, the results suggest that male brains
are structured to facilitate connectivity
between perception and coordinated action,
whereas female brains are designed to

facilitate communication between analytical
and intuitive processing modes" (p. 823).
However, not all researchers found the
strength and meaning of these results all that
compelling (Hänggi et al., 2014; Joel and
Tarrasch, 2014; but see Ingalhalikar et al.,
2014a). Key results are depicted in
Figure 10.15. This study did not include
any cognitive variables, but it suggested
how connectome data might be used to
investigate cognitive differences.

Limited cognitive data were included in
one compelling neuroimaging comparison
of males and females from the UK Biobank
consortium (Ritchie et al., 2018). The study
was based on 5,216 participants (2,750
females; 2,466 males) between forty-four
and seventy-seven years old who completed
structural MRI and resting-state fMRI. For
structural differences, "males had higher raw
volumes, raw surface areas, and white matter fractional anisotropy; females had higher

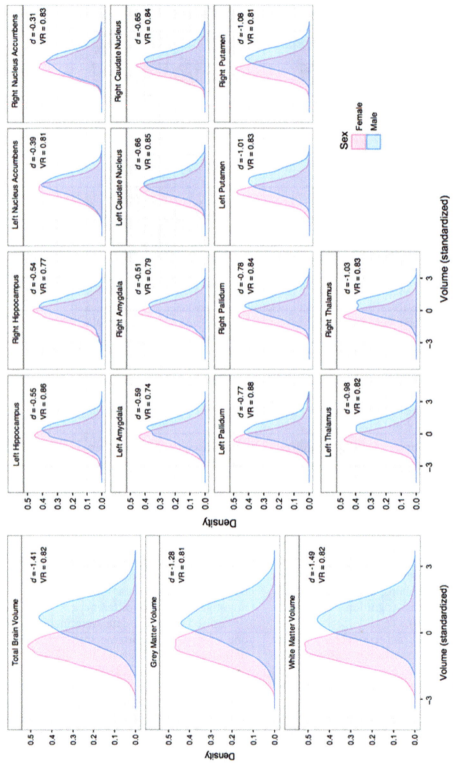

Figure 10.16 (left) Density plots of sex differences in overall brain volumes. (right) Subcortical structures. d = Cohen's d (mean difference); VR = variance ratio (variance difference). All mean differences were statistically significant at $p < 0.003$, after correction for multiple comparisons. All variance differences were significant at $p < 3.0 \times 10^{-25}$; From Ritchie et al. (2018).

raw cortical thickness and higher white matter tract complexity. There was considerable distributional overlap between the sexes. ... There was generally greater male variance across the raw structural measures. Functional connectome organization showed stronger connectivity for males in unimodal sensorimotor cortices, and stronger connectivity for females in the default mode network" (p. 2959) (see Chapter 5 for imaging terminology). These findings were generally consistent with previous studies. Figure 10.16 shows brain structural sex differences. As usual, there is considerable overlap between male and female distributions.

They also analyzed two cognitive variables: verbal-numeric reasoning and reaction time. Generally, they were weakly correlated to structural brain variables in the full sample, and there were no appreciable sex differences in these correlations. Nonetheless, an analysis of volume, surface area, and cortical thickness found some brain areas where the structural variable mediated a sex difference. As to the fundamental question of when any brain differences might originate, Wheelock and colleagues (2019) reported MRI male–female comparisons in fetuses. They found sufficient differences to conclude that "sexual dimorphism in functional brain systems emerges during human gestation" (p. 1), but more research is necessary to confirm this.

These neuroimaging studies show average differences in some brain parameters when groups of males and females are compared. However, a large-scale study by Daphna Joel and colleagues (2015) considered 1,400 autopsied human brains. Analyzing four data sets using different MRI approaches, they found considerable overlap between the distributions of females and males for gray matter, white matter, and connections evaluated. They concluded that the human brain is a mosaic: "this extensive overlap undermines any attempt to distinguish between a 'male' and a 'female' form for specific brain features" (p. 15471).

Adam Chekroud and colleagues (2016) challenged this conclusion. They showed that an individual's biological sex can be classified with greater than 90 percent accuracy "by considering the brain mosaic as a whole" (p. 1). They analyzed 1,566 individuals from the Brain Genomics Superstructure Project. They concluded, "The human brain may be a mosaic, but it is one with predictable patterns. ... Multivariate analyses of whole-brain patterns in brain morphometry can reliably discriminate sex" (p. 1).

The weight of evidence about brain differences between males and females and what they mean is still evolving, with some newer studies moving in a skeptical direction (Eliot et al., 2021; Zhang et al., 2021). It is too simplistic to characterize individuals as having a "male" or "female" brain, but there are sex differences that need to be understood.

10.5 Conclusions

We started this chapter asking if there were sex differences in intelligence. The weight of evidence indicates no appreciable difference for the g-factor. Other differences on verbal and numerical factors and on some specific mental abilities are small and likely have no impact on practical matters like the representation of males and females in various careers. An exception may be for spatial rotation, which on average shows a difference in favor of males. This difference, combined with males being more variable, results in more males at the upper end of the distribution of spatial rotation ability, an ability linked to success in STEM careers. The data also suggest that sex differences in interests along a things–people dimension may be more influential for career choice than cognitive differences, especially for men and women with equally high spatial rotation ability.

Any individual test, including the much-used matrix tests, will evaluate g and test-specific features, a point that Charles Spearman made more than a century ago. You can find a measure of general intelligence that is g and verbally loaded finding an advantage for females, or produce a measure of general intelligence that is g and loaded on visuospatial reasoning finding an advantage for males (Colom and García-López, 2002). Remember that counting is not measuring.

Men and women have somewhat different brains on average, and this appears to be observable in fetuses. These differences may or may not lead to somewhat different strengths and weaknesses in cognition, on average. To the extent that any male–female brain or cognition differences are based in biology and genetics, it is important to keep in mind that biology is not destiny and that the impact of genetics is probabilistic. Social and cultural influences can either accentuate or override these proclivities and trends. Humans are powerful learners, and learning itself influences brain organization.

As Diane Halpern underscored, the more you learn about sex differences in cognition, the less certain you are that you know the answers. Within very broad limits, biological and social influences produce tendencies toward certain cognitive and social behaviors. Once this is done, the behaviors themselves alter biological makeup and social status. Women, *on average*, are somewhat more verbal and somewhat less spatially oriented than men, but there are male chatterboxes and female combat helicopter pilots.

So, can information in this chapter inform social or educational policy? If, for example, women with high spatial or math ability prefer a people-oriented career in medicine or law to a thing-oriented STEM one, should we create policies to achieve a 50–50 goal for the sexes in STEM careers? Should we discourage women from the social science/health/education fields where they dominate to achieve 50 percent men? Should boys get remedial education classes in reading while girls train on spatial rotation skills? Such policies would be decisions based on social goals, but discussions about them would benefit from considering the existing data.

The conclusions for this chapter are not much different than those in the first edition of this book. It may be that further research

into sex differences in cognition and where they come from will not clarify this picture much more. It may be that research on sex differences will yield more meaningful results when it comes to understanding medical, psychological, and psychiatric conditions, disorders, and diseases and how to treat them (Bao and Swaab, 2010). Perhaps introducing gender data will be beneficial to such research. The good news is that we can ask about and discuss differences between the sexes with increasingly sophisticated data and analyses. The questions are legitimate for scientific investigations, and asking them does not disparage either sex nor imply inferior or superior judgments. This is a key message derived from the issues described and discussed in this chapter.

10.6 Summary

This chapter described the current scientific understanding of average sex differences in intelligence and cognitive abilities. For many of the questions addressed, there is a vast research literature and a weight of evidence that have been summarized here. On the key question, men and women do not differ on general intelligence (g), but there are differences on some mental abilities. Where do these differences come from? When do they arise? We discussed whether there are features of the brain involved and whether any of the differences are related to disparities between men and women in education and career choices. Can any of the sex difference data inform education or social policies? Addressing these questions can be contentious; some of the data are controversial, but an understanding of the data is required for reasonable discussions.

10.7 Questions for Discussion

10.1   Why is understanding the variance of a distribution of intelligence test scores necessary for addressing the question of whether one sex shows better performance on average than the other?

10.2 Do you have personal experiences consistent or inconsistent with any cognitive abilities that on average show some male/female differences?

10.3 Can you name any social/culture sex or gender role influences on the development of your own intelligence?

10.4 How would you critique evolutionary explanations of male/female differences in cognitive abilities?

10.5 What data would you use to refute the idea that there are male and female brains?

References

Abad, F. J., Colom, R., Rebollo, I., & Escorial, S. 2004. Sex differential item functioning in the Raven's Advanced Progressive Matrices: Evidence for bias. *Personality and Individual Differences*, 36, 1459–1470.

Arribas-Águila, D., Abad, F. J., & Colom, R. 2019. Testing the developmental theory of sex differences in intelligence using latent modeling: Evidence from the TEA Ability Battery (BAT-7). *Personality and Individual Differences*, 138, 212–218.

Bao, A. M., & Swaab, D. F. 2010. Sex differences in the brain, behavior, and neuropsychiatric disorders. *Neuroscientist*, 16, 550–565.

Bates, T. C., & Gignac, G. E. 2022. Effort impacts IQ test scores in a minor way: A multi-study investigation with healthy adult volunteers. *Intelligence*, 92, 101652.

Benbow, C. P., Lubinski, D., Shea, D. L., & Eftekhari-Sanjani, H. 2000. Sex differences in mathematical reasoning ability at age 13: Their status 20 years later. *Psychological Science*, 11, 474–480.

Benbow, C. P., Lubinski, D. J., & Stanley, J. C. 1996. *Intellectual talent: Psychometric and social issues.* Baltimore: Johns Hopkins University Press.

Berenbaum, S. A., & Resnick, S. 2007. The seed of career choice: Prenatal sex hormone effects. In Ceci, S. I., & Williams, W. M. (eds.), *Why aren't there more women in science: Top researchers debate the evidence.* Washington, DC: American Psychological Association.

Bernstein, B. O., Lubinski, D., & Benbow, C. P. 2019. Psychological constellations assessed at age 13 predict distinct forms of eminence 35 years later. *Psychological Science*, 30, 444–454.

Blinkhorn, S. 2005. Intelligence: A gender bender. *Nature*, 438, 31–32.

Bouchard, T. J. 1999. Genes, environment, and personality. In Ceci, S. J., & Williams, W. M. (eds.), *The nature–nurture debate: The essential readings.* Malden, MA: Blackwell.

Bouchard, T. J., Jr. 2014. Genes, evolution and intelligence. *Behavior Genetics*, 44, 549–577.

Brunner, M., Krauss, S., & Kunter, M. 2008. Gender differences in mathematics: Does the story need to be rewritten? *Intelligence*, 36, 403–421.

Campbell, D. T. 1975. The conflict between social and biological evolution and the concept of original sin. *Zygon*, 10, 234–249.

Casey, M. B., Nuttall, R., & Benbow, C. P. 1995. The influence of spatial ability on gender differences in mathematics college entrance test-scores across diverse samples. *Developmental Psychology*, 31, 697–705.

Ceci, S. J., & Williams, W. M. 2007. *Why aren't more women in science?* Washington, DC: American Psychological Association.

Chekroud, A. M., Ward, E. J., Rosenberg, M. D., & Holmes, A. J. 2016. Patterns in the human brain mosaic discriminate males from females. *Proceedings of the National Academy of Sciences of the United States of America*, 113, E1968.

Choi, J., & Silverman, I. 2003. Processes underlying sex differences in route-learning strategies in children and adolescents. *Personality and Individual Differences*, 34, 1153–1166.

Clayton, J. A., & Collins, F. S. 2014. Policy: NIH to balance sex in cell and animal studies. *Nature*, 509, 282–283.

Coleman, J. S. 1966. *Equality of educational opportunity.* Washington, DC: US Department of Health, Education, and Welfare, Office of Education.

Colom, R., & García-López, O. 2002. Sex differences in fluid intelligence among high school graduates. *Personality and Individual Differences*, 32, 445–451.

Colom, R., & Lynn, R. 2004. Testing the developmental theory of sex differences in intelligence on 12–18 year olds. *Personality and Individual Differences*, 36, 75–82.

Council on Graduate Schools. 2018. *Annual report on US graduate school enrollment and degrees for 2018.* Washington, DC: Council on Graduate Schools.

Court, J. H. 1983. Sex-differences in performance on Raven Progressive Matrices: A review. *Alberta Journal of Educational Research*, 29, 54–74.

Coutinho, M. J., & Oswald, D. P. 2005. State variation in gender disproportionality in special education: Findings and recommendations. *Remedial and Special Education*, 26, 7–15.

Dai, X., Ryan, J. J., Paolo, A. M., & Harrington, R. G. 1991. Sex differences on the Wechsler Adult Intelligence Scale–Revised for China. *Psychological Assessment*, 32, 282–284.

Deary, I. J., Irwing, P., Der, G., & Bates, T. C. 2007a. Brother–sister differences in the *g* factor in intelligence: Analysis of full, opposite-sex siblings from the NLSY 1979. *Intelligence*, 35, 451–456.

Deary, I. J., Strand, S., Smith, P., & Fernandes, C. 2007b. Intelligence and educational achievement. *Intelligence*, 35, 13–21.

Deary, I. J., Thorpe, G., Wilson, V., Starr, J. M., & Whalley, L. J. 2003. Population sex differences in IQ at age 11: The Scottish Mental Survey 1932. *Intelligence*, 31, 533–542.

Detterman, D. K. 2016. Education and intelligence: Pity the poor teacher because student characteristics are more significant than teachers or schools. *Spanish Journal of Psychology*, 19, 1–11.

de-Wit, L., Huygelier, H., van der Hallen, R., Chamberlain, R., & Wagemans, J. 2017. Developing the Leuven Embedded Figures Test (L-EFT): Testing the stimulus features that influence embedding. *Peer Journal*, 5, e2862.

Eliot, L., Ahmed, A., Khan, H., & Patel, J. 2021. Dump the "dimorphism": Comprehensive synthesis of human brain studies reveals few male–female differences beyond size. *Neuroscience and Biobehavioral Reviews*, 125, 667–697.

Ferriman, K., Lubinski, D., & Benbow, C. P. 2009. Work preferences, life values, and personal views of top math/science graduate students and the profoundly gifted: Developmental changes and gender differences during emerging adulthood and parenthood. *Journal of Personality and Social Psychology*, 97, 517–532.

Finnigan, K. M., & Corker, K. S. 2016. Do performance avoidance goals moderate the effect of different types of stereotype threat on women's math performance? *Journal of Research in Personality*, 63, 36–43.

Flore, P. C., Mulder, J., & Wicherts, J. M. 2018. The influence of gender stereotype threat on mathematics test scores of Dutch high school students: A registered report. *Comprehensive Results in Social Psychology*, 3, 140–174.

Geary, D. C. 2005. *The origin of mind: Evolution of brain, cognition, and general intelligence*. Washington, DC: American Psychological Association.

Geary, D. C. 2007a. An evolutionary perspective on sex differences in science and mathematics. In Ceci, S. J., & Williams, W. M. (eds.), *Why aren't there more women in science: Top researchers debate the evidence*. Washington, DC: American Psychological Association.

Geary, D. C. 2007b. Educating the evolving mind: Conceptual foundations for an evolutionary educational psychology. In Carlson, J. C., & Levine, J. L. (eds.), *Educating the evolving mind*. Charlotte, SC: Information Age Press.

Gelade, G. A. 2008. IQ, cultural values, and the technological achievement of nations. *Intelligence*, 36, 711–718.

Gur, R. C., & Gur, R. E. 2007. Neural substrates for sex differences in cognition. In Ceci, S. J., & Williams, W. M. (eds.), *Why aren't there more women in science: Top researchers debate the evidence*. Washington, DC: American Psychological Association.

Haier, R. J. 2007. Brains, bias, and biology: Follow the data. In Ceci, S. J., & Williams, W. M. (eds.), *Why aren't there more women in science: Top researchers debate the evidence*. Washington, DC: American Psychological Association.

Haier, R. J. 2017. *The neuroscience of intelligence*. Cambridge: Cambridge University Press.

Haier, R. J., & Benbow, C. P. 1995. Sex differences and lateralization in temporal lobe glucose metabolism during mathematical reasoning. *Developmental Neuropsychology*, 11, 405–414.

Haier, R. J., Jung, R. E., Yeo, R. A., Head, K., & Alkire, M. T. 2005. The neuroanatomy of general intelligence: Sex matters. *Neuroimage*, 25, 320–327.

Halpern, D. F. 1986. *Sex differences in cognitive abilities*. 1st ed. Hillsdale, NJ: Erlbaum.

Halpern, D. F. 2000. *Sex differences in cognitive abilities*. 2nd ed. Mahwah, NJ: Erlbaum.

Halpern, D. F. 2012. *Sex differences in cognitive abilities*. 3rd ed. New York: Psychology Press.

Halpern, D. F., Benbow, C. P., Geary, D. C., et al. 2007. The science of sex differences in science and mathematics. *Psychological Science in the Public Interest*, 8, 1–51.

Halpern, D. F., & Tan, U. 2001. Stereotypes and steroids: Using a psychobiosocial model to understand cognitive sex differences. *Brain and Cognition*, 45, 392–414.

Hänggi, J., Fövenyi, L., Liem, F., Meyer, M., & Jäncke, L. 2014. The hypothesis of neuronal interconnectivity as a function of brain size: A general organization principle of the human connectome. *Frontiers in Human Neuroscience*, 8, 915–915.

Hattori, K., & Lynn, R. 1997. Male–female differences on the Japanese WAIS-R. *Personality and Individual Differences*, 23, 531–533.

Hausmann, M., Slabbekoorn, D., Van Goozen, S. H., Cohen-Kettenis, P. T., & Gunturkun, O. 2000. Sex hormones affect spatial abilities during the menstrual cycle. *Behavioral Neuroscience*, 114, 1245–50.

Hedges, L. V., & Nowell, A. 1995. Sex-differences in mental test-scores, variability, and numbers of high-scoring individuals. *Science*, 269, 41–45.

Humphreys, L. G., & Lubinski, D. J. 1996. Assessing spatial visualization: An underappreciated ability for many school and work settings. In Benbow, C. P., & Lubinski, D. (eds.), *Intellectual talent: Psychometric and social issues*. Baltimore: Johns Hopkins University Press.

Hunt, E., & Madhyastha, T. 2008. Recruitment modeling: An analysis and an application to the study of male–female differences in intelligence. *Intelligence*, 36, 653–663.

Hunt, E., Pellegrino, J. W., Frick, R. W., Farr, S. A., & Alderton, D. 1988. The ability to reason about movement in the visual-field. *Intelligence*, 12, 77–100.

Hyde, J. S. 2007. Women in science: Gender similarities in abilities and sociocultural forces. In Ceci, S. J., & Williams, W. M. (eds.), *Why aren't there more women in science: Top researchers debate the evidence*. Washington, DC: American Psychological Association,.

Ingalhalikar, M., Smith, A., Parker, D., et al. 2014a. Reply to Joel and Tarrasch: On misreading and shooting the messenger. *Proceedings of the National Academy of Sciences of the United States of America*, 111, E638.

Ingalhalikar, M., Smith, A., Parker, D., et al. 2014b. Sex differences in the structural connectome of the human brain. *Proceedings of the National Academy of Sciences of the United States of America*, 111, 823–828.

Irwing, P., & Lynn, R. 2005. Sex differences in means and variability on the progressive matrices in university students: A meta-analysis. *British Journal of Psychology*, 96, 505–524.

Jensen, A. R. 1998. *The g factor: The science of mental ability*. Westport, CT: Praeger.

Joel, D., Berman, Z., Tavor, I., et al. 2015. Sex beyond the genitalia: The human brain mosaic. *Proceedings of the National Academy of Sciences of the United States of America*, 112, 15468–15473.

Joel, D., & Tarrasch, R. 2014. On the mispresentation and misinterpretation of gender-related data: The case of Ingalhalikar's human connectome study. *Proceedings of the National Academy of Sciences of the United States of America*, 111, E637.

Johnson, W., & Bouchard, T. J., Jr. 2005. The structure of human intelligence: It is verbal, perceptual, and image rotation (VPR), not fluid and crystallized. *Intelligence*, 33, 393–416.

Johnson, W., & Bouchard, T. J., Jr. 2007a. Sex differences in mental abilities: *g* masks the dimensions on which they lie. *Intelligence*, 35, 23–39.

Johnson, W., & Bouchard, T. J., Jr. 2007b. Sex differences in mental ability: A proposed means to link them to brain structure and function. *Intelligence*, 35, 197–209.

Johnson, W., Carothers, A., & Deary, I. J. 2008. Sex differences in variability in general intelligence: A new look at the old question. *Perspectives on Psychological Science*, 3, 518–531.

Johnson, W., Carothers, A., & Deary, I. J. 2009. A role for the X chromosome in sex differences in variability in general intelligence? *Perspectives on Psychological Science*, 4, 598–611.

Jung, R. E., Haier, R. J., Yeo, R. A., et al. 2005. Sex differences in N-acetylaspartate correlates of general intelligence: An H-1-MRS study of normal human brain. *Neuroimage*, 26, 965–972.

Kaufman, S. B., Reynolds, M. R., Liu, X., Kaufman, A. S., & McGrew, K. S. 2012. Are cognitive *g* and academic achievement *g* one and the same *g*? An exploration on the Woodcock–Johnson and Kaufman tests. *Intelligence*, 40, 123–138.

Kell, H. J., Lubinski, D., & Benbow, C. P. 2013. Who rises to the top? Early indicators. *Psychological Science*, 24, 648–659.

Kell, H. J., McCabe, K. O., Lubinski, D., & Benbow, C. P. 2022. Wrecked by success? Not to worry. *Perspectives on Psychological Science*, 17, 1291–1321.

Kimura, D. 1999. *Sex and cognition*. Cambridge, MA: MIT Press.

Law, D. J., Pellegrino, J. W., & Hunt, E. B. 1993. Comparing the tortoise and the hare: Gender differences and experience in dynamic spatial reasoning tasks. *Psychological Science*, 4, 35–40.

Levy, J., & Kimura, D. 2009. Women, men, and the sciences. In Sommers, C. H. (ed.), *Women and science*. Washington, DC: AEI Press.

Longman, R. S., Saklofske, D. H., & Fung, T. S. 2007. WAIS-III percentile scores by education and sex for US and Canadian populations. *Assessment*, 14, 426–432.

Lotze, M., Domin, M., Gerlach, F. H., et al. 2019. Novel findings from 2,838 adult brains on sex differences in gray matter brain volume. *Scientific Reports*, 9, 1671.

Lubinski, D. 2009. Exceptional cognitive ability: The phenotype. *Behavior Genetics*, 39, 350–358.

Lubinski, D. 2016. From Terman to today: A century of findings on intellectual precocity. *Review of Educational Research*, 86, 900–944.

Lubinski, D., Benbow, C. P., Webb, R. M., & Bleske-Rechek, A. 2006. Tracking exceptional human capital over two decades. *Psychological Science*, 17, 194–199.

Luders, E., Narr, K. L., Bilder, R. M., et al. 2008. Mapping the relationship between cortical convolution and intelligence: Effects of gender. *Cerebral Cortex*, 18, 2019–2026.

Lynn, R. 1999. Sex differences in intelligence and brain size: A developmental theory. *Intelligence*, 27, 1–12.

Lynn, R., & Irwing, P. 2004a. Sex differences on the Advanced Progressive Matrices in college students. *Personality and Individual Differences*, 37, 219–223.

Lynn, R., & Irwing, P. 2004b. Sex differences on the Progressive Matrices: A meta-analysis. *Intelligence*, 32, 481–498.

Lynn, R., & Mau, W. C. 2001. Ethnic and sex differences in the predictive validity of the scholastic achievement test for college grades. *Psychological Reports*, 88, 1099.

Makel, M. C., Kell, H. J., Lubinski, D., Putallaz, M., & Benbow, C. P. 2016. When lightning strikes twice: Profoundly gifted, profoundly accomplished. *Psychological Science*, 27, 1004–1018.

Malinowski, J. C. 2001. Mental rotation and real-world wayfinding. *Perceptual and Motor Skills*, 92, 19–30.

Mansour, C. S., Haier, R. J., & Buchsbaum, M. S. 1996. Gender comparisons of cerebral glucose metabolic rate in healthy adults during a cognitive task. *Personality and Individual Differences*, 20, 183–191.

Martínez, K., Madsen, S. K., Joshi, A. A., et al. 2015. Reproducibility of brain–cognition relationships using three cortical surface-based protocols: An exhaustive analysis based on cortical thickness. *Human Brain Mapping*, 36, 3227–3245.

Matarazzo, J. D., Bornstein, R. A., McDermott, P. A., & Noonan, J. V. 1986. Verbal IQ vs. performance IQ difference scores in males and females from the WAIS-R standardization sample. *Journal of Clinical Psychology*, 42, 965–974.

McCabe, K. O., Lubinski, D., & Benbow, C. P. 2020. Who shines most among the brightest? A 25-year longitudinal study of elite STEM graduate students. *Journal of Personality and Social Psychology*, 119, 390–416.

Mitchell, K. J. 2018. *Innate: How the wiring of our brains shapes who we are*. Princeton, NJ: Princeton University Press.

Muller, C. B., Ride, S. M., Fouke, J., et al. 2005. Gender differences and performance in science. *Science*, 307, 1043.

Murray, C. 2003. *Human accomplishment: The pursuit of excellence in the arts and sciences, 800 BC to 1950*. New York: HarperCollins.

Murray, C. 2020. *Human diversity: The biology of gender, race, and class*. New York: Twelve.

National Academy of Sciences, National Academy of Engineering, & Institute of Medicine. 2007. *Beyond bias and barriers: Fulfilling the potential of women in academic science and engineering*. Washington, DC: National Academies Press.

Park, G., Lubinski, D., & Benbow, C. P. 2008. Ability differences among people who have commensurate degrees matter for scientific creativity. *Psychological Science*, 19, 957–961.

Pokropek, A., Marks, G. N., & Borgonovi, F. 2021. How much do students' scores in PISA reflect general intelligence and how much do they reflect specific abilities? *Journal of Educational Psychology*, 114, 1121–1135.

Puts, D. A., McDaniel, M. A., Jordan, C. L., & Breedlove, S. M. 2008. Spatial ability and prenatal androgens: Meta-analyses of congenital adrenal hyperplasia and digit ratio (2D:4D) studies. *Archives of Sexual Behavior*, 37, 100–111.

Raven, J. 2000. The Raven's Progressive Matrices: Change and stability over culture and time. *Cognitive Psychology*, 41, 1–48.

Raven, J., & Raven, J. (eds.) 2008. *Uses and abuses of intelligence: Studies advancing Spearman and Raven's quest for non-arbitrary metrics*. New York: Royal Fireworks Press.

Ritchie, S. J., Cox, S. R., Shen, X. Y., et al. 2018. Sex differences in the adult human brain: Evidence from 5216 UK Biobank participants. *Cerebral Cortex*, 28, 2959–2975.

Sackett, P. R. 2021. Reflections on a career studying individual differences in the workplace. *Annual Review of Organizational Psychology and Organizational Behavior*, 8, 1–18.

Schmithorst, V. J. 2009. Developmental sex differences in the relation of neuroanatomical connectivity to intelligence. *Intelligence*, 37, 164–173.

Skårbrevik, K. J. 2002. Gender differences among students found eligible for special education. *European Journal of Special Needs Education*, 17, 97–107.

Snow, W. G., & Weinstock, J. 1990. Sex differences among non-brain-damaged adults on the Wechsler Adult Intelligence Scales: A review of the literature. *Journal of Clinical and Experimental Neuropsychology*, 12, 873–886.

Sommers, C. H. 2009. *The science on women and science*. Washington, DC: AEI Press.

Stanley, J. C. 1997. Varieties of intellectual talent. *Journal of Creative Behavior*, 31, 93–119.

Stanley, J., Keating, D. P., & Fox, L. H. 1974. *Mathematical talent: Discovery, description, and development*. Baltimore, MD: Johns Hopkins University Press.

Steele, C. M., & Aronson, J. 1995. Stereotype threat and the intellectual test-performance of African-Americans. *Journal of Personality and Social Psychology*, 69, 797–811.

Steele, C. M., & Aronson, J. 1998. Stereotype threat and the test performance of academically successful African Americans. In Jencks, C., & Meredith, P.

(eds.), *The black–white test score gap*. Washington, DC: Brookings Institution Press.

Stoet, G., & Geary, D. C. 2013. Sex differences in mathematics and reading achievement are inversely related: Within- and across-nation assessment of 10 years of PISA data. *PLoS ONE*, 8, e57988.

Stoet, G., & Geary, D. C. 2018. The gender-equality paradox in science, technology, engineering, and mathematics education. *Psychological Science*, 29, 581–593.

Strand, S., Deary, I. J., & Smith, P. 2006. Sex differences in Cognitive Abilities Test scores: A UK national picture. *British Journal of Educational Psychology*, 76, 463–480.

Su, R., & Rounds, J. 2015. All STEM fields are not created equal: People and things interests explain gender disparities across STEM fields. *Frontiers in Psychology*, 6, 189.

Su, R., Rounds, J., & Armstrong, P. I. 2009. Men and things, women and people: A meta-analysis of sex differences in interests. *Psychological Bulletin*, 135, 859–884.

Uttal, D. H., Meadow, N. G., Tipton, E., et al. 2013. The malleability of spatial skills: A meta-analysis of training studies. *Psychological Bulletin*, 139, 352–402.

Wai, J., Lubinski, D., & Benbow, C. P. 2005. Creativity and occupational accomplishments among intellectually precocious youths: An age 13 to age 33 longitudinal study. *Journal of Educational Psychology*, 97, 484–492.

Wai, J., Lubinski, D., & Benbow, C. P. 2009. Spatial ability for STEM domains: Aligning over 50 years of cumulative psychological knowledge solidifies its importance. *Journal of Educational Psychology*, 101, 817–835.

Wai, J., Lubinski, D., Benbow, C. P., & Steiger, J. H. 2010. Accomplishment in science, technology, engineering, and mathematics (STEM) and its relation to STEM educational dose: A 25-year longitudinal study. *Journal of Educational Psychology*, 102, 860–871.

Wainer, H., & Steinberg, L. S. 1992. Sex-differences in performance on the mathematics section of the Scholastic Aptitude Test: A bidirectional validity study. *Harvard Educational Review*, 62, 323–336.

Waller, D., Knapp, D., & Hunt, E. 2001. Spatial representations of virtual mazes: The role of visual fidelity and individual differences. *Human Factors*, 43, 147–158.

Wang, M.-T., Eccles, J. S., & Kenny, S. 2013. Not lack of ability but more choice: Individual and gender differences in choice of careers in science, technology, engineering, and mathematics. *Psychological Science*, 24, 770–775.

Wheelock, M. D., Hect, J. L., Hernandez-Andrade, E., et al. 2019. Sex differences in functional connectivity during fetal brain development. *Developmental Cognitive Neuroscience*, 36, 100632.

Wierenga, L. M., Bos, M. G. N., van Rossenberg, F., & Crone, E. A. 2019. Sex effects on development of brain structure and executive functions: Greater variance than mean effects. *Journal of Cognitive Neuroscience*, 31, 730–753.

Wilson, D. S., & Wilson, E. O. 2007. Rethinking the theoretical foundation of sociobiology. *Quarterly Review of Biology*, 82, 327–348.

Wilson, D. S., & Wilson, E. O. 2008. Evolution "for the good of the group." *American Scientist*, 96, 380–389.

World Economic Forum. 2015. *The global gender gap report*. Geneva: World Economic Forum.

Yang, C. F. J., Hooven, C. K., Boynes, M., Gray, P. B., & Pope, H. G. 2007. Testosterone levels and mental rotation performance in Chinese men. *Hormones and Behavior*, 51, 373–378.

Zaboski, B. A., Kranzler, J. H., & Gage, N. A. 2018. Meta-analysis of the relationship between academic achievement and broad abilities of the Cattell–Horn–Carroll theory. *Journal of School Psychology*, 71, 42–56.

Zhang, Y., Luo, Q., Huang, C. C., et al. 2021. The human brain is best described as being on a female/male continuum: Evidence from a neuroimaging connectivity study. *Cerebral Cortex*, 31, 3021–3033.

Zigerell, L. J. 2017. Potential publication bias in the stereotype threat literature: Comment on Nguyen and Ryan (2008). *Journal of Applied Psychology*, 102, 1159–1168.

CHAPTER 11

Intelligence and Aging

11.1 Introduction

Changes in intellectual ability over the adult years are complex and important to understand because they can inform social policies. There are 97 million people in the European Union at least sixty-five years old. Three out of 10 live alone, and only 9 out of 100 between sixty-five and seventy-five are economically active. In the United States, the number of people sixty-five or over is 48 million now, in 2023, and this number will rise to 98 million by 2060. In China, the estimate is 487 million people aged sixty-five or older by 2050. The number for Japan will be a quarter of its total population.

These values, however, can be misleading. Consider this question: how many people are sixty-five or older in the world – 10 percent, 30 percent, or 50 percent? Guess before you continue reading. The question has been surveyed in thirty-two countries around the world, and most people (58 percent) choose the second alternative, 24 percent choose the third alternative, and only 18 percent choose the correct answer, which is 10 percent. Therefore, 82 percent choose wrong answers. Issues about the "aging population" need to keep this perspective in mind when we consider aging effects on intelligence.

Intellectual abilities increase with age up to early adulthood, but after that, things are not so rosy (Cabeza et al., 2018; Fernández-Ballesteros et al., 2019; Tucker-Drob et al., 2019; Schaie, 2013). Some intellectual decline is inevitable with increased age, but there are considerable individual differences during aging. When people complete intelligence tests at age eleven and the same people are retested at age seventy, results show large variability in how their scores have changed during these six decades of life. If you take everyone with an IQ score of 110 at age eleven, when the same people are retested at age seventy, there will be a spread of scores. This phenomenon is known as the *great divergence*, and cross-sectional and longitudinal studies provide a wealth of evidence supporting the observation (Underwood, 2014).

As described in Chapter 1, the Scottish Council for Research in Education tested the intelligence of every eleven-year-old child attending school in 1932 ($N = 87,498$) using the Moray House Test; another cohort of eleven-year-old children was tested in 1947 ($N = 70,805$). The records of this countrywide study were rediscovered by Ian Deary in the 1990s (Deary, 2014, 2020). His research team recruited people still alive from the original testing groups for follow-up studies. This allowed examination of the stability of intelligence across the entire life span, from childhood to old age in an entire population. The follow-up studies with the largest number of participants were (1) the Lothian Birth Cohort born in 1921, age at retesting = 79, number of participants = 485; and (2) the Lothian Birth Cohort born in 1936, age at retesting = 70, number of participants = 1,017. Whereas test scores fluctuate over time, one key question was whether individuals retain their relative positions with respect to their peers as they age. The test–retest correlations in each cohort were $r = 0.73$ and $r = 0.78$, respectively. This indicated good stability: smarter kids tend to be the smarter seniors, even when changes in mean values are observed across the years. Remember that mean values and correlations tell different stories. These correlations also indicated individual changes consistent with the great divergence phenomenon.

An important issue was that the Scottish Mental Surveys relied on a single measure of intelligence (the Moray House Test). We have explained that single measures are not the best way for obtaining highly reliable estimates of the latent constructs of interest. It is much better to administer several tests and compute their shared variance. Another group did this in a longitudinal study (Rönnlund et al., 2015). They examined 262 men at age eighteen and reassessed the group again at age fifty, and then again at five-year intervals up to age sixty-five. Three intelligence measures were administered at age eighteen, four intelligence measures were administered at age fifty, and subsequently, working memory capacity was assessed with two other tasks. The data were analyzed with a refined statistical method for comparing latent variables (MGCFA, explained in Box 9.4). The measurement of the latent g-factor (at ages eighteen, fifty, fifty-five, sixty, and sixty-five) was stable over the nearly fifty-year period of study, consistent with observations of the Scottish studies, even though different tests were used. The key findings based on the latent g-factor were that (1) the stability of g was high (0.90) when variance specific to the administered tests was removed, (2) the correlation between g assessed at age eighteen and working memory assessed fifty years later was substantial (0.60), and (3) the concurrent correlation between g and working memory (at age sixty-five) was high (0.90).

Once again, we underscore that mean values and correlations tell different stories. Like height, intelligence increases in the first years of life, remains at similar values for decades, and then declines at old age. Those taller in childhood tend to be the tallest in old age. This same pattern is found for intelligence test scores.

Try to find another psychological trait showing the stability values we just enumerated for intelligence (you can use Google). The relative instability of many psychological traits has implications for understanding widespread human behavioral differences of interest to scientists, and perhaps for social policy (see Box 11.1).

Box 11.1 Cognitive Capacity (CC) and Psychosocial Maturity (PM)

Laurence Steinberg and colleagues (2009) published a seminal article in the journal *American Psychologist* addressing intricate legal issues associated with reports from the American Psychological Association regarding abortion and the death penalty for juveniles. The main question was, should we maintain a distinction between cognitive and psychosocial maturity in discussions of the legal status of adolescents?

Their analysis led them to conclude that the development of reasoning abilities (in structured situations) and of basic cognitive processes is complete at fifteen to sixteen years of age, on average. This conclusion is different for psychosocial features, such as impulsivity, sensation seeking, defining future goals, and susceptibility to peer pressure (Figure 11.1).

These researchers summarized results of the MacArthur Juvenile Capacity Study comprising 935 individuals (age range ten to thirty) who completed (1) the WASI for measuring intellectual ability (IQ), (2) five psychosocial maturity (PM) scales (risk perception, sensation seeking, impulsivity, resistance to peer influence, and future orientation), and (3) cognitive capacity (CC) measured by resistance to interference in working memory, digit span memory, and verbal fluency tasks.

The PM index combines the five measures, and lower scores indicate that individuals "characterize themselves as less likely to perceive dangerous situations as risky, more impulsive, more thrill seeking, more oriented to the immediate, and more susceptible to peer influence" (p. 590).

The correlation between IQ and the CC index was 0.46. Importantly, IQ scores were scaled by age, whereas the CC index was not. This latter index was calculated for different age bands (ten to

eleven, twelve to thirteen, fourteen to fifteen, sixteen to seventeen, eighteen to twenty-one, twenty-two to twenty-five, and twenty-six to thirty) controlling for participants' IQ scores. The correlation between the CC and PM indices was 0.15 (controlling for age).

The authors also represented the development of CC and PM controlling for IQ and household education. Age differences in PM were absent until mid-adolescence (fourteen to fifteen years) and became visible later. Differences were statistically significant between the sixteen- to seventeen-year-olds and those twenty-two or older and between the eighteen- to twenty-one-year-olds and those twenty-six and older. Results were sharply different for CC. Age differences were detectable before age sixteen to seventeen, but the upward trend stopped in this age band.

Interestingly, CC reaches adult levels well before the PM index. Adolescents reach adult levels of psychosocial maturity a decade after they reach adult levels of cognitive maturity. Nevertheless, even at twenty-four years of age, 50 percent of the adolescents fail to reach adult levels in both indices of mental maturity. This observation may be crucial in the courtroom. Having sound ways for the standardized assessment of cognitive and psychosocial maturity for individuals could be relevant for deciding punishment levels.

As the researchers note, scientific data can help provide guidelines for who should or should not be treated as adult: "the boundary between adolescence and adulthood should be drawn at a particular chronological age for one policy purpose and at a different one for another" (p. 592). In this regard, psychology research distinguishes two sociologically relevant decision-making contexts: those that allow for logical reflection and reasoned decision-making and those that do

Box 11.1 *(continued)*

not. Examples of the first category are medical and legal decision-making situations in which adults are present. Examples of the second category are emotional situations in which adults are absent. Juvenile delinquency (risky sex, purchase of alcohol and tobacco, risky driving) is an example of the latter.

Science alone cannot and should not dictate social policy, but data can help, especially if there is a solid weight of evidence from multiple studies.

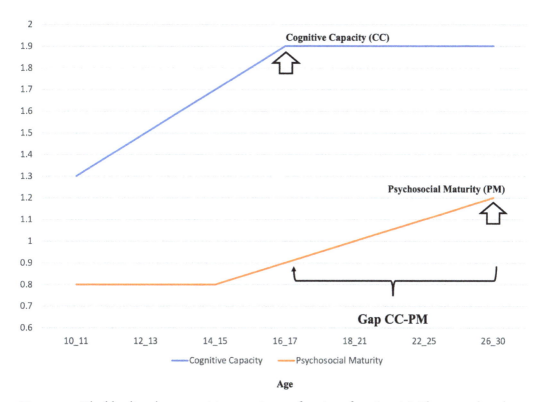

Figure 11.1 The blue line shows cognitive capacity as a function of age (*x*-axis). The orange line shows psychosocial maturity as a function of age. The *y*-axis is standard scores.
From Steinberg et al. (2009).

11.2 How Do We Study the Relationship between Intelligence and Aging?

Three basic research designs have been used in the study of aging. The simplest is the cross-sectional design (CSD), where the scientist takes measurements on people in different age groups, at roughly the same time. The positive aspects of CSDs are that they are relatively easy to conduct compared to other alternatives and they provide a picture of differences in the intellectual abilities of different age groups (cohorts) in the population, as exists at the time of testing. This information can be projected forward over short time spans. For instance, if we know the incidence of senile dementia in people seventy years and older as of 2022, and we know the number of people in the sixty-to-seventy age band, it is possible to estimate the incidence of senile dementia expected in 2032 and apply strategies for dealing with the problem.

However, CSDs pose problems of interpretation. It is difficult to obtain samples in which the participants differ only in age. For instance, elderly participants (seventy and over) will report having had less years of education, on average, than younger adults. This reflects changes in society, rather than psychological changes in the individuals, but it produces an unavoidable confound between any age and education effects. The general principle is that in cross-sectional

designs, age effects are confounded with cohort effects. If we compare twenty-year-olds to seventy-year-olds in 2022, we are comparing people in the 2002 birth cohort to people in the 1952 cohort. There is no way to tell whether any differences between these two groups are age effects (aging causes the difference) or cohort effects (see Box 11.2 for more details and examples).

Because cross-sectional studies measure each individual only once, there is also no way to distinguish between gradual and sudden changes in cognitive ability. To see this, consider two possible physiological events: (1) gradual deterioration of the brain and (2)

Box 11.2 The Cohort Effect

Studies of cohort effects have taken two forms. The first is typified by a classic study by Read D. Tuddenham (1948), some aspects of the SLS (Schaie, 2013), the normalization comparisons for Raven's Standard Progressive Matrices (Flynn, 2007a, 2012), and the European studies of military enlistees (Teasdale and Owen, 2000; Jokela et al., 2017). In the second form, two tests with norms established at different times are given to the same sample.

The first approach compares scores obtained on the same test by people of the same age, but from different cohorts, for example, registrants for the military in Denmark in 1958 and registrants in 1978. A generalization is then drawn to the larger population. This is the same-test-different-cohorts paradigm.

But cohorts over time may differ on relevant variables. For example, in 1917, the US Army developed the Army Alpha Test as a device for screening recruits; for a good history, see Warne et al. (2019). In World War II, the army used a successor test, the Army General Classification Test. To compare the two tests, the army gave a version of the World War I test to 768 World War II soldiers, selected to

represent the demographics of World War II enlistees. The median score of the World War II soldiers was approximately 1 standard deviation higher (using World War I norms) than the median score for World War I soldiers (this was an early example of what became known as the Flynn effect, described in Chapter 13).

Tuddenham (1948) pointed to several causes of the discrepancy, but he thought that the most important of these was that the 1945 soldiers had, on average, much more education than the 1917–1918 soldiers. He supported this conclusion by showing that World War I soldiers who were literate had test scores close to the scores of World War II soldiers. Figure 11.2 shows the general effect and the effect of introducing literacy as a covariate.

Drawing conclusions from the same-test-different-cohorts paradigm requires two assumptions. The first is that the test is a meaningful way to evaluate intelligence in different cohorts. This seems a reasonable assumption for fluid (Gf) tests, such as progressive matrix tests, given to the same overall population. It is also reasonable if the cohorts are not far apart in years, for cultures do not change that quickly. There are situations, however, where this assumption would not be reasonable. For instance, if the tests

Box 11.2 *(continued)*

were to be conducted in a developing country, it might be the case that, proportionately, more people in the more recent cohort would be accustomed to the testing paradigm due to dramatic increases in schooling and familiarity with testing. Such an argument is less tenable for a comparison of cohorts in a developed country. Also, tests of verbal and general knowledge have to be modified for the cohort involved. For example, in 1938, the term *gay bachelor* referred to an unmarried man who enjoys the company of women. In the twenty-first century, the term had come to acquire a rather different meaning. Any test involving cultural knowledge also faces the danger of being frozen in time or restricted to a particular cultural group. To maintain widespread applicability, commercial tests that evaluate crystallized knowledge (Gc) tend to assess knowledge that is held widely through the society. Common knowledge changes over cohorts, so Gc tests have to be changed accordingly. And this is what must happen to a good test, especially if it is commercialized.

The second assumption is that the cohorts are similar samples of a larger population. If this assumption does not hold, cohort effects can be mistakenly generalized to an entire population. In fact, few studies use random samples of the population for which generalization is intended. For instance, the 1992 US standardization sample for the Raven's Standard Progressive Matrices was entirely drawn from Des Moines, Iowa, a relatively small city. Des Moines was chosen because, on some statistical criteria, such as age distribution and distribution of ancestry groups, the city matched the United States as a whole. The problem with this approach is that matching solely on those variables that the investigator thinks are appropriate

leaves other measures free to vary. For instance, in the United States, educational standards vary widely across states and even across school districts within states. Was the quality of education in Des Moines equivalent to the typical quality of education in the United States at the time? Random sampling avoids such problems by equating statistical expectations for all covariates, not just for those the investigator feels to be important.

Tuddenham (1948) compared US military recruits from World War I to recruits from World War II. Recruitment procedures were not the same in the two wars, so is it valid to make an inference about changes in population intelligence? The SLS drew from a population of enrollees in a health care program, as described in the main text. To what extent is this population representative of the US population in general? Also, to what extent did the nature of the enrollees in the health care program change over the half-century lifetime of the project?

Because of questions like these, European military registration studies are particularly valuable. They involve repeated sampling of the same subpopulation, young men eligible for military service, over fairly brief time intervals. The fact that a cohort effect appears in several European studies of this nature is an important confirmation of secular gains in intelligence.

James R. Flynn (2016; Flynn and Shayer, 2018) relied on the second approach, in which two tests, with norms established at different times, were given to the same sample. This is the two-test-one-cohort design. Suppose that scores are higher, in terms of percentiles defined by the original standardization, on test 1 than on test 2. This implies that the standardization sample for test 1 had

Box 11.2 *(continued)*

lower abilities than the standardization sample for test 2. This argument depends on the two standardization samples being equally representative of the general population at the time that the standardization is done. Considerable care to account for cohort effects is taken in standardizing tests such as the Stanford–Binet and Wechsler tests, as these tests are widely used in clinical practice and to establish legal competency or qualification for special education programs.

An inflated IQ score can result from a raw score that has been compared to obsolete norms. Fatal consequences could result. Flynn's book (2007b) *What Is Intelligence?* includes details of a death penalty legal case. Flynn notes, "3 IQ points may be the difference between life and death. ... I have no expectation that psychologists in a normal clinical setting,

or a test publisher advising psychologists in such settings, are going to adjust obsolete IQ scores. But the court is not in that position" (pp. 192–195).

The raw score obtained by an individual on a standardized intelligence battery is valid if the battery is valid. The standardized score will be accurate if the examinee is compared to a representative sample within their cohort. If the examinee completes the Wechsler in 2022, but the norms were obtained in 2012, they will get a bonus of 3 IQ points because of the obsolescence of these norms. IQ gains over time on this intelligence battery amount to 0.30 points per year since 1947 (Flynn, 2007b; Trahan et al., 2014).

Cohort effects are not just statistical concerns; they can have profound real-world consequences.

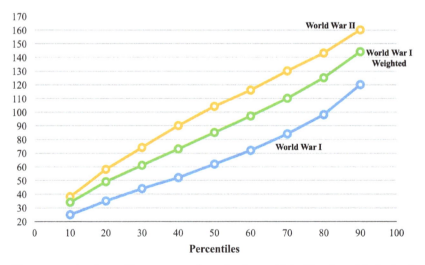

Figure 11.2 A cohort effect on scores from the Army Alpha Test (*y*-axis) obtained by World War I soldiers (blue), World War II soldiers (orange), and literate World War I soldiers (green, weighted). Scores are shown by increasing percentiles (*x*-axis). The same raw score corresponds to different percentiles, depending on the cohort. For instance, percentile 80 corresponds to raw scores 98 (World War I), 125 (World War I weighted), and 143 (World War II).
From Tuddenham (1948).

sudden damage due, for instance, to stroke. Either could produce slowed decision-making. In a cross-sectional design, we would find that, on average, decision-making processes slowed with increased age. However, because different people would be evaluated at each age, we would have no way of knowing how much of the difference was due to gradual deterioration, which affected everyone, and how much was because the older groups would include more people who had suffered sudden brain damage.

In a longitudinal design, the same people are studied across several time intervals. Therefore, instead of studying age differences, here we study age-related changes. The contrast between cross-sectional and longitudinal results is informative. Intelligence test scores show considerably less drop over the adult years in longitudinal studies than they do in cross-sectional studies. This is in part due to the confounding of age and cohort effects in the cross-sectional design.

Longitudinal studies are expensive and time consuming. They are also prone to the recruitment/attrition effects described in Chapter 9, because people with lower test scores at baseline are more likely to decline participation or to quit the study than people with high test scores. This is particularly true if the study involves multiple measures, thus requiring a considerable investment in time on the part of the study participants. Nonrandom attrition effects will be present. Unless allowance is made for this effect, aging may appear to be less debilitating than it actually is (Madhyastha et al., 2009).

The participants in a longitudinal study come from just one cohort. Therefore, the effects of general changes in society are mixed with the effects of aging. As an example, the percentage of women in the Genetic Studies of Genius (Terman, 1925) who followed professional careers was high for its time, but was much lower than the percentage of women following professional careers in the Study of Mathematics Precocious Youth (SMPY) study (Stanley et al., 1974), who were surveyed fifty years

later (Lubinski, 2016). Was this due to any psychological difference between women born around 1910 and those born around 1960, or was it due to the much greater career opportunities for young women in the 1980s than in the 1930s?

The gold standard is the cohort-sequential design, like the Seattle Longitudinal Study (SLS; Schaie, 2013). In this design, scientists recruit people of different ages at the start, follow them as in a longitudinal study, and in addition, periodically recruit new participants and follow them as well (see Table 11.1). This allows for a separate evaluation of cohort (age differences) and aging (age changes) effects and provides for longitudinal studies of different cohorts.

However, cohort-sequential designs are very difficult to implement. Aside from the expense, the biggest problem is ensuring comparability of the samples at each phase of recruitment. In the SLS, participants were recruited from people enrolled in a health maintenance organization. If enrollments in this organization have changed over the years of the study, recently recruited participants will not be comparable in all ways to earlier-recruited participants. There are also problems of selective attrition, as is the case in a longitudinal design.

All these problems make the study of aging difficult, but not impossible. To establish a weight of evidence, the effects of aging are discussed in this chapter from the three approaches presented in Chapter 2 and detailed in Chapters 3, 4, 5, and 6: (1) psychometric, (2) information processing, and (3) biological.

11.3 Psychometrics

Many psychometric studies of changes in intelligence with age rely on the distinction between fluid (Gf) and crystallized (Gc) intelligence. Two important conclusions based on a weight of evidence are shown in Figure 11.3: (1) Gf peaks in the mid-twenties and then falls fairly rapidly, with

Table 11.1 Cohort-sequential design used in the Seattle Longitudinal Study

1956 (T1)	1963 (T2)	1970 (T3)	1977 (T4)	1984 (T5)	1991 (T6)	1998 (T7)	2005 (T8)
S1 (N = 500)	S1 (N = 302)	S1 (N = 163)	S1 (N = 130)	S1 (N = 97)	S1 (N = 75)	S1 (N = 38)	S1 (N = 26)
	S2 (N = 997)	S2 (N = 419)	S2 (N = 333)	S2 (N = 225)	S2 (N = 163)	S2 (N = 111)	S2 (N = 74)
		S3 (N = 705)	S3 (N = 337)	S3 (N = 224)	S3 (N = 175)	S3 (N = 127)	S3 (N = 93)
			S4 (N = 609)	S4 (N = 293)	S4 (N = 203)	S4 (N = 136)	S4 (N = 106)
				S5 (N = 629)	S5 (N = 427)	S5 (N = 266)	S5 (N = 186)
					S6 (N = 693)	S6 (N = 406)	S6 (N = 288)
						S7 (N = 719)	S7 (N = 421)

Note. Grand total N = 4,852. S = sample. T = time.

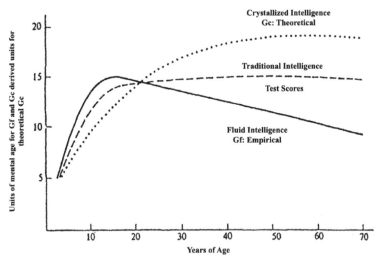

Figure 11.3 Illustration of life-span (*x*-axis) curves for fluid (Gf) and crystallized (Gc) intellectual abilities (*y*-axis). Traditional intelligence refers to combining both into a *g*-factor (Cattell, 1987).

the fall accelerating in old age (sixty-five plus), and (2) Gc peaks around age thirty and then is maintained at a surprisingly constant level until old age (Cattell, 1987; Horn, 1985; Horn and Noll, 1997). These findings support the importance of the Gf–

Gc distinction for aging research, rather than collapsing Gf and Gc into a single traditional general intelligence factor (g).

John B. Carroll, a psychometrician we met in Chapter 3 when presenting the Cattell–Horn–Carroll (CHC) psychometric model of intelligence, addressed intelligence changes across the life span using data sets from about twenty countries. In his classic encyclopedic book *Human Cognitive Abilities: A Survey of Factor Analytic Studies* (Carroll, 1993), he wrote, "At one time in the planning of the present work ... I had the intention of trying to glean from factor-analytic studies information from which I could derive growth curves for cognitive abilities. Several problems arose in any attempt to do this (so) I decided to abandon the attempt ... research on the change of cognitive abilities in adulthood and old age has been dogged by many methodological problems, including not only the cross-sectional vs. longitudinal controversy, but also the problem of how to deal with differences between cohorts and the problem of exactly what is to be measured, and how (speed versus level)" (p. 664).

We discussed in Chapter 3 that Carroll's expansion of the Cattell–Horn Gf–Gc model included crystallized and fluid intelligence as two of several broad second-stratum abilities, below g but less specialized than primary-level abilities such as vocabulary and language comprehension. Other important second-stratum abilities are visual and auditory reasoning, short-term memory, and long-term memory retrieval. John Horn had relied heavily on cross-sectional studies to draw his conclusions. As we have seen, such studies are confounded by cohort effects. This point is particularly troubling because the cohort effect is stronger for measures of Gf than for measures of Gc (Flynn, 2016; Schaie, 2013).

To avoid such a confound effect, results of a cross-sequential study would be of interest, and an important one was done with the Woodcock–Johnson test batteries, generated from the CHC three-stratum model of intelligence. They include tests suitable for early childhood to adulthood. The test batteries include measures for Gf, Gc, and several other second-stratum abilities comprising the CHC model. They also contain a measure of broad cognitive ability, which is analogous to the general intelligence (third stratum) factor in Carroll's (1993) extension of the Gf–Gc model. Because the different subtests have been calibrated using item response theory (see Chapter 3), they provide a score that can be treated as a linear scale of the trait underlying each subtest. This makes it possible to talk about differences in the rate of change of various measures of intelligence (Hunt, 2007).

The Woodcock–Johnson test was revised and renormed in 1990 to produce the WJ-R test (McGrew et al., 1991). The renorming was based on an approximate probability sample of the US population, containing about 6,500 cases. Subsequently, J. J. McArdle and colleagues (2002) contacted and retested 1,200 of the original participants. Chronological ages at the time of first testing ranged from two to ninety years. The interval between testing times ranged from one to five years. Figure 11.4 depicts factor scores for general intelligence (g), Gf, and Gc as a function of age.

Regarding general cognitive ability (g), there is a peak in the early twenties and a gradual decline thereafter, although there are considerable individual differences. Some participants in their seventies have scores that would be considered high for a twenty-five-year-old. As shown in Figure 11.4, the developmental patterns for Gf and Gc are substantially different. Also note the individual differences seen in the short lines that connect the same person's scores on the two testing occasions. Table 11.2 includes the specific values computed by McArdle et al. (2002) for different cognitive abilities. It shows the peak age and the rate of decline. To illustrate the rate of change, if, on retest, Chen has increased 10 W units from the initial score, he can now perform tasks with 75 percent success that were performed with 50 percent success on the initial test. If Chen has declined by 10 units from the initial score, this means he now performs tasks with 25 percent

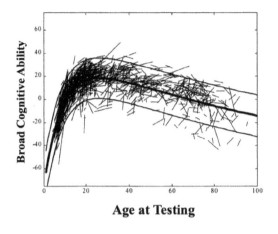

Age at Testing

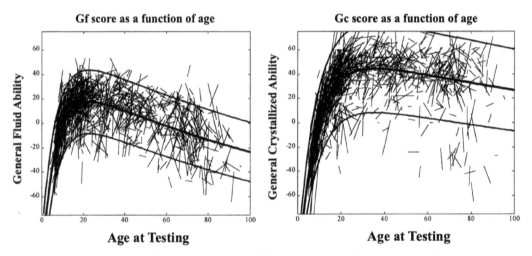

Figure 11.4 Age (*x*-axis) and change scores (*y*-axis) for *g* (top), Gf (left), and Gc (right). The short lines connecting two points indicate scores for the same individual at the two times of testing. Heavy lines show quartiles (McArdle et al., 2002).

success that were formerly performed with 50 percent success. This relationship holds for any 10-point difference on the W scale across age (McArdle et al., 2002, p. 119). All the second-stratum broad abilities, except Gc, behave very much like Gf, cresting in the mid-twenties and then declining throughout the adult years. The Gc exception is important because Gc is a better predictor of some workforce performance than Gf (Schmidt, 2014). McArdle et al.'s (2002) data support the CHC model as an appropriate way to understand the influence of age on intelligence.

Warner Schaie (Schaie, 1996, 2005; Schaie and Willis, 2010; Borghesani et al.,

2013) has reported data from the SLS that might complement the findings of McArdle et al. (2002). Figure 11.5 shows the cognitive abilities measured in the SLS, along with the findings obtained in the cross-sectional and longitudinal comparisons for reasoning, spatial orientation, processing speed, numerical ability, verbal comprehension, and verbal memory. Cross-sectional results show decline in reasoning, spatial orientation, and processing speed at age forty-six. Verbal memory begins to show decline at age thirty-nine. Numerical ability and verbal comprehension show a smoother developmental pattern. The average difference between the youngest and the oldest cohorts

Table 11.2 Ages at peak value, rate of decline, and annual change (in standard W units) within two age bands for cognitive factors evaluated on the WJ-R battery

			Annual change	
Cognitive ability	Age at peak (years)	Age at deceleration (years)	Ages 2–19 (years)	Ages 20–75 (years)
General intelligence	26	52	4	−0.3
Fluid reasoning	23	45	5.5	−0.5
Comprehension	36	71	7	−0.01
Long-term retrieval	18	36	2.4	−0.4
Short-term memory	24	48	4.8	−0.3
Processing speed	25	50	6.6	−0.6
Auditory processing	23	45	3.8	−0.3
Visual processing	24	49	3.8	−0.4
Quantitative ability	29	58	9.7	−0.3
Academic knowledge	30	60	6.6	−0.3
Reading and writing	26	52	15.2	−0.2

Note. From McArdle et al. (2002).

is equivalent to 2 standard deviations. Longitudinal results reveal significant decline for processing speed and numerical ability at age sixty. Reasoning, spatial orientation, and verbal memory decline at age sixty-seven. Verbal comprehension shows significant decline at age eighty-one.

Combining cross-sectional and longitudinal designs, results showed that between twenty-five and eighty-eight years of age, (1) verbal comprehension declines 0.4 standard deviations, (2) reasoning and spatial orientation decline 0.8 standard deviations, (3) verbal memory declines 1.1 standard deviations, (4) processing speed declines 1.2 standard deviations, and (5) numerical ability declines 1.5 standard deviations. Therefore, the downward trend with increased age ranges between less than half a standard deviation for verbal comprehension to one and a half standard deviations for numerical ability.

We see that the McArdle et al. (2002) study and the SLS studies are consistent in showing that some Gc measures are more resistant to the aging process than measures of inductive reasoning (Gf). However, the studies are not consistent about precisely when the declines occur. The McArdle et al. study indicates declines starting in the mid-twenties for Gf and in the mid-thirties for Gc. The SLS data show declines starting considerably later. Age trends in the WJ-R renorming were confounded with cohort effects, which would lead to an overestimation of the deleterious effects of aging. In the SLS, some participants have dropped out over time, so there is a bias toward continued participation by people with higher initial cognitive abilities. This would result in an understatement of the age effect.

There also is debate about the extent to which the pervasiveness of the general intelligence factor (g) increases with advanced

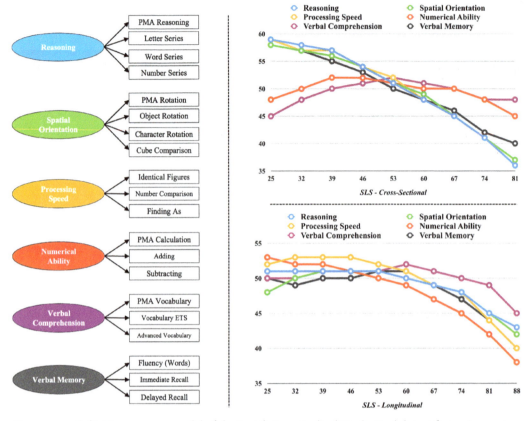

Figure 11.5 (left) Measurement model of the Seattle Longitudinal Study. Each latent factor (reasoning, spatial orientation, processing speed, numerical ability, verbal comprehension, and verbal memory) is measured by at least three tests, which is highly recommended from a psychometric standpoint. (right) Changes in latent variables over time from the (top) cross-sectional and (bottom) longitudinal comparisons. The *x*-axis is age; the *y*-axis is standard scores in T units.
Adapted from Schaie (2005).

age. Technically, this is called the *dedifferentiation hypothesis*. The hypothesis predicts that the correlations between measures of different types of intelligence, such as verbal and visuospatial reasoning, will increase with age.

Good arguments can be advanced for this hypothesis. Declines in intelligence are usually associated with declines in health. Injuries that influence the brain should have widespread deleterious effects on cognitive performance, although their differential impact may change dramatically for different premorbid intelligence levels (Santarnecchi et al., 2015). Nonpathological reductions in the prefrontal cortex and other areas associated with the working memory system,

which are typical of aging, are associated with lower fluid intelligence scores (Kramer et al., 2006). Reduction in the capabilities of the working memory system should have a pervasive effect on almost all cognitive functions, which should lead to an increased influence of individual differences in *g* on performance. Nevertheless, the cognitive training literature fails to substantiate this presumption, because improvements in the working memory system fail to impact *g* (Protzko, 2017).

The evidence for the dedifferentiation hypothesis is mixed (Juan-Espinosa et al., 2002; Schaie, 2005; Tucker-Drob and Salthouse, 2008; de Frias et al., 2007). However, a comprehensive meta-analysis comes to a robust conclusion: a general factor of cognitive aging (g-aging) strengthens with advancing adult age (Tucker-Drob et al., 2019). Their key research question was whether individual differences in longitudinal changes are related across different cognitive abilities. For addressing this question, the researchers identified twenty-two data sets, eighty-nine effect sizes, and more than 30,000 individuals. Conclusion: 60 percent of the variation in cognitive changes was shared across cognitive abilities. However, this was the average value, because changes increased with age, from 45 percent at age thirty-five to 70 percent at age eighty-five. Therefore, the dedifferentiation hypothesis was supported, even when the probable presence of dementia was carefully controlled.

These were the main conclusions from the meta-analysis:

1. Longitudinal individual differences in changes of distinguishable intellectual abilities are strongly correlated among themselves. If inductive reasoning decreases, for example, the remaining abilities also tend to decrease. Changes are coupled.
2. The amount of variance in cognitive ability levels explained by the general intelligence factor (g) was 56 percent, closely similar to the amount of shared variance in rates of change, which was 60 percent.
3. Cognitive decline may operate along a similar general dimension as does cognitive development in general.

These findings were interpreted within the conceptual framework proposed by Manuel Juan-Espinosa and colleagues (2002), what they call the *in-differentiation hypothesis*: the structure of life-span changes in cognitive abilities may be invariant in much the same way that the structure of changes in human anatomy are invariant. They proposed an anatomical metaphor: "As the human skeleton, there is a basic structure of intelligence that is present early in life. This basic structure does not change at all, although, like the human bones, the cognitive abilities grow up and decline at different periods of life" (p. 406).

The results reported by Tucker-Drob and colleagues (2019) qualify the classic models of adult intellectual development. It remains valid to say that there is a pattern of average mean decline in verbal abilities and steep mean declines in fluid abilities. Nevertheless, all abilities show substantial and uniform loadings on a common factor of age change (g-aging). See Box 11.3 for possible implications of this finding.

11.4 Information Processing

The general picture with respect to changes in information processing with age is straightforward: cognition gets slower (mental speed decreases) and the working memory system functions less well. In this section, we present evidence of cognitive slowing and reduced capacity. In the next section, we present data about how the brain is involved.

11.4.1 *Processing Speed*

Cognitive slowing with age can be quantified using reaction time and perceptual decision tasks (Chapter 4). Many studies show this trend, but one of the most sound was conducted by Geoff Der and Ian Deary (2006). They surveyed a large sample of the UK population ($N = 7,414$), finding marked slowing of simple (SRT) and choice (CRT) reaction times with age. Choice reaction time, which requires more complex decisions, was more sensitive to aging (Figure 11.6). Reaction times slow and results also become more variable with aging (recall the "great divergence" from Section 11.1). These were their four main conclusions: (1) SRT fails to show any increase until age fifty (longer RT is a slower time); (2) CRT increases across the whole life span, and the values become more variable with age; (3) the correlation between SRT and CRT was 0.67; and (4) the results are consistent with previous research using another

Box 11.3 Experience Accumulates with Age, and It May Count for Solving Problems in Everyday Life

According to average group data, if around age forty-five, people become slower, more easily distracted, and poorly focused, then, (1) how is it that humans in their fifties and sixties occupy most of the important leadership positions in our society? (2) why are people advised to go to experienced physicians when physicians do not become experienced until their forties? and (3) why do we routinely trust our lives to commercial airline captains, most of whom are in their fifties and sixties?

Accumulated experience requires long-term investments, and this seems to count in real life. The typical test of crystallized intelligence (Gc) evaluates how well a person knows information in their society. Although the elderly may have trouble dealing with the sorts of novel, out-of-context problems included in IQ tests (and especially tests of Gf), people in midlife and beyond do quite well in dealing with realistic decision-making (Marsiske and Margrett, 2006; Sternberg, 2003).

On the other hand, the elderly do not do particularly well in made-up experimental situations, especially when they are under time pressure to make decisions. This could be due in part to general cognitive slowing (as discussed in the next section), in part to the need to keep several factors in mind when making a complex decision (working memory), and in part to older individuals' tendency to stress accuracy rather than speed in decision-making situations.

Why is there a difference? Laboratory decision-making tasks are constructed to reveal the process of decision-making, in situations where all participants have little prior experience with the problem at hand. Outside the laboratory, decision-making is heavily influenced by the possession of relevant information (Gigerenzer, 2000; Klein, 1998, 2009), and people use heuristics (shortcuts) that are discouraged in laboratory studies (Klein, 2009). In everyday settings, the information-processing burden shifts from working memory to long-term memory and retrieval processes. The latter processes are relatively resistant to age-related decline until quite late in life.

The heuristics (shortcuts) used outside the laboratory rely on the decision maker's experience with similar decision-making situations, which is precisely what mature decision makers would have. This sort of reasoning has been extended to what some call wisdom. The intuitive idea is that as you age, you acquire wisdom – or, at least, you should.

How do we distinguish wisdom from practical problem solving? Practical problem solving and decision-making involve problems that the individual faces in the here and now, while wisdom refers to thinking about broader, ephemeral issues, often involving the course of society. A group of German researchers developed the Berlin wisdom theory, in which wisdom is defined as knowing the rules and meaning of life (Brugman, 2006). Sternberg (2003) has defined wisdom as knowing how to apply creativity and intelligence for the common good by balancing personal and societal interests. Gerontologists claim that wisdom is more typical of the old than of the young. That may be, but we would like to have clearer definitions and a better analysis of individual differences within the aged population before unequivocally associating age with wisdom.

Finally, leaders in our society, physicians, and commercial airline captains in their fifties and sixties hardly represent the general population of humans at

> **Box 11.3 (continued)**
>
> these ages. They likely would excel in typical tests of crystallized intelligence (Gc) assessing how well they know their society. But standardized Gc tests cannot be successfully completed without having the general ability required for selecting the relevant information within the database stored in long-term memory. Gc is not an isolated cognitive ability; it belongs to a network of cognitive abilities substantially correlated (Kovacs and Conway, 2019a, 2019b). It is not enough to have the knowledge – the critical factor is being aware of when to use the knowledge. This is why psychologists coined the concept of crystallized ability instead of referring simply to knowledge stored in long-term memory.

British sample (900 adults, mean age fifty-six). The correlation between general intelligence (g) and SRT was r = −0.31, whereas the correlation between CRT and g was r = −0.49 (Der and Deary, 2003). The higher g is, the faster are the SRT and CRT values (i.e., greater processing speed, especially for CRT).

There are two broad ways to consider the implications of age-related slowing (Hartley, 2006; Salthouse, 1996). First, cognitive slowing may produce problems in specific aspects of cognition. For example, allocating attention to concurrent tasks (like driving a car while attending to the GPS's verbal instructions and visual display) is possible only if multiple information inputs can be evaluated reliably at once. Diminished cognitive processing speed might weaken the ability to keep up with multiple inputs from the surrounding world. Second, diminished processing also serves as a marker for the general state of the nervous system. Therefore, diminished processing with age is important both in itself and as an indicator of general cognitive health. Either of these possibilities indicates why cognitive slowing may be one of the core factors for the age-related decline in intelligence.

Figure 11.7 presents results from a genetically informative Swedish twin study of aging in which changes in cognition were assessed with and without statistical controls for changes in processing speed. Controlling for processing speed virtually eliminates the change in cognition from age fifty to sixty-five and slows, but does not eliminate, the decline after sixty-five: "age trajectories for general cognitive ability changed from modest decline before age 65 to lack of changes before age 65 when speed variance was removed" (Finkel and Pedersen, 2004, p. 341). Nevertheless, statistically removing the contribution of processing speed does not change reality: older people are slower, on average. This fact of life (and, more specifically, of the nervous system) impacts general intelligence levels and everyday lives as people grow older.

The pattern shown in Figure 11.7 indicates that with increased age, genetic variance associated with processing speed becomes more important for explaining the genetic variance of general intelligence (g). These researchers concluded that genetic influences on g "act primarily via an indirect path through processing speed. ... We highlight the increasingly important roles that processing speed and the genetic influences on processing speed play in longitudinal age changes in general intelligence" (p. 342).

11.4.2 Working Memory

As people age, there is a breakdown in the working memory system. This is shown by decreased performance in short-term memory span tasks, greater susceptibility to interruption or distraction, and poorer control on dual tasks, such as simultaneously

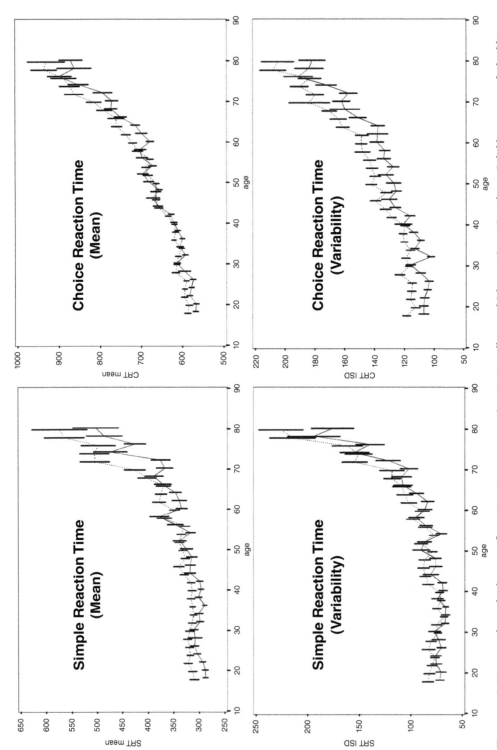

Figure 11.6 Means and standard errors of reaction time measures (y-axis; in milliseconds) by age (x-axis) and sex. Solid line = men; dashed line = women. The results for men and women have been offset slightly on the x-axis so that the overlap of the standard error bars can be seen more clearly.
From Der and Deary (2006).

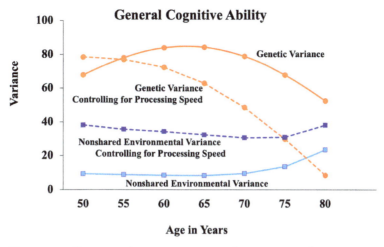

Figure 11.7 Cognitive changes in genetic and environmental variance
(*y*-axis) with age (*x*-axis), both with and without controlling for processing
speed, as estimated by a latent growth curve model (Finkel and Pedersen,
2004).
Courtesy of Deborah Finkel.

monitoring a stream of visual and auditory
signals (Hoyer and Verhaeghen, 2006;
Kramer et al., 2006; Salthouse et al., 2003).
These measures are in general those that are
most closely related to psychometric measures of *g* and Gf. On one hand, the failure
in working memory functioning appears to
be general, rather than associated with a
particular component of the working memory system (Salthouse, 1996, 2005, 2012;
Cowan, 2017). On the other hand, as discussed in Chapter 4, executive updating of
information in short-term memory may play
an important role.

Timothy Salthouse and colleagues (2003)
asked if executive function may account for
age-related intellectual decline. Is this
decline attenuated when individual differences in executive functioning are controlled? For answering this question, data
were obtained from 261 adults (age range
eighteen to eighty-four) who completed a
comprehensive battery of tests, listed in
Table 11.3.

Figure 11.8 shows their results for the
relationship between age and four cognitive
tests. All, except vocabulary, decline after
age thirty. No surprises here – it is the same
pattern seen in the RT studies. But what

happened to the executive measures with
age? Table 11.3 shows examples of performance across age for the neuropsychological
executive function, inhibition of (prepo-
tent) responses that automatically capture
your attention, updating, and time-sharing
tasks. The correlations with age (age *r*) also
are shown. The findings from this comprehensive study led to four main conclusions:

1. The executive function, updating, and
    time-sharing constructs showed very
    high correlations ($r > 0.85$) with fluid
    reasoning (Gf). Therefore, they might
    be tapping shared cognitive processes.
    The implication is that it does not make
    sense to control for executive function-
    ing differences and see if this attenuates
    intellectual decline. Recall that dimin-
    ished cognitive performances across age
    are coupled (Tucker-Drob et al., 2019).

2. The specifics of executive functioning
    are much less important than usually
    assumed within the neuropsychological
    literature. Typical executive function-
    ing measures tap cognitive processes
    shared by standardized measures of
    intelligence (Ruiz Sánchez de León
    et al., 2019).

Table 11.3 Average performance scores across age bands for executive functioning (EF) measures

EF measure	Age band (years)			Age r
	18–39	40–59	60–84	
Neuropsychological executive measures				
Wisconsin card sort*	20	24	37	0.21
Hanoi*	26	30	28	0.06
Verbal fluency	13	14	11	−0.15
Figural fluency	19	18	15	−0.41
Connections*	1.4	1.5	2.1	0.41
Inhibition of prepotent responses measures				
Stroop*	16	18	22	0.51
Reading distraction*	120	118	142	0.18
Anti-cue*	834	987	1,328	0.48
Stop signal*	720	726	732	0.05
Updating measures				
Keep track	72	70	64	−0.23
Digit monitoring	94	94	92	−0.04
Matrix monitoring	76	76	68	−0.26
2 back*	3.2	4.2	5.8	0.38
Time-sharing				
Tracking with paired associates	3.3	2.9	1.6	−0.49
Driving with n-back 2*	5.7	7.9	9.9	0.37
Connect and count*	62.9	59.8	70.5	0.15

Note. From Salthouse et al. (2003). An asterisk indicates where higher values equal worse performance. Age r is correlation between age and each measure.

3. The updating construct is more closely related to fluid reasoning ($r = 0.93$) than to memory ($r = 0.72$), speed ($r = 0.79$), and vocabulary ($r = 0.30$). This might support a distributed network involved for fluid intelligence and executive control. High-complexity cognitive problems evoke widespread activations encompassing the prefrontal cortex, but also subcortical and occipitotemporal regions (Martínez et al., 2011; Colom et al., 2016; see also Chapters 4 and 5).

4. The executive function measures usually administered in neuropsychological research may not qualify as distinguishable psychological constructs beyond the classic standardized measures of intelligence: "one needs to be cautious in postulating the existence of a new construct without relevant empirical evidence that it is distinct from constructs that have already been identified" (Salthouse et al., 2003, p. 589).

Unfortunately, this last point is a recurrent problem in psychological research generally (Judge et al., 2002). Using a new label like executive functioning (EF) for designating a set of tasks is not proof that we are tapping different cognitive processes than available measures like intelligence tests do not tap. Discriminant validity must be demonstrated before accepting the supposed new construct (Chapter 2). At this construct level, the general factor of intelligence (g) and EF

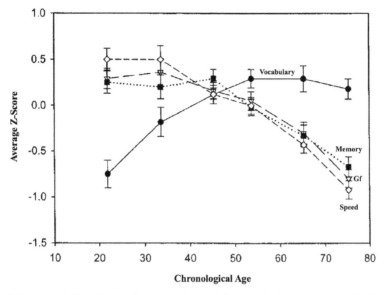

Figure 11.8 Standardized z-scores (y-axis) for vocabulary, memory, fluid reasoning (Gf), and processing speed across age (x-axis). Means and standard deviations (vertical brackets at each data point) are shown. The correlations between age and these scores were 0.36, −0.45, −0.47, and −0.58, respectively (Salthouse et al., 2003).

are almost identical with respect to the pattern of individual differences. This observation is corroborated in the lesion mapping study of intelligence and executive function described in Chapter 5 (Barbey et al., 2012).

11.5 Biology

We have discussed a set of selected findings from psychometric and information-processing perspectives regarding the relationship between intelligence and aging. As described in previous chapters, cognitive processes necessary to perform experimental tasks and their relationship to scores on standardized tests of intelligence are also associated with gene expression and brain structure and function. Therefore, a biological perspective is worthwhile for a more complete coverage of age effects.

Elliot Tucker-Drob and Daniel Briley (2014) analyzed the issue of the genetic and nongenetic contributions to intelligence differences across the life span. They conducted a comprehensive meta-analysis of longitudinal twin and adoption studies. It

included 150 combinations of time points and measures from fifteen longitudinal samples (4,548 MZT pairs, 7,777 DZT pairs, 34 MZ adopted pairs, 78 DZ adopted pairs, 141 adoptive sibling pairs, and 143 nonadoptive sibling pairs). The age range was from six months to seventy-seven years. Their analyses addressed four questions about the stability of intelligence:

1. To what extent are genetic and nongenetic factors stable over time?
2. To what extent does the stability of genetic and nongenetic factors change across the life span?
3. To what extent do genetic and nongenetic factors underlie changes in the stability of cognitive abilities?
4. What are the other moderators of phenotypic, genetic, and nongenetic stability?

Results showed that the stability of intelligence increases from effect sizes of 0.30 in early life to 0.6 at age ten and 0.7 at age sixteen, then reaches an asymptote (0.78) (Figure 11.9, left). These were the main

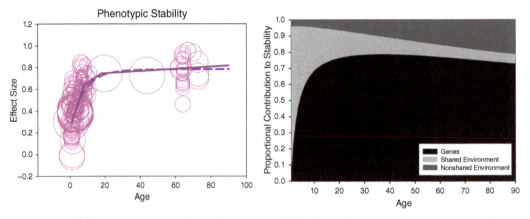

Figure 11.9 (left) Age (*x*-axis) trends in phenotypic stability (*y*-axis). Data points are represented as dots enclosed in circles. Larger circles indicate higher weight in the analysis and greater precision of the data point. The key message here is that cognitive differences have low stability in very early life ($d = 0.30$) but values increase during development ($d = 0.6$ by age ten and older, 0.7 by age sixteen). (right) Proportional genetic, shared environmental, and nonshared environmental contributions (*y*-axis) to the stability of intelligence across the life span (*x*-axis).
From Tucker-Drob and Briley (2014).

findings regarding the contribution of genetic and nongenetic factors to stability:

1. While the contribution of genetic factors is null in early life, their relevance increases through child development and stabilize at an asymptote effect size of 0.65 in adulthood.
2. Shared environment contributes moderately to stability in childhood (effect size 0.24), and this contribution fades to zero values by middle adulthood.
3. Nonshared environment makes a very small contribution, although it increases somewhat with advanced age.
4. By late childhood, the genetic contribution to the stability of intelligence was 75 percent, whereas nonshared environment contributes 20 percent (Figure 11.9, right; see also Figure 6.3).

The authors argue that these results support transactional models of cognitive development underscoring gene–environment correlation and interaction. If shared environmental influences accumulate their influence on cognition over time through variables like social class or school quality, depending on children's genotypes, then the

estimates of the contribution of shared environment to stability should decrease with increased age, and this is what was found. Furthermore, the gene–environment correlation should lead to increases with age in the genetic contribution to stability, and this too was supported in the meta-analysis: "As children increasingly select and evoke differential levels of stimulation on the basis of their genotypes over time, genetic stability will increase. These considerations together indicate that estimates of genetic influence are likely to reflect environmentally mediated mechanisms" (p. 971).

This is the same conclusion James Flynn (2018, p. 81) reached: "We must distinguish between environment losing predictive potency and environment losing causal potency. ... As children age, what really happens is that genes and environment become better and better correlated; and that environment retains its full causal potency in the process." This line of age-related research shows the importance of genetic and nongenetic factors for intelligence, but it is clear that both components are melded in the brain (Colom, 2016).

This brings us to the question: are developmental brain changes related to intelligence? A number of longitudinal MRI studies address this question. Typical dependent measures are cortical thickness (CT) and cortical surface area (CSA). A weight of evidence is emerging about the dynamic nature of these variables during development.

Two studies addressed simultaneous changes in intelligence and the brain using longitudinal data from children and adolescents representative of the general population and collected by the National Institutes of Health. In the first study, Francisco J. Román and colleagues (2018) analyzed the concurrent changes in both variables, whereas in the second study, Eduardo Estrada and colleagues (2019) addressed this key question: are brain changes a consequence of intelligence changes, or the other way around?

In their study, Román and colleagues (2018) considered an age range between six and twenty-one years. Cortical and intelligence changes were measured at three time points separated by approximately two years. The general factor of intelligence (g) was estimated at the latent level from a test battery (the optimal way), and cortical changes were correlated with g at different points in time. Figure 11.10 (left) depicts age changes in CT, CSA, and g. Each individual is represented as a trend line, with a dot standing for each of the three time measures.

Because g was calculated at the latent level, it was possible to estimate the maturational changes within the age range of interest: from time 1 to time 2, the increase was equivalent to $d = 0.60$ (9 IQ points); from time 2 to time 3, $d = 0.48$ (7 IQ points); and from time 1 to time 3, $d = 0.98$ (15 IQ points). Intellectual changes were greater at younger ages than at older ages. Figure 11.10 (middle) shows that the changes in g scores correlate with changes in CT and CSA: the greater the age is, the smaller are the changes. In the final stage of their analyses, brain structural changes were related to the distribution of intelligence scores as individuals age (Figure 11.10, right). Cortical thinning was present in individuals with lower intelligent levels at approximately age ten.

For brighter individuals, this thinning process happened only in late adolescence, about age seventeen. Thinning was statistically associated with intelligence from age ten to age fourteen; individuals with lower intelligence scores in this developmental stage showed greater thinning. Regarding CSA, expansion along the full range of intelligence scores for younger individuals (age eight) was observed. This cortical expansion stopped at age nine to ten. Surface area contraction was observed at age eleven and was more pronounced in brighter individuals. Later, surface area changes were almost absent. All surface area changes observed across the developmental periods were independent of the intelligence values.

One important feature of this study was the use of confirmatory factor analysis for demonstrating that the estimates of g were comparable across age. This means that the scores tap the construct of interest regardless of age. Not surprisingly, individuals were increasingly smarter (higher g scores), but the upward trend slowed with age. The pattern for g was also observed for CT and CSA: older individuals showed smaller changes in cortical and intelligence measures. In early adolescence, individuals with the highest g scores showed CT preservation, whereas individuals with the lowest g scores showed thinning. Moreover, the thinning and surface contraction processes lasted longer in individuals with lower intelligence scores.

The relationships reported by Román et al. (2018) cannot distinguish, however, whether intelligence modulates cortical changes, or the other way around. In the second developmental study we are highlighting, Estrada and colleagues (2019) addressed this question by identifying the developmental sequence using repeated measures of the same individuals across time. Three hypotheses were considered:

1. Developmental intellectual changes follow cortical maturation.
2. Cortical maturation follows intellectual changes.
3. Intellectual changes and cortical maturation influence each other over time.

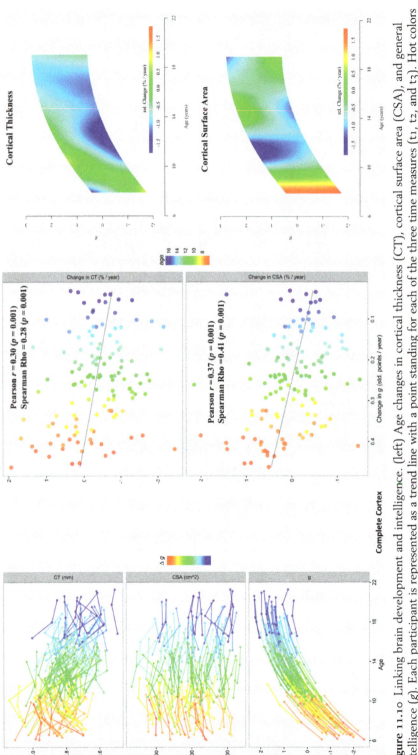

Figure 11.10 Linking brain development and intelligence. (left) Age changes in cortical thickness (CT), cortical surface area (CSA), and general intelligence (g). Each participant is represented as a trend line with a point standing for each of the three time measures (t₁, t₂, and t₃). Hot colors denote abrupt increases in intelligence (Δg). (middle) Correlation between changes in g (horizontal axis, reversed) and changes in cortical thickness (CT) and cortical surface area (CSA) (vertical axis). Each individual is represented by one point, and the age of the first evaluation is represented by the color of these points. (right) Relative change rates in cortical thickness (CT) and cortical surface area (CSA) over time for the different levels of g (general cognitive ability). Cold colors denote thinning and surface contraction, whereas hot colors denote thickening and surface expansion. The continuous trend goes upward because g increases during this time period. The image shows how this change relates to thickness and surface values for individuals with different g scores across time. Results for CT reveal prolonged thinning (cold colors) for less-intelligence individuals (from ten to eighteen years of age), whereas this thinning process becomes visible only in late adolescence (seventeen years) for more-intelligence individuals. Results for CSA show great expansion (hot colors) in childhood for all individuals regardless of their intellectual levels. Surface contraction (cold colors) is revealed at eleven years of age, especially for brighter individuals. This contraction process continues for a longer period of time in less intelligent individuals (Román et al., 2018).

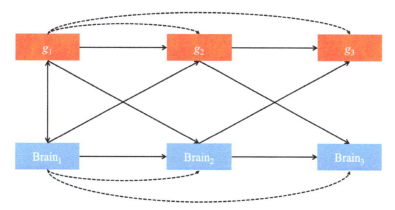

Figure 11.11 Theoretical cross-lagged model. This model is used to evaluate whether the predictive values for g (g_1 to Brain$_2$, g_2 to Brain$_3$) are greater than those for brain (Brain$_1$ to g_2, Brain$_2$ to g_3) across time (subscripts 1, 2, and 3). Latent change score (LCS) models add rates of change (not shown in the figure) across time for both variables. Therefore, successive levels on the variables of interest plus rates of change in these variables across time are modeled.

The key finding, based on 430 children and adolescents (age range six to twenty-two years and representative of the general population), was that the magnitude or rate of the change in both intellectual and cortical development at any time point predicts later changes, but the previous levels of these variables do not predict later changes. The authors concluded this based on a sophisticated approach for modeling age-related relationships between variables, called latent change score (LCS) models. Figure 11.11 depicts a simplified version of what LCS does. These were the main conclusions:

1. Intellectual changes and cortical changes were not predicted by previous levels of g, CT, or CSA.
2. Intellectual changes were predicted by previous change in CT and surface area. Individuals showing less cortical thinning and surface contraction during the previous interval showed greater increase in g.
3. Individuals who increased more in g during the previous interval showed greater subsequent cortical thinning.
4. Change in surface area was not predicted by any previous change.

5. Individuals who increased more in g during the previous interval showed less subsequent increase in g.
6. Individuals who experienced greater cortical thinning during the previous interval showed smaller subsequent thinning.

These findings underscore the highly dynamic nature of the relationship between cognitive and cortical development by showing that changes in intellectual abilities and CT predict their subsequent changes.

The finding that intellectual changes predict cortical changes may be consistent with the challenge hypothesis discussed in Chapter 2. Brain development and degeneration would influence intelligence; this seems to be the default hypothesis (Wendelken et al., 2017), but Estrada et al.'s (2019) research suggested that the influence may also go the other way around: less thinning predicts larger subsequent intellectual gains, and larger intelligence gains predict greater subsequent thinning.

These results also are related to results from the longitudinal analysis of the Lothian Birth Cohort 1936. Sherif Karama and colleagues (2014) found that intelligence levels measured at age eleven explain more than two-thirds of the relation

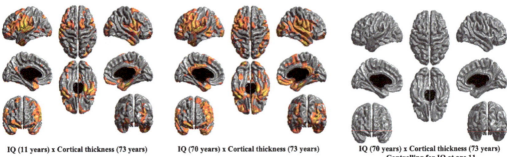

IQ (11 years) x Cortical thickness (73 years)   IQ (70 years) x Cortical thickness (73 years)   IQ (70 years) x Cortical thickness (73 years)
Controlling for IQ at age 11

Figure 11.12 Intelligence and cortical thickness. (left) Regions showing significant correlations between intelligence scores measured at age eleven and cortical thickness measured at age seventy-three. (middle) Regions showing significant correlations between intelligence scores measured at age seventy and cortical thickness measured at age seventy-three. The similarity between the left and middle panels is striking. (right) Correlations between IQ at age seventy and thickness at age seventy-three vanish when IQ at age eleven is statistically controlled (Karama et al., 2014).

between intelligence assessed at age seventy and CT measured at age seventy-three: "each person's rank order in terms of cortical thickness at age 11 may be mirrored at age 73 despite the fact that the cortex is known generally to thin slowly through adulthood" (p. 558). Figure 11.12 shows their results based on 588 individuals assessed at age eleven (left panel) and at age seventy (middle panel). The similarity of the brain maps is striking, but, even more surprising, all the correlations shown in the middle panel for age seventy disappeared (became statistically nonsignificant) after controlling for the intelligence of these same individuals at age eleven (right panel).

Why do intelligence levels at age eleven account for the association between intelligence and CT in old age? Three alternatives were suggested:

1. The lifetime association may be due to genetic factors. The same genetic factors may affect intelligence in childhood and old age. Genes that influence cortical growth might also affect cortical maintenance across the entire life span.

2. Individuals with higher intelligence levels from the outset keep doing more intellectually challenging activities as they age. Coping with these challenges may contribute to preserving brain integrity in old age.

3. There are reciprocal dynamic associations between intelligence and brain structural differences. Greater CT may lead to higher intelligence. Higher intelligence would promote (1) seeking stimulating environments that evoke cortical positive changes and (2) healthier lifestyle choices beneficial for brain structural maintenance.

The second and third alternatives are consistent with Estrada et al.'s (2019) finding that changes in intellectual abilities predict subsequent cortical changes. Children and adolescents involved in more cognitively demanding activities may accelerate their natural process of cognitive development and, therefore, achieve a higher rate of change. In turn, this faster rate of change may lead to changes in the cortex. In any case, a key feature of these studies is their focus on individual differences. Not everyone ages the same way on the same schedule (Tucker-Drob, 2019; Box 11.4).

Two final points on the biology of aging and intelligence are due. First, meta-analyses of the fade-out effect observed after ending programs aimed at enhancing intelligence suggest that the dissipation of the gains is an inevitable consequence of returning to a cognitively impoverished environment (Protzko, 2015). Alternatively, the fade-out effect may result from the failure of the

Box 11.4 Individual Differences in Healthy Aging

It is likely that individual differences in age-related intellectual decline result from a complex interplay between genetic and nongenetic factors. Three mechanisms, embedded in these interactions, have been proposed: (1) reserve, (2) maintenance, and (3) compensation (Figure 11.13). Roberto Cabeza and colleagues (2018) provide tentative formal definitions of these mechanisms:

1. *Reserve*: cumulative improvement of neural resources for reducing the effects of neural decline related with aging. This accumulation occurs before the brain is affected by age-related processes. Education is a good example. Intellectual engagement across the life span is also important. Individual differences in reserve would manifest as trait-like effects, meaning that they must be general.
2. *Maintenance*: preservation of neural resources. Its efficacy depends on the magnitude of decline and the efficacy of repair. Whereas *reserve* refers to augmenting resources beyond their current level, *maintenance* involves recovering previous higher levels.
3. *Compensation*: cognition-enhancing recruitment of neural resources in response to cognitive requirements. Compensation may involve (1) upregulation (enhancement of cognitive performance by boosting a neural process in response to task demands), (2) selection (qualitative differences in the cognitive processes – and related brain networks – engaged by older and younger individuals), and (3) recruitment of additional processes (older adults show more bilateral brain activity than younger adults).

According to this model, reserve involves the accumulation of brain resources across the life span (how much brain power you have), maintenance relates to the preservation of these resources by recovery and repair (how well you keep your brain power), and compensation implicates the deployment of these resources for coping with task requirements (when and how you use the brain power you have and how much you keep). These fine-grained distinctions might be important for prevention and intervention programs, although the coupling of the components of cognitive decline increases with age (Tucker-Drob et al., 2019).

To which extent do genetic factors influence the healthy aging process? Matt McGue and colleagues (2014) used data from a sample of 12,714 twins from the Danish Twin Registry (mean age fifty-nine, range forty to eighty). Each completed a self-report assessing six lifestyle variables usually related to aging: smoking (packs/year), drinking (drinks/week), diet, physical activities, social activities, and intellectual activities. The results revealed that these lifestyle variables are substantially heritable. The values of the estimated genetic contributions were 0.69 (smoking), 0.45 (drinking), 0.32 (diet), 0.45 (physical activities), 0.47 (intellectual activities), and 0.38 (social activities). The contribution of shared environmental factors ranged from 0 (smoking) to 0.06 (diet and intellectual activities). They concluded, "Lifestyle, rather than being extrinsic and distinct from genetic factors, is at least in part a manifestation of qualities that are intrinsic to the individual. Lifestyles are shaped by individual decisions that are likely guided by heritable personality and ability traits" (p. 780).

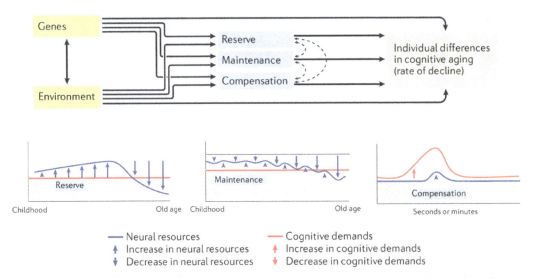

Figure 11.13 (top) Illustration of how reserve, maintenance, and compensation might mediate the effects of interacting genetic and nongenetic factors. (bottom) Reserve and maintenance involve an increase in neural resources. The recruitment of neural resources for compensation reduces the gap between task demands and available resources (Cabeza et al., 2018).

enhancement programs to evoke and preserve actual changes in the individual's brain: "such changes may be achieved only when the intellectual enhancement is fast enough (the amount of improvement during a given period of time is sufficiently large) to trigger reorganization in the cortex" (Estrada et al., 2019, p. 1349).

Second, the relationship between intellectual and brain development is a dynamic problem, more like taking a video instead of a picture (Viviano et al., 2017). This approach is what consortia like the Brain Initiative (Alivisatos et al., 2012) and ENIGMA (Thompson et al., 2014) are trying to achieve (for a description of several consortia, see Khundrakpam et al., 2021). As an example, Brouwer and colleagues (2020) analyzed longitudinal MRI data of 10,163 individuals (age range four to ninety-nine years). Their main goal was to identify common genetic variants (Chapter 6) related to rates of brain growth and atrophy across the life span. In their preliminary report, they identified five genome-wide significant loci and fifteen genes associated with brain structural changes. These genes have functional expressions in brain tissue in utero and continue to influence the brain after birth. This

report was the first GWAS research exploring the probable influence of common genetic variants on brain structural changes across the life span. According to the authors, the genetic effects influence individual differences in development and aging of brain structures; the same genetic variants identified in early brain development were crucial for brain structural changes later in life; our genetic architecture is associated with the dynamics of human brain structure throughout life.

11.6 Heathy Aging

Research findings discussed in this chapter refer to healthy aging. While this term has no precise definition (Fernández-Ballesteros and Sánchez-Izquierdo, 2019), it typically refers to two things:

1. The individual must not have experienced serious health problems that impact cognition. These include, but are not limited to, cardiovascular problems and dementia-producing diseases, such as Alzheimer's disease and the later stages of Parkinson's disease.

Dementias may unfold gradually, so any population of apparently healthy elderly people will include some predementia cases. This should be considered in aging research studies.

2. To stay healthy, an individual should remain engaged with society. Intelligence, like physical fitness, does not exist in a vacuum; it requires maintenance (Flynn, 2012). Studies of healthy young adults have shown that people vary in intellectual engagement and that those who engage with the world are, in general, more intelligent than those who do not, where intelligence is measured by psychometric and laboratory studies (Ackerman, 1996; Ackerman and Beier, 2001). Whether engagement causes intelligence or intelligence causes engagement is hard to say. There are probably reciprocal influences.

Longitudinal studies of the elderly, such as the SLS and the Swedish study of aging twins, have shown that the downward trend is more pronounced as we grow older (Schaie and Willis, 2010; Wetherell et al., 2002; Schaie, 2013; Finkel and Pedersen, 2004). Withdrawal and heightened anxiety are statistically associated with lower intelligence. This becomes a serious problem as people age, because the frequency of threatening life events increases with advancing age. Loss of a job is far more threatening to a person of fifty or sixty than to one of twenty or thirty. Loss of a spouse becomes more likely with advancing age. How we handle such events may both tell a good deal about someone's present intelligence – in the broader concept rather than in the narrower psychometric sense (Chapter 2) – and have implications for maintaining intelligence following trauma.

Finally, we underscore that chronological age may or may not be one reasonable proxy of intellectual age. Research findings from the Dunedin Longitudinal Study indicate that individuals of the same chronological age vary in their biological age, as defined by declining integrity of multiple organ systems (pulmonary, periodontal, cardiovascular, renal,

hepatic, and immune function): "before midlife, individuals who were aging more rapidly were less physically able, showed cognitive decline and brain aging, self-reported worse health, and looked older" (Belsky et al., 2015, p. 4104). This finding was extended in subsequent research from the same group. In one study, an index of brain age was estimated from brain scans for 869 individuals with a mean chronological age of forty-nine years (Elliott et al., 2019). However, individual differences in the brain age index ranged from twenty-four to seventy-two years (Figure 11.14). Individuals with greater values were characterized by lower intellectual functioning in early childhood and later in adulthood.

11.7 Summary

The study of age differences and age changes in intelligence has produced four well-established findings:

1. Fluid reasoning decreases over the adult life span, while crystallized abilities remain stable or even increase slightly until people are into the retirement years.
2. There is a generalized slowing of cognition with age. It is measurable as early as the forties and can influence mental abilities outside of the laboratory as we approach the seventies and beyond.
3. There is a similar decrease in the functioning of the working memory system over the life span. As is the case with general slowing of reaction time, the degree of decrease accelerates after the mid-sixties.
4. Laboratory studies may underestimate the effectiveness of older individuals when they are faced with familiar problems, especially in situations in which they have developed expertise and are not under time constraints. In such situations, people in their forties and older may outperform younger individuals.

Understanding the interactions between intelligence, development, and the aging brain is important and intricate. The

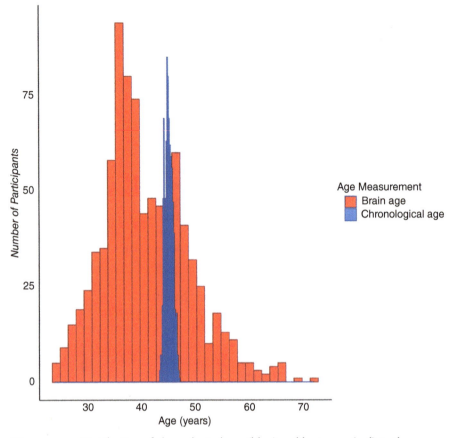

Figure 11.14 Distribution of chronological age (blue) and brain age (red) in the Dunedin Study sample. The mean chronological age was forty-five years, with a very small range because members were born the same year. By contrast, brain age variations for the same people were considerable (twenty-four to seventy-two years) (Elliott et al., 2019).

relationships are dynamic and multivariate and differ among individuals as they age. Research advances require coordinated efforts from the psychometric, information-processing, and biological approaches, especially in multicenter consortia. Unraveling the complexities is challenging, but not impossible, especially as new methods and technologies emerge.

11.8 Questions for Discussion

11.1   How do different aspects of intelligence change in older adults?

11.2   What is the relationship between cognitive capacity and psychosocial maturity?

11.3   What is an example of the cohort effect?

11.4   What is the general factor of age, and what does it influence?

11.5   What is the relationship between intelligence and wisdom?

References

Ackerman, P. L. 1996. A theory of adult intellectual development: Process, personality, interests, and knowledge. *Intelligence*, 22, 227–257.

Ackerman, P. L., & Beier, M. E. 2001. Trait complexes, cognitive investment, and domain knowledge. In Sternberg, R. J., & Grigorenko, E. L. (eds.), *The psychology of abilities, competences, and expertise*. Cambridge: Cambridge University Press.

Alivisatos, A. P., Chun, M., Church, G. M., et al. 2012. The brain activity map project and the challenge of functional connectomics. *Neuron*, 74, 970–974.

Barbey, A. K., Colom, R., Solomon, J., et al. 2012. An integrative architecture for general intelligence and executive function revealed by lesion mapping. *Brain*, 135, 1154–1164.

Belsky, D. W., Caspi, A., Houts, R., et al. 2015. Quantification of biological aging in young adults. *Proceedings of the National Academy of Sciences of the United States of America*, 112, E4104–E4110.

Borghesani, P. R., Madhyastha, T. M., Aylward, E. H., et al. 2013. The association between higher order abilities, processing speed, and age are variably mediated by white matter integrity during typical aging. *Neuropsychologia*, 51, 1435–1444.

Brouwer, R. M., Klein, M., Grasby, K. L., et al. 2020. Dynamics of brain structure and its genetic architecture over the lifespan. *bioRxiv*, 2020.04.24.031138.

Brugman, G. M. 2006. Wisdom and aging. In Schaire, K. W., & Willis, S. (eds.), *Handbook of the psychology of aging*. New York: Elsevier.

Cabeza, R., Albert, M., Belleville, S., et al. 2018. Maintenance, reserve and compensation: The cognitive neuroscience of healthy ageing. *Nature Reviews Neuroscience*, 19, 701–710.

Carroll, J. B. 1993. *Human cognitive abilities: A survey of factor-analytic studies*. New York: Cambridge University Press.

Cattell, R. B. 1987. *Intelligence: Its structure, growth, and action*. Amsterdam: North-Holland.

Colom, R. 2016. Advances in intelligence research: What should be expected in the XXI century (questions and answers). *Spanish Journal of Psychology*, 19, 1–8.

Colom, R., Chuderski, A., & Santarnecchi, E. 2016. Bridge over troubled water: Commenting on Kovacs and Conway's process overlap theory. *Psychological Inquiry*, 27, 181–189.

Cowan, N. 2017. The many faces of working memory and short-term storage. *Psychonomic Bulletin and Review*, 24, 1158–1170.

Deary, I. J. 2014. The stability of intelligence from childhood to old age. *Current Directions in Psychological Science*, 23, 239–245.

Deary, I. J. 2020. *Intelligence: A very short introduction*. Oxford: Oxford University Press.

de Frias, C. M., Lovden, M., Lindenberger, U., & Nilsson, L. G. 2007. Revisiting the dedifferentiation hypothesis with longitudinal multi-cohort data. *Intelligence*, 35, 381–392.

Der, G., & Deary, I. J. 2003. IQ, reaction time and the differentiation hypothesis. *Intelligence*, 31, 491–503.

Der, G., & Deary, I. J. 2006. Age and sex differences in reaction time in adulthood: Results from the United Kingdom Health and Lifestyle Survey. *Psychology and Aging*, 21, 62–73.

Elliott, M. L., Belsky, D. W., Knodt, A. R., et al. 2019. Brain-age in midlife is associated with accelerated biological aging and cognitive decline in a longitudinal birth cohort. *Molecular Psychiatry*, 26, 3829–3838.

Estrada, E., Ferrer, E., Román, F. J., Karama, S., & Colom, R. 2019. Time-lagged associations between cognitive and cortical development from childhood to early adulthood. *Developmental Psychology*, 55, 1338–1352.

Fernández-Ballesteros, R., Benetos, A., & Robine, J.-M. 2019. *The Cambridge handbook of successful aging*. Cambridge: Cambridge University Press.

Fernández-Ballesteros, R., & Sánchez-Izquierdo, M. 2019. Are psycho-behavioral factors accounting for longevity? *Frontiers in Psychology*, 10, 2516.

Finkel, D., & Pedersen, N. L. 2004. Processing speed and longitudinal trajectories of change for cognitive abilities: The Swedish Adoption/Twin Study of Aging. *Aging Neuropsychology and Cognition*, 11, 325–345.

Flynn, J. 2007a. The latest thinking on intelligence. *Psychologist*, 20, 356–357.

Flynn, J. R. 2007b. *What is intelligence? Beyond the Flynn effect*. New York: Cambridge University Press.

Flynn, J. R. 2012. *Are we getting smarter? Rising IQ in the twenty-first century*. Cambridge: Cambridge University Press.

Flynn, J. R. 2016. *Does your family make you smarter? Nature, nurture, and human autonomy*. Cambridge: Cambridge University Press.

Flynn, J. R. 2018. Reflections about intelligence over 40 years. *Intelligence*, 70, 73–83.

Flynn, J. R., & Shayer, M. 2018. IQ decline and Piaget: Does the rot start at the top? *Intelligence*, 66, 112–121.

Gigerenzer, G. 2000. *Adaptive thinking: Rationality for the real world*. Oxford: Oxford University Press.

Hartley, A. 2006. Changing role of the speed of processing construct in the psychology of aging. In Birren, J. E., & Schaie, K. W. (eds.), *Handbook of the psychology of aging*. 6th ed. Burlington, MA: Elsevier.

Horn, J. L. 1985. Remodeling old models of intelligence. In Wolman, B. B. (ed.), *Handbook of intelligence: Theories, measurements, and applications*. New York: John Wiley.

Horn, J. L., & Noll, J. 1997. Human cognitive capabilities: Gf–Gc theory. In Flanagan, D. P., Genshaft, J. L., & Harrison, P. L. (eds.), *Contemporary intellectual assessment: Theories, tests, and issues*. New York: Guilford Press.

Hoyer, W. J., & Verhaeghen, P. 2006. Memory aging. In Birren, J. E., & Schaie, K. W. (eds.), *Handbook of the psychology of aging*. 6th ed. Burlington, MA: Elsevier.

Hunt, E. B. 2007. *The mathematics of behavior.* Cambridge: Cambridge University Press.

Jokela, M., Pekkarinen, T., Sarvimäki, M., Terviö, M., & Uusitalo, R. 2017. Secular rise in economically valuable personality traits. *Proceedings of the National Academy of Sciences of the United States of America,* 114, 6527–6532.

Juan-Espinosa, M., Garcia, L. F., Escorial, S., et al. 2002. Age dedifferentiation hypothesis: Evidence from the WAIS III. *Intelligence,* 30, 395–408.

Judge, T. A., Erez, A., Bono, J. E., & Thoresen, C. J. 2002. Are measures of self-esteem, neuroticism, locus of control, and generalized self-efficacy indicators of a common core construct? *Journal of Personality and Social Psychology,* 83, 693–710.

Karama, S., Bastin, M. E., Murray, C., et al. 2014. Childhood cognitive ability accounts for associations between cognitive ability and brain cortical thickness in old age. *Molecular Psychiatry,* 19, 555–559.

Khundrakpam, B. S., Poline, J., & Evans, A. 2021. Research consortia and large-scale data repositories for studying intelligence. In Barbey, A., Karama, S., & Haier, R. J. (eds.), *The Cambridge handbook of intelligence and cognitive neuroscience.* New York: Cambridge University Press.

Klein, G. 1998. *Sources of power: How people make decisions.* Cambridge, MA: MIT Press.

Klein, G. 2009. *Streetlights and shadows.* Cambridge, MA: MIT Press.

Kovacs, K., & Conway, A. R. A. 2019a. A unified cognitive/differential approach to human intelligence: Implications for IQ testing. *Journal of Applied Research in Memory and Cognition,* 8, 255–272.

Kovacs, K., & Conway, A. R. A. 2019b. What is IQ? Life beyond "general intelligence." *Current Directions in Psychological Science,* 28, 189–194.

Kramer, A. F., Fabiani, M., & Colcombe, S. J. 2006. Contributions of cognitive neuro-science to the understanding of behavior and aging. In Birren, J. E., & Schaie, K. W. (eds.), *Handbook of the psychology of aging.* 6th ed. Burlington, MA: Elsevier.

Lubinski, D. 2016. From Terman to today: A century of findings on intellectual precocity. *Review of Educational Research,* 86, 900–944.

Madhyastha, T. M., Hunt, E., Deary, I. J., Gale, C. R., & Dykiert, D. 2009. Recruitment modeling applied to longitudinal studies of group differences in intelligence. *Intelligence,* 37, 422–427.

Marsiske, M., & Margrett, J. A. 2006. Everyday problem solving and decision making. In Birren, J. E., & Schaie, K. W. (eds.), *Handbook of the psychology of aging.* 6th ed. Burlington, MA: Elsevier.

Martínez, K., Burgaleta, M., Román, F. J., et al. 2011. Can fluid intelligence be reduced to "simple" short-term storage? *Intelligence,* 39, 473–480.

McArdle, J. J., Ferrer-Caja, E., Hamagami, F., & Woodcock, R. W. 2002. Comparative longitudinal structural analyses of the growth and decline of multiple intellectual abilities over the life span. *Developmental Psychology,* 38, 115–142.

McGrew, K. S., Werder, J. K., & Woodcock, R. W. 1991. *Woodcock–Johnson technical manual.* Allen, TX: DLM Teaching Resources.

McGue, M., Skytthe, A., & Christensen, K. 2014. The nature of behavioural correlates of healthy ageing: A twin study of lifestyle in mid to late life. *International Journal of Epidemiology,* 43, 775–782.

Protzko, J. 2015. The environment in raising early intelligence: A meta-analysis of the fadeout effect. *Intelligence,* 53, 202–210.

Protzko, J. 2017. Effects of cognitive training on the structure of intelligence. *Psychonomic Bulletin and Review,* 24, 1022–1031.

Román, F. J., Morillo, D., Estrada, E., et al. 2018. Brain–intelligence relationships across childhood and adolescence: A latent-variable approach. *Intelligence,* 68, 21–29.

Rönnlund, M., Sundström, A., & Nilsson, L.-G. 2015. Interindividual differences in general cognitive ability from age 18 to age 65 years are extremely stable and strongly associated with working memory capacity. *Intelligence,* 53, 59–64.

Ruiz Sánchez de León, J., Colom, R., & Quiroga, M. Á. 2019. Intelligence and executive function: Can we reunite these disparate worlds? In McFarland, D. J. (ed.), *General and specific mental abilities.* Cambridge: Cambridge Scholars.

Salthouse, T. A. 1996. The processing-speed theory of adult age differences in cognition. *Psychological Review,* 103, 403–428.

Salthouse, T. A. 2005. Relations between cognitive abilities and measures of executive functioning. *Neuropsychology,* 19, 532–545.

Salthouse, T. 2012. Consequences of age-related cognitive declines. *Annual Review of Psychology,* 63, 201–226.

Salthouse, T. A., Atkinson, T. M., & Berish, D. E. 2003. Executive functioning as a potential mediator of age-related cognitive decline in normal adults. *Journal of Experimental Psychology: General,* 132, 566–594.

Santarnecchi, E., Rossi, S., & Rossi, A. 2015. The smarter, the stronger: Intelligence level correlates with brain resilience to systematic insults. *Cortex,* 64, 293–309.

Schaie, K. W. 1996. *Intellectual development in adulthood: The Seattle longitudinal study.* Cambridge: Cambridge University Press.

Schaie, K. W. 2005. *Developmental influences on adult intelligence: The Seattle longitudinal study.* 1st ed. New York: Oxford University Press.

Schaie, K. W. 2013. *Developmental influences on adult intelligence: The Seattle longitudinal study*. 2nd ed. New York: Oxford University Press.

Schaie, K. W., & Willis, S. L. 2010. The Seattle Longitudinal Study of Adult Cognitive Development. *ISSBD Bulletin*, 57, 24–29.

Schmidt, F. L. 2014. A general theoretical integrative model of individual differences in interests, abilities, personality traits, and academic and occupational achievement: A commentary on four recent articles. *Perspectives on Psychological Science*, 9, 211–218.

Stanley, J., Keating, D. P., & Fox, L. H. 1974. *Mathematical talent: Discovery, description, and development*. Baltimore: The Johns Hopkins University Press.

Steinberg, L., Cauffman, E., Woolard, J., Graham, S., & Banich, M. 2009. Are adolescents less mature than adults? Minors' access to abortion, the juvenile death penalty, and the alleged APA "flip-flop." *American Psychologist*, 64, 583–594.

Sternberg, R. J. 2003. *Wisdom, intelligence, and creativity synthesized*. Cambridge: Cambridge University Press.

Teasdale, T. W., & Owen, D. R. 2000. Forty-year secular trends in cognitive abilities. *Intelligence*, 28, 115–120.

Terman, L. M. 1925. *Genetic studies of genius*. Stanford, CA: Stanford University Press.

Thompson, P. M., Stein, J. L., Medland, S. E., et al. 2014. The ENIGMA Consortium: Large-scale collaborative analyses of neuroimaging and genetic data. *Brain Imaging and Behavior*, 8, 153–182.

Trahan, L. H., Stuebing, K. K., Fletcher, J. M., & Hiscock, M. 2014. The Flynn effect: A meta-analysis. *Psychological Bulletin*, 140, 1332–1360.

Tucker-Drob, E. M. 2019. Cognitive aging and dementia: A life-span perspective. *Annual Review of Developmental Psychology*, 1, 177–196.

Tucker-Drob, E. M., Brandmaier, A. M., & Lindenberger, U. 2019. Coupled cognitive changes in adulthood: A meta-analysis. *Psychological Bulletin*, 145, 273–301.

Tucker-Drob, E. M., & Briley, D. A. 2014. Continuity of genetic and environmental influences on cognition across the life span: A meta-analysis of longitudinal twin and adoption studies. *Psychological Bulletin*, 140, 949–979.

Tucker-Drob, E. M., & Salthouse, T. A. 2008. Adult age trends in the relations among cognitive abilities. *Psychology and Aging*, 23, 453–460.

Tuddenham, R. D. 1948. Soldier intelligence in World Wars I and II. *American Psychologist*, 3, 54–56.

Underwood, E. 2014. Starting young. *Science*, 346, 568–571.

Viviano, R. P., Raz, N., Yuan, P., & Damoiseaux, J. S. 2017. Associations between dynamic functional connectivity and age, metabolic risk, and cognitive performance. *Neurobiology of Aging*, 59, 135–143.

Warne, R. T., Burton, J. Z., Gibbons, A., & Melendez, D. A. 2019. Stephen Jay Gould's analysis of the Army Beta Test in *The Mismeasure of Man*: Distortions and misconceptions regarding a pioneering mental test. *Journal of Intelligence*, 7.

Wendelken, C., Ferrer, E., Ghetti, S., et al. 2017. Frontoparietal structural connectivity in childhood predicts development of functional connectivity and reasoning ability: A large-scale longitudinal investigation. *Journal of Neuroscience*, 37, 8549–8558.

Wetherell, J. L., Reynolds, C. A., Gatz, M., & Pedersen, N. L. 2002. Anxiety, cognitive performance, and cognitive decline in normal aging. *Journals of Gerontology, Series B*, 57, P246–P255.

CHAPTER 12

Intelligence in the World

12.1 Introduction

Ancestry and country differences in intelligence test scores are a matter of heated discussion. This chapter addresses delicate issues. Even when the data are relatively clear, discussion about their interpretation and meaning easily slips into dispute. We wrote this chapter to shed light and insight instead of lightning and thunder. Our focus is the world because the issues are significant across the globe (Hunt, 2012; Jones, 2016; Rindermann, 2018). To maintain the global focus, we are not detailing data within the United States beyond what is summarized in Box 12.1. As we do throughout this book, we emphasize key research because we agree with James Flynn's (2018, p. 128) view on this subject: "There will be bad science on both sides of the debate. The only

antidote I know for that is to use the scientific method as scrupulously as possible."

We begin with four basic points that frame the contents of this chapter:

1. Although the meaning of intelligence may vary across societies, there is a common core defined by abilities such as comprehension of verbal arguments, manipulating numbers, understanding basic logical principles, or orientation in space. This likely is why the general ability factor (g) is ubiquitous among cultures (Warne and Burningham, 2019).

2. Intelligence tests are assessments of the cognitive abilities and skills required for success mainly in industrial and postindustrial societies. Information provided by test scores is relevant in these

Box 12.1 Intelligence Differences among Ancestry Groups in the United States

This is a controversial topic with competing interpretations of research findings that often depend on arcane technical statistical issues and cherry-picked studies to support one view or another. To avoid this morass, we summarize the evidence and the conclusions of the American Psychological Association (APA) task force report on intelligence published in 1996 (Neisser et al., 1996) and on a related review paper authored by several intelligence researchers published in 2012 (Nisbett et al., 2012).

The APA report on intelligence was published in response to the debate evoked by the publication of *The Bell Curve* in 1994: "The debate was characterized by strong assertions as well as by strong feelings. Unfortunately, those assertions often revealed serious misunderstandings of what has (and has not) been demonstrated by scientific research in this field.... The report presented here has the unanimous support of the entire task force (Ulric Neisser, Gwyneth Boodoo, Thomas Bouchard, A. Wade Boykin, Nathan Brody, Stephen Ceci, Diane Halpern, John Loehlin, Robert Perloff, and Susana Urbina)" (Neisser et al., 1996, p. 77).

Regarding group differences characterized by ancestry, this report considered Americans of European ancestry (AEA), Americans of African ancestry (AAA), Americans of Hispanic ancestry (AHA), Americans of Asian ancestry (AASA), and Native Americans (NA). After reviewing the available empirical evidence, the task force concluded that AEA and AASA show similar average intelligence levels (IQs of 100 and 98, respectively). AHA and NA averages were lower (no values were provided), and the AAA group had the lowest average IQ score (85): "The cause of that

differential is not known; it is apparently not due to any simple form of bias in the content or administration of the tests themselves. ... At present, no one knows what causes this differential" (Neisser et al., 1996, p. 97).

To illustrate what the APA report summarized, Figure 12.1 shows average scores of the AEA, AAA, and AHA groups obtained in the standardization of the WAIS-III (Lange et al., 2006) and in the NLSY79 (Herrnstein and Murray, 1994). Importantly, the information shown in Figure 12.1 is only *average* values, and other potentially relevant information is not shown, including time periods and socioeconomic backgrounds

The conclusion of the APA report regarding the average intelligence test score difference among US ancestry groups generally is consistent with educational assessments and occupational performance differences within the country. IQ scores and educational data reveal a consistent ordering of the major ancestry groups in the United States: the AASA group is slightly higher than the AEA group, the AHA group is about 0.5 to 0.7 deviation units (d) lower than the AEA group, and the AAA group is 0.8 to 1.1 d units lower than the AEA group.

The next review paper was published sixteen years after the 1996 APA report (Nisbett et al., 2012). It was authored by a self-selected independent group of researchers (Richard Nisbett, Joshua Aronson, Clancy Blair, William Dickens, James Flynn, Diane Halpern, and Eric Turkheimer). In contrast to the 1996 report, they cautioned, "We do not claim to represent the full range of views about intelligence" (p. 2). Regarding the average intelligence difference between AEA and AAA, this group of scientists highlighted the gain of approximately 5 IQ points in the AAA group between 1972 and 2002 (Dickens and Flynn, 2006). A similar

Box 12.1 *(continued)*

finding was reported by Murray (2007), who concluded that the average difference between AEA and AAA may be shrinking. The estimated average intelligence of AASA was equivalent to the value for AEA. No evidence was provided for AHA and NA. They also noted that group average differences may be greater in lower-SES than in higher-SES bands (Nisbett et al., 2012).

Here are four points to consider based on the 1996 APA report and the 2012 review paper:

1. The claim that intelligence test scores are irrelevant or inaccurate in predicting minority group performance is contradicted by the available evidence.
2. SES is a statistical abstraction (O'Connell, 2019) that depends on physical and social variables within the environment. It might not be the sole explanation for ancestry differences in intelligence, because SES is confounded with intelligence. Moreover, since intelligence is heritable to a substantial degree (Chapter 6), it is unclear if an effect of SES on children's intelligence is due to social or biological inheritance or, most probably, some combination of the two.
3. The idea that environment and culture explain 100 percent of group differences in average intelligence is a reasonable hypothesis. Whether genes might also contribute to group differences is one of the most controversial questions in the social sciences. At this point, we just do not know how the two views may interact. According to some behavioral geneticists, the question may be unanswerable because of the complexity of disentangling genetic and nongenetic variance for any group difference, even with current DNA methods (Turkheimer, 2019). Other behavioral geneticists are more optimistic (Plomin, 2018). If genes are shown to be involved, the implications could be positive if identifying genes leads to understanding neurobiological mechanisms of intelligence that can be modified to minimize average group differences and maximize individual potential – a prospect with ethical considerations of its own (Haier, 2017). There also likely could be extreme misuses of such information, so public discussion and education by scientists are critical for clarifying what the data mean and do not mean.
4. Bold statements about average group differences and what causes them make for dramatic debates (Murray, 2020; Ceci and Williams, 2018). The issues are complex, and oversimplification does not help. Answers can come only from more high-quality research and a willingness to follow the data and engage opposing views with respectful discussion, rather than rancor and accusations of bad faith.

---

societies, as underscored by James Flynn (2018, p. 129): "IQ tests provide priceless data about the cognitive development of parents and their children." Intelligence test scores predict success in other societies to the extent that the tests tap the cognitive skills relevant for success in them. And this relevance is best decided on a case-by-case basis.

3. Evaluating the abilities and skills required by postindustrial societies is important in itself. Learning about intelligence as it is expressed in Los Angeles, Berlin, Madrid, or Beijing will

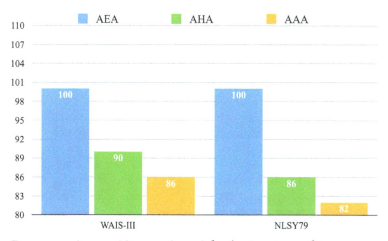

Figure 12.1 Average IQ scores (*y*-axis) for the Americans of European ancestry (AEA), Americans of Hispanic ancestry (AHA), and Americans of African ancestry (AAA) groups (*x*-axis). Scores are from the standardization of the WAIS-III and the National Longitudinal Study on Youth 1979 (NLSY79).
Adapted from Lange et al. (2006) and Herrnstein and Murray (1994).

have more impact on the human condition than learning about the expression of intelligence in small groups living in remote locations.

4. All cognitive abilities and skills ultimately are linked by neurobiology (Haier, 2017). However, it is important to distinguish between genetic influences, which establish probabilities for potential, and life experiences (including social and cultural ones), which impact the extent to which potential is realized. There is nothing easy about this. Estimating the size of the genetic contribution to variance in a trait within a particular society may not help in understanding the trait (Hur and Bates, 2019).

With these points in mind, here is a fundamental question: do ancestry groups have real meaning? On one hand, it has been fashionable for some time to claim that such groups are solely the result of arbitrary or subjective social constructions. It is argued that because the definition of a group varies from time to time and place to place, the group cannot have a distinct biological basis. Therefore, ancestry differences are not amenable to scientific study because the basic concept is ill defined (Fish, 2002; Smedley and Smedley, 2005).

On the other hand, people identify with ancestry groups, and such groups are correlated with a clustering of social and biological variables (Murray, 2020). Socially, for instance, seven out of ten residents of the United States who identify themselves as Hispanic are Catholic. Only two out of ten non-Hispanic residents are Catholic. Therefore, Catholicism is statistically associated with identity, even though there are Hispanics who are not Catholic and Catholics who are not Hispanic.

The same reasoning applies to genetic variation. While it is true that within-group genetic variation is greater than between-group variation, the amount of genetic variation between groups is statistically sufficient to make accurate probability assignments for some broad ancestry groups. For example, in the United States, this is true for the three largest groups, East Asian, African, and European (Bamshad et al., 2004). There is also a high level of agreement between self-identification and assignment of a person to clusters based on similarities of their genomes (Tang et al., 2005).

Although ancestry is a fuzzy concept, research advances indicate that there are identifiable genetic differences among populations across the globe (Green et al., 2015) (updates about this can be found at www.internationalgenome.org/). James Flynn (2018, p. 128) offers an example that invites all us to keep open minds: "We no longer hear much from those who once proposed a fourth argument: that all races share so many genes in common that it would be absurd to look for genetic differences (note: even this argument assumes the question is subject to investigation; they just think the answer is as obvious as height differences between Watusi and pigmies). We share 99% of our genes with Bonobo Chimpanzees. That 1% makes a huge difference in cognitive capacity: one hundredth of 1% might make a huge difference between socially identified groups." This is a reasonable basis for research.

Knowledge about covariation between ancestry identification and important social outcomes like health can help identify variables related to causal relationships. For instance, it is empirically the case that high blood pressure is more common in African Americans than in non-African US communities. What is it in the African American genetic constitution, social condition/practice, or interaction that leads to high blood pressure? Such questions are reasonable for investigation by the biomedical sciences (Bamshad et al., 2004). Similarly, if there is a discrepancy in intelligence test scores between the African American and other US communities (Box 12.1), seeking to understand the reasons for and implications of the discrepancy is legitimate for the psychological sciences generally motivated to remedy or alleviate problems. Accurate understanding is the best basis for efficient problem solving.

Unfortunately, information about any average group differences can be used to justify prejudice and discriminatory practices. For this reason, there is a point of view that researchers should avoid investigating group differences because nothing good could come of it. Even with the potential

for some undesired consequences, however, we think it is better to have accurate information than not (Cofnas, 2020; Ceci and Williams, 2018). People who use scientific information for hateful purposes should not be rewarded a veto power for deciding what kind of research is acceptable (Haier, 2020). Open discussions, skepticism, and constructive confrontation of competing ideas and interpretations of data are key aspects of science in democratic societies: "Nothing will be gained by systematic sanctions that protect ignorance. … Suppressing free inquiry is by its nature an expressive of contempt for truth by power" (Flynn, 2018, p. 5). In short, more research is better than less.

Nonetheless, the concept of ancestry, ethnicity, or race remains at the core of much of the controversy about average differences in intelligence test scores. Charles Murray (2020) discusses whether current scientific data support the perspective that these are social constructs with no biological meaning. After reviewing evidence for and against this view, he argued that they have some biological meaning and concluded the following:

- Human populations are genetically distinctive in ways that correspond to self-identified ethnicity.
- Evolutionary selection pressure since humans left Africa has been extensive and mostly local.
- Continental population differences in (genetic) variants associated with personality, abilities, and social behavior are common.

These conclusions are still a matter of debate, but the sentiment expressed in his view deserves consideration: "The mix of nature and nurture? That is not the issue. The differences themselves are facts. People around the world are similar in the basics and different in the details. We connect through the basics. We live with and often enjoy the differences. … No population is free of defects nor possessed of all the virtues. … There will be no moral or legal justification for treating individuals differently because of the population to which they belong" (pp. 201–202).

This chapter focuses on average population differences across the globe and how they are related to intelligence as assessed by standardized tests. However, it is crucial to keep in mind that what we know about individuals may not apply to populations, and evidence valid for populations may not apply to individuals or even subgroups within populations. Countries, for instance, compose a convenient category for analysis, but there are great differences within countries related to regions and states or even socioeconomic levels. Across countries, social groups showing equivalent incomes share a greater number of features than groups of different social levels within countries (Rosling et al., 2018).

12.2 The Intelligence of Nations

Wealth is not distributed uniformly across nations. Generally, European and North American countries are best off. Japan, Korea, and China are rapidly catching up. Sub-Saharan Africa lags behind so far. The Middle East and South Asian nations, along with South and Central America, fall in between. Neither the genetic composition nor the socioeconomic histories of these regions are the same. To what extent is the unequal distribution of wealth across nations related, at least in part, to average differences in cognitive abilities, like it is for individual citizens? As global economies develop, this becomes an important question, but not a new one. Herodotus speculated about different temperaments of peoples in the fifth century BCE. From the seventeenth century until modern times, economists have discussed the topic under the general term *human capital*. This term refers collectively to the capabilities of individuals in the workforce. Rates of literacy and the percentage of people with engineering degrees are examples. It is generally

agreed that human capital is critical in determining the economic potential of a state or nation. Politicians are more likely to justify taxes to support schools by claiming that a more educated population has a greater economic potential than a less educated one than by claiming that the more educated have a moral and philosophical advantage. An important PISA report takes this view: "without the right education, people will languish on the margins of society, countries will not be able to benefit from technological advances, and those advances will not translate into social progress" (Schleicher, 2019, p. 10).

Economists discussing human capital seldom use the word *intelligence*. Generally, they are interested in the distribution of developed skills in a society and the economic implications of the distribution. How those skills came to be generally are of less concern. Economists are also aware that the word *intelligence* can have connotations of fixed, genetically determined constraints on mental capacity. From this perspective, anyone interested in influencing policies concerning human capital, often a euphemism for intelligence, is well advised to avoid using the word. Suggesting that intelligence is genetic and limits the socioeconomic development of nations or ancestry groups is not a welcomed point of view, as illustrated in Box 12.2.

Nevertheless, scientists look at the international distribution of intelligence test scores and ask whether it is related to the distribution of wealth, at least in part. The psychologist Richard Lynn and the political scientist and economist Tatu Vanhanen addressed this question in a series of publications. Their work has resulted in controversial findings, considerable skepticism, and much debate. Before summarizing these studies and the skepticism that surrounds them, let us consider the problems faced by any empirical investigation of intelligence differences at the national level, building on the discussion in Chapter 9.

Box 12.2 The James Watson Affair

James Watson (1928–) is one of the world's most famous living scientists. In 1962, he, along with his colleague Francis Crick, received the Nobel Prize for their discovery of the structure of the DNA molecule. Their work sparked the explosion of discoveries in molecular biology that continue to this day. Subsequently, Watson left his appointment at Harvard to head the Cold Spring Harbor Laboratory, which he built into a major scientific institution. Late in his career, Watson was a leading figure in the successful effort to catalog the human genome.

In 2007, Watson, who is an outspoken person, wrote an autobiography, *Avoid Boring People*. He was scheduled to give several public talks in England. Just prior to the talks, he gave an interview to the *Sunday Times* (October 14, 2007). The following sentences are taken verbatim from the article:

1. He is inherently gloomy about the prospect of Africa.
2. All our social policies are based on the fact that their intelligence is the same as ours, whereas all the testing says not really.
3. His hope is that everyone is equal, but "people who have had to deal with black employees find that this is not true."
4. There is no firm reason to anticipate that the intellectual capacities of peoples geographically separated in their evolution should prove to have evolved identically.
5. Our wanting to reserve equal powers of reason as some universal heritage of humanity will not be enough to make it so.

A huge public furor ensued. The *New York Times* published a lengthy discussion of subsequent debates between Watson's supporters and critics

(December 1, 2007). Many of Watson's speaking engagements were canceled abruptly. He was removed as director of the Cold Spring Harbor Laboratory. He became radioactive in most scientific and social circles. Many honors and honorary titles were revoked.

In 2008, we (RH, RC, and Earl Hunt) were invited to a three-day seminar on "improving the brain" that had been arranged by the Cold Spring Harbor Laboratory in honor of Watson's retirement. The event was unexpected given the press coverage of his forced resignation. Standing in front of twenty-five scientists who had traveled from different countries to attend the event, Watson discussed the incident recounted here, with visible bitterness. He felt that his remarks had been taken out of context, which is often the case when media address population differences in psychological traits, but especially when intelligence is involved.

The incident illustrates the extreme sensitivity of any discussion of ancestry differences in intelligence, especially by a geneticist. Watson's remarks as printed (the *London Times* has stood behind its accuracy) seem to be like the famous Rorschach ink blot test in clinical psychology. Many people, including longtime friends and colleagues, saw nothing but racism in his remarks.

Given these strong sentiments, it is hardly surprising that policy makers, funders of scientific research, and researchers have backed away from studies about ancestry differences in intelligence. This has inhibited understanding and rational discussion, especially by those who hold an agnostic position on heredity–environment issues.

In 2019, Watson reiterated his arguments from a decade before. In a TV documentary (*Decoding Watson*), he declared, "My views have not changed since 2007. I would like for them to have

Box 12.2 *(continued)*

changed, that there be new knowledge that says that your nurture is much more important than nature. But I haven't seen any knowledge. And there's a difference on the average between blacks and whites on IQ tests. I would say the difference is, it's genetic."

Should Watson have been more careful expressing such a controversial opinion? Lee J. Cronbach (1975) published an excellent cautionary article half a century ago titled "Five Decades of Public Controversy over Mental Testing." In it we find an insightful observation still valid: "The ... academic is ill trained to cope with the media and the public. In his normal life he speaks to a captive, note-taking audience. He writes for

archives where ... the reader can be trusted to weigh sentences in context. But the public reads the headlines and the snappy quotes, and only half remembers even them. ... We may rail against the journalist for relaying what we said instead of what we meant to say, but mind-reading is not his job. We may rail against the public for not studying the text we put before them. But the writer in public is the servant, not the master, of the reader" (p. 12). Hopefully, scholars will heed this message before engaging the public, which is now much easier because of the worldwide and instant availability of social networks like Twitter, Instagram, and Facebook; blogs; and podcasts.

12.3 Methodological Issues

Estimating the distribution of intelligence in nations requires (1) a representative sample of the residents in each nation and (2) the use of tests that are valid for the nations compared. These are two stringent requirements to discuss further. First, representative national samples are hard to obtain in the most developed countries. The problem is virtually unsolvable in low-income countries, especially if these countries are subject to political unrest and unstable governments. It is often much easier to approximate a probability sample in a selected segment of the population, such as schoolchildren (often the most easily accessible population) or, in some countries, registrants for military service. However, there are three problems: (1) the accessible populations may differ across countries, (2) the estimates will apply only to the segment of the population involved, and (3) the recruitment procedures that construct the accessible samples from the general population may differ across

nations. The problem is particularly acute in the case of schoolchildren.

For example, consider the following contrast. In the United Kingdom, essentially all children go to school, and from time to time, the government has given well-standardized cognitive tests to all students. In Congo, large parts of the country are cut off from government control, and schooling outside the major cities is haphazard. National intelligence estimates based on studies of schoolchildren in the United Kingdom and Congo might not be comparable. This example is extreme, but differential recruitment also can occur in middle- and high-income nations. For example, South Korea reported that 42 percent of PISA participants were girls, whereas France reported 53 percent (Rindermann, 2007). Any comparison between the PISA results from Korea and France would be confounded by sex differences in sample recruitment.

The second stringent requirement, obtaining comparable and appropriate intelligence tests, is equally challenging. The

Box 12.3 General Intelligence (*g*) Is a Universal Phenomenon, but It Might Have Different Interpretations

Russell Warne and Cassidy Burningham (2019) tested whether general intelligence (*g*) is a cross-cultural phenomenon beyond Western industrialized societies. They analyzed ninety-seven samples from thirty-one non-Western countries (total number of individuals = 52,340). This was the key finding: the *g*-factor obtained from the samples explained 46 percent of the variance, essentially matching the value found in Western countries (Canivez and Watkins, 2010). The authors concluded, "Because these datasets originated in cultures and countries where *g* would be least likely to appear if it were a cultural artifact, we conclude that general cognitive ability is likely a universal human trait" (p. 263). The apparent universality may be related to the principle of the indifference of the indicator discussed in Chapter 3; that is, a *g*-factor emerges irrespective of the content of test items or the format of the test (Spearman, 1927).

Even the presence of *g* in non-Western cultures, however, does not mean that the *g*-factor can be used for comparing cognitive ability across countries. As noted in Chapter 9, this would require checking measurement invariance across nations "to ensure that test items and tasks function in a similar way for each group" (Canivez and Watkins, 2010, p. 266). This point suggests that the *g* obtained in different countries might not be interpreted in a straightforward fashion. Remember, counting is not measuring. The authors acknowledge this issue: "there is no guarantee that the *g* found in one sample is the same as the *g* in another sample" (p. 266). The design of the SLATINT project described in the main text remains optimal for country comparisons.

One final point Warne and Burningham (2019) stressed: cultural beliefs about human intelligence are anecdotes, not data. Mountains of anecdotes cannot replace objective empirical assessment of differences on standardized intelligence tests: "the same logic that researchers use to argue that a folk belief regarding intelligence provides evidence of the nature of intelligence could also be used to argue that widespread cultural beliefs in elves, goblins, or angels provide evidence of the existence of supernatural beings" (p. 238).

tests are weighted toward evaluating abilities and skills that are relevant to the developed world, and any test necessarily is partly an evaluation of test-taking skills. One has to be open to the possibility that, outside the middle- and high-income world, people may not have acquired the abilities and skills relevant to industrial and postindustrial societies or even to the test-taking paradigm itself. You might think that a potential strategy to address this problem is to focus on the latent *g*-factor, but this is not straightforward, as discussed in Box 12.3.

Despite these two major potential problems, good cross-national research is possible, as demonstrated by a comprehensive project coordinated by the Brazilian psychologist Carmen Flores-Mendoza (Flores-Mendoza et al., 2018). This research was designed to obtain knowledge about six South American countries. The Study of Latin American Intelligence (SLATINT) project is unique because it goes beyond the usual practice of administering school knowledge tests, such as those included in the PISA surveys. These researchers considered intelligence data because they are the best single predictor of school performance differences. In the preface to their book, they state, "Human capital deserves to be studied and understood. Not only from the point of view of education, but also from a psychological variable, strongly related to school performance, commonly known as intelligence."

In all, there were 4,282 students with a mean age of 14.5 years enrolled in sixty-six schools from Belo Horizonte (Brazil), Bogota (Colombia), Lima (Peru), Mexico City, Rosario (Argentina), and Santiago (Chile). Between 2007 and 2011, they completed the same test battery, including ten cognitive measures (the Standard Progressive Matrices [SPM] among them), a short version of the PISA assessment, and a comprehensive sociodemographic questionnaire.

This was the key research problem they addressed: why do the countries of the region show suboptimal human capital despite their socioeconomic growth? A paper by Flores-Mendoza and colleagues (2015) reported the raw scores for the SPM and for PISA for each country. Combining SPM and PISA scores, the countries were ranked highest to lowest: Peru, Mexico, Chile, Argentina, Brazil, and Colombia; the average was 89. A more recent review found a different ranking from highest to lowest: Chile, Argentina, Mexico, Peru, Colombia, and Brazil; the average was 85 (Lynn and Becker, 2019). Not surprisingly, the different sources of information used in different studies may lead to different rankings, which is another cautionary note.

The SLATINT authors underscored the relevance of intelligence for addressing the changing nature of human capital in the modern world, where some nations are better positioned than others. The mean IQ computed for the six SLATINT countries was 89: "If 90 is the IQ breakpoint for a nation achieving reasonable development (Whetzel and McDaniel, 2006), the performance of our sample indicates that the region is close to achieving relative social wellbeing. ... Policy-makers should know (and recognize) that raising the educational level of the population involves intervention on the intellectual level of the population" (Flores-Mendoza et al., 2018, pp. 107–109).

The SLATINT project took will, persistence, and resilience. After initial analyses, Flores-Mendoza wrote to Earl Hunt "Some of my colleagues fear that our data will be taken by other researchers to reaffirm existing discrimination." She was under pressure and asked for support from colleagues outside the region. This is part of the personal communication he wrote back to her:

I believe that SLATINT (Study of Latin American Intelligence) is an extremely important project for both scientific and social reasons. The project represents careful data collection using identical testing procedures, in several countries. There is also a substantial amount of demographic data, e.g. educational background, on participants. Such information is extremely important because, in addition to its scientific interest, this information can serve to plan educational, health, and economic policies in the countries involved. For instance, high scores, relative to other countries, can be used as arguments to encourage investments that increase national economic development. Low scores can be used to underscore the importance of further educational and health interventions. Improving the cognitive resources of a country's population is (a) possible and (b) highly important for the economic and social wellbeing of the country. However, policy makers will not solve a problem unless they realize that it exists and how important it is. Accordingly, it is important that SLATINT be completed because both high and low scores can be used to guide policies that will benefit the countries involved.

He continued: "There is another reason the reporting of SLATINT is important. Other people, using far less competent procedures, are going to report data for Latin America. Latin American psychologists and educators do not want others to present data their way. Whatever the data show, it is important that facts be established accurately. That way we deal with a problem as it exists, not as someone thinks it might exist. You and the SLATINT team can and should present your data honestly, and in the context of multivariate analyses that establish the relation between your findings and other socioeconomic variables."

Despite pressure from her social environment, Flores-Mendoza and colleagues (2018) published a monograph in which they stated, "Authorities sometimes make questionable decisions and demonstrate inabilities in identifying and acting on the needs of the population they represent. ... This project is based on 120 years of differential psychology research. Differential psychology is the largest division of psychological science, but it is unfortunately ignored by education policies" (preface and "Final Words").

12.4 Opening Pandora's Box

The modern discussion about IQ differences among countries began with the publication of Richard Lynn and Tatu Vanhanen's (2002) *IQ and the Wealth of Nations*. Four years later, they published *IQ and Global Inequality*, which extended the analysis beyond economic development (Lynn and Vanhanen, 2006). Six years later, they published *Intelligence: A Unifying Construct for the Social Sciences* (Lynn and Vanhanen, 2012a). In 2019, Richard Lynn and David Becker updated previous findings in their book *The Intelligence of Nations*.

Lynn and Vanhanen contend that intelligence contributes to national wealth directly and by its interaction with education. Their argument is that people with higher intelligence tend to be more educable and that, in addition to the benefits of education, people with higher levels of intelligence tend to be better thinkers and hence more able to deal with or solve problems posed by complex societies. They obtained estimates of IQs for nations across the globe by searching various publication sources. All the technical caveats about individual IQ scores are amplified for estimating average national IQs. Many of the studies considered by these researchers had small and nonrepresentative samples of their respective populations or used unreliable measures of the construct of interest. For good reasons, these limitations fueled widespread criticisms of this research concept. Before presenting these criticisms in detail, here is a summary of what they reported using their latest analyses (Lynn and Becker, 2019).

At the continent level, the average values were Americas (fifty-two countries), mean IQ 79 (range 48–99); Africa (fifty-two countries), 69 (range 45–87); Europe (thirty-nine countries), 95 (range 82–102); Asia (forty-five countries), 86 (range 43–106); and Oceania (eleven countries), 87 (range 78–99). The mean IQ for all countries studied (183 countries) was 82 (range 43–106). Whereas the averages are themselves problematic and raise reasonable doubts, the low end of the ranges indicates why these data are not taken seriously by many scientists.

The original Lynn and Vanhanen database of national IQ scores had major deficiencies beyond small, unrepresentative samples and equating disparate tests used in different countries. In some cases, their criteria for including a study can only be described as a blunder. For instance, their data point for Equatorial Guinea was a national IQ of 59. This was taken from a report that clearly stated that the IQ 59 figure referred to children in a school for the developmentally disabled in Madrid (Fernández-Ballesteros et al., 1997). Moreover, Earl Hunt and Werner Wittmann (2008) conducted separate analyses of the data for developed and developing countries, as presented in Lynn and Vanhanen's 2002 book. They found greater inaccuracies for low-income countries than for middle- and high-income ones. Additional reviews found more problems with selectivity of the data sets (Wicherts et al., 2010a, 2010b, 2010c). For example, one reanalysis raised the estimate of the median sub-Saharan African IQ of Lynn's estimate from 67 to 82, close to the average value observed in the United States (approximately 85) (Wicherts et al., 2010b). This indicated that Lynn and Vanhanen's research was biased toward underestimation of countries in sub-Saharan Africa (Wicherts et al., 2010a, 2010c).

Even with all the criticism, it is worth considering whether these average IQ estimates for nations have any validity at all. One way to check is to use an alternative

Table 12.1 Correlations between Lynn and Vanhanen (2012a) IQs and 2012 PISA scores

	PISA language	PISA math	PISA science	PISA everyday problem solving
IQ 2012	0.80	0.86	0.82	0.84
PISA language		0.96	0.98	0.91
PISA math			0.98	0.90
PISA science				0.91

test like PISA that also taps intelligence (general learning ability), as demonstrated by Pokropek and colleagues (2021). As explained in previous chapters, PISA exams are completed by students at age fifteen, and they tap knowledge and skills in language, mathematics, and science. They also periodically assess everyday problem-solving skills in addition to language, math, and science. For illustrative purposes, we computed the correlation between 2012 IQ values and 2012 PISA scores from participating countries. The results in Table 12.1 show that (1) the 2012 IQs estimated by Lynn and Vanhanen are highly correlated to PISA scores (r values between 0.80 and 0.86) and (2) PISA test scores are highly correlated among themselves (r values between 0.90 and 0.98), consistent with a common factor of learning ability.

The national IQ estimates also are related to numerous indicators beyond PISA scores. For instance, Lynn and Vanhanen (2006) considered the relation between the IQ values and five measures of the "quality of human conditions" (the QHC Index). The measures used were gross national income per capita (corrected for local purchasing power), the adult literacy rate, the fraction of the population enrolling in tertiary education (college/university), life expectancy, and an index of democratization developed by Vanhanen from previous work. The QHC Index can be considered as a measure of how much citizens of a country can participate in political life. The highest scores on the index (40 and higher) are from Belgium, Denmark, the Netherlands, and

Switzerland. Most large European and North American countries scored in the mid-30s. Dictatorships that enforce a one-party state, such as China, Cuba, Kazakhstan, and North Korea, scored zero.

The correlations between the national IQ estimates and the five individual components of the QHC were $r = 0.68$ for gross national income, 0.64 for adult literacy, 0.75 for tertiary education, 0.77 for life expectancy, and 0.57 for democratization. Given this level of correlation with the individual variables, it is inevitable that there will be a high correlation of IQ with the composite QHC Index, and there is: $r = 0.80$. Lynn and Vanhanen (2012b) published a summary of 244 educational, cognitive, economic, political, demographic, sociological, epidemiological, geographic, and climatic correlates of their national IQ estimates. Table 12.2 shows examples for the different categories. These correlations underscore the same point: on a national basis, an average IQ estimate is a good predictor of socioeconomic indicators of quality of life.

The authors noted three basic conclusions from this summary: (1) the range and heterogeneity of these correlates help validate the national IQ estimates; (2) there are presumably positive feedback loops among the variables – IQs might contribute to per capita income, education, and health in the first place, but these might contribute to IQs: "IQ is a determinant of income, and income is a determinant of IQ through its positive effects on nutrition, health and education" (Lynn and Vanhanen, 2012b, p. 230); and (3) research findings at the

Table 12.2 Selected correlations with national IQ estimates

Correlate	Number of countries	Correlation (r)
Math, science, literacy	108	0.91
Academic publications	139	0.87
GDP per capita	185	0.78
Economic growth, 1950–1990	185	0.44
Income inequality	148	−0.51
Economic freedom	165	0.52
Savings	129	0.48
Adult literacy	187	0.74
Homicide, 1990s	116	−0.25
Corruption, 1980–2003	132	−0.60
Democracy, 1950–2004	183	0.56
Life expectancy, 2002	192	0.75
Suicide	85	0.54
Fertility	192	−0.73
Maternal age (years)	172	0.29
Atheism	137	−0.60
Human Development Index	176	0.78
War	186	−0.22
Temperature	192	−0.63
Infectious disease	184	0.89

Note. From Lynn and Vanhanen (2012b).

individual level are reflected at the national level. The explanatory power of intelligence extends from the individual to the nation because nations are aggregates of individuals (Box 12.4). This is more evidence that social sciences should not ignore the pervasive relevance of the intelligence construct.

Whether or not we accept that the national IQ data are valid, one key implication of the overall picture is to consider ways to increase average intelligence levels in countries with low average scores. For example, Lynn (2018, p. 266) wrote, "The most promising ways for increasing intelligence in low-scoring countries are by improvements in nutrition, health, and education. During recent decades, these improvements have been taking place in a number of economically developing countries and contributed to raising intelligence, while from the mid-1990s. IQs have declined in several economically developed countries including France, Norway, Denmark, Australia, Britain, Sweden and the Netherlands. It is likely that these trends will continue leading to a reduction of global inequality." And Earl Hunt (2012, p. 303) wrote, "Virtually all the developing nations of the world are striving to obtain the benefits of the postindustrial world. The extent to which they will succeed will depend very much on their ability to develop and use those cognitive abilities that are essential in industrial and postindustrial societies. ... National indicators of intelligence are indicators of national differences in the ability to use the cognitive tools that are required to participate in modern industrial and postindustrial societies."

Real life for individuals and for populations is multivariate and colinear. Many factors contribute to the complexities of intelligence and wealth. However, not all factors are expected to be equally relevant. We need methods for disentangling collinearity. Box 12.5 shows two simple approaches taken by political scientist and social policy analyst Charles Murray. The next section shows more complicated approaches.

12.5 Cognitive Capitalism

We now turn to another approach that deals with collinearities and ambiguities. The

Box 12.4 Hive Mind: An Economist Considers Intelligence Seriously

The economist Garett Jones (2016) published *Hive Mind: How Your Nation's IQ Matters So Much More than Your Own*. His main point was that "the most important productive asset in each nation is the human mind. While standardized (cognitive) tests can't tell us everything about how productive the mind is, the tests can tell us more than you might think" (p. 2).

The evidence that Jones reviewed led him to conclude that nations with the highest cognitive scores are eight times more prosperous than nations with the lowest scores. From this perspective, finding ways for increasing the average cognitive level of nations is a vital goal: "A whole nation of people who tend to be good at standardized cognitive tests is world-changing. ... Millions of small cognitive contributions are what create each

nation's collective intelligence, each nation's hive mind" (pp. 7–12).

According to Jones, national prosperity is influenced by high cognitive ability for the following reasons:

1. High cognitive scores are associated with greater savings.
2. High cognitive scores are associated with cooperation.
3. High cognitive scores are associated with market-oriented policies.
4. High cognitive scores are associated with successful management of highly productive team-based technology.

Jones estimates that IQ-type scores "explain about half of everything across countries. ... The hive mind is no single-cause theory of prosperity. It's just a story that almost no one else is talking about" (p. 14).

German psychologist Heiner Rindermann (2007, 2018) has conducted investigations of national cognitive skills, wealth, and well-being using multivariate statistical analyses. He has reported a large general factor for "national cognitive skill" calculated across countries, conceptually similar to the *g*-factor of intelligence. The strength of the general national factor is noteworthy; the range of loadings on other factors is from 0.97 to 1.0. This might be surprising because some of these tests were intended to cover specific topics, such as reading or mathematics. On a cross-national basis, if children read well, they also do math well (and vice versa). Ranking countries on his measure of national cognitive skill resulted in almost the same ranking as reported by the IQ averages from Lynn and Vanhanen. Generally, the lowest scores are from sub-Saharan Africa; intermediate scores are in South Asia, the Middle East, North Africa, and South America. The highest scores are found in northeast Asia, Europe, and North America.

Rindermann (2018) has reported a number of correlations between this general cognitive factor and indices of economic and political well-being. For instance, the correlation to gross domestic product per capita (GDP/c) was 0.60, close to the Lynn–Vanhanen estimate. Other indicators show similar trends. The correlation with an index of economic freedom was 0.52 and with economic growth 0.44. For those who like food for thought (the readers of this book are first in this alimentary line), the correlation between national cognitive skill and the homicide rate was −0.23, and the correlation between cognitive skill and the rate of solved homicide cases was 0.32.

Because longitudinal education data were available, Rindermann was able to conduct "causal" analyses, in which he computed the extent to which a social or psychological variable measured at time 1 predicted social and psychological variables at time 2. His analysis relied on a technique called *cross-lagged analysis* (we saw an example in

Box 12.5 "Simplicity, Agent Starling"

In the 1991 movie *Silence of the Lambs*, this was the advice of Hannibal Lecter, an incarcerated serial killer, for the young detective Clarice Starling for finding another serial killer sought by the FBI. Psychological research can be complicated technically, but sophisticated statistical models are not necessarily better than simple comparisons. Here are two examples of straightforward approaches that provide valuable and easy-to-understand information.

The first example comes from part II of *The Bell Curve* (Herrnstein and Murray, 1994). It includes a short introduction explaining how the authors overcame some usual difficulties in prediction research. They were using the data set of the National Longitudinal Survey of Labor Market Experience of Youth (NLSY79, N = 12,686). First, they defined cognitive classes by dividing the sample into percentiles within the distribution of IQ scores: very low (fifth percentile), low (sixth to twenty-fifth), normal (twenty-sixth to seventy-fourth), high (seventy-fifth), and very high (ninety-fifth). Then they predicted a variety of socially relevant behaviors using IQ and SES (defined by parental education, income, and occupational prestige) as competing predictors (here is how IQ predicts social behavior controlling SES, or here is how SES predicts social behavior controlling IQ).

The authors' solution was elegant, simple, and straightforward – they reported the IQ predictive values for two subgroups of the NLSY with the same educational level: (1) high school diploma and (2) bachelor's degree. This shows if IQ makes a difference regardless of education, avoiding complicated computations.

They saw education as a special case: "One of the most common misuses of regression (predictive) analysis is to introduce an additional variable that in reality is mostly another expression of variables that are already in the equation" (p. 124). Including education and IQ into the same regression equation is statistically and conceptually dangerous because (1) the amount of education is related to both parental SES (r = 0.50) and personal intelligence (r = 0.64); (2) the education effect, regardless of personal intelligence, might be discontinuous; (3) closely related variables may generate interference (and wrong results) because of their high collinearity; and (4) equating education and intelligence at the same level means that educational and intelligence differences may have independent results over the outcome of interest, which is highly arguable.

Here is the second example. In his short monograph *Income Inequality and IQ*, Charles Murray (1998) addressed the question, what causes income inequality? He tried to discern if sociological factors were better than personal factors for predicting this inequality (or the other way around). The method for obtaining evidence was based in a simple and convincing way to assess the importance of intelligence for occupational prestige, keeping all family background factors equal.

The method was based on the analysis of siblings. The approach is powerful because it allows for controlling the network of variables associated with the family environments in which children are raised. The idea is to compare full siblings who have grown up in the same family but who show different IQ scores on a standardized test. Murray selected the siblings from the NLSY and classified them into five groups:

1. siblings with very high IQ scores (N = 128, mean IQ = 125 [SD = 5.6]);
2. siblings with high IQ scores (N = 326, mean IQ = 114 [SD = 2.7]);

Box 12.5 *(continued)*

3. reference group (N = 1,074, mean IQ = 99 [SD = 5.9]);
4. siblings with low IQ scores (N = 421, mean IQ = 86 [SD = 2.5]);
5. siblings with very low IQ scores (N = 199, mean IQ = 74 [SD = 5.4]).

Then he systematically compared one sibling with an IQ value within the normal range and another sibling within one of the remaining IQ groups, creating a simple index of occupational prestige. Based on 1,074 sibling pairs, these were the prestige values for the different IQ bands: very low IQ (25), low IQ (33), average IQ (43), high IQ (47), and very high IQ (54). The pattern was striking: siblings exposed to the same sociological family circumstances during childhood are linearly ordered in the occupational prestige they achieve in their adulthood (after leaving their families of origin) according to their personal intelligence level, as assessed by a standardized test.

These two examples show that sometimes there is no need to compute a complicated statistical regression or path analysis to identify valuable and easy-to-understand relationships. But more complex analyses do help verify such results. For example, using a much more sophisticated approach that included polygenic scores, McGue and colleagues (2020) reported results highly consistent with the main conclusion of this siblings simple comparison. These were their three key findings: (1) most individuals are educationally and occupationally mobile, (2) this mobility is predicted by offspring–parent differences in skills and genetic endowment, and (3) the relationship of offspring skills with the observed mobility does not change by parent social background. These researchers wrote, "Many people are likely to directly observe social mobility within their own families" (p. 845). Families are far from comprising homogeneous sets of individuals.

Figure 11.11), which made it possible to assess standardized path coefficients. These statistics can be interpreted as measuring the extent to which a raise of 1 standard deviation unit in a predicting variable will cause a raise, in standard deviation units, of a predicted variable. Figure 12.2 depicts an example of this kind of cross-lagged analysis.

The path coefficient between cognitive ability (cognitive human capital), measured in 1970, and wealth, measured in 2010, was 0.31. This indicates that raising academic achievement by one deviation unit will "pay off" by increasing wealth by 31 percent. This is no minor prediction. On the other hand, the path coefficient from wealth in 1970 to cognitive ability in 2010 was a meager 0.05, indicating that national wealth does not necessarily influence future national cognitive ability. More intensive efforts to use wealth to improve education might change this.

Figure 12.3 shows an average general path model derived from extensive sets of cross-nation analyses computed by Rindermann (2018) that he used to suggest several thought-provoking points:

1. An index of cultural background is a better predictor of adult education than an index of evolutionary background (0.60 vs. 0.15).
2. Adult education predicts school quality (0.72). Adult education also predicts cognitive ability differences among nations, both directly (0.19) and indirectly via school quality (0.72 × 0.42).

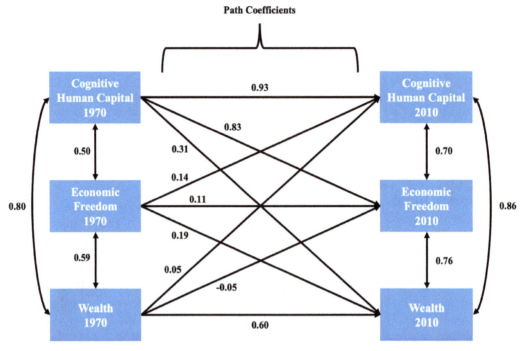

Figure 12.2 Cross-lagged effects of human capital–related variables (left) assessed in 1970 that (right) influenced/caused them when retested in 2010. Arrows show putative causal direction; numbers indicate strength of relationship pathways between variables (Rindermann, 2018).

3. Evolutionary background predicts cognitive ability better than cultural background (0.30 vs. 0.20). However, considering indirect effects, cultural background is a better predictor than evolutionary background (0.50 vs. 0.37).
4. Cultural background also is better than evolutionary background for predicting top cognitive ability level (0.54 vs. 0.33).
5. Cognitive and cultural factors predict institutional and societal features. Government effectiveness is predicted by top ability level (0.49) and culture (0.32). The average cognitive level of the society also predicts, indirectly, government effectiveness (0.40).
6. Economic freedom and rule of law predict GDP (0.13 and 0.27), but the stronger effect is for top cognitive ability (0.60).

The relationships shown in Figure 12.3 are intriguing. The idea of a general factor of national cognitive skill is important since it may have a major relationship to key social outcomes. The point of this kind of research is to identify possible causal relationships in complex systems that could provide information useful to policy makers interested in enhancing specific outcomes on a national level. This potential usefulness is why we include it in this chapter, even though it must be clearly stated that these data do not yet form a weight of evidence from independent research groups. Nonetheless, we repeat: controversy and debate require openness to discussion and encouragement of high-quality research.

12.6 What Makes Nations Intelligent?

The simple answer is, we do not know. Nations are aggregates of individuals, so variables relevant for understanding individual differences in intelligence might also be relevant for understanding national differences. Figure 12.4 illustrates a general model of intelligence development at the individual and national levels proposed by

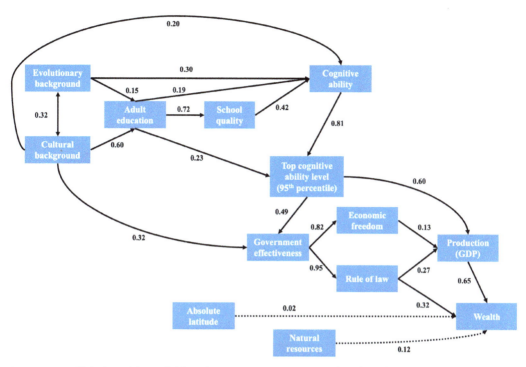

Figure 12.3 Global wealth model based on average cross-nation data for relevant variables. Arrows indicate direction of "causality," and numbers indicate strength of relationship between variables. Dotted lines show statistically nonsignificant values (Rindermann, 2018).

Earl Hunt (2012) in an invited address to the Association for Psychological Science. For the individual (Figure 12.4A), the potential for cognitive ability contained in the genome is developed by interactions with the physical environment, which establishes information-processing capacities. Interactions between information-processing capacity and the social environment develop cognitive skills. These include general ability and special cognitive skills. Depending on the society, individuals may be able to use their cognitive skills to alter the physical and social environments.

For the national level (Figure 12.4B), the genetic potentials for cognition contained in the gene pool are developed into information-processing capacities by the physical environment and then into cognitive abilities by the social environment. Because populations, including nations, operate on a timescale greater than a single person's lifetime, feedback mechanisms leading from social changes, including the adoption of the cognitive skills of other

societies, can change information-processing capacities and even the gene pool.

Models like the ones we have described in this chapter make many assumptions. For example, treating countries as homogeneous sets of individuals can be misleading and hide relevant insights. These were scores for high-, medium-, and low-scoring countries in the Science PISA survey: Singapore (556), China (518), Germany (509), United States (496), France (495), Spain (493), Argentina (475), Chile (447), Thailand (421), and Brazil (401). The mean score for the seventy countries assessed in this survey was 493. However, if we look at the eighteen regions within Spain, for example, we will find scores ranging from 473 to 519. Eleven of these regions show scores above the mean, ten regions show scores above France, four regions show scores above Germany, and one region shows scores above China. The same picture will emerge for the remaining countries, and therefore, it would be crucial to look for the factors that

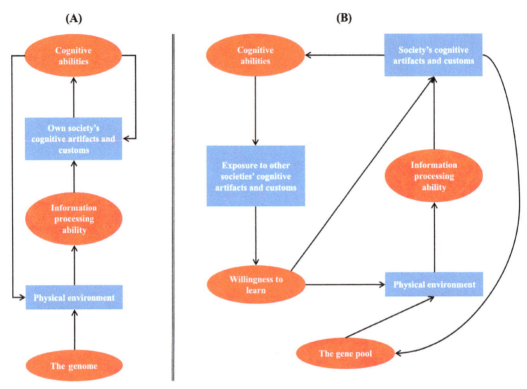

Figure 12.4 (A) Model of the development of individual general cognitive ability. The potential for cognitive ability contained in the genome is developed by interactions with the physical environment, which establishes information-processing capacities. Interactions between information-processing capacity and the social environment develop cognitive skills. These include general ability and special cognitive skills. Depending on the society, individuals may be able to use their cognitive skills to alter the physical and social environments. (B) Model of the development of cognitive skills at the national level. The genetic potentials for cognition contained in the gene pool are developed into information-processing capacities by the physical environment and then into cognitive abilities by the social environment. Because populations, including nations, operate on a timescale greater than a single person's lifetime, feedback mechanisms leading from social changes, including the adoption of the cognitive skills of other societies, can change both information-processing capacities and possibly even the gene pool (Hunt, 2012).

produce these differences within and across countries. Countries, like families, are not homogeneous entities.

The evidence reviewed in this chapter suggests that some nations have greater average cognitive resources than others and that this can make a difference for their prosperity. The concept of national intelligence and the data about it are controversial. Some critics reject any idea about national IQ out of hand, either because they see IQ scores as irrelevant or because they consider the weaknesses in some of the data as fatal flaws. Ad hominem criticism should not distract from the potential relevance of

the issues raised. This research is in early stages and is complicated and intricate but still worthy of discussion as various models are proposed and tested. As always, keep an open mind, along with a healthy and respectful skepticism.

12.7 Summary

Research suggests that nations differ in their cognitive resources, a key aspect of human capital. These differences might influence the economic and social well-being of societies. National indicators of intelligence are dynamic instead of static indicators of national differences in the ability to use the cognitive tools that are required to participate in the modern world. By "dynamic," we mean that the cognitive resources estimated for a country thirty years ago possibly were different than those estimated now and may be different thirty years from now. Indeed, some data suggest ways to improve average national intelligence. Responding to the challenges of doing so will be among the most complex and socially relevant global issues for the coming decades.

This chapter presents data that deserve consideration. Can science help policy makers improve the average intelligence of low-scoring countries? This might depend on the most intriguing question of all: can intelligence be increased? This may be the ultimate purpose of all intelligence research, as Haier (2017, 2021) proposed – and it is the focus of the final chapter of this book.

12.8 Questions for Discussion

12.1 Does the concept of national intelligence make sense to you?

12.2 Why is research on national intelligence controversial?

12.3 Are there any group differences that you think should not be discussed in public?

12.4 What are key methodological issues for evaluating group difference data?

12.5 Why could gaps in country average IQs be an issue of national well-being and prosperity?

References

Bamshad, M., Wooding, S., Salisbury, B. A., & Stephens, J. C. 2004. Deconstructing the relationship between genetics and race. *Nature Reviews Genetics*, 5, 598.

Canivez, G. L., & Watkins, M. W. 2010. Investigation of the factor structure of the Wechsler Adult Intelligence Scale – Fourth Edition (WAIS-IV): Exploratory and higher order factor analyses. *Psychological Assessment*, 22, 827–836.

Ceci, S. J., & Williams, W. M. 2018. Who decides what is acceptable speech on campus? Why restricting free speech is not the answer. *Perspectives on Psychological Science*, 13, 299–323.

Cofnas, N. 2020. Research on group differences in intelligence: A defense of free inquiry. *Philosophical Psychology*, 33, 125–147.

Cronbach, L. J. 1975. 5 decades of public controversy over mental testing. *American Psychologist*, 30, 1–14.

Dickens, W. T., & Flynn, J. R. 2006. Black Americans reduce the racial IQ gap: Evidence from standardization samples. *Psychological Science*, 17, 913–920.

Fernández-Ballesteros, R., Juan-Espinosa, M., Colom, R., & Calero, M. D. 1997. Contextual and personal sources of individual differences in intelligence. In Carlson, J. S., Kingma, J., & Tomic, W. (eds.), *Advances in cognition and educational practice: Reflections on the concept of intelligence*. London: JAI Press.

Fish, J. M. 2002. *Race and intelligence: Separating science from myth*. Mahwah, NJ: Erlbaum.

Flores-Mendoza, C., Ardila, R., Rosas, R., et al. 2018. *Intelligence measurement and school performance in Latin America*. New York: Springer Science +Business Media.

Flores-Mendoza, C., Mansur-Alves, M., Ardila, R., et al. 2015. Fluid intelligence and school performance and its relationship with social variables in Latin American samples. *Intelligence*, 49, 66–83.

Flynn, J. R. 2018. Academic freedom and race: You ought not to believe what you think may be true. *Journal of Criminal Justice*, 59, 127–131.

Green, E. D., Watson, J. D., & Collins, F. S. 2015. Human Genome Project: Twenty-five years of big biology. *Nature*, 526, 29–31.

Haier, R. J. 2017. *The neuroscience of intelligence*. Cambridge: Cambridge University Press.

Haier, R. J. 2020. Academic freedom and social responsibility: Finding a balance. *Intelligence*, 82, 101482.

Haier, R. J. 2021. Are we thinking big enough about the road ahead? Overview of the special issue on the future of intelligence research. *Intelligence*, 89, 101603.

Herrnstein, R. J., & Murray, C. 1994. *The bell curve: Intelligence and class structure in American life*. New York: Free Press.

Hunt, E. 2012. What makes nations intelligent? *Perspectives on Psychological Science*, 7, 284–306.

Hunt, E., & Wittmann, W. 2008. National intelligence and national prosperity. *Intelligence*, 36, 1–9.

Hur, Y.-M., & Bates, T. 2019. Genetic and environmental influences on cognitive abilities in extreme poverty. *Twin Research and Human Genetics*, 22, 297–301.

Jones, G. 2016. *Hive mind: How your nation's IQ matters so much more than your own*. Stanford, CA: Stanford Economics and Finance.

Lange, R. T., Chelune, G. J., Taylor, M. J., Woodward, T. S., & Heaton, R. K. 2006. Development of demographic norms for four new WAIS-III/WMS-III indexes. *Psychological Assessment*, 18, 174–181.

Lynn, R. 2018. The intelligence of nations. In Sternberg, R. J. (ed.), *The nature of human intelligence*. Cambridge: Cambridge University Press.

Lynn, R., & Becker, D. 2019. *The intelligence of nations*. London: Ulster Institute for Social Research.

Lynn, R., & Vanhanen, T. 2002. *IQ and the wealth of nations*. Westport, CT: Praeger.

Lynn, R., & Vanhanen, T. 2006. *IQ and global inequality*. Augusta, GA: Washington Summit.

Lynn, R., & Vanhanen, T. 2012a. *Intelligence: A unifying construct for the social sciences*. London: Ulster Institute for Social Research.

Lynn, R., & Vanhanen, T. 2012b. National IQs: A review of their educational, cognitive, economic, political, demographic, sociological, epidemiological, geographic and climatic correlates. *Intelligence*, 40, 226–234.

McGue, M., Willoughby, E. A., Rustichini, A., et al. 2020. The contribution of cognitive and noncognitive skills to intergenerational social mobility. *Psychological Science*, 31, 835–847.

Murray, C. A. 1998. *Income inequality and IQ*. Washington, DC: AEI Press.

Murray, C. 2007. The magnitude and components of change in the black–white IQ difference from 1920 to 1991: A birth cohort analysis of the Woodcock–Johnson standardizations. *Intelligence*, 35, 305–318.

Murray, C. 2020. *Human diversity: The biology of gender, race, and class*. New York: Twelve.

Neisser, U., Boodoo, G., Bouchard, T. J., Jr. et al. 1996. Intelligence: Knowns and unknowns. *American Psychologist*, 51, 77–101.

Nisbett, R. E., Aronson, J., Blair, C., et al. 2012. Intelligence: New findings and theoretical developments. *American Psychologist*, 67, 130–159.

O'Connell, M. 2019. Is the impact of SES on educational performance overestimated? Evidence from the PISA survey. *Intelligence*, 75, 41–47.

Plomin, R. 2018. *Blueprint: How DNA makes us who we are*. Cambridge: MIT Press.

Pokropek, A., Marks, G. N., & Borgonovi, F. 2021. How much do students' scores in PISA reflect general intelligence and how much do they reflect specific abilities? *Journal of Educational Psychology*, 114, 1121–1135.

Rindermann, H. 2007. The *g*-factor of international cognitive ability comparisons: The homogeneity of results in PISA, TIMSS, PIRLS and IQ tests across nations. *European Journal of Personality*, 21, 667–706.

Rindermann, H. 2018. *Cognitive capitalism: Human capital and the wellbeing of nations*. Cambridge: Cambridge University Press.

Rosling, H., Rosling, O., & Rönnlund, A. R. 2018. *Factfulness: Ten reasons we're wrong about the world – and why things are better than you think*. New York: Flatiron Books.

Schleicher, A. 2019. *PISA 2018: Insights and interpretations*. Paris: Organisation for Economic Cooperation and Development.

Smedley, A., & Smedley, B. D. 2005. Race as biology is fiction, racism as a social problem is real: Anthropological and historical perspectives on the social construction of race. *American Psychologist*, 60, 16–26.

Spearman, C. 1927. *The abilities of man; their nature and measurement*. New York: Macmillan.

Tang, H., Quertermous, T., Rodriguez, B., et al. 2005. Genetic structure, self-identified race/ethnicity, and confounding in case-control association studies. *American Journal of Human Genetics*, 76, 268–275.

Turkheimer, E. 2019. Genetics and human agency: The philosophy of behavior genetics – introduction to the special issue. *Behavior Genetics*, 49, 123–127.

Warne, R. T., & Burningham, C. 2019. Spearman's *g* found in 31 non-Western nations: Strong evidence that *g* is a universal phenomenon. *Psychological Bulletin*, 145, 237–272.

Watson, J. D. 2007. *Avoid boring people: Lessons from a life in science*. Oxford: Oxford University Press.

Whetzel, D. L., & McDaniel, M. A. 2006. Prediction of national wealth. *Intelligence*, 34, 449–458.

Wicherts, J. M., Dolan, C. V., Carlson, J. S., & van der Maas, H. L. J. 2010a. Raven's test performance of sub-Saharan Africans: Average performance, psychometric properties, and the Flynn effect. *Learning and Individual Differences*, 20, 135–151.

Wicherts, J. M., Dolan, C. V., & van der Maas, H. L. J. 2010b. The dangers of unsystematic selection methods and the representativeness of 46 samples of African test-takers. *Intelligence*, 38, 30–37.

Wicherts, J. M., Dolan, C. V., & van der Maas, H. L. J. 2010c. A systematic literature review of the average IQ of sub-Saharan Africans. *Intelligence*, 38, 1–20.

Enhancing Intelligence

13.1 Introduction

All the preceding chapters have led to this one fundamental question: can intelligence be increased? It is a simple question, but what exactly does it mean? As discussed in Chapters 2 and 3, from a scientific standpoint, intelligence can mean an assessment score (from a reliable and valid standardized test), a broad factor (like verbal, visuospatial, or perceptual ability), and the general factor common to all mental abilities (the g-factor). The measured performance on any given cognitive test results from the contribution of g, the specific ability tapped by the test, and the specific skills required for such a test. Therefore, when we observe an increase in the measures we administer, the change can be at the test, the ability, or the g level. An increase can be small, albeit statistically significant, or large enough to have a measurable effect on a relevant outcome variable like educational achievement or job performance. An increase can be temporary or long-lasting. In this chapter, we mean something potentially more interesting than an increase in IQ scores, something that is more permanent, and something that impacts *g*. As you were reading other chapters, perhaps you considered questions like the following:

- Is there anything I can do to be more intelligent?
- Can intelligence be increased beyond a person's genetic potential?
- Is there a theoretical limit on just how smart any individual can become?
- Do children and adults have an inner genius that can be unlocked?

The desire to enhance intelligence dramatically is as ancient as alchemy. So far, this goal is just as elusive as turning lead to gold – but is it even possible that any of these questions can be answered in the affirmative?

Intelligent humans have solved all kinds of complex problems: conquering some diseases, discovering the structure of DNA, building machines to see inside the brain and other machines to break atoms apart, building skyscrapers and city infrastructure, sending people to the moon and exploring other planets remotely, communicating instantaneously around the world with cell phones, and even explicating the dynamics of the first nanoseconds of the big bang. So why not the problem of increasing intelligence itself? What would the world be like if everyone were smarter? It certainly would be another giant leap for humankind.

If for no other reason, increasing intelligence should be a priority for alleviating social problems associated with IQs under 85, which describes 16 percent of the population based on the assumptions of a normal distribution (as discussed in Figure 1.1). At this level, individuals may not have the requisite cognitive abilities for gainful employment in a modern world or necessary to navigate everyday life with appropriate competence. In the United States, for example, about 53 million people, including about 14 million children, have IQ scores under 85. Surely the prospect of increasing intelligence for these people motivates serious consideration (Haier, 2017, 2021).

We do not underestimate the enormity of what would be required – the complexity of intelligence at many interacting levels is the theme of this book. With that in mind, this chapter explores possibilities and also challenges you to think about their implications.

13.1.1 *Sources of Confusion*

There is considerable confusion about what increasing intelligence means. There is the common conflating of intelligence, the *g*-factor, and IQ or other test scores. They are different aspects of the universe of mental abilities. An increase in one does not mean the others also increase. For example, IQ points do not have an absolute meaning; they only can be interpreted relative to other people's scores (IQ is on an interval scale). This is unlike units of distance or weight; they have absolute meaning (they are on a ratio scale). If you weigh yourself before and after a weight-loss program, the result does not depend on the weight change in other people. If you take an IQ test when you are sick and unable to concentrate, the score likely will be a bad estimate of your actual intelligence; if you retake it when you are well, the score will be a better estimate, but the increased score cannot be interpreted as an increase in your intelligence. Similarly, research that compares intelligence test scores before and after an intervention hypothesized to increase intelligence faces this fundamental problem (Haier, 2014). Moreover, increases in IQ scores may or may not be attributed to the *g*-factor; the increases could be related to other, more specific intelligence factors, to test-taking skills, or to social factors (Estrada et al., 2015). All are related, however, to each other and to biological factors, as illustrated in Figure 13.1. Where in this field of influences and interactions might interventions increase intelligence? Increasing *g* is an especially important focus because it alone accounts for the most variance among people (about 50 percent). All the other factors make smaller contributions, even though their relevance is not zero (Flynn, 2012, 2018; Hunt, 2012).

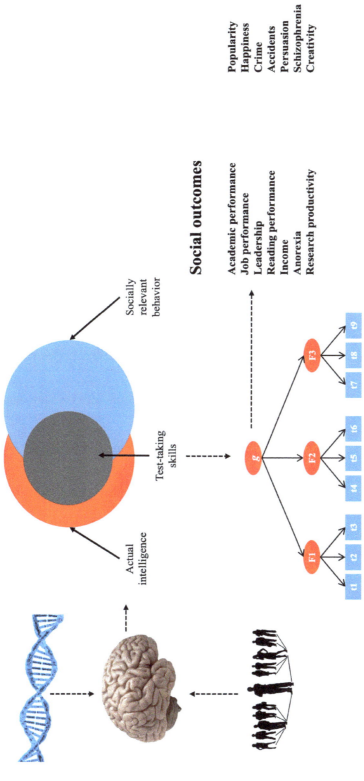

Figure 13.1 Interplay between brain, tests, and social outcomes. (left) All aspects of intelligence result from genome–environment interplay that takes place in the brain. (middle) Intelligence and abilities assessed by standardized tests (test-taking skills) are not synonymous (Chapter 2). Tests are the basis for psychometric models of intelligence (Chapter 3), and they tap socially important facets of the intelligence construct (Chapter 8). Improving any mental ability may or may not evoke changes elsewhere in the system, especially to *g*.

377

The strongest implication of the psychometric models of intelligence described in Chapter 3 is that measurable changes in lower-order abilities due to interventions will fail to impact higher-order abilities such as the g-factor. The arrows in the hierarchical psychometric model depicted in Figure 13.1 (hierarchical structure at the bottom) go downward, which implies that any intervention effects seen at lower levels will not move upward. This argument was made by John Protzko (2017) to explain why short-term cognitive training programs fail to show convincing and reliable effects on high-level abilities such as fluid intelligence: "the structure of intelligence may make it impermeable to changing subfactors in the hopes of upward effects" (p. 1030). In fact, no long-lasting increases to g, based on multiple measures, have been demonstrated in replicated intervention studies to date once the intervention ends.

So, the simple question of whether intelligence can be increased (enhanced, boosted, improved) can mean different things because the broad construct includes a large set of latent abilities and skills that vary in importance for explaining (in a statistical sense) individual differences in cognition. In the next sections, we discuss research that reports improvements in specific abilities in the broad sense, but whether the common general factor (g) can be increased in the long term is another matter.

13.1.2 Does Compensatory or Early Childhood Education Boost Intelligence?

An important question is whether increases in IQ scores after an intervention are temporary or long-lasting. Perhaps the most earnest attempts to permanently raise intelligence are based on decades-old observations that disadvantaged children on average scored lower on tests of mental ability, including IQ tests. By the 1960s, educators and social scientists were convinced that test score gaps resulted from unequal educational opportunities due in large part to ethnic/racial discrimination. Given the stark reality of unequal opportunities, this was a

reasonable attribution. The logical approach to eliminating test score gaps, especially IQ scores, was to provide compensatory education to blunt direct and indirect effects of unequal opportunity. The expectation was that permanent IQ increases would result.

Before the popular US federally funded Head Start program began, a number of pilot demonstration projects reported large gains in IQ scores for disadvantaged children after compensatory interventions involving curriculum enrichment and cognitive stimulation. Most projects reported IQ gains around 5 points, but some reported IQ gains up to almost 20 points, and one claimed a 30-point increase among low-IQ children (Page, 1972). These early reports fueled optimism that early intensive compensatory education was a realistic solution to the serious gap problem.

As discussed in Chapter 7, a comprehensive evaluation of these early compensatory efforts found no convincing evidence that they boosted IQ scores (Jensen, 1969). Most of the gains diminished substantially after the compensatory program ended, with virtually no long-lasting gains reported. Jensen's conclusion was summarized in a famous opening sentence: "compensatory education has been tried and it apparently has failed" (p. 2). From a scientific perspective, this blunt evaluation undermined the widely held fundamental assumption of unequal opportunity being the primary cause of average score differences. To make the conclusion more unsettling, Jensen proposed that IQ scores might be less malleable than thought because intelligence is influenced by genetics.

Many educators and social scientists at the time found Jensen's evaluation and speculation offensive. Criticism was so intense that intelligence research, IQ testing, and behavioral genetic research related to intelligence became radioactive, with repercussions to this day. In the history of psychology, Jensen's (1969) report may be the most infamous ever published. But it still is the starting point for anyone interested in this topic. "Jensenism" is still vilified, mostly for things Jensen never wrote or said (Cronbach, 1975a, 1975b; Johnson, 2012). Here is what Jensen concluded at the end of his long and detailed report: "Diversity rather than uniformity of approaches and aims would seem to be the key to making education rewarding for children of different patterns of ability. The reality of individual differences thus need not mean educational rewards for some children and frustration and defeat for others" (p. 117).

Writing about the controversy, Lee J. Cronbach (1975a, pp. 4–5) noted, "Jensen was right about the failure of compensatory efforts [but] buried within the paper was recognition that intensive small-scale programs often succeed, and in actuality Jensen's position was a call for intervention of effective educational procedures. . . . Even the most hereditarian position does not hold that ranks in performance will remain stable when the initially low-ranking children are treated specially." Few educators would disagree with these statements.

On her revisiting of Jensen's (1969) report, Wendy Johnson (2012, p. 119) wrote, "Though Jensen's presentation and the ensuing controversy were focused on the US, the issues involved were and still are clearly relevant throughout the world. . . . The evidence he presented stands to this day and has not been substantively refuted."

As explained in Chapter 7, the weight of evidence favors Jensen's conclusion that compensatory education does not boost IQ in the long run, although the evidence of temporary gains should not be ignored and requires additional research to understand the fade-out effect (Box 7.5). Nonetheless, if intelligence is as malleable as some people believe (and as assumed by the blank slate view; Pinker, 2002), early childhood interventions should have results clearly indicating lasting IQ increases. So far, they do not, but there is other evidence of some malleability.

13.2 Evidence of Malleability from the Environment

13.2.1 Secondary Education Effects for Individuals

Chapter 7 discussed research showing intelligence changes related to a number of environmental or nongenetic variables. We will not repeat them all here, but we will reiterate compelling data about education from a meta-analysis and from a longitudinal study that showed that regular education is associated with higher intelligence scores.

The meta-analysis by Ritchie and Tucker-Drob (2018) included more than 616,000 individuals from forty-two independent data sets. The average effect size of one year of education was 3.4 IQ points, but this was not shown to be cumulative, nor was it shown to be a *g* effect because required data were unavailable.

The longitudinal research report was based on 1,009 individuals at ages twenty, fifty-six, and sixty-six; it revealed an education effect of small gains in intelligence after age twenty (Kremen et al., 2019). Their results supported reverse causation, meaning that individuals with greater baseline intelligence obtain higher educational levels and occupational sophistication. Regarding the leveling-off period after age twenty, the authors suggested, "There is still substantial brain development during childhood and adolescence. Additional education likely provides an enriching environment that promotes brain and cognitive development: however, by age 20 there would be much less subsequent brain development. . . . It may also suggest that a reduction in later-life cognitive decline and dementia risk actually might begin with improving earlier educational quality and access" (p. 2025).

Whether these gains in test scores attributed to education have any practical benefit remains to be determined. Changes in IQ scores of 4 points or less could be due to measurement error inherent in the test, especially in children, and fluctuations in IQ scores are not evidence of changes in *g*. Some components of an IQ score may be more sensitive to environmental changes, whereas the *g*-factor might be less malleable, although this is unresolved by the weight of evidence so far.

13.2.2 The Flynn Effect for Generations

Perhaps the most intriguing example of intelligence increases is the so-called Flynn

Change trajectories

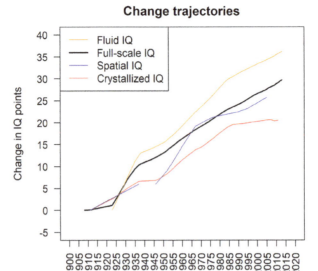

Figure 13.2 Generational change in IQ points (*y*-axis) on
four measures of intelligence from 1909 to 2013 (*x*-axis)
based on a comprehensive meta-analysis. Fluid IQ is highly
related to the *g*-factor and shows the Flynn effect
(Pietschnig and Voracek, 2015).
Courtesy of J. Pietschnig.

effect, which describes a slow but substantial increase in IQ scores over generations
(Flynn, 2012, 2018; Hegelund et al., 2021).
It is a major example of a cohort effect (see
Box 11.2). Many studies show this robust
trend worldwide, although there is some
evidence that the effect may be diminishing
(Bratsberg and Rogeberg, 2018; Pietschnig
and Gittler, 2015; Pietschnig and Voracek,
2015). Nonetheless, the Flynn effect begs
two key questions related to enhancement:
is it a *g* effect, and what causes it?

For the first question, a comprehensive
meta-analysis by Jakob Pietschnig and
Martin Voracek (2015) included 271 independent samples (4 million individuals from
thirty-one countries) recruited and analyzed
between 1909 and 2013. As shown in
Figure 13.2, the Flynn effect was found for
four estimates of intelligence, ranked here
by effect size (*d*) of the change: fluid intelligence, *d* = 0.41 (equivalent to 6 IQ points);
spatial intelligence, *d* = 0.30 (4.5 IQ points);
full-scale IQ, *d* = 0.28 (4.2 IQ points); and
crystallized intelligence, *d* = 0.21 (3 IQ

points). Since fluid intelligence is closely
associated with the *g*-factor, this evidence
indicates a *g* effect in addition to an effect
on other factors, but there is not yet a compelling weight of evidence about whether a
g effect continues.

As far as causes, there are many candidate
explanations of the Flynn effect, such as
reduced family size, increased schooling,
greater availability of technology, better
nutrition, and better health care, to name a
few. Virtually all of them are related to envi-
ronmental/social factors because the effect
occurs over far too a short a time for evolu-
tionary/genetic processes, although there is
some disagreement about this (Mingroni,
2004, 2007; Woodley, 2011). There likely is
more than one cause, and the causes may
vary from place to place. One important
finding is that infant development quotients
(quantifications based on behaviors) in the
first two years of life show a generational
increase of 3.7 points per decade (Lynn,
2009). Similarly, an increase of 3.9 points
per decade was observed in preschool

children (aged four to six) (Wongupparaj et al., 2015). These gains are approximately the same as the Flynn effect for adults. They favor (but do not prove) causal explanations focused on the first years of life, such as better prenatal and early postnatal nutrition and health care. If this is so, a cost-effective way to increase intelligence for millions of children is already known and getting attention in many countries.

Actual intelligence increases (not due to test artifacts) of 3 or 4 points might have little if any practical importance for individuals. Small average increases for a population, however, could be consequential at the extremes of the normal distribution, moving more people into the higher end and more out of the lower end. There might also be substantial social impacts. Based on the normal distribution of IQ scores, Haier (2017) pointed out that in the United States, approximately 51 million people have IQs under 85. Surely there is overlap between this population and the population of people living in poverty, so increasing intelligence might alleviate at least some consequences of poverty (see Haier, 2017, section 6.6).

Herrnstein and Murray (1994) presented an interesting example using data from *The Bell Curve*: "We cannot predict exactly what would happen if the mean IQ rose from 100 to 103, but we can describe what does happen to the statistics when the NLSY sample is altered so that its subjects have a mean of 103. The poverty rate falls by 25%. So does the proportion of males ever interviewed in jail. High school dropouts fall by 28%. Children living without their parents fall by 20%. Welfare recipiency, both temporary and chronic, falls by 18%. Children born out of wedlock drop by 15%. The incidence of low-weight births drops by 12%. Children in the bottom decile of home environments drop by 13%. Children who live in poverty for the first three years of their lives drop by 20%" (p. 367). Imagine what might result if a population average were to increase by a full standard deviation (15 points). This is why we think increasing intelligence is a crucial goal.

Some educational interventions and commercial schemes already have claimed success for increasing intelligence with the intent to minimize average population differences or increase a person's intelligence. These efforts typically conflate IQ and other test scores with intelligence and the *g*-factor. So far, none of the claims are established by independent replication, a cornerstone of the scientific method. Several researchers also have claimed to increase intelligence with varied techniques, and their evidence is discussed in the next section.

13.3 What Does Not Work (So Far) from Laboratory Studies

One of the most popular claims about increasing intelligence originated with a simplistic study that reported increased IQ scores after listening to a musical piece by Mozart. We describe this study and its debunking in Box 13.1. The Mozart effect should be a warning to well-intentioned researchers about the difficulties of demonstrating intelligence increases with simple experimental manipulations. Such claims require skepticism and a high standard of evidence. It is instructive to scrutinize some of the most popular claims to illustrate the importance of sound research design and independent replication for establishing a weight of evidence.

13.3.1 *Memory Training*

A flurry of interest came from a report in another prestigious scientific journal, *Proceedings of the National Academy of Sciences of the United States of America*. This report claimed that training on a difficult working memory task increased fluid intelligence, the construct most synonymous with *g* (Jaeggi et al., 2008). The authors concluded, "The finding that cognitive training can improve [fluid intelligence] is a landmark result because this form of intelligence has been claimed to be largely immutable" (p. 6832). Self-proclaimed landmark status usually is a signal for skepticism.

Box 13.1 Mozart and IQ

A common theme among efforts to increase intelligence is the notion that simple, even passive, interventions can be successful. No better illustration of this is the Mozart effect. In 1993, a short report in a prestigious scientific journal, *Nature*, claimed an increase of 8 IQ points after college students listened to a specific Mozart composition (Sonata for Two Pianos in D Major) for ten minutes (Rauscher et al., 1993). This is about half a standard deviation and would be regarded as a major increase. The effect lasted only ten to fifteen minutes, but showing this apparent degree of malleability was sufficient proof for some people that intelligence is not fixed and that IQ can be increased easily. Mozart music sales had a dramatic resurgence, and this study often was used to validate school music programs (a good thing, but for the wrong reason). This striking finding was widely reported and readily accepted by some people as fact. After all, it was published in *Nature*, and surely the expert peer reviewers were convinced of the results and eager to publish this scientific breakthrough. Researchers familiar with the science of human intelligence, however, were skeptical.

There were many serious flaws in this study, and at this point in the book, you should be able to spot them. The sample was small, three different visuospatial reasoning subtests of the Stanford–Binet battery were used separately for each condition, and their scores were converted to estimates of full-scale IQ points in an effort to make them comparable. There was no IQ testing prior to the experiment or information about the participants' musical experience. (For a detailed critique, see Haier [2017, pp. 139–143].)

The Mozart effect sparked hundreds of studies (not including countless high school student science fair reports), but meta-analyses found no supporting evidence (Pietschnig et al., 2010; Chabris, 1999). There was no consistent independent replication. Nevertheless, to this day, listening to Mozart is still widely believed to increase IQ. It does not.

There is, however, some evidence of a relationship between active musical training and cognitive ability. The evidence is inconsistent. At best, there may be only a weak relationship (Bigand and Tillmann, 2022; Sala and Gobet, 2020). Nevertheless, an interesting report of longitudinal data (from the Scottish study described in Chapter 1) cautiously supports a weak relationship (Okely et al., 2022).

This "landmark" result that apparently overturned contrary results from hundreds of previous studies was based on eight or nine students in each of four training groups and another four control groups. The actual increase in reasoning test scores after training was an additional two to five correctly answered items, compared to one for the controls (more training was apparently related to more improvement). These changes are not IQ scores but correctly answered items, which were converted into estimated IQ changes.

These small increases in small samples might be interesting if the research design had not been flawed, especially with a visuospatial measurement of fluid intelligence that was confounded with the visuospatial aspect of the training task (for a detailed analysis, see Haier, 2017, pp. 143–150). The eagerness to prove that intelligence is easily malleable drove widespread news coverage, grant money, and pitches to educators to incorporate memory training in schools, often by purchasing expensive commercial programs.

As expected, intelligence researchers were skeptical. Criticisms of this study evoked the embarrassing failure of cold fusion, a single physics experiment conducted by researchers naive to technical issues of low-level heat measurement. They claimed they had proved that unlimited cheap energy creation was possible, but their result could never be replicated. Not surprisingly, independent studies of memory training failed to replicate the original claims of increased fluid intelligence, and independent meta-analyses report little evidence for any effect (Redick, 2019; Melby-Lervåg et al., 2016; Thompson et al., 2013; Wiemers et al., 2019; Gobet and Sala, 2022).

13.3.2 *Computer Games*

There are ubiquitous claims that playing computer games improves specific cognitive abilities like attention, processing speed, and memory, but evidence is mostly unconvincing (Unsworth et al., 2015; Moreau, 2022). These are abilities related to the intelligence construct, and some researchers and gaming companies make the connection. Claims about increasing intelligence with computer gaming have narrowed because there are no independently replicated data to support such a claim, and, perhaps more importantly, at least one company (Lumosity) paid a $2 million fine after being charged by the US Federal Trade Commission for advertising claims about improving cognition that were not supported by scientific evidence.

An early example that playing computer games specifically increased intelligence is instructive. It is a study from the University of California, Berkeley (Mackey et al., 2011). This was a study of disadvantaged children (aged seven to ten years old) exposed to computer games that focused either on matrix reasoning ($N = 17$) or on speed of processing information ($N = 11$). The games were played during school for an hour each day for twelve days on average over an eight-week period. The posttraining changes on selected intelligence subtests results showed a large increase of 9.9 IQ points estimated from subtest scores. Remarkably, some students showed apparent increases over 20 IQ points. The authors concluded, "Counter to widespread belief, these results indicate that both fluid reasoning and processing speed are modifiable by training" (p. 582). They further concluded that these results gave hope for closing achievement gaps with "a mere 8 weeks of playing commercially available games" (p. 587). These are bold statements, given that the sample sizes were small and the training period limited. There are other problems with this study (detailed in Haier, 2017, pp. 150–153), so you might not be surprised to learn that there is no replication study, even by the original researchers.

There is another claim about computer game playing and increasing IQ scores published in a top journal, *Scientific Reports* (Sauce et al., 2022). Like the last example, this study was widely reported in popular media with sensationalized headlines like "Video Games Can Boost Children's Intelligence." The study was well designed and had a large sample of more than 5,000 children. Measures of each child's screen time, included gaming, were self-reported at an initial testing and again two years later. There were many analyses, but the main result was an apparent increase of 2.55 IQ points attributed to gaming time after two years, compared to a 1.8-point increase attributed to general screen time. Do you think such changes justify the headlines?

There is debate about whether playing computer games is useful for decreasing the effects of normal aging or whether they result in lasting improvements in cognitive abilities like memory and concentration (Simons et al., 2016; Bediou et al., 2018; Anguera et al., 2013). As far as increasing the core of intelligence, however, we are not aware of any compelling data from independent or replicated studies.

There are many interactive aspects to intelligence and its development. The complexities indicate that any single, simple approach for increasing intelligence would be unlikely to have a major impact. Another interesting example is in Box 13.2, which explores educational consequences based on whether a person believes that intelligence is fixed or malleable.

These examples demonstrate flaws in research design and analyses that are

Box 13.2 Mind-set Theory and Practice

This example is different from the others discussed in Section 13.3. Whereas Mozart, memory training, and gaming studies used intelligence measures as dependent variables to assess changes after an intervention, the mind-set approach uses a person's opinion about whether intelligence is fixed or malleable. On the basis of this implicit theory of mind, thought to be a stable personal characteristic, people are divided into two groups: those with a fixed mind-set and those with a growth mind-set. The main claims are that (1) having a growth mind-set results in better academic performance and that (2) children can be taught to have a growth mind-set.

The basic idea is that what students believe about their brains has profound effects on their motivation, learning, and school achievement (Dweck, 2008). This approach has had thirty years of enormous popularity in education and among parents. Numerous best-selling books and research papers tout how a simple intervention – teaching that abilities like intelligence are malleable – can unlock greater academic performance, especially among disadvantaged children. Foundations have invested millions bringing it to schools. School systems have spent millions on mind-set growth programs.

However, according to a comprehensive review based on two large meta-analyses, it does not work (Sisk et al., 2018). The first meta-analysis focused on the relationship between mind-set type and academic achievement. It included 162 independent samples, 273 effect sizes, and a total of 365,915 students. The second focused on whether mind-set interventions increased academic achievement. It included thirty-eight independent samples, forty-three effect sizes, and a total of 57,155 students. Given the impact mind-set has had on concepts about intelligence and education, it is instructive to present details of this comprehensive analysis.

The effect sizes were minuscule for both analyses, as assessed by Cohen's d (see Chapter 9). In meta-analysis 1, most effect sizes were not different from zero, and no moderator variables like SES accounted for the pattern of results. In meta-analysis 2, the average difference between students getting the mind-set intervention and control students was 0.08, far smaller than Cohen's criterion of 0.2 for even a small effect. Based on all the analyses, the authors concluded, "Some researchers have claimed that mind-set interventions can 'lead to large gains in student achievement' and have 'striking effects on educational achievement.' Overall, our results do not support these claims. ... Mind-set interventions on academic achievement were non-significant for adolescents, typical students, and students facing situational challenges (transitioning to a new school, experiencing stereotype threat). However, our results support claims that academically high-risk students and economically disadvantaged students may benefit from growth-mind-set interventions. ... Although these results should be interpreted with caution because (a) few effect sizes contributed to these results, (b) high-risk students did not differ significantly from non-high-risk students, and (c) relatively small sample sizes contributed to the low-SES group" (p. 568).

After noting limitations of their analyses, in the final paragraph, the authors conclude, "However, from a practical perspective, resources might be better allocated elsewhere than mind-set interventions. Across a range of treatment types, Hattie, Biggs, and Purdie (Hattie et al., 1996) found that the meta-analytic

Box 13.2 *(continued)*

average effect size for a typical educational intervention on academic performance is 0.57. All meta-analytic effects of mind-set interventions on academic performance were < 0.35, and most were null. The evidence suggests that the 'mindset revolution' might not be the best avenue to reshape our education system" (p. 569).

There is a general argument that even minuscule effects are better than no effects and that even a tiny effect can be aggregated across a school system to show some improvement. Nonetheless, after decades of applications worldwide, research shows that mind-set theory apparently does not elucidate anything new or important about intelligence, nor does it fulfill its revolutionary promise to increase academic achievement merely by teaching students that intelligence and other mental abilities are easily malleable, an assertion that so far is contrary to the weight of evidence.

---

relatively common if researchers have little or no psychometric background. Ackerman and Hambrick (2020) have provided twenty-one specific do and don't guidelines for laboratory studies of intelligence. For example, *don't* disregard the reliability of measures, *don't* use samples too small to detect differences in correlations, and *don't* confuse statistically significant with meaningful; *do* adjust correlations for restriction of range, *do* prespecify expected correlations, *do* take account of underlying heterogeneity of the intelligence construct, and *do* use multiple tests to assess underlying ability factors. If adopted, these guidelines will help minimize misleading results and unwarranted conclusions.

13.4 What Might Work from Biology

There are claims that biological interventions can increase intelligence, most notably with drugs or dietary supplements. Drug use to enhance cognitive performance is not uncommon, especially among students before exams. However, any drugs or supplements that affect the brain and promise enhanced intelligence should be viewed with caution and skepticism. There is no compelling weight of evidence for their effectiveness. The same is true for electrical devices that stimulate brain activity, as described in Box 13.3. The rest of this section explains some approaches that might be effective in the near term and perhaps in the long run.

13.4.1 *Embryo Selection*

Two biological parents both with high IQ increases the likelihood of their biological offspring having high IQ too, although regression to the mean statistically predicts that the offspring's high IQ would be lower than the IQ of the parents. Having a smart mate is not a guarantee, just an increased statistical probability with considerable variance. The same is true for two tall parents and the height of their offspring. Richard Herrnstein (1971, p. 58) explained the issue this way: "To predict the IQ of the average offspring in a family: (1) average the parents' IQs, (2) subtract 100 from the result, (3) multiply the result of (2) by 0.8 (the heritability value), and (4) add the result of (3) to 100. Given a mother and father with IQs of 120, their average child will have an IQ of 116. Some of their children will be brighter and some duller, but the larger the family, the more nearly will the average converge onto 116. With parents averaging an IQ of 80, the average child will have an IQ of 84. . . . The amount of regression for a

Box 13.3 Zapping Your Brain

It sounds a bit like science fiction, but there is some evidence that electrical stimulation of the brain can enhance performance on some specific cognitive tasks. Researchers have explored four ways to do this.

The first is transcranial magnetic stimulation (TMS). TMS uses a wand-like device containing a metal coil to produce magnetic field pulses when electricity is applied in short bursts. When the wand is placed over a part of the scalp, neurons in the cortex underneath undergo depolarization, with a resulting increase in cortical excitation or inhibition, depending on the TMS parameters. TMS can test whether a part of the cortex is involved in a particular cognitive task. TMS allows for the experimental manipulation of cortical activity to see how task performance may change. Inducing neuron activation may result in better performance; deactivation may result in poorer performance. Although TMS has been used for about twenty years, there is not yet a consistent weight of evidence as to whether it reliably improves normal cognitive performance (Luber and Lisanby, 2014), and we are not aware of TMS manipulations that increase intelligence.

The second technique is called transcranial direct current stimulation (tDCS). It uses a nine-volt battery to generate a weak current between electrodes on the scalp. The resulting shocks are mild and barely detectable. Similar to TMS, tDCS can increase or decrease neuronal excitability under the electrodes, and effects on cognitive performance can be tested. So far, early studies suggesting improved cognitive performance have not been replicated (Horvath et al., 2015a, 2015b). In fact, one study suggested a decrease in performance on the WAIS-IV (Sellers et al., 2015), and another study reported a small increase in matrix reasoning (Brem et al., 2018), but so far, there is no compelling weight of evidence.

The third technique uses transcranial alternating current stimulation (tACS) instead of direct current. tACS can be more focused on particular brain areas than tDCS. One early report ($N = 20$) suggested that tACS to the left frontal lobe resulted in faster solution times for difficult matrix reasoning problems (Santarnecchi et al., 2013). This was important because the change was assessed with a measure of time on a ratio scale. Somewhat similar results were reported in another study ($N = 28$) that suggested that test scores on fluid intelligence tests improved (Pahor and Jausovec, 2014). These are provocative studies, but there is not yet a weight of evidence (Antal et al., 2022). Nonetheless, there are anecdotal reports that some gamers have built their own home-made versions of tACS devices to enhance gaming performance. Even if you think zapping your brain with mild shocks sounds intriguing, we do not recommend such experimentation.

The fourth technique is deep brain stimulation (DBS). This is the conceptual equivalent of a pacemaker for the heart or an implant into brain control areas for movement to alleviate shaking due to Parkinson's disease (Lozano and Lipsman, 2013). Fortunately, this cannot be done at home. There is exploratory work on DBS to control depression and other brain disorders, and a number of studies suggest possible enhancement of learning and memory (Suthana and Fried, 2014; Widge et al., 2019; Hescham et al., 2020). We are not aware of any attempts to use DBS to enhance intelligence, but it is intriguing to think about using the results of neuroimaging studies of intelligence to select targets for the implantation of electrodes. These could be multiple brain sites from the P-FIT framework or multiple sites based on connectivity hubs discussed in Chapter 5.

Box 13.3 *(continued)*

For now, research on intelligence using these techniques is in early stages, and results are tentative. Nevertheless, they are worth mentioning because they illustrate a possible experimental way to explore brain changes that might influence general problem-solving ability.

---

trait depends on the heritability – with high heritability, the regression is smaller. Also, for a given trait the regression is greater at the extremes of a population than at its center."

Looking for an intelligent mate is quite common and widely practiced throughout the world, especially by upper social classes. The phenomenon is well documented by scientific research on assortative mating. The search is often implicit; for example, attendance at a highly selective university or employment at a hi-tech company or medical facility maximizes the chances of finding a highly intelligent mate. Sometimes the strategy is more explicit; fertility clinics may prefer donors with advanced education, assuming higher intelligence.

There is an emerging strategy that could maximize the probability of having children with high intelligence. Recall from Chapter 6 that genome-wide association studies (GWAS) can be used to create polygenic scores from DNA that predict IQ scores or the related variable of educational attainment. So far, these nascent predictions are relatively weak, but they are expected to become better (Cheesman et al., 2020). There are commercial companies (like Genomic Prediction) that work in conjunction with fertility clinics to make polygenic score predictions on a number of traits in embryos not yet brought to term so that parents can choose among the embryos. Currently, most of the predictions are for negative traits or for the risk of disorders like schizophrenia. The point is for parents to avoid embryos with a statistical risk of potential serious problems.

Parents also may be informed about embryos with low polygenic scores for intelligence so they can minimize the probability of having a child with compromised intelligence. As a matter of ethical sensitivity, identifying the embryos with the statistical probability of having high intelligence is not now typically done, at least in Western countries. Given that the ability to make high intelligence predictions is likely to improve, parents may demand to know the full rank order of all their embryos on any trait. There is a reasonable concern, however, that it is possible that selecting for high intelligence will bring along risk for other problems associated with the same genes, so much more needs to be learned for safe applications and informed ethical discussions (Turkheimer, 2019; Pagnaer et al., 2021; Turley et al., 2021; Tellier et al., 2021).

13.4.2 *Genetic Engineering*

We are more than two decades into the twenty-first century, and already powerful technology is available to edit the genome. One of the most discussed is CRISPR (clusters of regularly interspaced short palindromic repeats), which allows the cutting and pasting of DNA segments to replace or repair bits of DNA that cause medical problems. This technique was first developed late in the twentieth century, and by about 2015, it became clear that improvements and variations of the technique could expand the scope of possibilities for genetic engineering, including for complex traits like intelligence, at least theoretically. Jennifer Doudna, from the University of California, Berkeley, received the 2020 Nobel Prize for her role in developing this technology after the first insight obtained from basic research by Francisco Mojica (Mojica and Rodriguez-Valera, 2016).

There are many challenges to make gene editing techniques viable, safe, and effective. On one hand, as discussed in Chapter 6, there are experts in the field who think the challenges are so complex that using genetic engineering for increasing intelligence may not ever be possible. They point to the incredibly complex cascade of events from DNA to protein formation to brain development to synapses to circuits to phenotypic expression. Interactions among these processes, some with environmental influences, and the effects of random fluctuations make interventions with predictable results a remote prospect (Turkheimer, 2016; Mitchell, 2018). On the other hand, there is the view that science is not to be underestimated in the long run by the complexity of problems. Where some see nightmares of complexity, others see challenges to overcome step by incremental step (Haier, 2017, 2021). Progress may be slow, but inexorable. What do you think?

It might take decades, but once more details are known about the processes illustrated in Figures 6.11 and 6.16, it may well be possible to craft interventions at strategic points that would result in brain development optimized for increased general problem-solving ability. This might first come as a genetic editing treatment for genetic disorders that cause low intelligence (Arnadottir et al., 2022) and that might give insights to other ways to modify the genetic influence on the core *g*-factor or even the genetic influence on specific cognitive abilities like math, visuospatial, or verbal ability (Sisodiya et al., 2007; Paul et al., 2022). It might come from research on boosting learning and memory in patients with Alzheimer's disease. Science fiction? Perhaps for now, but there is a worldwide effort among countless neuroscientists to discover more about basic brain mechanisms (Grasby et al., 2020). These efforts could address questions about increasing intelligence and how they might be answered.

13.4.3 *Drugs and an IQ Pill*

Even if editing the genome proves too complex to be practical for increasing general or specific mental abilities, a better understanding of the events depicted in Figures 6.11 and 6.16 will ultimately form the basis of a new discipline called the *molecular biology of intelligence*. A major systematic effort to learn how intelligence-relevant genes are expressed and function is already under way in the Netherlands, led by Professor Danielle Posthuma (https://ctg.cncr.nl/). This effort is conceptually identical to understanding gene expression and function in medical disorders of all kinds. Once relevant DNA/genetic influences are identified and understood on a molecular basis, a next formidable step is developing drugs or other interventions to modify key points in the neurobiological systems to attain an outcome that can look like a treatment, a cure, or prevention. In the case of intelligence, an outcome conceivably could be a brain optimized for better problem solving in a way that looks like enhanced intelligence. This could come about by a drug that changes synaptic function or brain circuit efficiency or increases information-processing capacity by stimulating gray matter volume or white matter connectivity and function. Many other possibilities could be developed when details of the systems are better understood.

No such drug exists now. Many drugs claim to give better attention and even enhanced performance on IQ tests. But there is not any weight of evidence that supports such claims for intelligence. Nonetheless, researchers in some countries see a scientific race already under way for intelligence enhancement to maintain or advance technological innovations and economic growth that depend in part on intelligence resources (human capital). Consider, for example, how many people are in the top 2 percent of the IQ score normal distribution. Table 13.1 shows some examples of country intelligence gaps, assuming that the distribution is the same everywhere (it may not be). Table 13.1 is based solely on 2 percent of 2020–2021

Table 13.1 Population estimates of selected countries and how many people are estimated to be in the top 2 percent of IQ based on the normal distribution

	Approx. population	Approx. top 2% of IQ
China	1.4 billion	28,000,000
India	1.3 billion	26,000,000
United States	332 million	6,640,000
Brazil	230 million	4,600,000
Russia	147 million	2,940,000
Japan	125 million	2,500,000
Mexico	130 million	2,600,000
Germany	83 million	1,660,000
Iran	83 million	1,660,000
United Kingdom	68 million	1,360,000
France	67 million	1,340,000
Italy	60 million	1,200,000
South Korea	52 million	1,040,000
Spain	47 million	940,000
Poland	38 million	760,000
Canada	38 million	760,000
Netherlands	18 million	360,000
Sweden	10 million	200,000
Israel	9 million	180,000
Singapore	5.8 million	116,000
World	7.8 billion	156,000,000

population estimates independent of estimated average country IQ and educational differences discussed in Chapter 12. Clearly, smaller countries have a numerical disadvantage that sometimes can be offset by education, immigration, and other policies. Winning the race for enhanced intelligence could have profound international impact (Jones, 2016).

Notice that the same numbers in the top 2 percent also would be found for the bottom 2 percent of IQ scores in each country. Large numbers of people in this category are at risk for many social problems that could be alleviated with successful ways to increase intelligence. The goal of increasing intelligence in as many people as possible, throughout the entire distribution, should be considered a priority for research. In the long run, intelligence may be what matters most for individuals and for countries.

13.5 Summary and Conclusion

The most profound question about intelligence is whether it can be increased in any person or population by any means. There is evidence of small increases in intelligence test scores for general education; some education intervention programs; and environmental factors like better nutrition and health care for infants and young children, thought to account for the Flynn effect. Interventions designed to increase fluid intelligence, however, have not fared well. Although there are claims that some interventions have done so already, the initial evidence typically was weak, and the claims failed to survive independent replication. In fact, bold claims that intelligence is largely malleable are based on weak evidence, especially with respect to conflating intelligence, IQ scores, and the *g*-factor. Changes in test scores for specific cognitive abilities and skills are one thing, but assuming that they change general problem-solving ability is not warranted so far. The evidence indicates that enhancing intelligence is not that easy once basic needs for health and brain development are met and environmental toxins are minimized or eliminated.

Arthur Jensen (1997, p. 80) pointed out that the apparent difficulty in enhancing intelligence might be adaptive from an evolutionary perspective: "An overly plastic nervous system, with its functions shaped too easily by the environment, would put the organism's adaptive capacity at risk of being wafted this way or that by haphazard experiences. ... General mental ability is elastic rather than plastic in its temporary deviations from its biologically programmed trajectory."

In the twenty-first century, the concept of an elastic *g*-factor underscores new excitement about the possibility of increasing intelligence, perhaps dramatically, based on research advances applying neuroscience and DNA methods. These nascent investigations of intelligence are forming the foundation for a molecular neurobiology of intelligence. With such research, we may find ways to influence or modify the sequences of multiple events that result in a brain optimized for increased general problem-solving and reasoning ability. Imagine a world where everyone was considerably smarter. Human beings and their societies are complex, and enhanced intelligence certainly would not solve all problems, but we can start to think about what might be different.

Understanding the cascade of events from DNA to genes to proteins to neurons/synapses to brain development to brain circuits to brain functions to individual differences in reasoning ability (notwithstanding many steps and processes in between these, along with environmental interactions) may be the most formidable challenge in science, and one with the most profound consequences. We think it will be possible to enhance intelligence by targeting the *g*-factor, although for now, we do not know how to do so. It is a goal that should be pursued with vigor. To paraphrase President John F. Kennedy in a speech about going to the moon, given at Rice University on September 12, 1962, we choose this goal "not because [it] is easy but because [it] is hard, because that goal will serve to organize and measure the best of our energies and skills, because that challenge is one that we are willing to accept, one we are unwilling to postpone, and one which we intend to win."

Perhaps as you finish reading this book, you might consider this goal of intelligence research in a new light.

13.6 Questions for Discussion

13.1 Do you agree with the data that indicate that early childhood education has little, if any, lasting effect on intelligence?

13.2 Do you have any personal experiences trying to improve your cognitive abilities by listening to classical music, improving your memory, or playing computer games?

13.3 If you were to use a fertility clinic, would you insist on knowing the full range of polygenic score predictions for intelligence (or educational attainment) from a donor?

13.4 Would you consider genetic engineering if it increased the probability of having more intelligent children?

13.5 If there was an IQ pill, how much would you pay to use it for yourself or for your children?

13.6 After reading this book, would you consider a career related to intelligence research?

References

Ackerman, P. L., & Hambrick, D. Z. 2020. A primer on assessing intelligence in laboratory studies. *Intelligence*, 80, 101440.

Anguera, J. A., Boccanfuso, J., Rintoul, J. L., et al. 2013. Video game training enhances cognitive control in older adults. *Nature*, 501, 97–101.

Antal, A., Luber, B., Brem, A. K., et al. 2022. Non-invasive brain stimulation and neuroenhancement. *Clinical Neurophysiology Practice*, 7, 146–165.

Arnadottir, G. A., Oddsson, A., Jensson, B. O., et al. 2022. Population-level deficit of homozygosity unveils CPSF3 as an intellectual disability syndrome gene. *Nature Communications*, 13, 705.

Bediou, B., Adams, D. M., Mayer, R. E., et al. 2018. Meta-analysis of action video game impact on perceptual, attentional, and cognitive skills. *Psychological Bulletin*, 144, 77–110.

Bigand, E., & Tillmann, B. 2022. Near and far transfer: Is music special? *Memory, and Cognition*, 50, 339–347.

Bratsberg, B., & Rogeberg, O. 2018. Flynn effect and its reversal are both environmentally caused. *Proceedings of the National Academy of Sciences of the United States of America*, 115, 6674–6678.

Brem, A. K., Almquist, J. N., Mansfield, K., et al. 2018. Modulating fluid intelligence performance through combined cognitive training and brain stimulation. *Neuropsychologia*, 118, 107–114.

Chabris, C. F. 1999. Prelude or requiem for the "Mozart effect"? *Nature*, 400, 826–827.

Cheesman, R., Hunjan, A., Coleman, J. R. I., et al. 2020. Comparison of adopted and nonadopted individuals reveals gene–environment interplay for education in the UK Biobank. *Psychological Science*, 31, 582–591.

Cronbach, L. J. 1975a. 5 decades of public controversy over mental testing. *American Psychologist*, 30, 1–14.

Cronbach, L. J. 1975b. Balanced presentation of controversy: Reply. *American Psychologist*, 30, 938–939.

Dweck, C. S. 2008. *Mindset: The new psychology of success*. New York: Ballantine Books.

Estrada, E., Ferrer, E., Abad, F. J., Román, F. J., & Colom, R. 2015. A general factor of intelligence fails to account for changes in tests' scores after cognitive practice: A longitudinal multi-group latent-variable study. *Intelligence*, 50, 93–99.

Flynn, J. R. 2012. *Are we getting smarter? Rising IQ in the twenty-first century*. Cambridge: Cambridge University Press.

Flynn, J. R. 2018. Reflections about intelligence over 40 years. *Intelligence*, 70, 73–83.

Gobet, F., & Sala, G. 2022. Cognitive training: A field in search of a phenomenon. *Perspectives on Psychological Science*, 18, 125–141.

Grasby, K. L., Jahanshad, N., Painter, J. N., Colodro-Conde, L., Bralton, J., et al. 2020. The genetic architecture of the human cerebral cortex. *Science*, 367, eaay6690.

Haier, R. J. 2014. Increased intelligence is a myth (so far). *Frontiers in Systems Neuroscience*, 8, 34.

Haier, R. J. 2017. *The neuroscience of intelligence*. Cambridge: Cambridge University Press.

Haier, R. J. 2021. Are we thinking big enough about the road ahead? Overview of the special issue on the future of intelligence research. *Intelligence*, 89, 101603.

Hattie, J., Biggs, J., & Purdie, N. 1996. Effects of learning skills interventions on student learning: A meta-analysis. *Review of Educational Research*, 66, 99–136.

Hegelund, E. R., Teasdale, T. W., Okholm, G. T., et al. 2021. The secular trend of intelligence test scores: The Danish experience for young men born between 1940 and 2000. *PLoS ONE*, 16, e0261117.

Herrnstein, R. 1971. I.Q. *Atlantic Monthly*, September.

Herrnstein, R. J., & Murray, C. 1994. *The bell curve: Intelligence and class structure in American life*. New York: Free Press.

Hescham, S., Liu, H., Jahanshahi, A., & Temel, Y. 2020. Deep brain stimulation and cognition: Translational aspects. *Neurobiology of Learning and Memory*, 174, 107283.

Horvath, J. C., Forte, J. D., & Carter, O. 2015a. Evidence that transcranial direct current stimulation (tDCS) generates little-to-no reliable neurophysiologic effect beyond MEP amplitude modulation in healthy human subjects: A systematic review. *Neuropsychologia*, 66, 213–236.

Horvath, J. C., Forte, J. D., & Carter, O. 2015b. Quantitative review finds no evidence of cognitive effects in healthy populations from single-session transcranial direct current stimulation (tDCS). *Brain Stimulation*, 8, 535–550.

Hunt, E. 2012. What makes nations intelligent? *Perspectives on Psychological Science*, 7, 284–306.

Jaeggi, S. M., Buschkuehl, M., Jonides, J., & Perrig, W. J. 2008. Improving fluid intelligence with training on working memory. *Proceedings of the National Academy of Sciences of the United States of America*, 105, 6829–6833.

Jensen, A. R. 1969. How much can we boost IQ and scholastic achievement? *Harvard Educational Review*, 39, 1–123.

Jensen, A. R. 1997. The puzzle of non-genetic variance. In Sternberg, R. J., & Grigorenko, E. (eds.), *Intelligence, heredity, and environment*. Cambridge: Cambridge University Press.

Johnson, W. 2012. How much can we boost IQ? An updated look at Jensen's (1969) question and answer. In Slater, A., & Quinn, P. (eds.), *Developmental psychology: Revisiting the classic studies*. Thousand Oaks, CA: SAGE.

Jones, G. 2016. *Hive mind: How your nation's IQ matters so much more than your own*. Stanford, CA: Stanford Economics and Finance.

Kremen, W. S., Beck, A., Elman, J. A., et al. 2019. Influence of young adult cognitive ability and additional education on later-life cognition. *Proceedings of the National Academy of Sciences of the United States of America*, 116, 2021.

Lozano, A. M., & Lipsman, N. 2013. Probing and regulating dysfunctional circuits using deep brain stimulation. *Neuron*, 77, 406–424.

Luber, B., & Lisanby, S. H. 2014. Enhancement of human cognitive performance using transcranial magnetic stimulation (TMS). *Neuroimage*, 85, 961–70.

Lynn, R. 2009. What has caused the Flynn effect? Secular increases in the development quotients of infants. *Intelligence*, 37, 16–24.

Mackey, A. P., Hill, S. S., Stone, S. I., & Bunge, S. A. 2011. Differential effects of reasoning and speed training in children. *Developmental Science*, 14, 582–590.

Melby-Lervåg, M., Redick, T. S., & Hulme, C. 2016. Working memory training does not improve performance on measures of intelligence or other measures of "far transfer": Evidence from a meta-analytic review. *Perspectives on Psychological Science*, 11, 512–534.

Mingroni, M. A. 2004. The secular rise in IQ: Giving heterosis a closer look. *Intelligence*, 32, 65–83.

Mingroni, M. A. 2007. Resolving the IQ paradox: Heterosis as a cause of the Flynn effect and other trends. *Psychological Review*, 114, 806–829.

Mitchell, K. J. 2018. *Innate: How the wiring of our brains shapes who we are*. Princeton, NJ: Princeton University Press.

Mojica, F. J., & Rodriguez-Valera, F. 2016. The discovery of CRISPR in archaea and bacteria. *FEBS Journal*, 283, 3162–3169.

Moreau, D. 2022. How malleable are cognitive abilities? A critical perspective on popular brief interventions. *American Psychologist*, 77, 409–423.

Okely, J. A., Overy, K., & Deary, I. J. 2022. Experience of playing a musical instrument and lifetime change in general cognitive ability: Evidence from the Lothian birth cohort 1936. *Psychological Science*, 33, 1495–1508.

Page, E. B. 1972. Miracle in Milwaukee: Raising the IQ. *Educational Researcher*, 1, 8–16.

Pagnaer, T., Siermann, M., Borry, P., & Tsuiko, O. 2021. Polygenic risk scoring of human embryos: A qualitative study of media coverage. *BMC Medical Ethics*, 22, 125.

Pahor, A., & Jausovec, N. 2014. The effects of theta transcranial alternating current stimulation (tACS) on fluid intelligence. *International Journal of Psychophysiology*, 93, 322–331.

Paul, M. M., Dannhäuser, S., Morris, L., et al. 2022. The human cognition-enhancing CORD7 mutation increases active zone number and synaptic release. *Brain*, 21, 3787–3802.

Pietschnig, J., & Gittler, G. 2015. A reversal of the Flynn effect for spatial perception in German-speaking countries: Evidence from a cross-temporal IRT-based meta-analysis (1977–2014). *Intelligence*, 53, 145–153.

Pietschnig, J., & Voracek, M. 2015. One century of global IQ gains: A formal meta-analysis of the Flynn effect (1909–2013). *Perspectives on Psychological Science*, 10, 282–306.

Pietschnig, J., Voracek, M., & Formann, A. K. 2010. Mozart effect–Shmozart effect: A meta-analysis. *Intelligence*, 38, 314–323.

Pinker, S. 2002. *The blank slate: The modern denial of human nature*. New York: Viking.

Protzko, J. 2017. Effects of cognitive training on the structure of intelligence. *Psychonomic Bulletin and Review*, 24, 1022–1031.

Rauscher, F. H., Shaw, G. L., & Ky, K. N. 1993. Music and spatial task performance. *Nature*, 365, 611.

Redick, T. S. 2019. The hype cycle of working memory training. *Current Directions in Psychological Science*, 28, 423–429.

Ritchie, S. J., & Tucker-Drob, E. M. 2018. How much does education improve intelligence? A meta-analysis. *Psychological Science*, 29, 1358–1369.

Sala, G., & Gobet, F. 2020. Cognitive and academic benefits of music training with children: A multilevel meta-analysis. *Memory and Cognition*, 48, 1429–1441.

Santarnecchi, E., Polizzotto, N. R., Godone, M., et al. 2013. Frequency-dependent enhancement of fluid intelligence induced by transcranial oscillatory potentials. *Current Biology*, 23, 1449–1453.

Sauce, B., Liebherr, M., Judd, N., & Klingberg, T. 2022. The impact of digital media on children's intelligence while controlling for genetic differences in cognition and socioeconomic background. *Scientific Reports*, 12, 7720.

Sellers, K. K., Mellin, J. M., Lustenberger, C. M., et al. 2015. Transcranial direct current stimulation (tDCS) of frontal cortex decreases performance on the WAIS-IV intelligence test. *Behavioural Brain Research*, 290, 32–44.

Simons, D. J., Boot, W. R., Charness, N., et al. 2016. Do "brain-training" programs work? *Psychological Science in the Public Interest*, 17, 103–186.

Sisk, V. F., Burgoyne, A. P., Sun, J. Z., Butler, J. L., & Macnamara, B. N. 2018. To what extent and under which circumstances are growth mind-sets important to academic achievement? Two meta-analyses. *Psychological Science*, 29, 549–571.

Sisodiya, S. M., Thompson, P. J., Need, A., et al. 2007. Genetic enhancement of cognition in a kindred with cone–rod dystrophy due to *RIMS1* mutation. *Journal of Medical Genetics*, 44, 373–380.

Suthana, N., & Fried, I. 2014. Deep brain stimulation for enhancement of learning and memory. *Neuroimage*, 85, 996–1002.

Tellier, L. C. A. M., Eccles, J., Treff, N. R., et al. 2021. Embryo screening for polygenic disease risk: Recent advances and ethical considerations. *Genes*, 12, 1105.

Thompson, T. W., Waskom, M. L., Garel, K. L. A., et al. 2013. Failure of working memory training to enhance cognition or intelligence. *PLoS ONE*, 8, e63614.

Turkheimer, E. 2016. Weak genetic explanation 20 years later: Reply to Plomin et al. (2016). *Perspectives on Psychological Science*, 11, 24–28.

Turkheimer, E. 2019. Genetics and human agency: The philosophy of behavior genetics – introduction to the special issue. *Behavior Genetics*, 49, 123–127.

Turley, P., Meyer, M. N., Wang, N., et al. 2021. Problems with using polygenic scores to select embryos. *New England Journal of Medicine*, 385, 78–86.

Unsworth, N., Redick, T. S., McMillan, B. D., et al. 2015. Is playing video games related to cognitive abilities? *Psychological Science*, 26, 759–774.

Widge, A. S., Zorowitz, S., Basu, I., et al. 2019. Deep brain stimulation of the internal capsule enhances human cognitive control and prefrontal cortex function. *Nature Communications*, 10, 1536.

Wiemers, E. A., Redick, T. S., & Morrison, A. B. 2019. The influence of individual differences in cognitive ability on working memory training gains. *Journal of Cognitive Enhancement*, 3, 174–185.

Wongupparaj, P., Kumari, V., & Morris, R. G. 2015. A cross-temporal meta-analysis of Raven's Progressive Matrices: Age groups and developing versus developed countries. *Intelligence*, 49, 1–9.

Woodley, M. A. 2011. Heterosis doesn't cause the Flynn effect: A critical examination of Mingroni (2007). *Psychological Review*, 118, 689–693.

Epilogue: A Final Word

E.1 Ten Key Points

Many of the facts discussed in this book may come as a surprise or even a shock, given that there is so much misinformation about intelligence in high school courses, among teaching faculties in colleges and universities, in the workplace, and in the media. It is quite all right to be skeptical, and we encourage it for every page of this book, but the weight of evidence presented in each chapter supports the following basic statements, which have been replicated across decades and reinforced by recent research:

1. Intelligence can be defined for scientific investigations.
2. Standard measures of intelligence provide quantitative assessments of individual differences for rigorous statistical analyses despite acknowledged limitations.
3. Standard intelligence tests have the required reliability and validity for scientific study.
4. Standard tests of mental ability, including IQ tests, are not biased against any population when administered and interpreted properly.
5. The general factor of intelligence (g), especially when assessed by a diverse battery of mental ability tests, is the single most predictive construct in psychology for a wide variety of real-world relevant social outcomes.
6. Measures of g are related to a number of quantifiable brain features that appear to have developmental sequences.
7. Individual differences in intelligence are influenced by genetics, although details at the molecular level are just beginning to be investigated. Nongenetic factors are also relevant, but there also is no clear model of the

mechanics of their impact (although the mechanisms must be biological and ultimately influence the brain).

8. The sources of average population differences for IQ and other measures of mental ability remain unknown, but science is making progress disentangling purported influences.

9. So far, there is no proven way to increase the general ability to integrate cognitive abilities (the *g*-factor), but the possibility is at an exciting frontier of neuroscience and molecular genetic research.

10. More intelligence is no guarantee of being a better person in any sense, but perhaps enhanced intelligence will help our species solve long-standing global problems.

This is a short list that is to the point. Much of what we know about intelligence is represented in this book, but each chapter also shows how much we have yet to learn. There are still psychometric issues, and the goal of a ratio scale intelligence test remains elusive. Neuroimaging is verifying the roles that brain structure and function play in cognitive differences. The demonstration that not all brains work the same way potentially has profound implications for education and social policy, especially if neuroimaging data can predict outcomes as well as or better than currently available standardized intelligence measures. Applying this kind of neuroscience information has barely been attempted anywhere. Perhaps the most challenging information to process is the rapidly advancing genetic research. When it comes to complex traits like intelligence, some see genetics as limiting, but the fact is that genes are probabilistic, not deterministic. Moreover, understanding genes and how they function can unlock tremendous opportunities to facilitate and enhance rather than limit human behavior. As genetic studies, especially based on DNA technology, begin to answer old questions about intelligence, an informed public is vital for ethical discussions about how new knowledge about what intelligence is and where it comes from can be used to make life better.

It is that spirit that motivates every sentence in this book, and writing it has been a daunting challenge. On one hand, we are obligated to provide an objective summary of key research and a balanced presentation of core issues that depend in part on how research findings are interpreted. On the other hand, some controversies do not require balance, because the weight of research evidence strongly supports one position over others. This book is based on our understanding of the evidence that has evolved over seventy-five years of our combined research experience and the expertise we have developed. We have our biases and are probably not completely correct in the way we evaluate all the facts, but we are open to changing our minds if new, compelling data become available or new theories are proposed that better explain the data. That is the fundamental way science works.

E.2 A Final Challenge for You

Reading a book on intelligence is also a daunting challenge. The first chapter began with a necessary voyage to the past to understand the current state of intelligence research and as a foundation for informed thinking. Quite a lot of technical knowledge must be learned to appreciate what the data say and what they do not say. Statistics and psychometrics have their own language, as does cognitive psychology, neuroimaging, and genetics, not to mention the complex ways data are analyzed beyond basic statistics, especially in molecular biology.

Working through all the chapters is not for the faint of heart. But, at the end of this book, we hope you can say you learned interesting and important information about intelligence. From the day we agreed to update and write this edition of Hunt's book, we aimed to give you a sense of the excitement researchers have as they integrate diverse findings into the big picture for understanding human intelligence. We have given you much to think about, and we hope we have kindled your imagination. Perhaps you are motivated to work on some of the problems we have presented. It would be terrific if you accepted the challenges of such a career to explore the mysteries of intelligence and what a deeper understanding of this fundamental human ability might mean for the future.

At the end of the preface, we quoted Douglas Detterman on the importance of intelligence research. We choose to end this book reiterating that quote:

> Intelligence is the most important thing of all to understand, more important than the origin of the universe, more important than climate change, more important than curing cancer, more important than anything else. That is because human intelligence is our major adaptive function and only by optimizing it will we be able to save ourselves and other living things from ultimate destruction.
>
> (Detterman, 2016, p. v)

Reference

Detterman, D. K. 2016. Was intelligence necessary? *Intelligence*, 55, v–viii.

Index

Numbers: bold = table or box; *italics* = figure

Small words ("and," "at," "but," "in," "versus") *are* alphabetized.

Printed in the USA
CPSIA information can be obtained
at www.ICGtesting.com
CBHW071644080624
9776CB00010B/1033